Third Edition

PRINCIPLES OF ELECTRONIC INSTRUMENTATION

Third Edition

PRINCIPLES OF ELECTRONIC INSTRUMENTATION

A. James Diefenderfer

Late of California State University, Fullerton

Brian E. Holton

Rutgers University

THOMSON

BROOKS/COLE

Australia • Canada • Mexico • Singapore • Spain
United Kingdom • United States

For permission to use material from this text contact us by:
Phone: 1-800-730-2214
Fax: 1-800-730-2215
Web: www.thomsonrights.com

Cover reprinted with permission of Motorola.

Printed in the United States of America.

Principles of Electronic Instrumentation

ISBN 0-03-074709-0

7 8 9 10 11 12   07 06 05 04 03 02

Preface

The goal of this text is to provide scientists and engineers with the appropriate background in electronics and instrumentation so that they, and not a manufacturer, can select the instrumentation most appropriate to a measurement or control problem. In writing the third edition, there has been an earnest attempt to update material, to maintain the emphasis of the first two editions and, of course, to eliminate any errors in the previous edition. The approach presents electronics in a way that allows students to use their gained knowledge in real, working circuit applications; that is, the material strikes a balance between verbal and mathematical descriptions and gives real circuits as examples. The early portions of the book are the most quantitative. This approach is modified in later chapters to a more descriptive one, because we have found this can be done more effectively once the earlier, quantitative development has been established.

More than two decades have passed since the United States placed men on the moon. The pride generated by this accomplishment has largely subsided. Most people have failed to realize that the space program provided dramatic "spin-offs," the results of which we enjoy today. The principal beneficiary has been electronics. Many of the electronic devices and assemblies in common use today had their genesis in the space program. When the first edition of this text was written, only a few of these developments were evident. Today, however, we can appreciate the plethora of devices and systems spawned by that era. LCD watches, calculators, microprocessors and microcomputers are a few examples of the devices that have an impact on our daily lives as a result of the space program. In short, we are in the midst of an electronics revolution. Few, if any, individuals are current in state-of-the-art electronics; it is changing as you read this preface. It falls upon the reader to continually update him- or herself. This textbook provides the basis for accomplishing this task.

One's horizons in electronics are limited only by one's imagination. It was Albert Einstein who said, "Imagination is more important than knowledge." And this couldn't be truer than in the field of electronics and measurement. Twenty years ago it would have been inconceivable to consider diagnosis of an instrument's ills by long-distance telephone or the possibility of sending copies of documents via the telephone. Yet today these are possible with microcomputer-based instruments communicating over long distances.

APPROACH

Knowledge of electronics comes from building on a base of principles. The text starts with a chapter on direct circuits and a second on capacitors and inductors. Mastery of the material in the first two chapters is essential to further understanding and use of electronics. Chapter Three introduces alternating current circuits using resistors, capacitors and inductors. It is by far the most mathematically complex chapter, using phasor notation to present and analyze the circuitry. Further analysis of alternating current circuitry follows in Chapter Four. Chapter Five introduces diodes and their applications including use in voltage regulators and power supplies, and for the first time in the text useful circuits are provided. Chapter Six introduces test equipment and measuring—the tools of electronics. Given the hands-on nature of the course, it might be useful to study this chapter earlier in the course and then return to it after the reader is acclimated to more of the basics contained in Chapters 1 through 4.

Chapter Seven is an extensive collection of useful transducers used in the laboratory to measure physical quantities. In a sense, this chapter is the heart of the text. Chapter Eight involves transistors. Here the material has been rewritten and updated from the second edition to enhance its usefulness in real-world electronics. While transistors still have their uses today, transistor amplifiers are more or less outdated and have been kept to a minimum of useful circuitry. Chapter Nine provides a discussion of operational amplifiers—the building blocks of a majority of analog instrumentation circuits. This chapter has been rewritten to provide some simple rules to follow when designing and analyzing op-amp circuits as opposed to the more complex, less intuitive approach toward the material discussed in the second edition and in other textbooks. The formal approach is provided in Appendix B. As an interface between the world of digital electronics and that of analog, Chapter Ten concentrates on waveform generators.

The essentials of digital electronics are covered in Chapter Eleven and further developed in Chapter Twelve. Although this text does not delve deeply into microprocessors, Chapter Thirteen gives an overview of microprocessor basics. The gateway between the analog world in which we take measurements and the digital world in which we record and analyze these data is covered in Chapter Fourteen. Chapter Fifteen presents noise.

Although every course is different, we recommend that the course material follow the essential order of the text. The exceptions follow. Much of the material in Chapter Six, "Test Equipment and Measuring," can only be thoroughly understood after the foundational work of Chapters 1–4. It could be covered earlier, especially if a laboratory is part of the course. Chapter Seven, "Transducers," could be covered anywhere in the course. In fact, we seriously considered making it the first chapter of the text.

The text has fifteen chapters, which probably contain too much material to be covered in a one-term course but does fit comfortably into two terms. For a one-term course certain material could be reduced or left out without seriously affecting the quality of instruction. Chapter Thirteen, "Microprocessor Basics," and Chapter Fifteen, "Noise," could be left to the student to learn outside the course. Transistors covered in Chapter Eight could be cut back to the essentials, which include the common collector emitter follower, the EET follower and the transistor switch.

ANCILLARIES

The following ancillaries are available with this text:

Instructor's Manual with Full Solutions This manual contains summarized answers and fully worked solutions to all of the problems in the text.

Laboratory Manual Contains approximately 30 experiments, including Motorola 6811 microcontroller and circuit simulation software examples.

A SPECIAL THANKS

I have been fortunate to have had the assistance of many individuals in preparing this text. The reviewers who gave me an enormous number of useful suggestions were particularly helpful. They opened my eyes in many ways. A special thank you to

John R. Amend
Montana State University

David J. Cooke
Chaminade University of Honolulu

Jai N. Dahiya
Southeast Missouri State University

D. Scott Davis
Naval Postgraduate School

Richard Denton
Dartmouth College

James J. Donaghy
Washington & Lee University

Rick Feinberg
Western Washington University

Eric Gantz
University of Minnesota

Jim Holler
University of Kentucky

Earl J. Kirkland
Cornell University

Peter Kissinger
Bioanalytical Systems

Dan Reger
University of South Carolina

Michael S. Spritzer
Villanova University

Bruce R. Thomas
Carleton College

Ray von Wandruszka
University of Idaho

Richard Wolfson
Middlebury College

and an extra special thanks to Harry L. Pardue from Purdue University who has helped critique all three editions of this text. I would like to thank the professionals at Saunders College Publishing, particularly Kate Pachuta who drove me to write a third edition and to Chiara Puffer who has had more patience with me than could be expected and has done a brilliant job with the enormous task of coordinating my work, reviewing the manuscript and producing this text. I would also like to thank Elm Street Publishing Services, especially Martha Beyerlein and Andrew Roe. Personally, I would like to thank Professor Joe Pifer who established the Rutgers University course, "Modern Instrumentation," which was a driving force behind my knowledge base. Professors Mark Croft and Haruo Kojima, both from Rutgers University, who spurred my interest in this fascinating subject, Bob Darling for his skilled mentorship and willingness to share his extraordinary electronics knowledge and James O'Dowd for lighting that first LED. Finally, a thank you to the students in Modern Instrumentation at Rutgers University in the summers of 1991 and 1992 who class tested drafts of the text.

Brian Holton

Contents

Chapter 7 **Transducers 127**

Chapter 8 **Transistors 155**

Chapter 9 **Operational Amplifiers 183**

Chapter 14   Digital and Analog I/O   331

Chapter 15   Noise   359

1

Direct Current Circuits

1-1 INTRODUCTION

Today most of us have a reasonable understanding of electricity. When we wish to make use of electrical measurement or control, however, this level of understanding is not enough. We must also understand and use quantitative laws of electricity. These laws are not new, but you may not have encountered them before or have forgotten how they are used. This chapter on direct current (DC) circuits will provide a foundational insight into the understanding and application of electrical circuit laws. The qualitative ideas of potential difference, charge, current and resistance will be developed first. Then the fundamental simple circuit laws will be presented with examples and solutions to ensure mastery. Keep in mind that although this chapter is presented in a quantitative manner, you should also strive for a qualitative understanding of the material.

1-2 CURRENT

A basic property of the electron is its charge (-1.603×10^{-19} coulombs). The movement of electrons from one point to another is called *current, I,* and has the units of charge per unit time, or in Standard International (SI) units, coulombs per second. The SI unit for current is the ampere (A), named after André M. Ampère (1775–1836), with one ampere being equivalent to one coulomb (C) per second(s):

$$1 \text{ A} \equiv 1 \text{ C/s.}$$

The abbreviations and symbols for electrical quantities used throughout the text are listed on the front and back endsheets of this edition.

The current flowing through a 100-watt (W) light bulb is about 1 A, but for most electronic instrumentation circuits, the ampere is a rather large unit. It is much more common to speak of milliamperes ($1 \text{ mA} = 1 \times 10^{-3}\text{A}$), microamperes ($1 \text{ }\mu\text{A} = 1 \times 10^{-6} \text{ A}$), nanoamperes ($1 \text{ nA} = 1 \times 10^{-9} \text{ A}$) or picoamperes ($1 \text{ pA} = 1 \times 10^{-12} \text{ A}$).

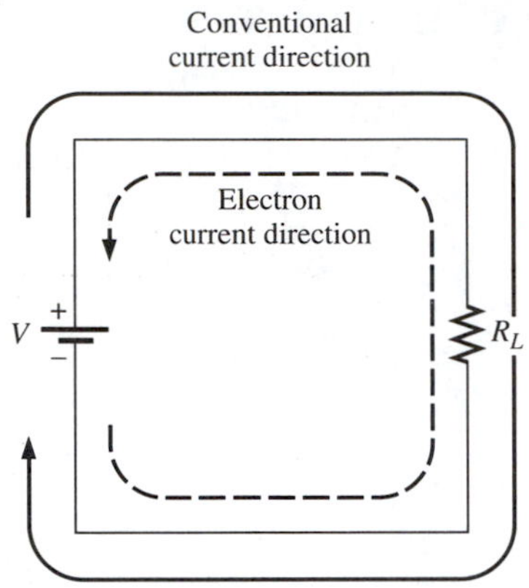

Conventional
current direction

Electron
current direction

V

R_L

FIGURE 1.1 **Diagram showing directions of conventional and electron currents.**

Historically (and somewhat erroneously), current flow in metals was believed to result from the movement of positive charges. Because of this, the direction of *conventional current* is generally understood to flow in the opposite direction of the actual *electron current*. This difference in direction between electron current flow and the conventional current flow can be understood by reference to Figure 1.1. Although we know the electrons are moving throughout the circuit, you may want to put this knowledge to the back of your mind while analyzing circuits and think in terms of conventional current flow.

1-3 POTENTIAL DIFFERENCE

Current flow in a conductor is caused by a difference in electric potential between the two ends of the conductor. Because electrons are free to move about in a conductor, the electrons, which are negatively charged, move from a point of less positive potential to one of more positive potential. Thus, conventional current flow, opposite in direction from electron flow, is from a more positive potential to a less positive potential. The conductor provides the conduit through which the current can move. A mechanical analogy of current flow through a potential difference may be found in a waterfall, where the moving water (current) falls from a higher to a lower potential (in this case gravitational rather than electric potential).

When a charge moves through a potential difference, work is done. The work done per unit charge in moving across the potential difference is measured in units of joules per coulomb. The SI unit of potential difference is the *volt* (V), with one volt being equivalent to one joule (J) per coulomb:

$$1 \text{ V} \equiv 1 \text{ J/C}.$$

Many people refer to potential difference and voltage interchangeably. For the purist this is incorrect, but the practice has become so common that the terms will be used interchangeably in this text.

Ground

Because the work done in moving a charge is measured by a *difference* in potential, not an absolute measure, the voltage must always be measured relative to some reference point. We may properly express the potential across an electric component as some number of volts, but often we speak of the voltage at only one point in a circuit. In such situations, it is assumed that the reference point is at *ground.* By definition, ground is a point (or conductor) in the circuit considered to be at zero volts, and all other voltages are measured with respect to it. Figure 1.2 shows three common schematic symbols for a ground connection.

Under strict definition, ground is exactly that: the body of the earth. It is an electrical "sink"; that is, it can accept or supply any reasonable amount of charge without changing its electrical characteristics. In actual practice, a true earth ground can be obtained by a solid connection to a cold water pipe. Frequently, though, the establishment of such a reference point is unnecessary, and a common (or reference) connection to the metal chassis of the instrument suffices because it should, for safety and electrical noise isolation

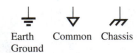

Earth   Common   Chassis
Ground

FIGURE 1.2 **Commonly used symbols for ground connections. Although it is specific per application, such distinction is often not adhered to, and the three are frequently used interchangeably.**

reasons, be connected to the ground prong of the wall outlet, which in turn is connected to earth. Sometimes the zero voltage point of a circuit is isolated from an actual earth ground, as in the case of battery-powered circuits. Zero voltage in such a circuit may be referenced to the negative terminal of the battery or some other voltage point. Unless otherwise noted, however, when we speak of ground in this book, it is assumed to be zero volts within the circuit in question, which is not necessarily the same as that of the earth.

If the metal case of an instrument powered by line voltage is not connected to the third prong of the input power cord, the case represents a potentially lethal danger to a user. Most circuits that plug into the wall, input this line voltage to a transformer (Chapter Four) which isolates the line voltage from the rest of the circuit. The simplest transformer has two inputs for the neutral and hot connections and two outputs. One output serves as a common in the circuit and the other a supply with a voltage referenced to the common. However the isolation by the transformer of the input line voltage and the output voltage implies that not only will the power supply lines be detached, but also that the line ground will be isolated from the transformer common. In fact, the neutral of the power line is not necessarily connected to the earth ground represented by the third prong. Therefore the common in the circuit can be many volts higher than the earth ground and if the common is connected to the metal case of the instrument and the earth ground is not, then an electric potential will probably exist between case and earth ground. A number of people have been shocked, some severely, in designing and using instruments that do not connect the earth ground to the instrument case.

Batteries

A battery is a common source of voltage and power. Table 1.1 lists a number of battery types. Because the potential is developed by a chemical reaction, batteries have finite lifetimes, limited by the quantity of chemicals available for the reaction and the integrity of the materials from which they are constructed. In some cases, the chemical reactions are such that the battery voltage changes with use. If a battery is to be used to supply a reference voltage, a flat discharge curve (i.e., constant voltage throughout the life of the battery)

TABLE 1.1*   *Electrical Characteristics of Batteries*

	Zinc-Carbon	Zinc-Chloride	Alkaline	Magnesium	Mercury-Oxide	Silver-Oxide	Divalent Silver-Oxide	Lithium
Energy output, watt-hours per in.^{3}	2	3	2 to 3.5	4	6	8	14	8 to 15
Nominal cell voltage	1.5	1.5	1.5	2.0	1.35 or 1.4	1.50	1.5	2.8
Practical drain rates	100 mA/in.^{2}	150 mA/in.^{2}	200 mA/in.^{2}	200 to 300 mA/in.^{2}				
Impedance, R_i (internal resistance)	low	low	very low	low	low	low	low	less than 1 Ω
Shape of discharge curve	sloping	sloping	sloping	fairly flat	flat	flat	flat	flat

*From *Electronics,* J. Lyman: V. 48; Ap. 3, 1975, p. 75.

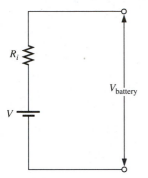

FIGURE 1.3   Equivalent circuit of a battery.

is essential, but most applications do not require such a restriction. On a cost-per-work basis, the battery is one of the most expensive power sources available.

Because a battery cannot produce an infinite amount of current, a model for battery behavior uses an internal resistance in series with an ideal voltage source (zero resistance) as indicated in Figure 1.3. The exhaustion of the battery with use, accompanied by a decrease in the available voltage at its terminals, is represented as an increase in the internal resistance as current is drawn from the battery. This model correctly suggests that measurement of a battery voltage is best performed with a load on the battery, i.e. when current is being drawn from it.

1-4   RESISTANCE

The electron movement in the circuit of Figure 1.1 is controlled by both the potential difference and the resistance, R_L, in the external circuit. The higher the resistance, the lower the current for a given potential difference. The resistance of a conductor depends upon the mobility of the electrons in the material. On a microscopic level, the resistance to current flow can be related to the number of collisions the electron makes with the lattice of atoms in the wire. The larger the number of collisions, the more resistive the material is said to be.

The best conductors of electricity have the lowest resistance, R. Resistance, expressed in ohms (Ω), is related to the resistivity, ρ, of the material by the equation

$$R = \rho L/a \tag{1-1}$$

where L and a represent the length and cross-sectional area of the conductor, respectively. L has units of m, a of m^2, and ρ of $\Omega - m$. The resistivities of some common conductors are presented in Table 1.2. From the table it is apparent that the resistivities of carbon and nichrome are much higher than those of other materials on the list. For this reason, and one of economics, resistors are often made of carbon or nichrome. Although silver has the lowest resistivity of any of the common materials, it is generally not used for wiring because of its relatively high cost. The most commonly used material is copper. Table 1.3 contains the approximate resistances for various standard sizes of copper wire. Note that wire is available in solid and stranded varieties. Stranded wire is commonly used for "hook-ups" between pieces of equipment because it is more flexible than a single strand or solid wire of equivalent diameter.

TABLE 1.2   Resistivity of Various Materials Commonly Used in Electrical Circuits

Material	Resistivity (ρ) in Ω m
Silver	1.6×10^{-8}
Copper	1.7×10^{-8}
Gold	2.4×10^{-8}
Aluminum	2.7×10^{-8}
Tungsten	5.7×10^{-8}
Nichrome	100×10^{-8}
Carbon	$3,500 \times 10^{-8}$
Glass	10^{12}
Teflon	10^{13}
Mica	10^{14}

1-5   OHM'S LAW

If a potential difference, V, is applied across a conductor with resistance, R, then a current I will flow through the conductor. The relationship of current, voltage and resistance in a conductor is described by Ohm's law and has the form:

$$V = IR \tag{1-2}$$

TABLE 1.3   Resistance of Copper Wire*

AWG[1] Size	Usual Number of Strands	Nominal Diameter per Strand (Inch)	Resistance per 1000 ft. (Ohms)
24	solid	0.0201	28.4
24	7	0.0080	28.4
22	solid	0.0254	18.0
22	7	0.0100	19.0
20	solid	0.0320	11.3
20	7	0.0126	11.9
18	solid	0.0403	7.2
18	7	0.0159	7.5
16	solid	0.0508	4.5
16	19	0.0113	4.7

[1]American Wire Gauge

*From *Components Handbook*, John F. Blackburn, Editor; Radiation Laboratory Series, V. 17, McGraw-Hill Book Company, Inc., New York, 1949.

which states that the voltage drop V across a conductor with resistance R is proportional to the current through the conductor. The constant of proportionality is the resistance of the conductor. The SI unit for resistance is the *ohm*, defined as being equal to one volt per ampere. The ohm is a rather small unit. Other commonplace SI units are the kilohm (1 kΩ = 1 $\times$ 10$^3\Omega$) and the megohm (1 MΩ = 1 $\times$ 10$^6\Omega$).

EXAMPLE 1.1

In the circuit shown in Figure 1.4 a current of 2.5 mA flows when V = 10.0 V and R = 4 kΩ; this is found using Ohm's law (Equation 1-2):

Solution:

$$I = V/R = 10.0 \text{ V}/4 \times 10^3\Omega = 2.5 \times 10^{-3} \text{ A} = 2.5 \text{ mA}.$$

(Note that V/kΩ = mA and V/MΩ = μA, for quick reference in mental calculations.)

Note that the standard symbol for a resistor is shown in the figure. Other examples of Ohm's law calculations are found in practice problems at the end of this chapter.   ∎

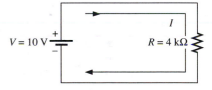

FIGURE 1.4   Simple resistance circuit.

1-6 POWER

When current flows through a resistive element, the electrons suffer inelastic collisions with the conductor atoms, losing some of their energy in heat. In other words, current flowing through a resistor causes a temperature increase in the resistor. This means that energy or power is expended. The amount of power, P, is computed from the definitions of current and voltage:

$$P(\text{J/s}) = V(\text{J/C}) \times I(\text{C/s}). \tag{1-3}$$

The SI unit for power is the *watt* (W) and is defined as one joule per second or one volt − ampere. Alternatively, by substitution from Ohm's law, power can be expressed as

$$P = V^2/R \qquad \text{(1-4a)}$$

or

$$P = I^2R \qquad \text{(1-4b)}$$

Resistors are manufactured with specific ratings stating the maximum rate at which they are capable of dissipating heat without the resistance value being changed significantly. Typical values of power ratings in resistors commonly available are 1/8, 1/4, 1/2, 1, 2, 5 and 10 W. Note that a resistor dissipating power at its maximum rating will get quite hot. A resistor of higher power rating than minimally required should be chosen but, because power, physical size and expense increase together, resistors with inordinately large power ratings should be avoided.

1-7 TYPES OF RESISTORS

Resistors are available in a wide variety of shapes, sizes and compositions. The most common types are carbon-film, carbon-composite, metal-film and wire-wound resistors. The carbon-film resistor is by far the most frequently used variety in electronic circuitry. It is readily available in power ratings from 1/8 W to 1 W, resistance values from 0.5 Ω to 10 MΩ and tolerances (possible variation of advertised resistance with actual resistance) of 5%, 10% or 20%. At the time of publishing, the 1/4 W, 5% tolerance, carbon-film resistor, the most generic variety in this family, is available for less than a penny each in quantities of 1000, and at a slightly higher price in smaller quantities.

For circuits requiring tolerances of 1%, metal-film resistors are a wise choice. These are commonly available with power ratings of 1/8, 1/4 and 1/2 W and resistance values from 4.02 Ω to 2.2 MΩ. Costs of metal-film resistors are about two-fold those of the carbon-film variety.

A standard resistor color code is used to signify the values of both carbon-film and most metal-film resistors. At first this may seem a bit awkward; would it not be simpler to simply write the value on the resistor? The answer is: for the experienced user, the color code indicates the value and tolerance of the component at a quick glance, regardless of the position of the resistor. Once accustomed to the color code, you will find it an efficient means of identifying resistors. Figure 1.5 shows the standard color coding for resistors.

Sometimes resistors of large power ratings are needed. Here we turn to wire-wound resistors which are readily available with ratings from 3 W to hundreds of watts. The wire-wound resistor consists of a high-resistivity metal such as nichrome, usually wrapped around a ceramic or air core, and housed in an aluminum, ceramic, or cement casing or left open to the air. Because they are generally rather large in physical size, the resistance values of these components, which commonly range from 0.1 Ω to 100 kΩ, are usually printed on the housings. Prices range from half a dollar to several

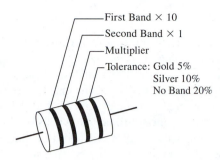

- First Band × 10
- Second Band × 1
- Multiplier
- Tolerance: Gold 5%
             Silver 10%
             No Band 20%

Silver	−2
Gold	−1
Black	0
Brown	1
Red	2
Orange	3
Yellow	4
Green	5
Blue	6
Violet	7
Gray	8
White	9

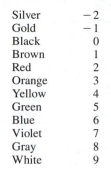

Value = (First Band) (Second Band) × 10$^{\text{(Multiplier)}}$ ± Tolerance
Example: Red, Violet, Orange, Gold = $27 \times 10^3 \, \Omega \pm 5\% = 27 \, \text{k}\Omega \pm 5\%$

FIGURE 1.5   Standard resistor code.

dollars per resistor. Figure 1.6 shows the relative physical sizes of resistors with a variety of power ratings.

Numerous other types of fixed-value resistors are easily available to meet just about any application. Precision resistors with tolerances of much less than 1% can be obtained, as can resistors with extremely high resistances. However, one should be cautioned to use specialty resistors only when such use cannot be avoided. Generally speaking, a circuit which requires many high tolerance resistors, or high wattage resistors, is more than likely a bad design. There are, of course, exceptions, particularly in instrumentation circuits. Circuits which require moderate precision typically use 1/4 W metal-film resistors.

In addition to the power, resistance and tolerance ratings, several other parameters such as moisture resistance, maximum operating voltage and temperature are sometimes important, particularly in precision circuits. Special resistors are available for which resistances change dramatically as functions of temperature (thermistors) or the amount of light falling on them (photo-resistors). These will be discussed in a later chapter on transducers.

1-8   KIRCHOFF'S EQUATIONS

A number of simple circuits can be treated mathematically with only Ohm's law; however, more complicated circuits require the use of Kirchoff's equations. The first of these, Kirchoff's voltage equation, states that the algebraic sum of the voltages in any closed circuit loop is zero:

$$\Sigma \, V = 0. \tag{1-5}$$

This law is a statement of the conservation of energy in a closed loop.

The second, Kirchoff's current equation, states that the algebraic sum of the currents into any node (junction) in the circuit is zero:

$$\Sigma \, I = 0. \tag{1-6}$$

where current entering the junction is considered positive and current leaving the junction is considered negative. This is a statement of conservation of charge. A few illustrations will clarify these laws and their applications.

Carbon Film

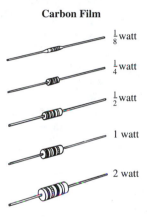

$\frac{1}{8}$ watt

$\frac{1}{4}$ watt

$\frac{1}{2}$ watt

1 watt

2 watt

Wire Wound

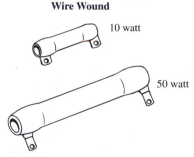

10 watt

50 watt

FIGURE 1.6   Various types of resistors with the different wattages, shown approximately one-half actual size.

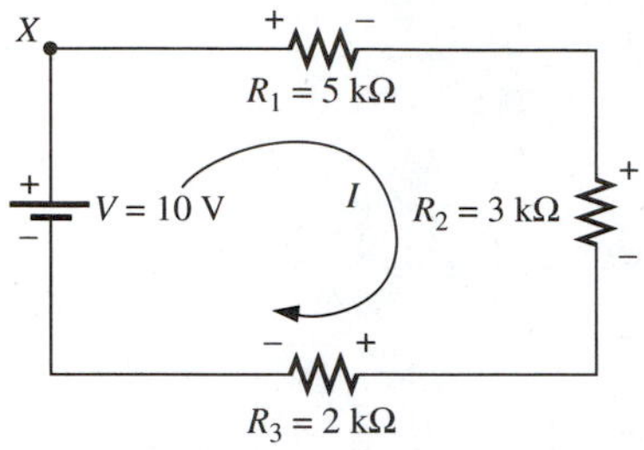

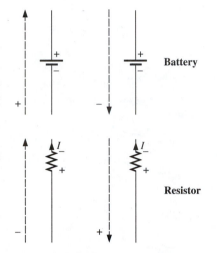

FIGURE 1.7 Series resistance circuit.

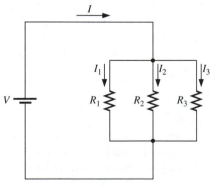

FIGURE 1.8 Sign convention for Kirchoff's equation. Dashed line indicates direction of analysis, solid line indicates current flow.

Resistors in Series

Note that the first law refers to the algebraic sum of the voltages around a loop. There are several "points of view" which can be used to ensure that the correct algebraic signs are used in the analysis. Consider the series resistor circuit in Figure 1.7. The direction of current flow is from the positive terminal of the battery, through the circuit, and back to the negative terminal. Note that the same current I flows through each resistor. As we move clockwise around the loop starting at point X, we add the voltage drops using the sign conventions shown in Figure 1.8. In the circuit of Figure 1.7 this is represented as

$$- IR_1 - IR_2 - IR_3 + V = 0. \tag{1-7}$$

Alternatively, if moving around the loop in a counter-clockwise direction starting at point X we find

$$- V + IR_1 + IR_2 + IR_3 = 0. \tag{1-8}$$

Clearly, the result is the same in each case and we have shown that the direction of analysis taken around the loop is irrelevant as long as we adhere to the conventions of Figure 1.8. Upon rearrangement of Equation 1-7 or 1-8

$$V/I = R_1 + R_2 + R_3 = R_s$$

or

$$R_s = \Sigma R_i \tag{1-9}$$

Hence the effective value of resistance of a *series* (connected one after the other) resistor circuit, R_s, is the sum of the individual resistances.

For the circuit of Figure 1.7, the current can be readily calculated as

$$I = V/R_s = 10 \text{ V} / (5.0 \text{ k}\Omega + 2.0 \text{ k}\Omega + 3.0 \text{ k}\Omega) = 1.0 \text{ mA}$$

Resistors in Parallel

Another arrangement of resistors, the parallel (connected side by side) circuit is shown in Figure 1.9. Note that the voltage is the same across all resistors, and accordingly the currents through the resistors will not be the same unless all resistors have the same value. The junction points (where the resistors are connected) are called nodes. Using Kirchoff's current law and designating individual currents in the same manner as the resistors, we can write for the upper node

$$\Sigma I = 0 = I - I_1 - I_2 - I_3 \tag{1-10}$$

since I flows into the upper node and I_1, I_2 and I_3 flow out of it. Applying Ohm's law to Equation 1-10 for each current through each resistor yields

$$I = (V/R_1) + (V/R_2) + (V/R_3) = V/R_p \tag{1-11}$$

where R_p is the effective resistance of the parallel network. This equation simplifies to

$$1/R_p = (1/R_1) + (1/R_2) + (1/R_3)$$

or

$$1/R_p = \Sigma (1/R_i) \tag{1-12}$$

which is the equation for equivalent resistance for resistors in parallel.

FIGURE 1.9 Parallel resistance circuit.

Note that for two resistors in parallel, R_1 and R_2, Equation 1-12 can be written as the product of the two resistance values divided by the sum, or

$$R_p = R_1R_2/(R_1 + R_2). \qquad \textbf{(1-12a)}$$

Examples of Kirchoff's Equations and Resistors in Series and Parallel

The equations developed above for series and parallel resistance networks can be used to simplify more complex resistance networks in a stepwise manner, as demonstrated in the example below.

EXAMPLE 1.2

Calculate the current, I, from the battery in the circuit shown in Figure 1.10A. Then calculate the voltage drop across each resistor in the circuit.

Solution:

Step One: Calculate the equivalent resistance for the six resistors. Figure 1.10B shows the result of combining the 6 Ω and 12 Ω resistors which are in parallel ($1/R_p = 1/6 + 1/12$, $R_p = 4$ Ω). Figure 1.10C shows the combination of the 6 Ω and 4 Ω resistors in series ($R_s = 4 + 6 = 10$ Ω). Figure 1.10D shows the combination of the two 10 Ω parallel resistors and Figure 1.10E shows the equivalent resistance of the circuit after combining the three series resistors shown in Figure 1.10D. The current from the battery is thus $I = V/R = 7.5$ V/15 Ω $= 0.5$ A.

Step Two: Find the currents. Assign current directions using the knowledge that current flows, in the external circuit, from the positive terminal to the negative terminal of the battery as shown in Figure 1.11, then define loops and apply Kirchoff's laws.

Loop 1: Kirchoff's voltage equation yields

$$7.5 \text{ V} = I5 + I_2 10 + I5$$

and since $I = 0.5$ A,

$$\text{then } I_2 = 0.25 \text{A.} \qquad \textbf{(1-13)}$$

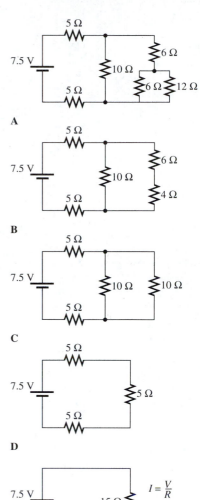

A

B

C

D

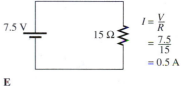

E

FIGURE 1.10   Steps in circuit simplification.

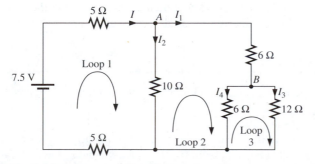

FIGURE 1.11   **Assigned current directions.**

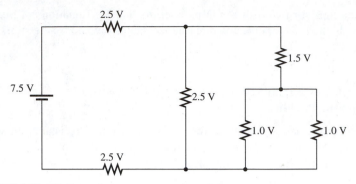

FIGURE 1.12   **Voltage drops across each resistor.**

Node A: Kirchoff's current equation yields

$$I = I_1 + I_2$$

and since $I = 0.5$ A and $I_2 = 0.25$ A, then

$$I_1 = 0.25 \text{ A.} \tag{1-14}$$

Node B: Kirchoff's current equation again gives

$$I_1 = I_3 + I_4. \tag{1-15}$$

Loop 3: Kirchoff's voltage equation gives

$$6 I_4 = 12 I_3, \text{ or } I_4 = 2 I_3. \tag{1-16}$$

Combining Equations 1-14, 1-15 and 1-16 yields

$$I_1 = 3 I_3 \text{ and } I_3 = (0.25/3) \text{ A and } I_4 = (0.5/3) \text{ A.}$$

Step Three: Evaluate the voltages. Knowing the four currents in the circuit, use Ohm's law to calculate the voltage drop across each resistor. The results are shown in Figure 1.12. ■

Note that for a circuit this complex, you can take many different routes to solve the simultaneous equations generated by applying Kirchoff's laws. The above analysis represents one way. Another way is to write *n* independent equations for *n* unknowns and then solve them using the method of determinants. This method is discussed in Appendix A.

EXAMPLE 1.3

Calculate the currents I_1, I_2 and I_3 in the circuit shown in Figure 1.13.

Solution:

Step One: Write down Kirchoff's current equation, the sum of the currents into a node equals the sum of the current leaving the node, for node *X*

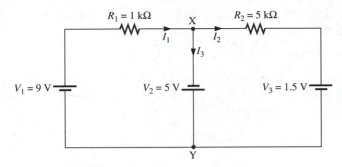

FIGURE 1.13   **Multiple battery-resistor circuit in Example 1.3.**

(or equivalently node Y). Note that the current directions were arbitrarily chosen.

$$I_1 = I_2 + I_3 \qquad\qquad \textbf{(1-16a)}$$

Step Two: Choose loops and apply Kirchoff's voltage law. Here we choose the first loop represented by V_1, R_1 and V_2 and the second loop V_2, R_2 and V_3. Using the clockwise direction, Kirchoff's law yields

$$V_1 - I_1R_1 + V_2 = 0 \qquad\qquad \textbf{(1-16b)}$$

and

$$-V_2 - I_2R_2 - V_3 = 0 \qquad\qquad \textbf{(1-16c)}$$

or

$$I_1 = (V_1 + V_2)/R_1 = (9\ \text{V} + 5\ \text{V})/1\ \text{k}\Omega = 14.0\ \text{mA}$$

and

$$I_2 = -(V_2 + V_3)/R_2 = -(5\ \text{V} + 1.5\ \text{V})/5\ \text{k}\Omega = -1.3\ \text{mA}.$$

Substituting values for I_1 and I_2 into Equation 1-16a gives $I_3 = 12.7$ mA. Notice that the assumed direction for I_2 was incorrect as indicated by the negative sign on the current value. As a check to our work, we can choose the outer loop of the circuit comprised of all components except V_2 and apply Kirchoff's voltage equation to see if it holds true. We use the same current directions as before and obtain

$$V_1 - I_1R_1 - I_2R_2 - V_3 = (?)0. \qquad\qquad \textbf{(1-16d)}$$

Substituting values yields

$$9\ \text{V} - (14 \times 10^{-3}\ \text{A})(1 \times 10^3\ \Omega)$$

$$- (-1.3 \times 10^{-3}\ \text{A})(5 \times 10^3\ \Omega) - 1.5\ \text{V} = (?)0$$

$$9\ \text{V} - 14\ \text{V} + 6.5\ \text{V} - 1.5\ \text{V} = 0$$

which verifies our previous result.   ■

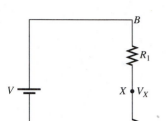

FIGURE 1.14   Voltage divider.

1-9   VOLTAGE DIVIDERS, POTENTIOMETERS AND CURRENT DIVIDERS

Voltage Divider

Consider the circuit shown in Figure 1.14. Since the current flowing in the circuit is given by $I = V/(R_1 + R_2)$ and the voltage drop across R_2 is $V_X = I R_2$, then the voltage at point X is given by

$$V_x = V \left(\frac{R_2}{R_1 + R_2} \right). \qquad (1\text{-}17)$$

The two resistors in series constitute a "voltage divider," because the voltage at the point between them is some fraction (determined by the ratio $R_2/(R_1 + R_2)$ of the battery voltage. This application of a resistor network is very common in electronics, since it is often necessary to control a voltage by reducing it in a predictable fashion. We will see many such uses in later chapters.

EXAMPLE 1.4

Using two resistors, one of which has a 10 kΩ resistance and a 12 V battery, design a circuit from which an 8 V potential can be derived.

Solution: Using Equation 1.17 and the two resistor voltage divider in Figure 1.14 we find that the ratio of V and V_X is

$$V_x / V = 8 \text{ V} / 12 \text{ V} = 2/3 = \frac{R_2}{R_1 + R_2}.$$

If we assign R_2 the value of 10 kΩ, then we find R_1 to be equal to 5 kΩ. Likewise if we assign R_1 the value of 10 kΩ, then R_2 would be 20 kΩ.   ∎

Potentiometer

The variable resistor, or *potentiometer* ("pot" for short), is a form of voltage divider. The variable resistor is best understood by examining its schematic symbol shown in Figure 1.15. Notice that it is a three-terminal device (three connections). A fixed resistance is located between terminals B and C, and terminal A is attached to a "wiper" which slides along the fixed resistor. In this way the resistance between A and C goes up as the wiper approaches terminal B, while the resistance between A and B goes down and vice-versa. The potentiometer finds use as a variable voltage divider in many applications in electronic instrumentation where it may be used to control some circuit parameter. A common example in a consumer instrument is the volume control of an audio amplifier. Figure 1.16 shows some common types of variable resistors.

The potentiometer can also be used as a two-terminal variable resistor by connecting it into a circuit using either terminals B and A, or terminals C and A. When used as a two-terminal variable resistor, the free terminal should be connected to the wiper to reduce stray signals from being picked up by the circuit. Although the concept of voltage dividers is described here

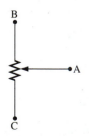

FIGURE 1.15   Variable resistor schematic symbol.

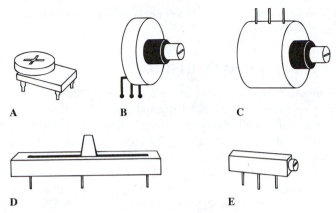

A   B   C

D   E

FIGURE 1.16 **Several types of variable resistors: (A) trimmer pot, (B) turn pot, (C) 10 turn pot, (D) slide pot, (E) 10 turn trim pot.**

in terms of resistors, the concept also applies to other types of circuit components.

Current Divider

Now consider the circuit in Figure 1.17 which shows a current source and two parallel resistors. The current source, shown schematically on the left as two interlocking circles, supplies a constant current. Until now we have discussed ideal voltage sources. An ideal voltage source supplies a constant voltage and is capable of producing any current demanded, whereas an ideal current source produces a constant current at any voltage demanded. Thus a 1 A current source would supply a 1 A current through a 10 Ω resistor at 10 V and through a 100 Ω resistor at 100 V.

The voltage supplied by the current source can be found given I, the current produced by the source and R_1 and R_2 as

$$V = [R_1R_2/(R_1 + R_2)] I \qquad \text{(1-17a)}$$

Therefore, the currents through R_1 and R_2 are given by

$$I_1 = V/R_1 = [R_2/(R_1 + R_2)] I \qquad \text{(1-17b)}$$

and

$$I_2 = V/R_2 = [R_1/(R_1 + R_2)] I \qquad \text{(1-17c)}$$

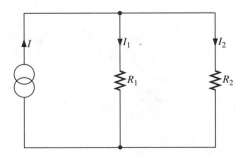

FIGURE 1.17 **Constant current source in current divider circuit.**

The current divider is useful but is not employed in electronic circuitry as often as the voltage divider.

1-10   THE MESH LOOP METHOD

A large number of techniques can be used to simplify the calculation of circuit parameters. Most are applied by specialty electronic designers, but several have rather wide applicability and general utility. These will be presented as an aid to the student because they do indeed simplify some otherwise complex calculations. The mesh loop method is one such technique.

Consider the circuit of Figure 1.18. If the currents in the three resistors are to be calculated by Kirchoff's equations, a set of three simultaneous equations is needed. If the mesh loop method described below is applied to the circuit, however, only two equations are needed because only one current is drawn for each loop shown in the figure. By calling the respective currents I_a and I_b and using Kirchoff's voltage equations,

$$V = I_a R_1 + R_2(I_a - I_b). \tag{1-18}$$

and

$$0 = I_b R_3 + R_2(I_b - I_a). \tag{1-19}$$

Rearranging these equations yields

$$V = (R_1 + R_2)\, I_a - R_2 I_b \tag{1-20}$$

and

$$0 = -R_2 I_a + (R_2 + R_3)\, I_b. \tag{1-21}$$

Rearrangement of Equation 1-21 yields, as an intermediate step,

$$I_b = R_2 I_a/(R_2 + R_3). \tag{1-22}$$

Equation 1-22 is now substituted into Equation 1-20, and the expression is reduced to its simplest form:

$$I_a = (V(R_2 + R_3))/(R_1 R_2 + R_1 R_3 + R_2 R_3). \tag{1-23}$$

This is inserted into Equation 1-22 to yield

$$I_b = (V\, R_2)/(R_1 R_2 + R_1 R_3 + R_2 R_3). \tag{1-24}$$

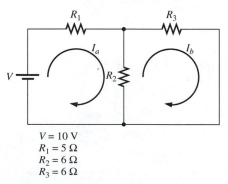

$V = 10\ \text{V}$
$R_1 = 5\ \Omega$
$R_2 = 6\ \Omega$
$R_3 = 6\ \Omega$

FIGURE 1.18   **Circuit for mesh loop method.**

It should be obvious that I_a is the current in R_1 and that I_b is the current in R_3. The current in R_2 is thus $(I_a - I_b)$.

A more vivid example of the utility of the mesh loop method is illustrated in the calculation of the currents in the Wheatstone bridge circuit shown in Figure 1.19. The three currents are identified in the figure as I_a, I_b and I_c. The voltage loop expressions for the three current loops are

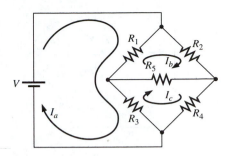

$$V = R_1(I_a - I_b) + R_3(I_a - I_c). \tag{1-25}$$

$$0 = R_1(I_b - I_a) + R_2I_b + R_5(I_b - I_c). \tag{1-26}$$

$$0 = R_3(I_c - I_a) + R_5(I_c - I_b) + R_4I_c. \tag{1-27}$$

FIGURE 1.19   **Loop method for Wheatstone bridge circuit.**

Collecting terms containing the same current,

$$V = I_a(R_1 + R_3) - I_bR_1 - I_cR_3 \tag{1-28}$$

$$0 = -I_aR_1 + I_b(R_1 + R_2 + R_5) - I_cR_5 \tag{1-29}$$

$$0 = -I_aR_3 - I_bR_5 + I_c(R_3 + R_4 + R_5). \tag{1-30}$$

If the values for the parameters shown in the diagram are used, the current values can be found by solving the simultaneous equations to be

$$I_a = 0.267 \text{ A}, I_b = 0.140 \text{ A}, \text{ and } I_c = 0.113 \text{ A}.$$

Moreover, if we number the individual currents through each resistor using the same scheme as we have for each component (current through R_1 is I_1, R_2 has I_2, etc., and identify I_0 as the current out of the battery, then

$$I_0 = I_a = 0.267 \text{ A}$$

$$I_1 = I_a - I_b = 0.127 \text{ A} \tag{1-31}$$

$$I_2 = I_b = 0.140 \text{ A}$$

$$I_3 = I_a - I_c = 0.154 \text{ A} \tag{1-32}$$

$$I_4 = I_c = 0.113 \text{ A}$$

$$I_5 = I_b - I_c = 0.027 \text{ A}. \tag{1-33}$$

These are, as expected, the same currents that would be found using only Kirchoff's equations; however, here we had to handle only three simultaneous equations instead of six.

1-11   THÉVENIN'S EQUIVALENT CIRCUIT THEOREM

While analyzing electrical circuits, we often wish to replace part (or all) of a circuit with a greatly simplified representation. For example, if we wished to compute the effect of different detector resistances in the Wheatstone bridge of the preceding example, we could tediously go through all of the calculations each time R_5 was changed. Thévenin's theorem permits the replacement of the bridge circuit (except for R_5) with a voltage source and a single resistance in series with it.

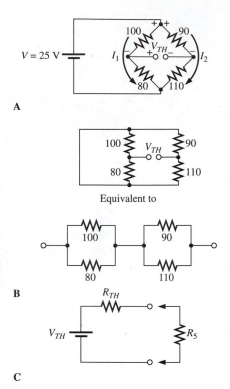

A

Equivalent to

B

C

FIGURE 1.20   **Thévenin's theorem applied to the Wheatstone bridge circuit.**

Thévenin's theorem states that a complex network of sources and resistances (or generalized impedances, which will be defined in the next chapter) can be replaced by a single equivalent source and a single equivalent resistance (or generalized impedance) in series with the source.

When this theorem is applied to the example cited, computation of the change in current flow due to change of detector resistance is reduced to a simple and straightforward calculation for a voltage with two resistances in series. Other examples of this theorem will be presented in later chapters.

The steps necessary to use Thévenin's theorem are as follows:

1. The element to be studied is "removed" from the circuit, and the resulting open ends of the circuit are labeled V_{TH}.
2. The value of V_{TH} is computed.
3. All voltage sources are short-circuited and all current sources are open-circuited.
4. The equivalent resistance of the circuit, R_{TH}, is calculated.
5. The connection of V_{TH} in series with R_{TH} is made and the element under study is reinserted into the series circuit for calculations.

These "rules" will become clear by use of an example, for which we will use the same Wheatstone bridge circuit as before.

Step 1. R_5 is removed and the open terminals are labeled V_{TH} (see Figure 1.20A). The polarity assigned is arbitrary and will be verified in the calculations.

Step 2. The evaluation of V_{TH} is performed using Kirchoff's laws:

$$0 = 25 - (100 + 80)I_1 \tag{1-34}$$

$$0 = 100I_1 + V_{TH} - 90I_2 \tag{1-35}$$

$$0 = 25 - (90 + 110)I_2 \tag{1-36}$$

The result is $V_{TH} = -2.64$ V. The minus sign means only that the arbitrary choice of polarity was incorrect.

Step 3. The voltage source is shorted out and R_{TH} is calculated (see Figure 1.20B):

$$R_{TH} = \frac{(100)\,(80)}{100 + 80} + \frac{(90)\,(110)}{90 + 110} = 93.94 \ \Omega$$

Note that when the source is shorted out, the resistors that were in series (R_1 and R_3; R_2 and R_4) become parallel combinations.

Step 4. The network is assembled in series as shown in Figure 1.20C, and the current through R_5 is calculated.

$$I_5 = \frac{V_{TH}}{R_{TH} + R_5} = \frac{-2.64}{93.94 + 2} = -0.027 \text{ A}$$

Note that the numerical value of the current is the same as that in the preceding calculations, but the sign is opposite. This is simply due to the incorrect choice of polarity of V_{TH} for this calculation. In fact, the current flow is in the same direction in both examples, as would be expected.

A practical use of Thévenin's theorem is found in the "loaded" voltage divider shown in Figure 1.21. When $R_L = \infty$,

$$V_0 = \frac{R_2}{R_1 + R_2} V \qquad \text{(1-37)}$$

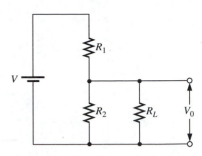

When R_L is finite, the above expression fails because current is drawn through R_L. Application of Thévenin's theorem offers a general solution to this problem. With R_L removed from the circuit,

$$V_{TH} = V_0 = \frac{R_2}{R_1 + R_2} V$$

FIGURE 1.21   The loaded voltage divider circuit.

When the battery is short-circuited, it can also be seen that

$$R_{TH} = \frac{R_1 R_2}{R_1 + R_2} \qquad \text{(1-38)}$$

Forming the Thévenin equivalent circuit with V_{TH}, R_{TH}, and R_L in series, we can now calculate $I = V_{TH}/(R_{TH} + R_L)$; then V_L, the voltage developed across R_L, is calculated as:

$$V_L = IR_L = \frac{\dfrac{VR_2 R_L}{R_1 + R_2}}{\dfrac{R_1 R_2}{R_1 + R_2} + R_L} = \frac{VR_2}{\dfrac{R_1 R_2}{R_L} + (R_1 + R_2)} \qquad \text{(1-39)}$$

When the first term in the denominator, $R_1 R_2 / R_L$, is small compared to the second term (that is, $R_L \gg R_1 R_2$), the voltage divider output approximates the "no load" condition. Table 1.4 presents selected values which illustrate this effect.

TABLE 1.4   Voltage Divider Performance

$V = 10.0\text{ V}$ $R_1 = R_2 = 5\text{ k}\Omega$ $\therefore V_0 = 5\text{ V}$		
$R_L(\Omega)$	$V_L(\text{V})$	$V_0 - V_L(\text{V})$
500	0.833	4.167
5 k	3.333	1.167
50 k	4.762	0.238
500 k	4.975	0.025
5 M	4.998	0.002
50 M	5.000	0.000

1-12 NORTON'S EQUIVALENT CIRCUIT THEOREM

Thévenin's theorem was used to replace a complicated circuit with a single voltage source and a resistance in series with that source. Norton's theorem replaces a complicated circuit with a *constant current source* and a single resistance in *parallel* with the current source. These replacement theorems are the essence of circuit modeling. Transistor modeling begins with a current source; thus, Norton's theorem is more useful for this purpose than Thévenin's.

The steps necessary to use Norton's theorem are as follows:

1. The element to be studied is removed and replaced by a wire. The current in that wire is the short-circuit or Norton's current, I_{NOR}.

2. The value of I_{NOR} is calculated.

3. All voltage sources are short-circuited and all current sources are open-circuited.

4. The equivalent resistance, R_{NOR}, of the circuit is calculated. (Note that R_{NOR} always equals R_{TH}.)

5. The connection of I_{NOR} in parallel with R_{NOR} is made. The element removed earlier is also placed in parallel.

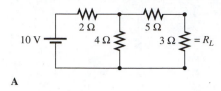

A

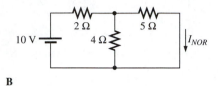

B

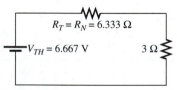

C

FIGURE 1.22 **Norton's theorem applied to a resistive circuit.**

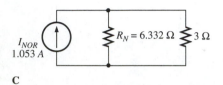

FIGURE 1.23 **Thévenin's equivalent circuit for the circuit of Figure 1.22.**

The use of Norton's theorem in this stepwise manner is illustrated with the circuit of Figure 1.22A. The element of interest is the 3 Ω resistor. Removing it and replacing it with a shorting wire results in the circuit of Figure 1.22B. The current in the 5 Ω resistor is I_{NOR}. This current may be calculated in two steps. First, the equivalent resistance of the circuit is used to calculate the current flowing through the battery:

$$R_{eq} = 2\ \Omega + \frac{5\ \Omega \cdot 4\ \Omega}{4\ \Omega + 5\ \Omega} = 4.222\ \Omega \tag{1-40}$$

$$I = \frac{V}{R_{eq}} = \frac{10\ \text{V}}{4.222\ \Omega} = 2.368\ \text{A} \tag{1-41}$$

The current just calculated is the current in the 2 Ω resistance. Using current division, the current in the 5 Ω resistance (I_{NOR}) can be calculated:

$$I_{NOR} = \left(\frac{4\ \Omega}{4\ \Omega + 5\ \Omega}\right)(2.368\ \text{A}) = 1.053\ \text{A} \tag{1-42}$$

The Norton resistance, R_{NOR}, is the resistance "seen" looking into the circuit from the place where the 3 Ω resistor was originally removed and with the voltage source shorted out:

$$R_{NOR} = \frac{2\ \Omega \cdot 4\ \Omega}{2\ \Omega + 4\ \Omega} + 5\ \Omega = 6.333\ \Omega \tag{1-43}$$

Finally, the Norton current source is placed in parallel with the Norton resistance. The element of interest is also added in parallel, as shown in Figure 1.22C. Current division is used to calculate the current in the 3 Ω resistance,

$$I = \left(\frac{6.333\ \Omega}{3\ \Omega + 6.333\ \Omega}\right)(1.053\ \text{A}) = 0.714\ \text{A} \tag{1-44}$$

The Thévenin equivalent for this circuit is given in Figure 1.23. The current in the Thévenin series circuit is the current through the 3 Ω resistor.

$$I = \frac{6.667\ \text{V}}{3\ \Omega + 6.333\ \Omega} = 0.714\ \text{A}$$

Thus, the two equivalent circuits (that is, equivalent to the complicated circuit we began with) are equivalent to each other, which is as it should be. Hence, the choice of which theorem to use depends entirely upon personal whim.

1-13 VOLTAGE AND CURRENT SOURCES

In modeling complicated processes, we frequently use a source of voltage or current. An *ideal voltage source* is one which has zero Thévenin resistance. This means that the voltage does not change as the load on the source is changed. A symbolic comparison between ideal and real voltage sources is depicted in Figure 1.24A. An *ideal current source* provides a specified current regardless of the load through which the current is driven. Thus, the

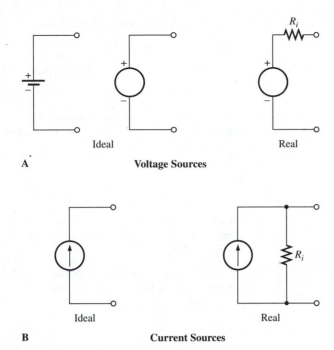

A   **Voltage Sources**

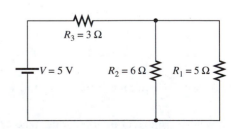

FIGURE 1.24   **Voltage and current sources.**

ideal current source has an infinite Norton resistance. The ideal and real current sources are symbolized in Figure 1.24B. In later chapters, time-varying voltage and current sources will be encountered. As long as their outputs are unaffected by the circuit load, they are ideal.

The mercury battery closely approximates an ideal voltage source when used correctly, in that the battery voltage remains constant. A simple constant current source can be constructed with a battery and a very large resistance in series with it. Within limits, the current is constant as long as the load resistance does not approach that of the constant current resistance.

1-14   QUESTIONS AND PROBLEMS

1.   What is the resistance of a carbon rod which is 1 cm in diameter and 10 cm long?

2.   A nichrome wire, 1 mm in diameter and 1 meter long, has what resistance?

3.   What is the maximum allowable current through a 10 kΩ, 10 W resistor? Through a 10 kΩ, 1/4 W resistor?

4.   What wattage rating is needed for a 100 Ω resistor if 100 V is to be applied to it? For a 100 kΩ resistor?

5.   Compute the current through R_3 of Figure A.

6.   Compute the currents through R_1 and R_2 of Figure A.

7.   The output of the voltage divider of Figure B is to be

FIGURE A   **Circuit for problems 5 and 6.**

$R_3 = 3\ \Omega$

$V = 5\ V$ $R_2 = 6\ \Omega$ $R_1 = 5\ \Omega$

measured with voltmeters whose effective resistances are 100 Ω, 1 kΩ, 50 kΩ, and 1 MΩ. What voltage will each indicate?

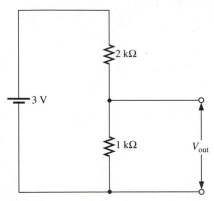

FIGURE B Circuit for problem 7.

8. The internal resistance of a 1.5 V zinc-carbon battery is found to be 1.5 Ω. What is the maximum current available?

9. A 12 V storage battery is rated at 70 A. What is its internal resistance?

10. A voltmeter whose resistance is 1000 Ω measures the voltage of a worn-out 1.5 V flashlight battery as 0.9 V. What is the internal resistance of the battery?

11. If the flashlight battery of the preceding problem had been measured with a voltmeter with a resistance of 10 MΩ, what voltage would have been read?

12. What is the effective resistance of the circuit in Figure C?

13. Suppose that a 25 V battery had been applied to the terminals of the circuit in Figure C. Calculate the current in the 10 Ω resistor.

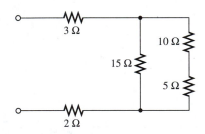

FIGURE C Circuit for problems 12 and 13.

14. Derive the conditions of balance for the Wheatstone bridge.

15. Compute the current through R_2 of the circuit in Figure D.

16. Using the following values for the circuit in Figure D, compute the current through R_3.

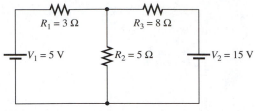

FIGURE D Circuit for problems 15, 16, and 17.

$$V_1 = 1.500 \text{ V} \quad V_2 = 1.082 \text{ V}$$

$$R_1 = 2787 \ \Omega \quad R_2 = 7213 \ \Omega \quad R_3 = 10 \ \Omega$$

17. Using the values of the preceding problem, compute the current through R_2. Also compute the current through R_2 when $R_3 = 10$ kΩ.

18. Using Thévenin's theorem, determine the effective voltage and the effective resistance of the circuit in Figure E with V_2 removed.

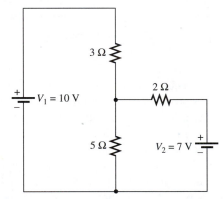

FIGURE E Circuit for problem 18.

19. Using Thévenin's theorem, determine the effective voltage and the effective resistance of the circuit in Figure F with R_5 removed.

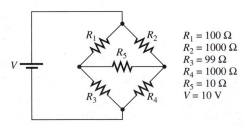

FIGURE F Circuit for problems 19, 20, 21 and 22.

20. What is the current through R_5 of the circuit of Figure F?

21. If $R_5 = 10$ kΩ instead of 10 Ω in Figure F, what would be the current through it?

22. Calculate the voltage across R_5 for the conditions cited in Problems 20 and 21.

23. In the circuit of Figure G, compute the current in the 3 Ω resistor and find the value of V_2.

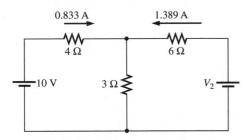

FIGURE G Circuit for problem 23.

24. In the circuit of Figure H, find the value of V_3 such that the current through the 10 Ω resistor is zero.

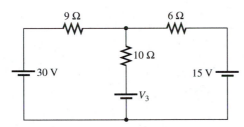

FIGURE H Circuit for problem 24.

25. Compute the current in each of the resistors in the circuit of Figure I.

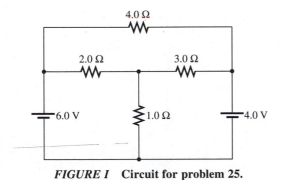

FIGURE I Circuit for problem 25.

26. Compute the currents labeled in the circuit of Figure J.

27. Compute the current through the 10 Ω resistor in the circuit of Figure K. Also compute the Thévenin voltage, the Thévenin resistance, and the Norton current when the 10 Ω

resistor is removed. Verify the equivalence of these circuits by comparing the current through the 10 Ω resistor in each case.

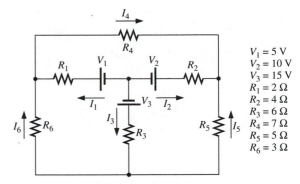

FIGURE J Circuit for problem 26.

$V_1 = 5$ V
$V_2 = 10$ V
$V_3 = 15$ V
$R_1 = 2\ \Omega$
$R_2 = 4\ \Omega$
$R_3 = 6\ \Omega$
$R_4 = 7\ \Omega$
$R_5 = 5\ \Omega$
$R_6 = 3\ \Omega$

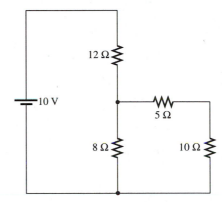

FIGURE K Circuit for problem 27.

28. Write the appropriate equations for the three currents designated in Figure L.

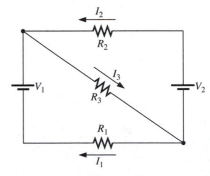

FIGURE L Circuit for problems 28 and 29.

29. Calculate the current through R_3 of the circuit in Figure L when $V_1 = 5$ V, $V_2 = 10$ V, $R_1 = 20\ \Omega$, $R_2 = 40\ \Omega$, and $R_3 = 2\ \Omega$.

Capacitors & Inductors

2-1 INTRODUCTION

In Chapter One we dealt exclusively with currents and voltages that do not vary with time. However, most electrical signals that we encounter in instrumentation circuitry vary with time. We shall now investigate the properties of time-dependent signals as well as their interactions with electrical components such as resistors, capacitors and inductors, and the transient phenomena which occur in DC circuits using these components.

Note that this and the next chapter are by far the most mathematically demanding chapters in the text. For the reader with little mathematical background, be aware that although the remainder of the text uses the principles presented in these chapters, a detailed knowledge of the derivations involved is not necessarily required for further understanding.

2-2 CAPACITANCE

A *capacitor* is an element commonly encountered in electronic circuitry. It has the property of being able to store charge, Q. The voltage, V, across the capacitor is proportional to the amount of charge stored, or

$$V = Q/C \qquad (2\text{-}1)$$

where the proportionality constant, C, is defined as the *capacitance*.

The unit of capacitance is the *farad* (F), which is equal to one coulomb per volt. [The farad is named after British physicist Michael Faraday (1791–1867).] More frequently encountered in circuit applications are the microfarad (1 μF = 1×10^{-6}F) and the picofarad (1 pF = 1×10^{-12}F). For historical reasons, the nanofarad (1 nF = 1×10^{-9}F), a perfectly acceptable SI unit, is less commonly used as a unit of capacitance and the picofarad is sometimes called a micro-microfarad ($\mu\mu$F).

Physically, a capacitor is best described as two thin, parallel plates of metal separated by an insulating (or dielectric) material. For a parallel-plate type capacitor, as shown in Figure 2.1A, the value of its capacitance depends

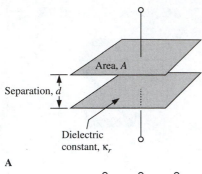

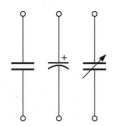

A

B

FIGURE 2.1 **The capacitor. (A) The physical representation. (B) The electronic symbols for non-polarized, polarized and variable capacitors.**

upon the area, *A,* of one plate, the distance, *d,* between the plates, the permittivity of free space, ϵ_O ($\epsilon_O = 8.85 \times 10^{-12}$ F/m), and a property called the *dielectric constant,* κ, of the medium (dielectric) between the plates:

$$C\ (F)\ =\ \kappa\epsilon_O\ (\text{F/m})\ A\ (\text{m}^2)/d(\text{m}) \tag{2-2}$$

Among materials commonly used in commercial capacitors, the dielectric constants vary from approximately 1.0 (air) to 11 (tantalum oxide).

EXAMPLE 2.1

The dielectric constant for polystyrene is approximately 2. If plates with an area of 1 cm^2 are used to sandwich a polystyrene sheet 0.1 mm thick, the nominal value of the capacitance can be calculated as follows:

Solution:

$$C = 2 \times 8.85 \times 10^{-12}\ (\text{F/m}) \times 1\ \text{cm}^2 \times 10^{-4}\ (\text{m}^2/\text{cm}^2)\ /$$

$$[0.1\ \text{mm} \times (10^{-3}\text{m/mm})]$$

$$= 18 \times 10^{-12}\text{F} = 18\ \text{pF}. \qquad \blacksquare$$

The circuit symbols for capacitors are shown in Figure 2.1B. Some idea of the capacitor's function can be derived from the figure. The gap suggests that the device acts as an open circuit. This is true when the capacitor is part of a DC circuit (after all transients have died, and the DC voltage has no time-dependent voltage component as will be explored later in the text).

Capacitors are commercially available in a wide variety of types and values. The value of the capacitance is determined by the dielectric constant of the material used, its thickness and the area of the plates. Some of the common dielectric materials are listed in Table 2.1, along with some of their properties. The largest values are generally those of electrolytic capacitors, which consist of a repeating sandwich of aluminum sheets and a conducting

TABLE 2.1 Commercial Capacitor Properties

Type of Capacitor	Dielectric Material	Range of Capacitance	Maximum Voltage
Variable	Air	5 to 500 pF	500 V
Ceramic	Barium titanate	1000 pF to 1 μF	2000 V
Oil	Paper in oil	0.01 to 1 μF	10000 V
Mica	Mica	100 to 5000 pF	10000 V
Film	Mylar, Teflon, Polystyrene, Polycarbonate	0.01 to 50 μF	1000 V
Electrolytic			
Tantalum	Tantalum oxide	0.01 to 3000 μF	*
Aluminum	Aluminum oxide	0.1 to 100,000 μF	*
Chip	Ceramic (see text)		

*The maximum voltage which can be applied to an electrolytic capacitor is very dependent upon the capacitance value. For example, at 100,000 μF, a voltage in excess of 3 V will cause a breakdown, while at 100 μF, the same type of capacitor can withstand 400 V.

paste, rolled into a cylinder for minimum size. A DC voltage is applied to form very thin layers of aluminum oxide, a dielectric material; as Equation 2-2 shows, the thinner the dielectric medium, the higher the capacitance value will be. This method of construction has the drawback that the capacitor can serve as an electrochemical cell generating hydrogen gas if the voltage source is connected incorrectly. Accordingly, the polarity of the voltage must be observed when the capacitor is used in a circuit. Electrolytic capacitors must be connected in circuits such that the lead marked with a " + " remains more positive than the other lead; otherwise hydrogen gas build-up may cause an explosion and the destruction of the capacitor.

Other dielectric materials often used in capacitors are mica, polyester/mylar, polypropylene and ceramics. Capacitors including these materials provide significantly lower values of capacitance than electrolytic types but offer several advantages, such as much smaller leakage than electrolytic capacitors and no polarity restrictions.

Another major variety of capacitor is the variable type in which air is used as the dielectric material between two alternate series of metal plates. The plates are fixed on separate mountings that can rotate with respect to each other, changing the area by which the two sets of plates overlap each other. Because the area in Equation 2-2 refers only to the overlapping area (and not the total area of the plate), the result of the rotation is a varying capacitance. Because of the low dielectric constant of air and the fact that the two sets of plates must be generously spaced in order to move between each other without touching, variable capacitors have very modest values of capacitance. Some major types of commercial capacitors are shown in Figure 2.2.

A fairly recent development in capacitor manufacture, that of small, surface-mount[1] "chip" capacitors, follows the sandwich idea of the electrolytic capacitors. In this case, the dielectric material is often a ceramic. The spacing is kept very small and the number of sandwiches is very large, to achieve large capacitance values. Chip capacitors can be made in a large range of capacitance values and with very compact sizes. The present range of commonly available sizes is from about 1 pF to 1 μF. Other capacitor types, such as electrolytic, are rapidly becoming available in small surface-mount configurations, although their values are relatively small compared to capacitors in the same family with axial (one lead at either end) or radial (both leads at the same end) leads.

Because no capacitor is perfect, the equivalent circuit of a real capacitor is often best approximated by an ideal capacitor along with two resistors and an inductor (discussed later in this chapter) as shown in Figure 2.3. Here R_p represents a leakage path for the stored charge between the plates. From the circuit it can be seen that the leakage current is approximately

$$I_{\text{leak}} \approx V_{\text{DC}}/R_p. \tag{2-3}$$

[1]Surface-mount technology is fairly new to the electronics industry. Such components are connected directly to the component side of a printed circuit (PC) board, rather than through holes which the component's leads are routed and connected to the board's other side.

Ceramic Capacitors

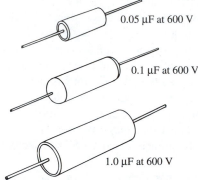

0.2 μF at 10 V

0.01 μF at 50 V

100 pF at 500 V

100 pF at 1000 V

Electrolytic Capacitors

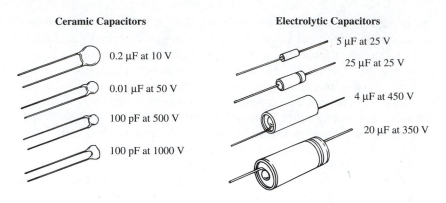

5 μF at 25 V

25 μF at 25 V

4 μF at 450 V

20 μF at 350 V

Mylar Capacitors

0.05 μF at 600 V

0.1 μF at 600 V

1.0 μF at 600 V

Variable Air Capacitor
(Butterfly)

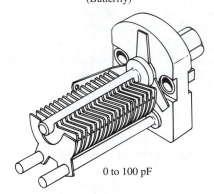

0 to 100 pF

FIGURE 2.2   Various types of capacitors.

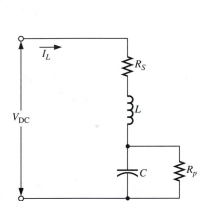

FIGURE 2.3   Equivalent circuit of a capacitor.

TABLE 2.2   Leakage Times of Some Dielectric Materials

Material	Ohms × Farads = s (approximate values for 25°C)
Teflon	2×10^6
Polystyrene	1×10^6
Polycarbonate	2×10^5
Mylar	1×10^5
Glass	1×10^3
Mica	to
Paper	1×10^5
Ceramic	
Electrolytic	1×10^1 to 1×10^3

Since the capacitance and leakage current are both directly proportional to the area of the capacitor, we see that R_p is inversely proportional to the capacitance. Manufacturers generally rate their capacitors by the product of R_p and C, in units of either ohms times farads or megohms times microfarads. If we convert this product to basic units, we find:

$$R(\Omega) \times C \, (\text{F}) = V/I \times Q/V$$

or in units

$$= V / (C/\text{s}) \times C/V = \text{s}.$$

This quantity is called the leakage time of the capacitor; the higher its value, the more capable the capacitor is of holding a charge stored on it. Table 2.2 gives some approximate leakage values for various materials commonly used in capacitors. We can readily judge, on this basis, that a teflon capacitor is much better than an electrolytic capacitor (in terms of leakage), since the leakage time of the former is 2×10^6 seconds while the latter is between 10 and 1000 seconds. Naturally, we find that the cost of a capacitor increases with its rated leakage time.

┌─ **C A U T I O N** ──────────────────┐

In a power supply, and in some other circuits, particularly high voltage circuits, a significant charge may remain on capacitors even though the line voltage has been disconnected for some time. Extreme caution must be used when
investigating such circuits in order to avoid electrical shocks or damage to test
equipment. Each capacitor should be shorted out with a relatively small resistor to remove the charge before attempting any modification or measurement.

└──────────────────────────────────┘

2-3   CAPACITORS IN SERIES AND IN PARALLEL

When several capacitors are connected in series, as in Figure 2.4, the equivalent capacitance, C_s, can be determined. Consider the circuit shown in Figure 2.5. The battery acts to separate charge, thus one unit of charge is taken
from the negative side of C_3 for every unit of charge that is placed on the
positive side of C_1. Further, the net charge on each isolated element in the
circuit, such as the one shown in the box, must equal zero since no charge is
introduced onto it. This suggests that the charge on each capacitor must be
equal or

$$Q_1 = Q_2 = Q_3. \tag{2-4}$$

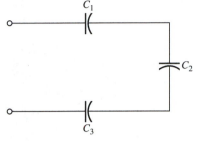

FIGURE 2.4   Series capacitor circuit.

From Kirchoff's laws we also know that the sum of the voltage drops across
each capacitor is equal to the potential of the battery. Using Equation 2-1,
$V = Q/C$, we write

$$V_b = V_1 + V_2 + V_3 = Q_1/C_1 + Q_2/C_2 + Q_3/C_3. \tag{2-5}$$

Since all the charges on the capacitors are equal, we can write

$$V_b/Q = 1/C_1 + 1/C_2 + 1/C_3 \tag{2-6}$$

or

$$C_s = 1/C_1 + 1/C_2 + 1/C_3 = \Sigma(1/C_i) \tag{2-7}$$

which states the equivalent resistance for a series capacitor circuit.

Likewise it can be shown that for capacitors in parallel, as in Figure 2.6,
the equivalent capacitance, C_p, is

$$C_p = C_1 + C_2 + C_3 = \Sigma C_i. \tag{2-8}$$

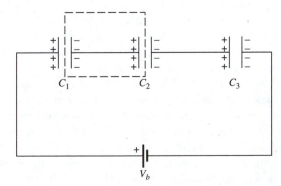

FIGURE 2.5   Series capacitor circuit showing charge distribution.

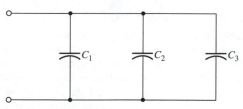

FIGURE 2.6 Parallel capacitor circuit.

A physical model which may aid in understanding the above results is to realize that stringing capacitors in series is much like increasing the separation of plates, *d,* in the parallel plate model of capacitance. This is so because the lower plate of the first capacitor is electrically the same as the upper plate of the second and so on. Because the magnitude of *C* is reciprocally related to the plate separation (see Equation 2-2), a reciprocal relation for series capacitors is obtained. Likewise, this model views the parallel arrangement as equivalent to increasing the area, *a,* of an equivalent single plate. Increasing the plate size increases the capacitance in direct proportion.

2-4 INDUCTANCE

The operation of inductors relies on the fact that a varying current in a coil of wire produces a varying magnetic field. The varying magnetic field induces a potential or electromotive force (EMF) in the coil. (A constant current produces a constant magnetic field, which does not induce an EMF.) The potential across the inductor is thus related to the rate of change of the varying current:

$$V_L = L(di/dt) \tag{2-9}$$

where V_L is the voltage across the inductor, *L* the inductance and *di/dt* is the rate at which the current changes. The unit of inductance is the henry (H), named after the American physicist Joseph Henry (1797–1878). A current changing at the rate of 1 A/s in a 1 H inductor will produce an EMF of 1 V. The henry is defined as

$$1 \text{ H} \equiv 1 \text{ V s/A}. \tag{2-10}$$

Notice that if the current is constant (*di/dt* = 0), there is no induced EMF. This is opposite to the behavior of a capacitor. Whereas the inductor acts as a short circuit (like a wire) to a direct current, a capacitor acts as an open circuit (infinite resistance) to a direct current. The values of commercial inductors commonly range from microhenries (1 μH = 1 × 10⁻⁶ H) through millihenries (1 mH = 1 × 10⁻³ H), into the 1 to 10 H range.

Because practical inductors (sometimes referred to as "chokes") are made of coiled wires, they have some resistance. Keeping the resistance low in high-power coils becomes a problem, and this generally limits the practical sizes of inductors. Inductors generally have either air, iron or ferrite cores. (The filling of the air core with iron or ferrite increases the inductance of an inductor just as increasing the dielectric constant of the dielectric increases the capacitance of a capacitor.) Schematic symbols for inductors are shown in Figure 2.7.

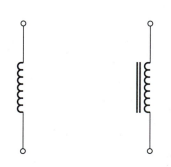

Generic   Iron-core

FIGURE 2.7 Electronic symbols for inductors.

A real inductor with its coils of wire often has significant capacitance as well as resistance. The equivalent circuit of a real inductor is shown in Figure 2.8. Commercially, inductors often have ratings of DC resistance and quality factor Q (which will be discussed later in the section on RLC circuits). Applications are less widespread than those of the capacitor, but include filters and oscillators; however, the inductor is very common in high frequency circuits.

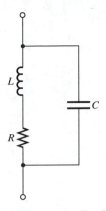

FIGURE 2.8   **Equivalent circuit of a real inductor.**

2-5   RC AND RL CIRCUIT TRANSIENTS

Resistor–Capacitor Circuits

Although capacitance behaves as an open circuit to direct current, initially current must flow to charge the capacitor. This initial response is called a transient, because it initially changes with time, but terminates to the steady-state, DC case. Consider the resistor–capacitor (RC) circuit of Figure 2.9. When the switch is closed, current flows to charge the capacitor. From Kirchoff's law,

$$V_b = v_R + v_C \qquad (2\text{-}11)$$

or

$$V_b = iR + Q/C \qquad (2\text{-}12)$$

where lower case letters are used for voltage and current to indicate that they are time-varying parameters.

We can differentiate this equation with respect to time, obtaining

$$0 = R(di/dt) + (1/C)(dq/dt)$$

and since current is given as the time rate of change of charge ($i = dq/dt$),

$$= RC(di/dt) + i.$$

This equation can be rearranged into the form

$$(1/i)di = -\,(1/RC)dt.$$

Integrating both sides,

$$ln(i) = -\,(1/RC)\,t + K$$

thus the current as a function of time is given by

$$i(t) = (e^K)(e^{-(t/RC)}) \qquad (2\text{-}13)$$

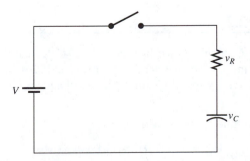

FIGURE 2.9   **RC charging circuit.**

where K is a constant of integration to be evaluated. This is accomplished by noting that at time $t = 0$, there is no charge on the capacitor ($Q = 0$) and therefore no voltage drop, from Equation 2-12, then

$$i(0) = V_b/R$$

and writing Equation 2-13 with $t = 0$ yields

$$i(0) = e^K$$

and therefore

$$e^K = V_b/R.$$

Substituting this into the exponential result above, we obtain our solution for the current flowing in the circuit as a function of time after the switch is closed:

$$i(t) = (V_b/R) \, e^{-(t/RC)}. \qquad (2\text{-}14)$$

Multiplying by R gives us a solution for the voltage across the resistor as a function of time

$$v_R(t) = V_b \, e^{-(t/RC)}. \qquad (2\text{-}15)$$

Using Equation 2-11 we can solve for the voltage across the capacitor during charging

$$v_C(t) = V_b(1 - e^{-(t/RC)}). \qquad (2\text{-}16)$$

These results indicate that, while the capacitor is charging, the current decreases exponentially, from the value it would have if the capacitor were short-circuited, to zero. The voltage across the capacitor, at the same time, increases from zero to the value of the voltage source.

Note that the above results give us a natural time unit for the charging process; the product RC (which we noted in Section 2-2 has the units of seconds) is called the *time constant* of the circuit. At time $t = $ RC seconds, the voltage across the capacitor has reached the value of $V_b\,(1 - (1/e))$ volts, or 63% of the applied voltage, while the current has decreased to 37% of its maximum value. The graph in Figure 2.10 depicts the transient signals as a function of time. It is calibrated in units of RC, and is therefore applicable to any RC circuit. Note that at any given time the sum of the voltage across the capacitor and the voltage across the resistor equals the applied voltage ($V_{applied} = V_R + V_C$), as expected from Kirchoff's law.

EXAMPLE 2.2

Consider the RC circuit shown in Figure 2.11. Calculate the voltage across the capacitor and the voltage across the resistor 0.02 seconds after switch SW1 is closed. Verify that the sum of these voltages is equal to the applied voltage.

Solution: The time constant RC is given by

$$RC = (2.7 \text{ k}\Omega) \, (5.0 \text{ μF}) = 0.0135 \text{ s}.$$

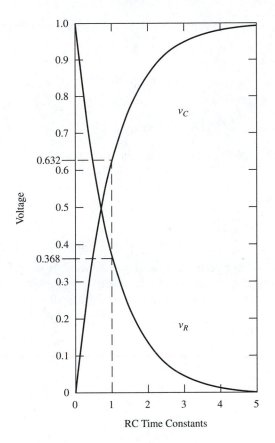

FIGURE 2.10   Response for the circuit of Figure 2.9.

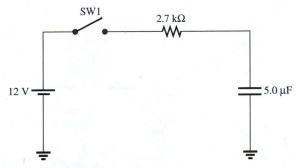

FIGURE 2.11   RC circuit in Example 2.2.

Using Equation 2-15 we find the voltage across the resistor to be

$$v_R(t) = V_b\, e^{-(t/RC)}$$

$$v_R(0.02) = 12\text{V}\; e^{-(0.02/0.0135)} = 2.73 \text{ V}.$$

The voltage across the capacitor can be calculated using Equation 2-16 as

$$v_C(t) = V_b(1 - e^{-(t/RC)})$$

$$v_C(0.02) = 12(1 - e^{-(0.02/0.0135)}) = 9.27 \text{ V}$$

and therefore

$$v_R(0.02) + v_C(0.02) = 2.73 \text{ V} + 9.27 \text{ V} = 12 \text{ V} = V_{applied}. \quad \blacksquare$$

A similar analysis yields the equations for a discharging capacitor circuit

$$i(t) = V_0/R \; e^{-(t/RC)} \tag{2-17}$$

$$v_R(t) = -V_0 \; e^{-(t/RC)} \tag{2-18}$$

$$v_C(t) = V_0 \; e^{-(t/RC)}. \tag{2-19}$$

Here, V_0, the voltage across the capacitor at time $t = 0$, replaces V_b because the discharging circuit has no applied voltage. A graphical representation of the voltage across the capacitor and resistor in a discharging RC circuit is shown in Figure 2.12.

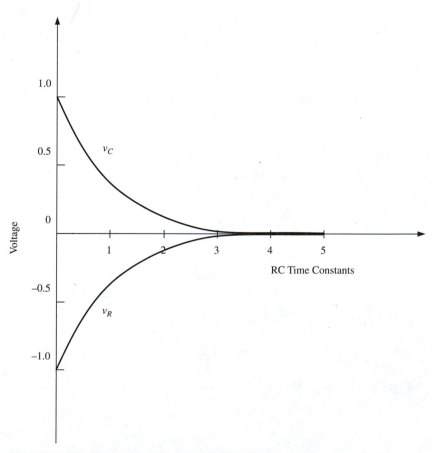

FIGURE 2.12   **Response of discharging RC circuit.**

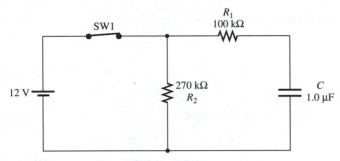

FIGURE 2.13   RC circuit in Example 2.3.

EXAMPLE 2.3

Consider the circuit shown in Figure 2.13. What is the voltage across the capacitor one second after switch SW1 is opened?

Solution:

Step 1: Use Kirchoff's equations to determine the initial voltage drop across the capacitor.

Using the loop defined by the battery and R_2, we find using Kirchoff's voltage equation that the voltage across R_2 must be 12 V, the same as the battery. With this knowledge, we analyze the second loop defined by R_1, R_2 and the capacitor, and find that the voltage across the capacitor must also be 12 V, with the top plate positive with respect to the lower plate.

Step 2: Open switch and apply Equation 2-19.

When the switch opens, the battery is essentially eliminated from the circuit and the capacitor discharges through the series combination of R_1 and R_2. Thus the voltage across the capacitor one second after the switch opens is given by

$$v_C(1) = 12 \text{ V } e^{-(1/((370 \text{ k}\Omega)(1.0 \text{ }\mu\text{F}))} = 8.3 \text{ V}$$

assuming the leakage resistance of the capacitor is large compared to the other resistance values.   ■

EXAMPLE 2.4

A 1.0 μF electrolytic capacitor has a typical leakage time equal to 100 s and that of a 1.0 μF polystyrene capacitor is 2×10^6 s. If both are charged to a voltage of 10 V and then disconnected from the charging circuit, how long will it take each capacitor to discharge to 5 V?

Solution: Capacitors with internal leakage constitute an RC circuit which is initially charged. The voltage drop across the capacitor is thus given by

$$v_C(t) = V_0 \, e^{-(t/RC)}$$

where RC is the leakage time in seconds and V_0 the voltage at time $t = 0$. Thus for our example we find that

$$v_C(t)/V_0 = e^{-(t/RC)}$$

Taking the natural log of both sides yields

$$\ln(v_C(t)/V_0) = -(t/RC)$$

or

$$t = -RC \left[\ln(v_C(t)/V_0)\right].$$

For the electrolytic capacitor we find

$$t = -100 \left[\ln (5/10)\right] = 69 \text{ s}$$

whereas for the polystyrene capacitor the time to fall to half the initial voltage is

$$t = -2 \times 10^6 \left[\ln (5/10)\right] = 1.4 \times 10^6 \text{ s}$$

or 16 days!   ∎

In a more general sense, we have described the response of an RC circuit to a step-function change in voltage, since the closing of a switch in series with a battery in a DC circuit is precisely that. If we combine the charging and discharging analysis of the circuit, we have the RC response to an on-off signal, or a square wave (in this case a signal that oscillates between V_b and 0 V). One half-cycle of the square wave begins as the charging RC circuit and the "off" half as the discharging RC circuit. This is a common circuit in modern electronics. Figure 2.14 shows this situation with different values of the RC time constant.

Resistor–Inductor Circuits

The response of a resistor–inductor (RL) circuit (Figure 2.15) to a step signal is similar to that of the RC circuit; however, there are important differences. Consider the RL circuit shown in the figure. The current goes from zero to a finite value at the instant the switch is closed; hence di/dt is very large at that time. The EMF generated in the inductor impedes the current flow (thus the

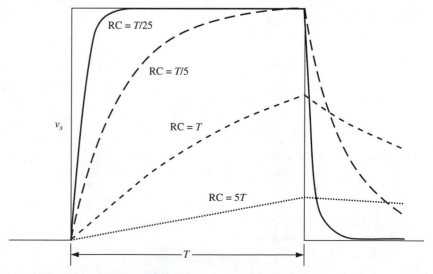

FIGURE 2.14   RC circuit responses to a step input.

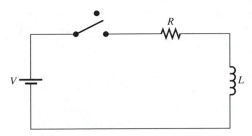

FIGURE 2.15   RL circuit.

common name, "choke," for inductors), but as current flow is impeded, *di/dt* decreases, until the current is constant (*di/dt* = 0). This is illustrated in Figure 2.16 which shows the exponential current growth in the circuit. The expression for current in the RL circuit in the "charging" state is

$$i(t) = V_b/R(1 - e^{-(tR/L)}).\qquad\text{(2-20)}$$

Here the time constant of the circuit is *L/R*. After one time constant passes, the current has risen to 63% of its final value, V_b/R. The voltages across the inductor and resistor after the switch is closed are shown in Figure 2.17. Note

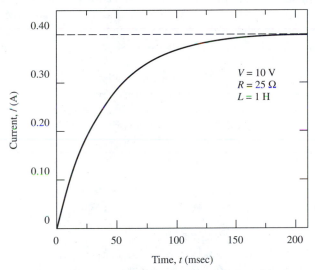

FIGURE 2.16   Current through RL circuit shown in Figure 2.15 using values given.

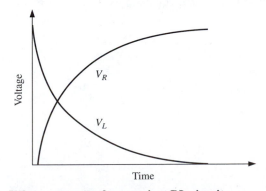

FIGURE 2.17   Voltage response for transient RL circuit.

that this is quite similar to the capacitor circuit, but opposite in that the voltage across the resistor is increasing exponentially in the RL circuit and decreasing exponentially in the RC circuit after a voltage is applied. Likewise the voltage across the inductor in the RL circuit is decreasing with time, whereas in the RC circuit the voltage across the capacitor is increasing with time.

Although both the RC and the RL circuits can be used to extract a somewhat similar type of behavior from a circuit, the RC circuit is used more often. Capacitors are generally easier to use than inductors and a larger variety of types and values is more readily available.

2-6   SIGNAL DIFFERENTIATION AND INTEGRATION

Before closing this chapter, a last look at the RC circuit is in order. The RC circuit of Figure 2.18, with proper selection of values of R and C for a given frequency range, is called a differentiation circuit, because its output voltage is the mathematical derivative of the input voltage, with respect to time, when certain circuit parameters are met. Note the schematic symbol for a square wave generator (or a *function generator* which can produce various forms of periodic signals) used as the input voltage to the RC circuit. Also in the figure are two common representations for the circuit of interest; all three are equivalent and used interchangeably throughout the text and the electronics industry. (A more detailed analysis of periodic signals is given in the next two chapters.)

Attention is called to this circuit for two reasons. First, a differentiated signal is desirable on numerous occasions. Second, the circuit designer should be aware of this inherent property when selecting an RC circuit for other purposes, or differentiation may inadvertently be incorporated into the design. Imagine a high fidelity amplifier design which differentiates the signal over a given range!

When the signal period, T_s, is much greater than the time constant of the circuit, RC, differentiation occurs. This is shown in Figure 2.19. Recall that the derivative of a constant voltage, V_o, is given by

$$dV_o/dt = 0 \qquad (2\text{-}21)$$

except at the transition points, where the derivative of the slope is very large (depending on the quality of the generator). The differentiator circuit finds many applications in electronic circuitry. A common use is that of an edge detector, where an incoming square wave is transformed into a very short pulse.

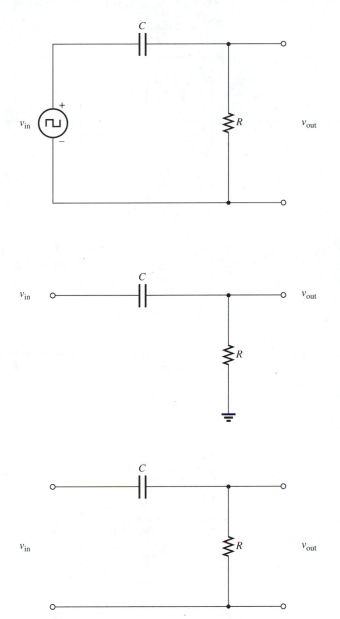

FIGURE 2.18   **Three schematic representations of the same RC differentiator circuit.**

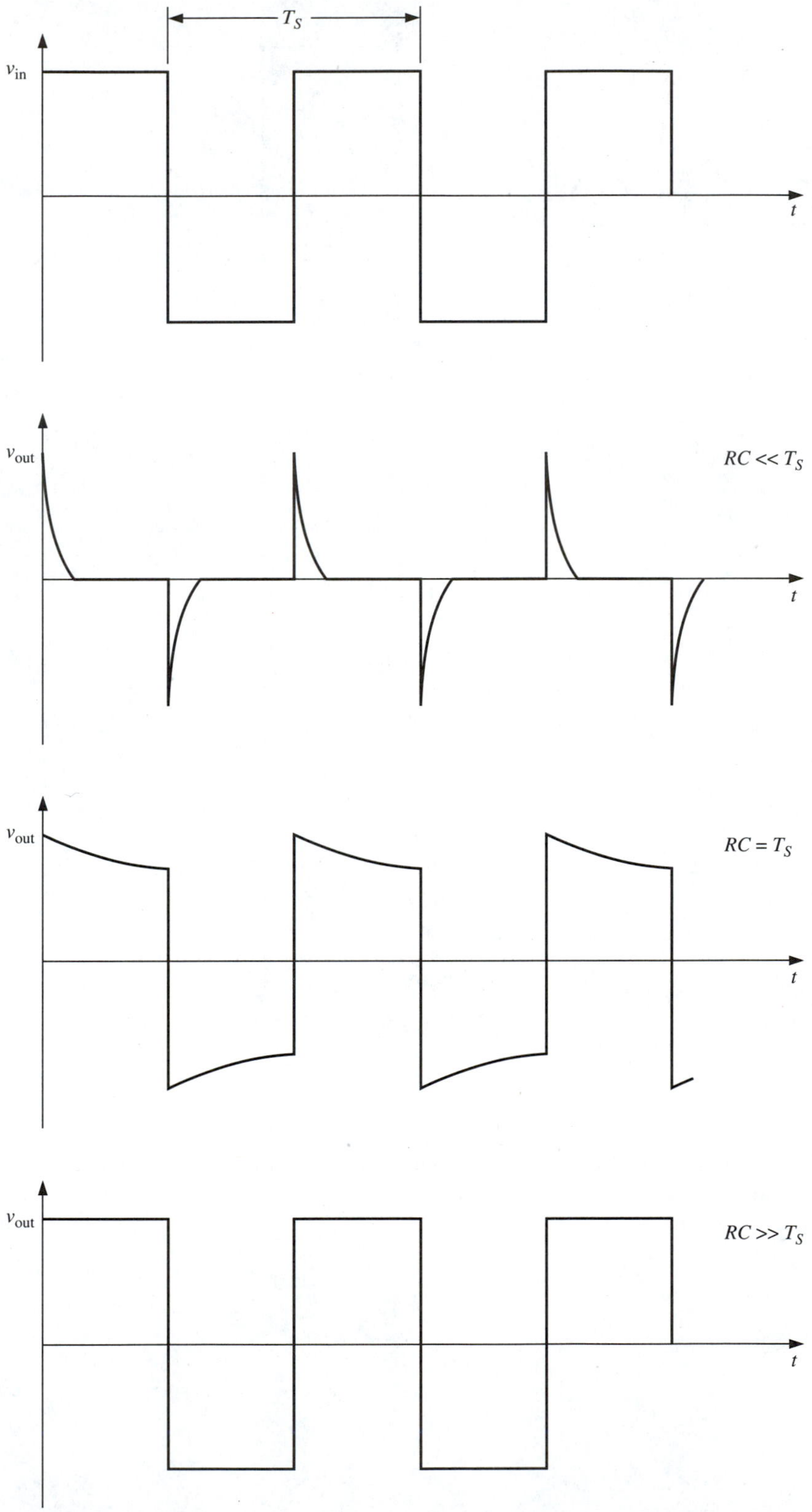

FIGURE 2.19 **RC differentiation of a square wave with different time constants, RC.**

The RC circuit of Figure 2.20 is called an integrator circuit for reasons similar to those above. Here, however, the output voltage is proportional to the integral of the input voltage (assuming $T_s \ll RC$), or for a square wave

$$v_{out} \propto \int V_o dt = V_o t + \text{constant}. \tag{2-22}$$

The output voltage of the circuit is shown in Figure 2.21.

Although the above relies on the exponential decay and growth of transients in RC circuits, the differentiation and integration performed by these combinations can be shown to be valid for more than just a special case of a DC signal. If a sine wave or triangular wave were applied to the inputs, the output voltage would indeed be the derivative or integral of the signal. The mathematical tools necessary for such a proof are presented in the next chapter.

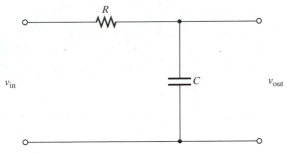

FIGURE 2.20   RC integrator circuit.

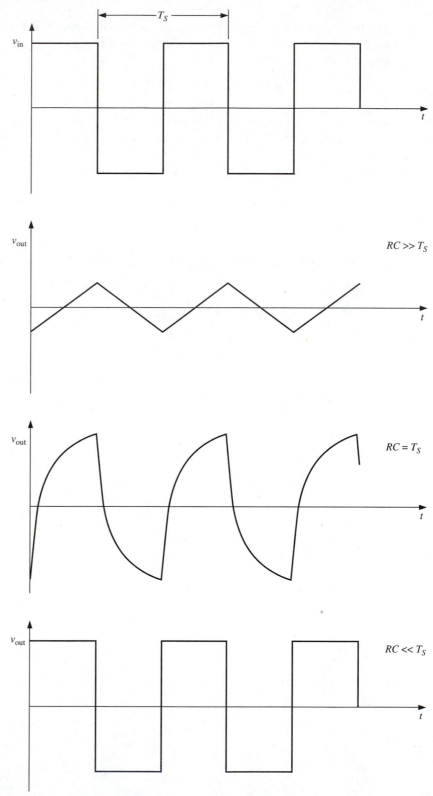

FIGURE 2.21 **RC integration of a square wave with different time constants, RC.**

2-7   QUESTIONS AND PROBLEMS

1.   Calculate the effective capacitance of the circuit shown in Figure A.

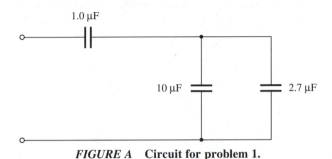

FIGURE A   Circuit for problem 1.

2.   Calculate the effective capacitance of the circuit shown in Figure B.

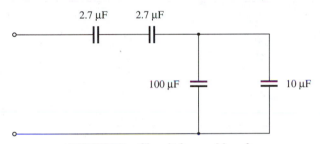

FIGURE B   Circuit for problem 2.

3.   Sketch v_{out} vs. time for the circuit shown in Figure C after the switch is closed.

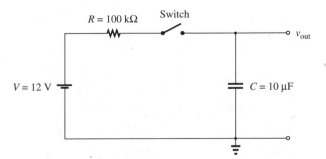

FIGURE C   Circuit for problems 3, 4 and 5.

4.   Given that the tolerance for the resistor is $\pm 5\%$ and that for the capacitor is -20% to $+80\%$, what would be the minimum and maximum values of v_{out} you would expect from the circuit of Figure C after a time of 1.0 second?

5.   An RC circuit is needed with a time constant of 1.000 second. Explain why the circuit shown in Figure C is a poor circuit if the capacitor is electrolytic.

6.   A designer wishes to replace the circuit shown in Figure D with an RC circuit that has the same voltage vs. time characteristics. Design an acceptable RC circuit.

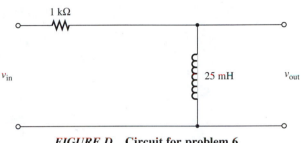

FIGURE D   Circuit for problem 6.

7.   In the circuit shown in Figure E, describe the current that flows from the battery to point A after the switch is closed. In what direction does it flow and what is its value?

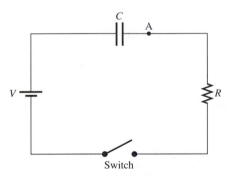

FIGURE E   Circuit for problem 7.

8.   A timing circuit may use the RC charging time to control a subsequent stage, as in Figure F. With $R = 1\ M\Omega$, what value of C is needed so that v_c equals 70 V at 10.0 seconds after the switch closes?

9.   If the subsequent stage referred to in problem 8 has an effective resistance to ground of 10 kΩ, how long after the

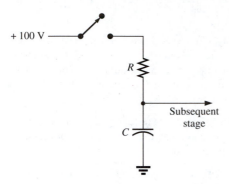

FIGURE F   **Circuit for problems 8 and 9.**

+ 100 V

R

C

Subsequent
stage

switch is reopened will it take to discharge C from full charge to 1 V under the conditions stated in problem 8?

10.   An electromagnet is operated at 50 A. Assuming an inductance of 100 H for the magnet's coil, what voltage will be developed when the switch is opened? When it is closed? Assume that the switch requires 1 millisecond to operate.

Alternating Current Circuits I

3-1 INTRODUCTION

Earlier we discussed time-independent DC voltages and DC voltages that vary as a function of time. In addition to these characteristics, a periodic, alternating characteristic of the electrical signal must be considered. This is called alternating current (AC). Pure direct current is defined as current that moves in one direction, whereas pure alternating current moves in one direction and then back in the other direction repeatedly. Many electrical signals have both AC and DC properties.

The simplest form of alternation is the sine wave, and in this chapter we will develop the characteristics of this signal extensively. We will investigate the properties of capacitors and inductors in AC circuits, introducing such terms as reactance, impedance and phasors.

3-2 THE SINE WAVE AND ITS PROPERTIES

The physical representation of a sine wave developed from a rotating vector is presented in Figure 3.1. This representation may be for either current or voltage. At a given time t, the instantaneous voltage, $v(t)$, is related to the maximum or peak voltage, V_p, by the equation

$$v(t) = V_p \sin (\omega t) \qquad \textbf{(3-1)}$$

where ω is the angular velocity of the rotating vector in radians per second and lower case v is used for the time-varying voltage. (For situations involving pure resistance, the expression for current is obtained by replacing $v(t)$ by $i(t)$ and V_p by I_p.) The time taken for one revolution or cycle is called the *period,* and is symbolized as T. Hence the *frequency, f,* in cycles per second is related to the period as

$$f = \frac{1}{T}. \qquad \textbf{(3-2)}$$

There are 2π radians (360°) in one cycle; hence the relationships

$$\omega = 2\pi f = \frac{2\pi}{T}. \qquad \textbf{(3-3)}$$

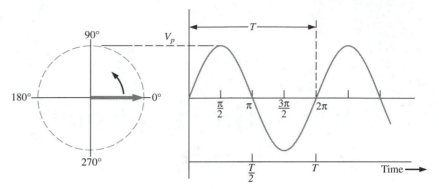

FIGURE 3.1   Rotating vector representation of a sine wave.

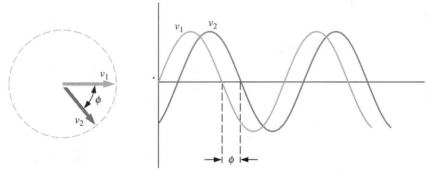

FIGURE 3.2   Representation of two sine waves of equal amplitude and frequency, with a phase angle between them.

The SI unit of cycles per second is the hertz (Hz), named after Heinrich Hertz (1857–1894). Common multipliers are kilohertz (1 kHz = 1×10^3 Hz), megahertz (1 MHz = 1×10^6 Hz) and gigahertz (1 GHz = 1×10^9 Hz).

Several signals having the same frequency and peak amplitude will not necessarily have precisely the same instantaneous amplitude, as illustrated in Figure 3.2. This difference is the result of the phase angle, ϕ, which is the angle between the two rotating vectors which define the signals. When we speak of the phase angle of a single sine wave, it is understood that its rotating vector is ϕ radians counterclockwise from zero (horizontal) at time $t = 0$. When the peak amplitude, V_p, frequency, f and phase angle, ϕ, of a sine wave are given, the wave is completely specified, that is, its instantaneous value at any time t can be calculated, and its equation is

$$v(t) = V_p \sin (\omega t - \phi). \tag{3-4}$$

EXAMPLE 3.1

Given that the voltage in Figure 3.3 is represented by $v(t) = V_p \sin (\omega t - \phi)$, find V_p, ω and ϕ from the graph.

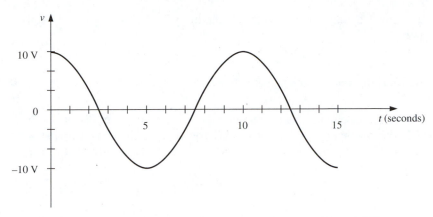

FIGURE 3.3   **AC signal represented by $v(t)$ = 10 V sin (0.63t + 1.575).**

Solution: First, we recognize that V_p is the maximum voltage attained by the signal, which in our case is 10 V. Second, we find the period of the signal by noting two identical points in the waveform which are separated by the period—say the points where the wave crosses the time axis on the downswing side. The first crossing occurs at t = 2.5 s and the second at t = 12.5 s, thus the period, T = 10 s. The frequencies f and ω are calculated using Equation 3-3 as

$$\omega = \frac{2\pi}{T} = \frac{2\pi}{10 \text{ s}} = 0.63 \text{ rad/s}$$

and

$$f = \frac{1}{T} = 0.10 \text{ Hz.}$$

To find ϕ, we write the signal in terms known as

$$v(t) = 10 \sin (0.63\, t - \phi)$$

and choose a convenient time where we can measure $v(t)$. Here we choose t = 0 s and from the graph find $v(t)$ = 10 V. Substituting into the above equation and solving for ϕ yields

$$10 = 10 \sin ((0.63\,(0)) - \phi)$$

$$\sin(-\phi) = 1$$

$$-\phi = \sin^{-1}(1)$$

There is only one solution to the above, but for other values like $\sin^{-1}(0.5)$, there are two solutions. Special care must be taken to choose the correct one. The maximum voltage is a wise choice as the point to evaluate, since there will be only one solution. The solution to the above is

$$\phi = -1.575 \text{ rad} = -90°$$

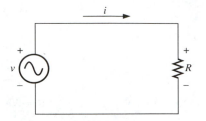

FIGURE 3.4   Resistive circuit with an AC signal applied to it.

To verify our solution we can check another point. Choosing $t = 10$ s gives us

$$v(10) = 10 \sin (0.63(10) + 1.575) = 10 \text{ V},$$

which is the value of the signal at that time.   ∎

A simple circuit with an alternating current is shown in Figure 3.4. Note that the schematic symbol for an AC source is a circle with a sine wave inside. Using Ohm's law, we can write the instantaneous current at time t as

$$i(t) = \frac{v(t)}{R} \tag{3-5}$$

or

$$i(t) = \left(\frac{V_p}{R}\right) \sin (\omega t - \phi) = i_p \sin (\omega t - \phi). \tag{3-6}$$

The circuit of Figure 3.3 raises an important question: How is the power computation made for a circuit with an alternating signal? Since the algebraic sum of currents over the complete cycle is zero for a sine wave (half is positive and half is negative), we resort to consideration of effective current, or *rms (root-mean-square)* current. To calculate an rms value, we square the instantaneous values, take the average (or mean), and obtain the square root. By methods of calculus,

$$I_{rms} = \sqrt{\frac{1}{T} \int_0^T i^2 \, dt}. \tag{3-7}$$

Substituting $i(t)$ from Equation 3-6, with ϕ taken to be zero, yields

$$I_{rms} = \sqrt{\frac{1}{2\pi} \int_0^{2\pi} I_p^2 \sin^2 \omega t \, dt}$$

$$= \left(\frac{I_p}{\sqrt{T}}\right) \left(\sqrt{\frac{T}{2}}\right) \tag{3-8}$$

which results in

$$I_{rms} = \frac{I_p}{\sqrt{2}} = 0.707 \, I_p. \tag{3-9}$$

The physical interpretation of the rms current is an AC current of peak value I_p which will dissipate the same amount of power in a resistor as a DC current of value I_{rms}.

Likewise, the rms voltage for a sine wave is given by

$$V_{rms} = \frac{V_p}{\sqrt{2}} = 0.707 \, V_p. \tag{3-10}$$

An additional way of describing the current is by the peak-to-peak reading, symbolized by I_{pp}. As the name implies, the peak-to-peak value is

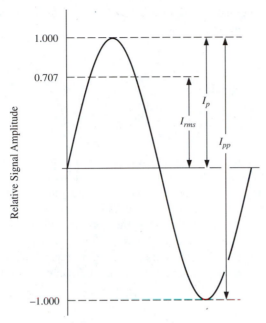

FIGURE 3.5   **Relationships of AC signal designations for a sine wave.**

the reading from the negative peak to the positive peak of the signal. Figure 3.5 shows the relationships of the peak current, I_p, the rms current, I_{rms} and the peak-to-peak current, I_{pp}. Caution must be exercised when describing the value of an alternating signal; common practice is to speak in terms of rms values, but it is wise to be specific.

To get some idea of the relative sizes of these quantities, consider the sine wave voltage available at a wall socket in the United States, $V_{rms} = 117$ V at 60 Hz. Translating the rms value into peak and peak-to-peak values yields $V_p = 165$ V and $V_{pp} = 330$ V.

The rms quantities for waveforms other than sine waves can be developed in the same manner by using Equation 3-7 with the signal value for $i(t)$. Several exercises are provided at the end of this chapter to emphasize the relationships.

3-3   REACTANCE, IMPEDANCE AND PHASORS

Before analyzing the AC response of capacitor and inductor circuits, we must review complex numbers and phasor notation. The mathematics may seem unwieldy at first, but it will soon become apparent how much simpler such representations make AC circuit analysis. A detailed description of phasors and complex numbers can be found in the appendices.

The wide use of capacitors in AC circuitry depends partially on the phase change in the signal which is introduced by the capacitor. Recall that $q = C v$. This equation can be differentiated to obtain

$$(dq/dt) = C(dv/dt). \tag{3-11}$$

Since current is defined as $i = (dq/dt)$ and $i(t) = I_p \sin(\omega t)$ for a sine wave with $\phi = 0^0$, we can write

$$C\left(\frac{dv}{dt}\right) = I_p \sin(\omega t) \tag{3-12}$$

which has the solution

$$v(t) = \left(\frac{I_p}{\omega C}\right) \sin(\omega t - 90°). \tag{3-13}$$

This result states that the voltage across the capacitor remains sinusoidal and, by comparison with Equation 3-6, lags 90° behind the current. (Note that the current and voltage are 90° out of phase, thus no power is dissipated by a capacitor, since $P = I V \cos \Theta$, where Θ is the phase angle between the current and voltage.) Put another way, the current leads the voltage by 90°. This relationship is illustrated in Figure 3.6A. Further comparison with Equation 3-6 suggests that the quantity $1/\omega C$ is equivalent to the "resistance" of the capacitor. This quantity is termed the capacitive reactance, X_C, and is defined as

$$|X_C| = \frac{1}{\omega C} = \frac{1}{2\pi f C}. \tag{3-14}$$[1]

Similarly it can be shown that the inductive reactance for an inductor is

$$|X_L| = \omega L = 2\pi f L \tag{3-15}$$

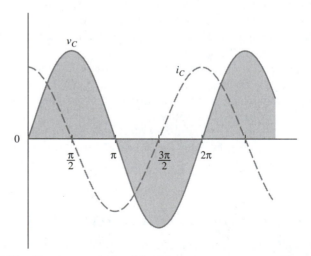

FIGURE 3.6A **Sine wave relationships between current and voltage in a capacitor.**

---

[1]Equation 3-14 states that X_C is a real, negative number. When we speak of magnitudes of X_C or of quantities derived from it, we mean $|X_C|$. This designation may be troublesome at first. There is no appropriate designation which avoids a sense of ambiguity for the reader, however. It is our feeling that the designation of X_C as above (real but negative number) is the most straightforward approach.

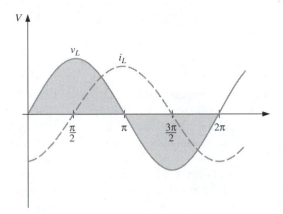

FIGURE 3.6B   **Sine wave relationships between current and voltage in an inductor.**

and that the *voltage* leads the *current* of an inductor by 90° (Figure 3.6B), opposite the case of the capacitor.[2]

The capacitive and inductive reactances have units of ohms (Ω). One of the most important characteristics of the reactance is the frequency dependence. At DC ($f = 0$ Hz), the reactance of the inductor is zero (a short circuit) whereas the reactance of the capacitor is infinite (an open circuit). As the frequency increases, the reactance of the inductor increases, and that of the capacitor decreases. At high frequencies the inductor has a very high reactance and the capacitor has a very low reactance.

Ohm's law can be applied to impedances of reactive elements if phase relationships are taken into account properly. To do this in a manageable way we introduce two different, but equivalent notations for impedance: the complex and the phasor representations. For those unfamiliar with complex number representation and phasors, a review is given in Appendix A.

The net resistance to current flow in circuits including reactive components is called *impedance*. Impedance, Z, is defined as

$$Z = R + jX \qquad \text{(3-16)}$$

where R represents the purely resistive portion, X represents the reactive portion of the overall circuit and $j = \sqrt{-1}$. Thus Z may be purely resistive if $X = 0$; it may be purely reactive if $R = 0$; or it may be a mixture of both if neither R nor X is zero, in which case Z is a complex number. Note the following distinction: X is a real number, but Z is a complex number. For an ideal capacitor or inductor, the resistance is zero ($R = 0$). Therefore the impedances of a capacitor and an inductor, respectively, are given by

---

[2]One way to keep the voltage and current relationships straight is to remember the phrase "ELI the ICE man." Here ELI signifies that the voltage, E, leads the Current, I, in an inductor, L, and vice-versa for the capacitor part—ICE. This phrase was passed on to one of the authors by Matthew Brinkman.

$$Z_C = \frac{1}{j\omega C} = \frac{-j}{\omega C} \qquad (3\text{-}17)^3$$

$$Z_L = j\omega L. \qquad (3\text{-}18)$$

In the rectangular coordinate system (real numbers on the abscissa and imaginary ones on the ordinate), this notation accounts for the 90° phase differences between current and voltage. Figure 3.7 depicts the respective impedances of an inductor, capacitor and resistor on such a coordinate system. This type of representation is called a phasor diagram. A phasor is a vector on this complex plane rotating counter-clockwise. Note that such representation deals with only amplitude and phase, but is nonetheless vital in AC circuit analysis, as we shall see.

The polar representation of a phasor $\mathbf{V}$ (here we introduce the bold face to signify the vector nature of the phasor) with magnitude V (not bold faced), and phase angle ϕ, can be written in rectangular coordinates as

$$\mathbf{V} = V_{\text{real}} + V_{\text{imaginary}} \qquad (3\text{-}19)$$

$$\mathbf{V} = V \cos \phi + jV \sin \phi \qquad (3\text{-}20)$$

or in polar coordinates as

$$\mathbf{V} = V\underline{/\phi}. \qquad (3\text{-}21)$$

The magnitude and phase angle are given by

$$V = \sqrt{(V_{\text{real}})^2 + (V_{\text{imaginary}})^2} \qquad (3\text{-}22)$$

and

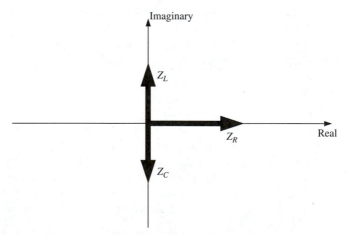

FIGURE 3.7   **Phasor diagram of impedances of resistor, capacitor and inductor.**

---

[3]For a capacitor C driven by a voltage $v(t) = e^{j\omega t}$, $v(t) = q/C$. Taking the derivative with respect to time gives $dv(t)/dt = i(t)/C$ or $j\omega C e^{j\omega t} = i(t)$ which implies $j\omega C\, v(t) = i(t)$ or $Z = v(t)/i(t) = 1/j\omega C$.

$$\phi = \tan^{-1}\left(\frac{V_{\text{imaginary}}}{V_{\text{real}}}\right) \tag{3-23}$$

It is worth reviewing multiplication and division using the polar coordinate notation Equation 3-21. Quite simply if A and B are phasors

$$A = A\underline{/\phi_A} \tag{3-24}$$

$$B = B\underline{/\phi_B}$$

then

$$A \times B = A \times B\underline{/\phi_A + \phi_B} \tag{3-25}$$

and

$$\frac{A}{B} = \frac{A}{B}\underline{/\phi_A - \phi_B}. \tag{3-26}$$

As stated before, Ohm's law is applicable to impedances, but now we write it as

$$v = iZ. \tag{3-27}$$

The equations for adding impedances in series and parallel are identical to those of resistors, with the exception that R is replaced with Z. Keep in mind, however, the vector nature as presented above, when adding and subtracting. Examples of such addition comprise the remainder of the chapter.

EXAMPLE 3.2

Using the above notation, calculate the impedance Z for a capacitor and an inductor.

Solution: First we define our voltage to be $V = V\underline{/\Theta}$ for both components.

In the capacitor, the current leads the voltage by 90° and in the inductor the current lags the voltage by 90° so

$$i_C = i\underline{/\Theta + 90°}$$

$$i_L = i\underline{/\Theta - 90°}.$$

Then by Ohm's law

$$Z_C = V/i_C = \frac{V\underline{/\Theta}}{i\underline{/\Theta + 90°}} = X_c\underline{/-90°} \tag{3-28}$$

$$Z_L = V/i_L = \frac{V\underline{/\Theta}}{i\underline{/\Theta - 90°}} = X_L\underline{/90°}. \tag{3-29}$$

∎

The following sections illustrate the use of phasors in the analysis of several AC circuits.

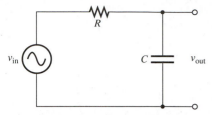

FIGURE 3.8   A series resistance-capacitance circuit (low pass).

3-4   AC ANALYSIS OF RC CIRCUIT

Low-Pass Filter

We may illustrate the frequency dependence of the capacitor by considering the *RC* circuit shown in Figure 3.8. The circuit impedance for the series circuit is

$$\mathbf{Z} = \mathbf{Z}_R + \mathbf{Z}_C = R + jX_C. \tag{3-30}$$

This expression in polar coordinate notation becomes

$$\mathbf{Z} = \sqrt{R^2 + X_C^2}\,\underline{/\tan^{-1}(X_C/R)} \tag{3-31}$$

and the magnitude of the impedance is therefore

$$Z = \sqrt{R^2 + X_C^2}. \tag{3-32}$$

If we allow the input voltage to assume a phase angle of 0° (since the phase angle is relative, we can assign it any angle we wish, and zero is a convenient reference point), then

$$\mathbf{V_{in}} = V_{in}\underline{/0^\circ} \tag{3-33}$$

and the current out of the voltage source is

$$i = \frac{\mathbf{V_{in}}}{\mathbf{Z}} = \frac{V_{in}\underline{/0^\circ}}{Z\underline{/\tan^{-1}(X_C/R)}} \tag{3-34}$$

or

$$i = \frac{V_{in}}{Z\underline{/-\tan^{-1}(X_C/R)}}. \tag{3-35}$$

The voltage across the capacitor is then

$$\mathbf{V_C} = \mathbf{V_{out}} = i\mathbf{Z}_C. \tag{3-36}$$

Using Equation 3-28 for the impedance of the capacitor yields

$$\mathbf{V_C} = \frac{V_{in}}{Z\underline{/-\tan^{-1}(X_C/R)}} \times X_c\underline{/-90^\circ} \tag{3-37}$$

or

$$\mathbf{V_C} = \left(\frac{V_{in}X_C}{Z}\right)\underline{/-\tan^{-1}(X_C/R) - 90^\circ}. \tag{3-38}$$

Therefore the phase angle is

$$\phi = -\tan^{-1}(X_C/R) - 90^\circ$$
$$= -\tan^{-1}(-1/\omega RC) - 90^\circ \tag{3-39}$$

and the magnitude of the output voltage across the capacitor is

$$V_C = V_{out} = \frac{V_{in}X_C}{Z}. \tag{3-40}$$

We can conveniently cast this expression as a ratio of output voltage to input voltage. This unit-less ratio is defined as the gain, A, of the circuit. If we bring the equation back to the original parameters, ω, R and C, then the expression for the gain becomes

$$A = V_{\text{out}} / V_{\text{in}} = X_C/Z$$

$$= \frac{(1/\omega C)}{\sqrt{R^2 + (1/\omega^2 C^2)}}$$

$$= \frac{1}{\sqrt{\omega^2 R^2 C^2 + 1}}. \qquad \text{(3-41)}$$

Thus Equations 3-40 and 3-41 can be used to evaluate the frequency performance of the circuit. An example of the gain and phase angle as a function of frequency is shown in Figure 3.9 using values of $R = 1$ kΩ and $C = 1$ μF. This circuit is called a low-pass filter because it passes low-frequency signals with less attenuation than high-frequency signals. Notice that these are log graphs.

A passive RC circuit such as this one does not have a sharp cutoff above a certain frequency; rather, the attenuation changes gradually. The frequency at which the magnitude of the impedances of both components is equal is called the *break-point frequency, f_B*. We can rearrange $|X_C| = 1/\omega C = R$ and insert $f = \omega/2\pi$ to obtain

$$f_B = 1/2\pi RC. \qquad \text{(3-42)}$$

We can write the gain at this frequency from Equation 3-41 as

$$A = R/\sqrt{2R^2} = 1/\sqrt{2} = 0.707. \qquad \text{(3-43)}$$

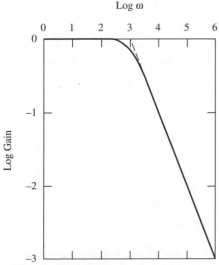

FIGURE 3.9A   Gain *versus* frequency for the circuit of Figure 3.8 (low pass).

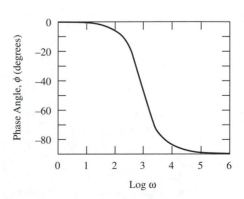

FIGURE 3.9B   Phase angle *versus* frequency for the circuit of Figure 3.8 (low pass).

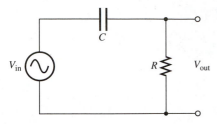

FIGURE 3.10   A series *RC* circuit (high pass).

This is often referred to as the 3 dB point, since it is the frequency at which the gain has fallen 3 dB (dB = $20 \log_{10}(A)$). The 3 dB point is often referred to when speaking about amplifiers. At this point the power out of the circuit is equal to half the power into the circuit. Notice as well that at this point the phase angle is $-45°$, midway in its range of $-90°$ to $0°$.

High-Pass Filter

If the output of a series *RC* circuit is taken across *R* as shown in Figure 3.10, then a high-pass filter circuit results. By analogy to the previous circuit

$$V_{out} = iZ_R = \frac{V_{in}}{Z} \angle -\tan^{-1}(X_C/R) \times R\angle 0°$$

$$= \frac{V_{in}R}{Z} \angle -\tan^{-1}\left(\frac{X_C}{R}\right).$$

$$(3\text{-}44)$$

The magnitude of the gain is

$$A = \frac{V_{out}}{V_{in}} = \frac{\omega RC}{\sqrt{\omega^2 R^2 C^2 + 1}} \qquad (3\text{-}45)$$

and the phase angle is

$$\phi = -\tan^{-1}(X_C/R) = \tan^{-1}(1/\omega RC). \qquad (3\text{-}46)$$

The response for this high-pass filter is shown in Figure 3.11 for a circuit with $R = 10 \text{ k}\Omega$ and $C = 1\mu\text{F}$.

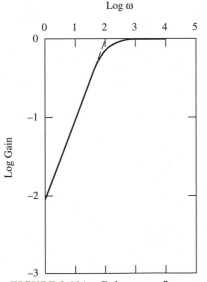

FIGURE 3.11A   Gain *versus* frequency for the circuit of Figure 3.10 (high pass).

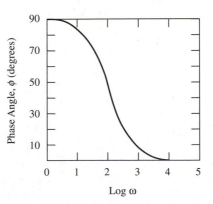

FIGURE 3.11B   Phase angle *versus* frequency for the circuit of Figure 3.10 (high pass).

Consider the possibility of inserting the low-pass filter in series with the high-pass filter. With the values used in the examples, the output would be only slightly attenuated between $\omega = 100$ rad/s and $\omega = 1000$ rad/s; at all other frequencies, significant attenuation would occur. This combination is a primitive example of a band-pass filter.

Similar analysis can be done to the *RL* circuit. As expected, if we replace the capacitor of a low-pass filter with an inductor, the circuit becomes a high-pass filter, and likewise for the high-pass filter. Such analysis will be left to the reader.

3-5   RLC RESONANT CIRCUIT

The *RLC* circuit shown in Figure 3.12 is commonly called a resonant circuit because energy is stored alternatively in the capacitor and the inductor. You may have already realized that the phase relationships of inductors and capacitors are such that they might cancel each other. At a particular frequency this is true. This frequency, f_0, is called the *resonant frequency* of the circuit. It is the frequency at which the following condition holds:

$$|X_C| = |X_L| \tag{3-47}$$

so that

$$\frac{1}{2\pi f_0 C} = 2\pi f_0 L \tag{3-48}$$

or

$$f_0 = \frac{1}{2\pi\sqrt{LC}} \tag{3-49}$$

Since the capacitive reactance is 180° out of phase with the inductive reactance, the reactances cancel each other at f_0. At other frequencies ($f \neq f_0$) the phasor diagram of Figure 3.13 applies when $|X_L| > |X_C|$; the Z vector lies below the R vector if $|X_L| < |X_C|$. The impedance is

$$Z = \sqrt{R^2 + (X_L - X_C)^2} \tag{3-50}$$

and the phase angle is

$$\phi = \tan^{-1}[(|X_L| - |X_C|)/R]. \tag{3-51}$$

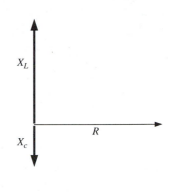

A

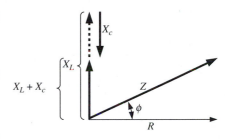

B

FIGURE 3.13   Impedance representation of an *RLC* circuit.

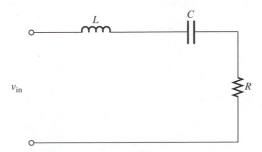

FIGURE 3.12   **A series *RLC* circuit.**

At $f = f_0$, Z by Equation 3-50 is a minimum and the current in the circuit is therefore at a maximum. Similarly, at f_0 the phase angle is 0°. For frequencies other than this, the impedance is greater than R and the current is smaller than its maximum value.

The *quality factor, Q,* of the circuit is defined as

$$Q = X_L/R = \omega_0 L/R. \tag{3-52}$$

Figure 3.14 shows the general shape of the V_{out} or current-vs.-frequency curve and illustrates how different values of Q affect this response. Note that the larger the value of Q, the sharper and narrower the peak; and the lower the value of Q, the more broad the peak. A higher value of Q indicates that the frequency response of the circuit is very select.

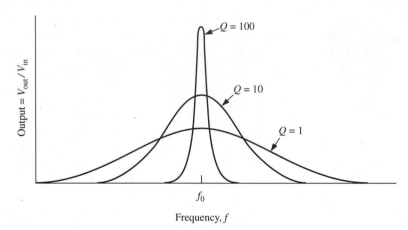

FIGURE 3.14 $V_{\text{out}}/V_{\text{in}}$ *versus* **frequency for several series *RLC* circuits with different *Q* values.**

Many other possible configurations exist that involve L, C and R. The treatment of these circuits is somewhat less straightforward, owing to the various impedance relationships. Below we present an example of an *RLC* circuit with actual values.

EXAMPLE 3.3

Consider the circuit shown in Figure 3.15. Using the values shown, calculate the impedance and phase angle at $f = 1000$ Hz.

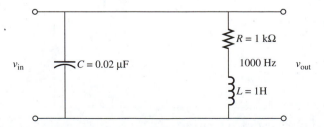

FIGURE 3.15   **A parallel resonant circuit.**

Solution: First, calculate the impedances of the individual components at $f = 1000$ Hz.

$$Z_C = \frac{-j}{\omega C} = \frac{-j}{2\pi fC} = \frac{-j}{2\pi(10^3)(2.0\times 10^{-8})}$$

$$= -j(8.0 \times 10^3) \; \Omega$$

and

$$Z_L = j\omega L = j2\pi fL = j(2\pi)(10^3)(1)$$

$$= j(6.3 \times 10^3) \; \Omega.$$

The impedance of the LR branch is the series combination of R and L or

$$Z_{LR} = (1 \times 10^3) + j(6.3 \times 10^3) \; \Omega.$$

Since the impedance of the C branch is in parallel with the LR branch

$$Z_{total} = \frac{(Z_C Z_{LR})}{(Z_C + Z_{LR})}$$

or

$$Z_{total} = \frac{[-j(8.0 \times 10^3)][(1.0 \times 10^3) + j(6.3 \times 10^3)]}{-j(8.0 \times 10^3) + (1.0 \times 10^3) + j(6.3 \times 10^3)}$$

$$= \frac{50.4 \times 10^6 - j\,8 \times 10^6}{1 \times 10^3 - j\,1.7 \times 10^3}$$

multiplying numerator and denominator by $(1 \times 10^3 + j\,1.7 \times 10^3)$ and dividing through gives the result

$$= (1.65 \times 10^4) + j\,(2.0 \times 10^4) \; \Omega.$$

Hence the magnitude of the impedance is

$$Z = \sqrt{(1.65 \times 10^4)^2 + (2.0 \times 10^4)^2} = 2.6 \times 10^4 \; \Omega$$

while the phase angle is

$$\phi = \tan^{-1}[(2.0 \times 10^4)/(1.65 \times 10^4)] = 50.5°.$$

If the same components were used to make a series resonant circuit, the calculations would yield different results. Then the impedance and phase angle would be

$$Z = 2.0 \times 10^3 \; \Omega$$

$$\phi = -59.5°. \qquad\qquad \blacksquare$$

Resonant circuits find use in oscillators, for which the oscillator frequency is controlled by the values of the components. Likewise, they are used in radio receiver circuits where tuned filters are employed to allow selection of signals of any required frequency.

3-6   QUESTIONS AND PROBLEMS

1.   Complete the following summary table.

	Resistance	Capacitance	Inductance
general v-i relationship			$v = L\dfrac{di}{dt}$
dc relationship		$I = 0$	
resistance or reactance	R		
impedance	$R + j0$		
v-i vector relationship		$V = \dfrac{-j}{\omega C} I$	

2.   A sine wave signal (20 V PP) is connected to a 10 kΩ resistor. Calculate the current through the resistor in terms of average, rms, and peak currents. What wattage rating should the resistor have?

3.   Determine the relationships among V_p, V_{rms}, and V_{PP} for a square wave of amplitude $\pm V_p$.

4.   Determine the relationships among V_p, V_{rms}, and V_{PP} for a triangular wave of amplitude $\pm V_p$, for which the description between 0 and $\pi/2$ is $v = \dfrac{V_p}{\pi/2}\, \omega t$.

5.   Compare the rms currents and power dissipations that would occur if the waveforms of problems 2, 3 and 4 were applied in turn to a 10 kΩ resistor; in all cases, $V_{PP} = 20$ V.

6.   Calculate the magnitude and phase angle of the impedance for the circuit of Figure A.

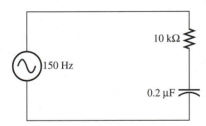

FIGURE A   Circuit for problems 6, 7, 8, 9 and 10.

7.   At what frequency will the phase angle be 30° for the circuit of Figure A?

8.   Using the circuit values of Figure A, plot v_{out}/v_{in} (where v_{out} is taken across the capacitor) versus the log of frequency. Also plot the log of v_{out}/v_{in} versus the log of frequency.

9.   Again using the circuit values of Figure A, make the same plots as in problem 8, but taking v_{out} across the resistor.

10.   With reference to the plots of problems 8 and 9, is there a frequency at which v_{out}/v_{in} is the same in both cases? Explain.

11.   Determine the resonance frequency for the series circuit of Figure B.

FIGURE B   Circuit for problems 11 and 12.

12.   Calculate the magnitude and phase angle of the impedance of the circuit in Figure B when it is operated at 2500 Hz.

13.   What value of L is necessary for resonance at 10,000 Hz for the circuit of Figure C when $R = 570\ \Omega$ and $C = 0.01\ \mu$F?

14.   Given that $L = 0.05$ H in the circuit of Figure C, and the values of R and C as in problem 13, calculate the phase angle and magnitude of the impedance of the circuit at 100 kHz.

15.   For the R, L and C values in problem 14, what are the Q values of the circuit of Figure C at 100 kHz, 1000 Hz, and 100 Hz?

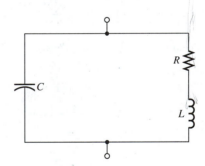

FIGURE C   Circuit for problems 13, 14 and 15.

16.   For the high-pass filter of Figure D, at what frequency is $v_{out}/v_{in} = 0.95$?

17.   At what frequencies will the circuit of Figure D act as a differentiator?

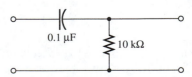

FIGURE D   Circuit for problems 16, 17, 18 and 19.

18. Interchanging R and C of Figure D produces a low-pass filter. At what frequency will $v_{out}/v_{in} = 0.95$?

19. At what frequencies will the RC circuit of problem 18 act as an integrator?

20. Design a low-pass filter that has $v_{out}/v_{in} = 0.5$ at 5 kHz.

21. For the circuit that you designed in problem 20, at what frequencies will v_{out}/v_{in} equal 0.01 and 0.99?

22. Consider an RLC series circuit in which $R = 1$ kΩ, $L = 1$ H and $C = 0.1$ μF. Calculate the resonance frequency, the impedance at resonance and the current at resonance when $V_{rms} = 10$ V. Also sketch X_L, X_C, ϕ and Z as functions of frequency.

23. Consider an RLC parallel circuit, as in Figure C, in which the component values are $R = 1$ kΩ, $C = 0.1$ μF, and $L = 1$ H. Sketch Z, I_{rms} and ϕ as functions of frequency.

24. Derive the following gain expression for the circuit of Figure E:

$$\left|\frac{v_o}{v_s}\right| = \frac{R_2}{R_2 + R_1(1 + \omega CR_2)}$$

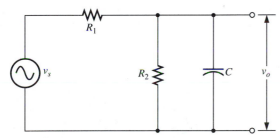

FIGURE E   Circuit for problems 24 and 25.

25. Under what frequency conditions will the circuit of Figure E behave as a simple voltage divider?

26. Derive the following gain expression for the circuit of Figure F:

$$\left|\frac{v_o}{v_s}\right| = \frac{R_2}{R_2 + R_1\left(\dfrac{1 + \omega R_2 C_2}{1 + \omega R_1 C_1}\right)}$$

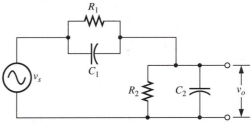

FIGURE F   Circuit for problems 26, 27 and 28.

27. Under what conditions of R and C will the gain of the circuit of Figure F be the same as that of a simple voltage divider?

28. Suggest a possible use for the circuit of Figure F.

29. Consider the circuit of Figure G, where $R_1 = 20$ Ω, $X_1 = 37.7$ Ω, $R_2 = 10$ Ω, and $X_2 = -53.1$ Ω. Compute the current and the phase angle between the current and the applied voltage.

30. Determine the values of L and C used in problem 29.

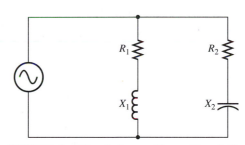

FIGURE G   Circuit for problems 29 and 30.

AC Circuits II

4-1 INTRODUCTION

In this chapter we will explore a variety of topics which are pertinent to AC circuit design and analysis. First we will introduce the transformer and its associated properties. Then we will discuss transmission lines, which carry electronic signals between components, and explore impedance matching and mismatching. We will conclude with an analysis of electronic waveforms, other than the already discussed sine wave, that find use in instrumentation.

4-2 TRANSFORMERS

A transformer is essentially two inductors wound in such a way that the energy from one coil is transferred to the other through changes in the magnetic flux. For power applications the coils are usually wound on an iron or ferrite core; when power is not a significant consideration, an air core is often used. The electronic symbol for the transformer is shown in Figure 4.1. An exemplary circuit, which will help define the parameters we will explore, is shown in Figure 4.2. An important property of a coil, or inductor, is that the inductance is directly proportional to the number of turns of wire in the coil, N. Since the induced EMF is also directly proportional to the inductance, it follows for a transformer consisting of two coils that

$$\frac{v_s}{v_p} = \frac{N_s}{N_p} \qquad (4\text{-}1)$$

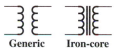

Generic   Iron-core

FIGURE 4.1 **Electronic symbols for transformers. Often the two symbols shown are used interchangeably.**

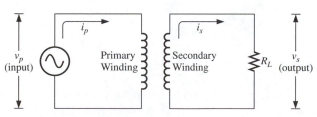

FIGURE 4.2 **A transformer circuit.**

where the subscript, s, indicates the secondary, or output, winding and the subscript, p, indicates the primary, or input, winding. The ratio N_s/N_p is called the turns ratio. When a transformer has a turns ratio greater than 1, it is called a *step-up transformer* since the output voltage, v_s, is always greater than the input voltage, v_p. A *step-down transformer* has a turns ratio of less than 1, with v_s being always less than v_p. A transformer with a turns ratio of 1 is often used as an *isolation transformer* to separate 120 V AC line voltage from having a direct, mechanical link to the apparatus being powered.

In an ideal transformer, no power is gained or lost as power is transferred from the primary winding to the secondary winding. (Real transformers do have some loss, but this loss is generally small if the transformer limit ratings are not exceeded.) It follows then that

$$P_s = P_p \tag{4-2}$$

which can also be written as

$$i_p v_p = i_s v_p \tag{4-3}$$

or

$$\frac{v_s}{v_p} = \frac{i_p}{i_s}. \tag{4-4}$$

The result, compared to Equation 4-1, shows that

$$\frac{i_p}{i_s} = \frac{N_s}{N_p}. \tag{4-5}$$

In other words, a step-up transformer which produces a higher voltage at the secondary than is applied to the primary, yields less current at the secondary than that which enters the primary. Conversely, a step-down transformer supplies more current to the secondary than that flowing into the primary.

If Equations 4-1 and 4-5 are multiplied, the result is

$$\left(\frac{v_s}{v_p}\right)\left(\frac{i_p}{i_s}\right) = \left(\frac{N_s}{N_p}\right)^2. \tag{4-6}$$

This is rearranged to give

$$\frac{(v_s/i_s)}{(v_p/i_p)} = \frac{Z_s}{Z_p} = \left(\frac{N_s}{N_p}\right)^2. \tag{4-7}$$

This result shows that the impedance of the secondary winding is different from that of the primary winding (with the exception of an isolation transformer, where $N_s = N_p$). The ratio of these impedances varies with the square of the turns ratio.

EXAMPLE 4.1

A commonly encountered step-down transformer yields 12.6 V (rms) at 1 A in the secondary when the primary is supplied with 120 V (rms). What are the values of the turns ratio, input impedance and output impedance?

Solution: The inverse of the turns ratio required to achieve this is

$$\frac{N_p}{N_s} = \frac{v_p}{v_s} = \frac{120}{12.6} = 9.5$$

which states that the primary winding has 9.5 times the number of loops of wire than that of the secondary. Since the secondary impedance at full load is calculated as

$$Z_s = \frac{v_s}{i_s} = \frac{12.6 \text{ V}}{1 \text{ A}} = 12.6 \; \Omega,$$

the impedance of the primary, Z_p, is

$$Z_p = Z_s(N_p/N_s)^2 = (12.6)(9.5)^2 = 1137 \; \Omega.$$

In order to produce 12.6 V at 1 A, the current flowing in the primary, i_p, is given by

$$i_p = \frac{120 \text{ V}}{1137 \; \Omega} = 106 \text{ mA.} \qquad \blacksquare$$

Transformers are available in an enormous variety of shapes and sizes. They are commonly used to transform incoming 120 V line voltage (either increase or decrease) to a value needed in a circuit. Transformers used in this application are called *power transformers*. Smaller transformers used for transforming other electronic signals are called *signal transformers*. Another common use for the transformer is as an isolation device. As mentioned earlier, a transformer with a turns ratio of 1 may seem rather useless until one realizes that such a device acts to physically separate the circuitry connected to the primary from the circuitry connected to the secondary.

Generally speaking, two parameters need to be specified when selecting a power transformer: the maximum current and voltage ratings of the secondary. For power transformers, the ratings generally, but not always, are given with the assumption that the primary of the transformer will be connected to 120 V line voltage. Secondary ratings range from hundreds of amperes down to milliamperes and from thousands of volts to one volt. Exceeding the current rating of a transformer will cause power losses in the transformer which result in heating of the iron core and windings, as well as distortion of the secondary voltage signal. Variable voltage output power transformers, often called *variacs,* are also available.

Signal transformers carry ratings of maximum current, but differ from power transformers in that the impedances of the primary and secondary windings are of prime consideration. This is due to the fact that the most common use of the signal transformer is that of matching impedances, as we shall discuss in the next section.

Transformers with multiple taps from their primary or secondary windings are very common, as we shall see in several applications in the next chapter. The most common variety of these is the *center-tap transformer,* which has one lead coming from the middle of the secondary coil. *Split transformers,* which consist of a single core with two (or more) separate primary and secondary windings, are also fairly common. Figure 4.3 shows the schematic symbols for a center-tap transformer and a split transformer.

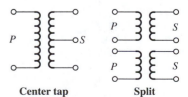

Center tap   **Split**

FIGURE 4.3   Common center-tap and split transformers. Primary windings are on left of diagram.

4-3 IMPEDANCE MATCHING AND POWER TRANSFER

The transfer of power from a source to a load is maximum when the impedance of the load equals that of the source. This statement is frequently called the *maximum power transfer theorem*. Consider the circuit of Figure 4.4, where R_s represents the source resistance and R_L represents the resistance of the load. We can write the power dissipated by the load as

$$P = I^2 R_L = (V/(R_s + R_L))^2 R_L. \tag{4-8}$$

To find the value of R_L that yields the maximum load power, we take the derivative of the above equation with respect to R_L, set it equal to zero, and solve for R_L. The derivative is given by

$$\frac{dP}{dR_L} = \frac{V^2 (R_s^2 - R_L^2)}{(R_s + R_L)^4} = 0 \tag{4-9}$$

and has the solution

$$R_L = R_s \tag{4-10}$$

which is what we stated above.

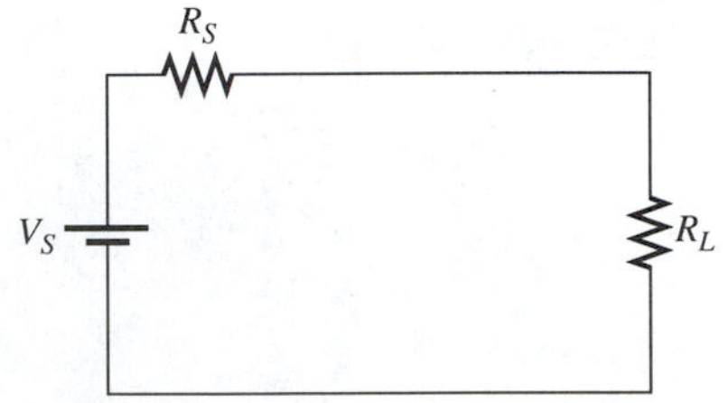

A numerical exercise will demonstrate this. Consider the same circuit in Figure 4.4 wherein the voltage source, V, and the impedance, R_s, represent the power source and R_L represents the load. The measured values of P_L for different values of R_L are given in Table 4.1. Note that the maximum power is indeed transferred from the source to the load when the source and load impedances are equal (matched). At this condition, 50% of the total power ($I^2 R_{\text{TOTAL}}$) is transferred to the load and 50% is wasted as heat in the source. Maximum power transfer at $R_L = R_s$ is, at first, not intuitive for most people. Common sense might say that maximum power transfer occurs when the current is at a maximum, so a small load resistance should yield maximum power transfer. But, we must keep in mind that the product of the current squared and the load resistance determines the power dissipated in the load resistor.

The transformer is extremely useful in matching the load and source impedances for maximum power transfer of AC signals. Consider the following example shown in Figure 4.5. A low impedance load of 100 Ω is to be matched to a high impedance source of 4.9 kΩ by placing the appropriate transformer between them. The proper turns ratio is

FIGURE 4.4 Circuit used to illustrate impedance matching.

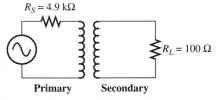

FIGURE 4.5 Matching transformer for high impedance source and low impedance load.

TABLE 4.1 *Power Transfer to Various Loads*

V_s (volts)	R_s (ohms)	R_L (ohms)	R_{TOTAL} (ohms)	I (amps)	P_L (watts)
100	40	5	45	2.222	24.69
100	40	10	50	2.000	40.00
100	40	20	60	1.667	55.56
100	40	40	80	1.250	62.50
100	40	80	120	0.833	55.56
100	40	120	160	0.625	46.88
100	40	200	240	0.417	34.72

$$\frac{N_s}{N_p} = \left(\frac{Z_s}{Z_p}\right)^{1/2} = \left(\frac{100}{4900}\right)^{1/2} = \frac{1}{7}.$$

Such a transformer would solve the impedance match problem, but let us carry on with the analysis. If the source voltage is 28 V (rms), then the voltage applied to the load is

$$v_s = v_p\left(\frac{N_s}{N_p}\right) = \frac{28}{7} = 4 \text{ V(rms)}.$$

The current supplied to the load is

$$i_s = \frac{v_s}{Z_s} = \frac{4 \text{ V}}{100 \text{ }\Omega} = 40 \text{ mA}$$

and the primary source current is

$$i_p = i_s\left(\frac{N_s}{N_p}\right) = \frac{40}{7} \text{ mA} = 5.7 \text{ mA}$$

or alternatively

$$i_p = \frac{v_p}{Z_p} = \frac{28 \text{ V}}{4900 \text{ k}\Omega} = 5.7 \text{ mA}.$$

Since the load is matched to the source, the power dissipated in the load equals the power dissipated in the source, or

$$P_p = i_p v_p = P_s = i_s v_s = 160 \text{ mW}.$$

This has the effect of appearing as though the matching transformer is not present. The transformer dissipates no power.

If, on the other hand, the matching impedance transformer had not been used, only a small fraction of the total power would have been transferred to the load; most of it would have been dissipated as heat in the large impedance of the source. In our example, the current through the circuit without the impedance matching transformer, as shown in Figure 4.6, would be

$$i = \frac{V}{(R_s + R_L)} = \frac{28 \text{ V}}{5000 \text{ }\Omega} = 5.6 \text{ mA}.$$

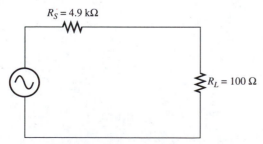

FIGURE 4.6   **High impedance source connected to low impedance load.**

The power dissipated by the load is then

$$P_L = i^2 R_L = (5.6 \text{ mA})^2 (100 \text{ } \Omega) = 5.6 \text{ mW}$$

whereas the power dissipated by the source resistance is

$$P_s = i^2 R_s = (5.6 \text{ mA})^2 (100 \text{ } \Omega) = 154 \text{ mW}.$$

Note as well that the voltage drop across the load is

$$V_L = i R_L = (5.6 \text{ mA})(100 \text{ } \Omega) = 0.56 \text{ V}$$

compared to 4 V with the impedance-matching transformer.

4-4 TRANSMISSION LINES

Although one consideration for impedance matching is maximum transfer of power, a related consideration is that of signal reflection. Consider a cable with two conductors, one held at ground and the other carrying an electronic signal. Here we consider two types of cables: the *coaxial cable* with a center conductor, surrounded by an insulating material which in turn is surrounded by a sheath of conductor (which is usually held at ground), and a *parallel cable* with two wires, side by side, embedded in an insulating material. A diagram of these two types of cable is shown in Figure 4.7 along with the dimensions of interest, *a, b* and *d*.

It can be shown that the formulae which relate the characteristic impedances of these two types of cables, or transmission lines, to their physical parameters are, for the parallel line

$$Z_o = \left(\frac{\mu \mu_o}{\pi^2 \epsilon \epsilon_o} \right)^{1/2} \ln \left(\frac{2d}{a} \right) \tag{4-11}$$

and that for the coaxial cable

$$Z_o = \left(\frac{\mu \mu_o}{\pi^2 \epsilon \epsilon_o} \right)^{1/2} \ln \left(\frac{b}{a} \right) \tag{4-12}$$

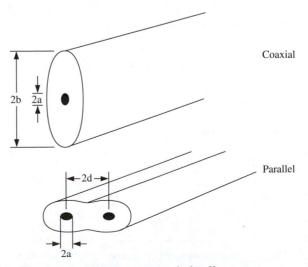

FIGURE 4.7 **Two types of common transmission lines.**

where μ_o is the permeability and ϵ_o the permittivity of free space, $(4\pi \times 10^{-7}$ H/m) and $(8.854 \times 10^{-12}$ F/m), respectively. The relative permittivity ϵ and relative permeability μ in the above equations are those of the insulating material between the conductors. Usually the permeability of this material is very close to 1. Notice that the characteristic impedance does not depend on length, although we are assuming no resistive losses here.

Consider a parallel wire transmission line with a conductor size of 20 AWG (radius = .0160 inch) separated by 0.10 inches in air. We can calculate the resistance of the cable per unit length using Equation 1-1, but this is negligible (assuming the cable is not very long). The reactance, however, is not negligible. Using Equation 4-11 we find

$$Z_o = \left(\frac{4\pi \times 10^{-7} \text{ H/m}}{\pi^2 \, 8.854 \times 10^{-12} \text{ F/m}} \right)^{1/2} \ln \left(\frac{2 \times 0.10''}{0.016''} \right)$$

$$= 300 \, (\text{H/F})^{1/2} = 300 \, \Omega.$$

Such an impedance is not uncommon in parallel wire transmission lines. Common values for coaxial cables are 50 Ω for electronic instrumentation and 75 Ω for communication applications.

Each of these cables has a capacitance C per unit length and an inductance L per unit length. It can be shown that the characteristic impedance of the transmission line is given by

$$Z_o = \left(\frac{L}{C} \right)^{1/2} \tag{4-13}$$

and that the signal propagates along the cable at a speed given by

$$v = \frac{1}{(LC)^{1/2}} \tag{4-14}$$

which in terms of permittivity ϵ and the permeability μ is

$$v = \frac{c}{(\epsilon \, \mu)^{1/2}}. \tag{4-15}$$

This is the same speed light would travel within the insulating medium. For RG-58/U cable, a common instrumentation line, the characteristic impedance is 50 Ω. The center conductor is a solid 20 AWG wire and the shielded outer conductor has a diameter of 0.116 inches. Typically, the value of the capacitance per unit length is 28.5 pF/ft. Using these figures we find the inductance per unit length from Equation 4-13 to be

$$L = Z_0^2 C = (50 \, \Omega)^2 (28.5 \times 10^{-12} \text{ F/ft})$$

$$= 7.1 \times 10^{-8} \text{ H/ft}.$$

The speed of signal propagation along such a line is then given by Equation 4-14 to be

$$v = \frac{1}{[(7.1 \times 10^{-8} \text{ H/ft})(28.5 \times 10^{-12} \text{ F/ft})]^{1/2}}$$

$$= 7.0 \times 10^8 \text{ ft/s} = 2.1 \times 10^8 \text{ m/s} \tag{4-16}$$

or about 2/3 times the speed of light.

When the transmission line is terminated by an impedance, Z, different than its characteristic impedance, Z_o, signal reflection occurs. A mechanical analogy of this phenomenon is a wave pulse traveling along a rope. If the rope has the same linear density all along its length, then the wave will traverse the rope undisturbed. However, if the long rope consists of two different, shorter ropes, each having a different linear density, the wave will be partially reflected at the connection (analogous to the electrical termination) and partially transmitted. This situation is identified as a *mismatched* line, or a mismatched impedance. The ratio of the amplitude of the original signal, A, to that of the reflection at the termination, A_o, is given by

$$A/A_o = K = (Z - Z_o)/(Z + Z_o) \qquad (4\text{-}17)$$

where K is called the reflection coefficient. Note that if the terminating end of the conducting line is shorted, $Z = 0$, then the reflection coefficient is -1 and the full incoming signal is reflected back, but inverted. For an open termination, $Z = \infty$, the reflection coefficient is equal to 1 and the full incoming signal is reflected, but not inverted. Using the rope analogy, the shorted situation is represented as a wave pulse traveling along a rope which has the terminating end connected to a solid wall. The open termination is akin to the end of the rope being free. In between these extremes, the reflected signal is less than the incoming signal, and a coefficient less than zero implies an inversion.

When the transmission line is terminated by an impedance equal to the characteristic impedance, then the reflection coefficient is equal to 0 and all of the signal is absorbed at the end of the line. Clearly this is the most desirable arrangement, but such design can become quite cumbersome. Luckily, however, it is not necessary to give careful attention to reflection and impedance matching in low-frequency circuits.

For clarification, consider the apparatus pictured in Figure 4.8. If we assume that the pulse generator and pulse counter are mismatched with the characteristic impedance of the cable which connects the two pieces, then when the pulse generator puts out a single pulse to be counted by the counter, the signal will travel down the line, be partially reflected at the counter and counted. The signal will then travel back down the line, be reflected by the generator and so on. This reflection will occur until the reflected pulse decays (the reflection coefficient is not exactly unity). Note that until the reflected signal is small enough that it is not detected by the counter, each reflection will be counted, resulting in erroneous experimental data. If this line is in a computer and the line is used to send digital signals to the disk drive, then the system will certainly not operate.

If we assume there are X reflections before the pulse is essentially gone, then the time over which the reflections occur is determined by the length of

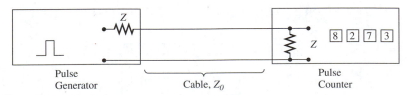

Pulse
Generator

Cable, Z_0

Pulse
Counter

FIGURE 4.8 **Pulse generator, cable and pulse counter.**

the cable and speed at which the signal propagates in the cable. For the RG-58/U cable analyzed above, the signal speed is about 2×10^8 m/s, and the time over which X reflections take place is calculated as

$$\text{time for signal to move length } L = t_{1/2} = \frac{L}{v}$$

$$\text{time for signal to move } X \text{ round trips} = t_X. = (2X)t_{1/2} = \frac{2\,XL}{v}$$

For a length $L = 2$ m, number of reflections $X = 20$ (poor matching), and $v = 2 \times 10^8$ m/s, we calculate the time over which the reflections occur to be about

$$t_X = \frac{2(20)\,(2 \text{ m})}{2 \times 10^8 \text{ m/s}} = 4 \times 10^{-7}\,\text{s} = 0.4\,\mu\text{s}.$$

For low-frequency signals, the reflections occur in such a short time that the original signal is still present. For example, a 1 kHz signal has a period of 1 ms, or a pulse width of 0.5 ms. Note how long this time is compared to the time over which the reflections are made in the above 2 m cable. For short cables and low frequencies, reflections are usually not a concern, but for long cables or high frequencies, reflections may be a concern. A good guideline is that for a cable of length L and velocity of propagation v, carrying a signal with frequency f, corresponding to a period T_o, termination is required if T_o is less than the time it takes for 10 reflections. In other words

$$\text{if } T_o = 1/f < 2(10)L/v \text{ then terminate.} \qquad \textbf{(4-18)}$$

Keep in mind that this is a general rule, not an equation derived from any exacting tolerances.

Termination is usually a simple matter of placing a resistor between the cable connections of the device in question. For example, a high-speed signal generator connected to a counter with a very large input impedance via a 50 Ω cable could be terminated by placing a 50 Ω resistor across the input terminals of the counter (where the cable connects). Computer peripherals, with multiple lines, are often terminated internally with a series of resistors across the inputs, but sometimes one must connect these resistors externally. Commercial devices which contain the appropriate resistors are aptly called *terminators*.

EXAMPLE 4.2

Will a 100 Mhz signal, traveling down a 2 m RG-58/U cable, need a termination?

Solution:

$$T_o = 0.01\,\mu\text{s} < \frac{20(2)}{2 \times 10^8} = 0.20\,\mu\text{s}$$

Yes, the line should be terminated.   ∎

4-5 COMPLEX WAVEFORMS

Many kinds of periodic waveforms exist. The rate at which a waveform repeats itself is one of the most easily found characteristics. A more detailed description of the waveform involves the use of a mathematical technique called the Fourier series, named after Jean Baptiste Joseph Fourier (1768–1830). The theorem underlying this technique states that any repetitive waveform can be represented as the sum of a series of sine and cosine waves of various frequencies and amplitudes. (Appendix A contains a brief introduction to the use of the Fourier series.) In general, a waveform $v(t)$, repeating in time can be represented as

$$v(t) = \frac{A_o}{2} + \Sigma A_n \cos n\omega t + \Sigma B_m \sin m\omega t. \tag{4-19}$$

The constants $A_o, A_1, \ldots, A_n, B_1, B_2, \ldots, B_m$ are evaluated to determine the amplitudes of the contributing sine and cosine terms. The frequencies, $n\omega t$ and $m\omega t$, are composed of a fundamental (n or $m = 1$) and higher harmonics (n or $m > 1$). For example, a sawtooth waveform is represented as

$$v(t)_{\text{sawtooth}} = \left(\frac{2 V_o}{\pi}\right) \left(\sin \omega t - \left(\frac{1}{2}\right) \sin 2\omega t + \left(\frac{1}{3}\right) \sin 3\omega t\right.$$

$$\left. - \left(\frac{1}{4}\right) \sin 4\omega t + \cdots\right) \tag{4-20}$$

$$= \left(\frac{2 V_o}{\pi}\right) \Sigma (-1)^n \left(\frac{1}{n+1}\right) \sin ((n+1)\omega t) \; [n = 0, 1, 2 \ldots].$$

This is illustrated in Figure 4.9. Note that the approximation improves as more terms are added to the series. The rapidly changing portions of the waveform, such as at the peak of the sawtooth, require the greatest number of terms for a good approximation. This is made evident when the series expansion for the square wave is examined:

$$v(t)_{\text{square}} = (4 V_o/\pi)\left(\sin \omega t + \left(\frac{1}{3}\right) \sin 3\omega t + \left(\frac{1}{5}\right) \sin 5\omega t\right.$$

$$\left. + \left(\frac{1}{7}\right) \sin 7\omega t) + \ldots\right)$$

$$= (4 V_o/\pi) \Sigma (1/n) \sin n\omega t \quad [n = 1, 3, 5, \ldots]. \tag{4-21}$$

Figure 4.10 shows that, even with the fourth term of the series added, the approximation to the square wave lacks sharpness at the corners. By plotting the amplitudes of the fundamental and harmonics, as shown in Figure 4.11, we can estimate the effect of adding more harmonic terms to the series. Notice that for both the square wave and the sawtooth wave, the amplitudes remain significant through the ninth harmonic; in contrast, for the triangular waveform (which has no rapidly changing portion), the amplitude is already quite small by the time the seventh harmonic is reached, and further terms will not greatly affect the accuracy of the approximation.

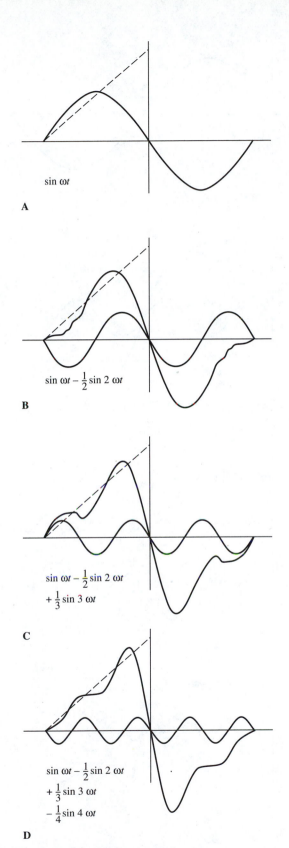

$\sin \omega t$

A

$\sin \omega t - \dfrac{1}{2} \sin 2\, \omega t$

B

$\sin \omega t - \dfrac{1}{2} \sin 2\, \omega t$
$+ \dfrac{1}{3} \sin 3\, \omega t$

C

$\sin \omega t - \dfrac{1}{2} \sin 2\, \omega t$
$+ \dfrac{1}{3} \sin 3\, \omega t$
$- \dfrac{1}{4} \sin 4\, \omega t$

D

FIGURE 4.9 **Fourier synthesis of a sawtooth wave.**

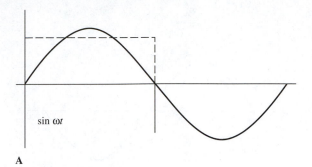

$\sin \omega t$

A

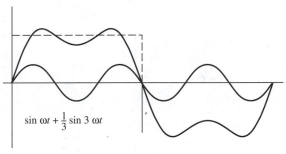

$\sin \omega t + \frac{1}{3} \sin 3 \omega t$

B

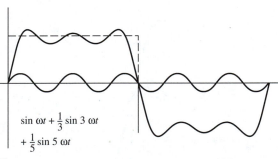

$\sin \omega t + \frac{1}{3} \sin 3 \omega t$
$+ \frac{1}{5} \sin 5 \omega t$

C

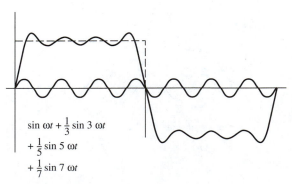

$\sin \omega t + \frac{1}{3} \sin 3 \omega t$
$+ \frac{1}{5} \sin 5 \omega t$
$+ \frac{1}{7} \sin 7 \omega t$

D

FIGURE 4.10 **Fourier synthesis of a square wave.**

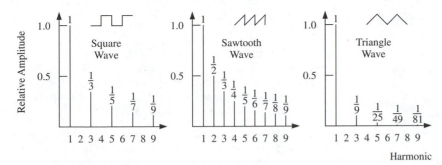

FIGURE 4.11   Fourier plot of various waveforms.

$$v(t)_{\text{triangle}} = (8\,V_o/\pi^2)\left(\sin \omega t + \left(\frac{1}{9}\right)\sin 3\omega t + \left(\frac{1}{25}\right)\sin 5\omega t\right.$$

$$\left. + \left(\frac{1}{49}\right)\sin 7\omega t\right) + \ldots)$$

$$= (8\,V_o/\pi^2)\,\Sigma\left(\frac{1}{n^2}\right)\sin(n\omega t) \quad [n = 1, 3, 5, \ldots]. \quad \textbf{(4-22)}$$

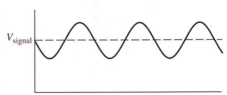

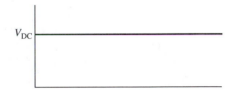

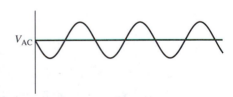

    To put it another way, sharply changing portions of even a low-frequency signal have high-frequency components. This means that these high-frequency components of the signal are affected differently by *RC* and *RL* configurations than are the lower or fundamental frequencies. Waveforms are rarely created electronically by adding a series of sine or cosine waves, but the recognition that even a simple wave is composed of high-frequency components is important. Neglecting this factor may result in a circuit behaving quite differently than expected. Consider, for example, a good quality high-pass filter which allows only sine waveforms above 2 kHz to pass, attenuating dramatically any signal below this frequency. If a square wave with frequency below the cutoff, say $f = 1$ kHz, is sent through the filter, the fundamental component will not pass through, but the higher order terms which represent the square wave will, resulting in a distorted wave leaving the filter. If the goal was to filter the signal entirely, the circuit has failed to do what was required.

    Finally, any alternating signal may have a DC component associated with it, as illustrated in Figure 4.12. The recognition of this feature is important, as it permits separation of the signal into its AC and DC components. In Fourier series representation, the leading term of the series, $A_o/2$ in Equation 4-19, indicates the DC component of the wave.

FIGURE 4.12   **The DC and AC portions of a signal voltage.**

4-6   QUESTIONS AND PROBLEMS

1.   In the circuit shown in Figure A, what current flows through the primary of the transformer?

2.   An audio signal generator has an output impedance of 600 Ω. To drive an 8 Ω speaker with maximum power transfer, an impedance-matching transformer is used between the

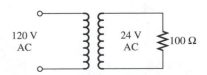

FIGURE A   **Circuit for problem 1.**

generator and the speaker. What are the specifications of such a transformer?

3. Design a transformer that converts 120 V AC line voltage to 5 kV AC. How can you limit the current from the output to 5 mA?

4. Given a transformer that outputs 24 V at 5 A when supplied by 120 V AC, what are the values of the turns ratio, input impedance and output impedance?

5. In the circuit shown in Figure B, what is the turns ratio of the transformer, what current flows through the primary of the transformer and what current flows through the secondary?

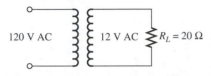

$$120 \text{ V AC} \qquad 12 \text{ V AC} \qquad R_L = 20 \, \Omega$$

FIGURE B Circuit for problems 5 and 6.

6. In the circuit of problem 5, use Kirchoff's voltage equation to analyze the loop defined by the transformer secondary and the 20 Ω resistor.

7. Assuming that the center-tapped transformer is wound in such a way that the center output is in the middle of the secondary coil, what are the voltage outputs of the circuit shown in Figure C? If the input is drawing 10 mA, what is the output current of the transformer if the load is connected across outputs A and B?

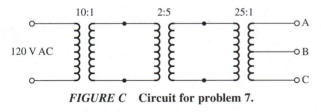

FIGURE C Circuit for problem 7.

8. Explain how and why a transformer can be used as an inductor.

9. A function generator with a 600 Ω output impedance can produce a signal with 50 V peak to peak. What is the maximum current that the generator can produce? If the generator is connected to a 600:50 Ω transformer, what is the maximum current that the generator can produce?

10. Given that the function generator in problem 9 is to drive a 50 Ω load, what are the maximum powers it can transfer to the load with and without the transformer?

11. A step-down transformer with no markings is measured to have a 1.6 Ω impedance on one side and a 40 Ω impedance on the other. Which side is the primary and which side is the secondary? If the input voltage is 120 V AC, what will the output voltage be?

12. Given a signal cable with a capacitance per unit length of 30.0 pF/ft and an inductance per unit length of 0.169 μH/ft, what is the speed of signal propagation along the wire?

13. Two meters of cable from problem 12 are connected between a pulse generator with a 75 Ω output impedance and a pulse counter with a 1 kΩ input impedance. If the output of the generator is a single 5 V pulse of duration 0.1 ns, what is the amplitude of the pulse when it reaches the detector? What is the voltage amplitude of the reflected pulse when it reaches the detector for the first time? The second time? If the pulse counter can count pulses which have a voltage amplitude between 0.3 and 7.0 V, how many pulses will the detector count?

14. How would you solve the reflection problem caused by mismatched impedances in problem 13?

15. A low-pass filter with a break-point frequency of 100 Hz is used to filter stray signal from a 90 Hz square wave signal. Describe the output of the filter. Is it the same as the 90 Hz square wave signal that entered the filter?

16. Calculate the Fourier components of the signal shown in Figure D.

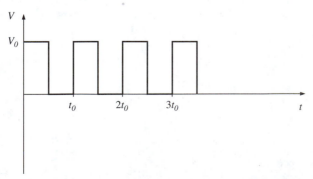

FIGURE D Circuit for problem 16.

Diodes and Selected Applications

5-1 INTRODUCTION: A PHYSICAL MODEL FOR SEMICONDUCTORS

Materials can be classified as insulators, semiconductors or conductors. We turn our attention to the semiconductor, which, as the word implies, is not quite a conductor. Table 5.1 lists the conductivities of some materials which are often incorporated into electronic devices. Conductance is a measure of the relative ease with which electrons move in a material. Metallic conductance involves movement of free electrons through the crystal lattice of the metal. The understanding of the movement of electrons in semiconductors requires the development of some ideas of bonding and structure in materials. What follows is a simplified physical picture which allows qualitative understanding of semiconductor action.

The most commonly encountered semiconductor materials are single crystals of silicon and germanium. The nucleus of a silicon atom has 14 protons which impart charge of $+14$ to the nucleus. In the space around the nucleus, 14 electrons are disposed in orbitals of varying energy. A simplified model of the atom considers that only the four outermost electrons enter into chemical bonding, the other ten being more tightly held by the nucleus. The "effective" nuclear charge is now $+4$. Such a situation is depicted in Figure 5.1. Note that a germanium atom may be treated in the same manner. In a single crystal of silicon or germanium, the bonding is said to be covalent; that is, a two-electron bond is formed, with one electron coming from each of two atoms. The crystal structure, of course, is three-dimensional, but a two-dimensional representation of a segment of crystal is given in Figure 5.2. Note that each atom is now surrounded by eight valence electrons, which is a minimum energy configuration.

Because each electron is rather strongly held, the conductance in a pure crystal is low. If sufficient energy is added to the crystal, some electrons overcome the attraction of their original nuclei and are free to move through the crystal. This results in an increase in conductance. The application of sufficient heat (thermal energy) causes this process to take place in an uncontrolled manner and leads to destruction of the device.

TABLE 5.1 Resistivity of Common Conductors, Semiconductors and Insulators

Material	Resistivity (ohm cm)
Conductors	
Aluminum	2.7×10^{-6}
Copper	1.7×10^{-6}
Gold	2.4×10^{-6}
Silver	1.6×10^{-6}
Tungsten	5.7×10^{-6}
Semiconductors	
Germanium	10^5
Selenium	10^7
Silicon	10^3 to 10^6
Insulators	
Beeswax	10^{14} to 10^{15}
Ceramics	10^{11} to 10^{14}
Ebonite	10^{15} to 10^{17}

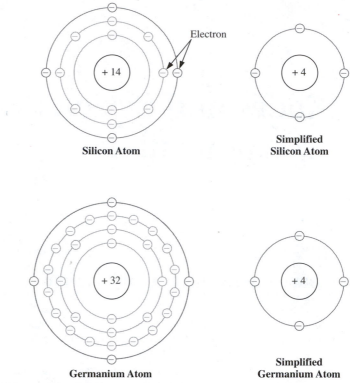

FIGURE 5.1   Descriptions of atoms of silicon and germanium.

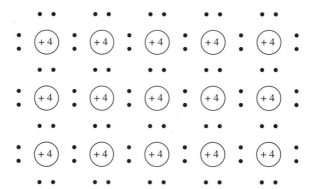

FIGURE 5.2   Two-dimensional representation of a portion of a crystal of silicon or germanium.

Suppose a small amount of some element is introduced into silicon crystal as an impurity. If the number of valence electrons for this impurity atom is not four, a change in the conductance of the crystal is observed. Suppose an atom of phosphorus, arsenic or antimony (each of which has five valence electrons) had been inserted into the lattice of Si or Ge. These atoms introduce valence electrons into a crystal that previously had only four at each lattice site. A crystal of this type is depicted in Figure 5.3. The net result is that a loosely bound electron is introduced into a structure where most electrons are more tightly bound. This is not to suggest that the rule of electro-

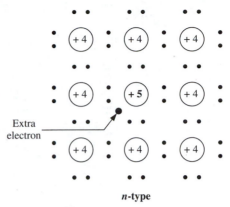

FIGURE 5.3   Two-dimensional representation of *n*-type silicon (or germanium) showing "impurity" atom.

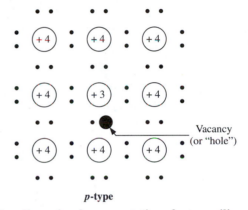

FIGURE 5.4   Two-dimensional representation of *p*-type silicon (or germanium) showing "impurity" atom.

neutrality has been violated. Since the new atom has an effective nuclear charge of +5, the five valence electrons achieve an electrical balance. The additional electron, however, has higher mobility than the others. Because the electron is free to move about, the place from which it came is called a *vacancy* or *hole*. Other electrons can and do fill the vacancy, but, in the process, create other holes. The only *particle* movement in a crystal is that of the electrons, even though the holes appear to move in a direction opposite to that of the electrons when a potential is applied. The holes are locked into the structure, but when one is filled by an electron, another hole is created, giving the impression of hole movement.

A crystal of either Si or Ge thus "doped" with an element such as P or As is termed an *n-type crystal* because the charge carrier is negative. If an element such as B, Al or In, with only three valence electrons, is inserted into the crystal, as depicted in Figure 5.4, a *p-type material* is formed. In this situation a hole, rather than an excess electron, is incorporated into the lattice. The *majority current carrier* in this type is the hole and the *minority current carrier* is the electron. The designations are reversed for the *n*-type material. The idea of hole "movement" is presented in Figure 5.5, where the

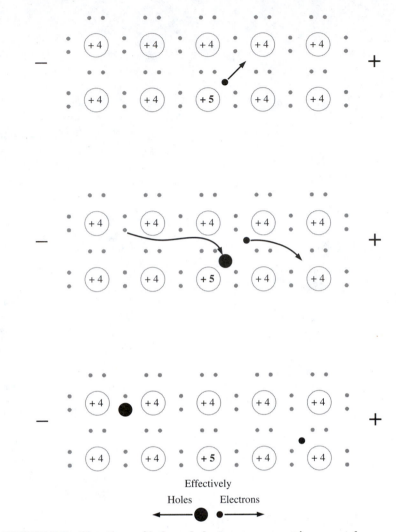

FIGURE 5.5   **Directions of hole and electron movement in a crystal.**

movement can be perceived as generation of electron-hole pairs with subsequent recombination at other sites in the crystal lattice. Even though only electrons leave and enter the crystal when an external voltage source is applied to the crystal, the process illustrated in Figure 5.5 imparts the effect of an overall movement of holes in one direction and electrons in the opposite direction. At room temperature, a certain number of electron-hole pairs are generated within the crystal as an electron is released from its bonding site, leaving behind the hole. These carriers, as well as those purposely added, constitute the movement or drift. The processes of generation and recombination occur continuously in the crystal.

5-2   THE *p-n* JUNCTION

When a crystal is made which contains both donor and acceptor atoms, a *p-n* junction is created. The situation at the instant the junction is formed with *n*-type and *p*-type material is shown in Figure 5.6A. The *n*-type material is "rich" in electrons and the *p*-type material is deficient in electrons. Likewise, the *p*-type material is rich in holes and the *n*-type material is deficient in holes. As the junction is formed, electrons diffuse across the junction as shown in Figure 5.6B. If no other processes than diffusion were active, ultimately there would be a complete uniformity of holes and electrons throughout the material. Fortunately, another process limits diffusion; an electrostatic potential or barrier retards the diffusion. The barrier is created because of a negative charge moving into the *p*-type material, leaving a positive charge in the *n*-type material. This situation is depicted in Figure 5.6C, where a depletion layer has been created by the diffusion of electrons and subsequent recombination with holes. In effect, a junction potential has been created which is opposite in sign to the designations of the materials. The voltage source, shown in Figure 5.6D, symbolizes this result. The junction also behaves as a capacitor. The depletion of charge carriers from the junction is equivalent to the separation of charges on a capacitor. The depletion zone serves as the dielectric medium. The magnitude of the junction capacitance is small, but it can be an important factor in high-frequency performance. The junction potential depends upon the material. At 25°C, the silicon junction potential is approximately 0.6 V, while that of germanium is 0.4 V.

If metallic contacts are added to the opposite ends of the crystal and a voltage is applied, the unidirectional character of the junction is clearly illustrated, as shown in Figure 5.7A. If the applied voltage (bias) is connected to agree with the junction potential polarity (Figure 5.7B), the barrier height is increased and essentially no current flows. (A current of a few microamperes flows because of minority carrier conduction.) This is termed *reverse bias.* If the polarity of the applied voltage is opposite that of the junction, the barrier is reduced and current flows (Figure 5.7C). This is called *forward bias.* This device is called a *diode,* and its circuit terminology is given in Table 5.2. Note that the *p*-side of the *p-n* junction is always the anode.

5-3   DIODE ACTION

Up to this point, we have dealt with *linear* devices, devices that obey Ohm's law. An increase in voltage across the device results in a proportional increase in current through the circuit and vice-versa. Diodes are *nonlinear* devices. They depart from the proportional action of the linear elements. An *ideal diode* would, in fact, act like a switch; it would be either on or off, conducting or non-conducting.

The current-voltage characteristics of an ideal diode and real diodes are shown in Figure 5.8. When the ideal diode is *forward-biased,* current flows through it with no resistance and thus there is no voltage drop across the diode. When the ideal diode is *reverse-biased,* no current flows—regardless of the voltage applied across it. On the other hand, a forward-biased real diode has a voltage drop across it which is dependent upon the current through the device. A good rule of thumb to use is that a forward-biased silicon diode (the most common diode variety) has an associated 0.6 V drop across its terminals. A reverse-biased real diode allows some current to flow, but for most applications this current can be neglected.

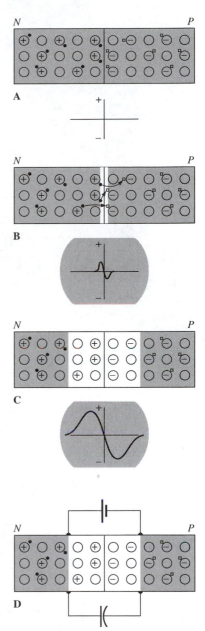

FIGURE 5.6   Schematic representation of the process of "building" a *p-n* junction.

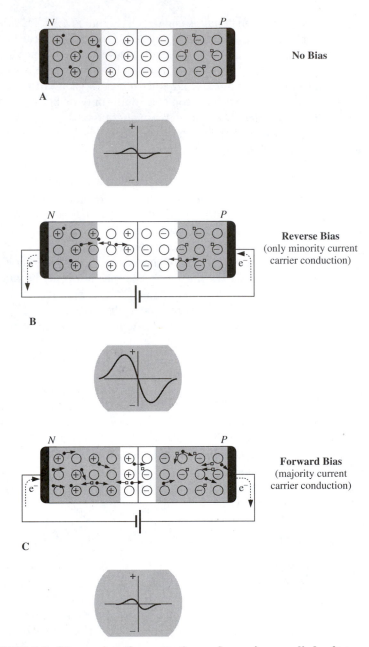

FIGURE 5.7 The *p-n* junction operation under various applied voltages.

Approximate testing of a diode which is not in a circuit can be conducted simply with a digital ohmmeter. The "resistance" is measured with the leads of the meter one way, then measured again with the leads reversed. A good diode will have a small "resistance" in one direction and a very large "resistance" in the other. The multimeter diode test indicates whether the diode is rectifying or not, but does not indicate the integrity of its other characteristics. Note that the meter passes a current through the diode, measures the

TABLE 5.2   Diode Designations

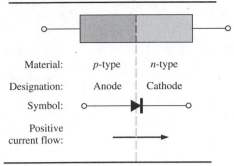

Material:	*p*-type	*n*-type
Designation:	Anode	Cathode
Symbol:		
Positive current flow:		

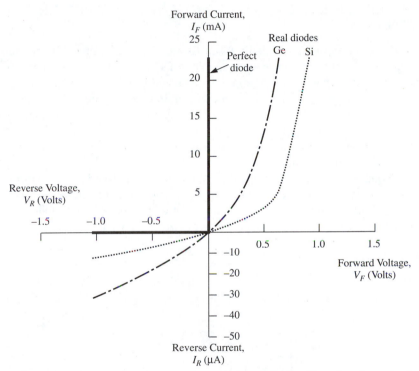

FIGURE 5.8   **Current-voltage characteristics of diodes. Note that the units for reverse current are μA while forward current is in mA.**

resulting voltage and displays the resistance ($R = V/I$). Since the diode is a nonlinear device, the resistance is a function of the current passing through it. Do not be surprised to get a different measurement using different resistance ranges. The important parameter is the difference in readings from one direction to the other. Table 5.3 illustrates some common diodes and their specifications.

The switching action of diodes is of fundamental importance in electronics. This sensitivity to the polarity of a signal finds a multitude of uses, some of which will be described in the following sections of this chapter.

TABLE 5.3   Diode Specifications

	Symbol	1N914B	1N4001	1N4002	1N4003	1N4004	1N4005	1N4006	1N4007	Units
Maximum ratings:										
Peak inverse voltage (PIV)	V_R	100	50	100	200	400	600	800	1000	volts
Average forward current	I_0	0.10	←			1.0			→	amps
Electrical characteristics:										
Forward voltage	V_F	1.0	←			1.1			→	volts
Maximum reverse current (25° C)	I_R	0.025	←			0.01			→	mA

5-4   RECTIFICATION AND SINGLE-VOLTAGE POWER SUPPLIES

A *DC power supply* is generally taken to mean a source of unidirectional voltage or current. The battery is a source of DC power, but it has a number of drawbacks, including expense, limited life and weight. The common carbon-zinc battery also suffers from producing a decreasing output voltage as power is taken from it. The mercury battery can be used as a power source of constant voltage, since its voltage remains essentially constant until just before it is completely discharged, but it has the drawback of producing only small amounts of current.

A much more reliable method of obtaining DC power is to transform, rectify, filter and regulate AC line voltage, as indicated in Figure 5.9. The principles of rectification, filtering and regulation are treated in the following sections of this chapter. Transformers were covered in Chapter Four.

Half-Wave Rectifier

When a source of alternating voltage is applied to a diode in series with a load, a half-wave rectifier is formed as shown in Figure 5.10. Figure 5.10A

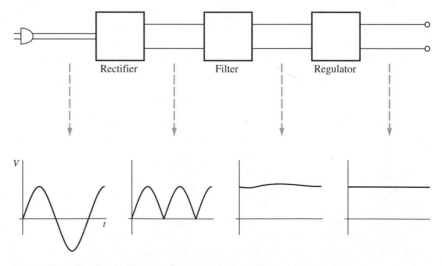

FIGURE 5.9   **Block diagram representation of a power supply.**

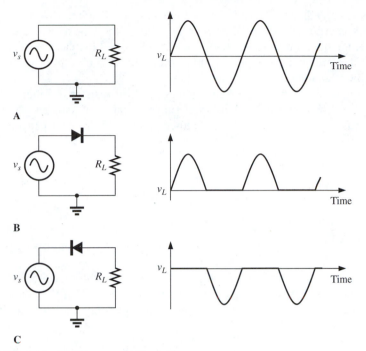

FIGURE 5.10   Half-wave rectifier action.

shows the voltage across R_L with no diode in the circuit. The voltage across the resistor is simply the voltage supplied by the power source. Figures 5.10B and 5.10C show rectified signals. Since the diode will only allow current to flow in the direction of the arrow, the orientation of the diode in Figure 5.10B shows rectification of the positive half of the signal. Likewise Figure 5.10C shows rectification of the negative half of the signal. Remember that a real diode has an associated voltage drop of about 0.6 V when it is conducting. The graph of V_L versus t in Figures 5.10B and 5.10C is therefore actually 0.6 V lower than that of Figure 5.10A.

In this application the diode is an AC to DC converter. Although the signal is not very smooth, it is, under strict definition, a DC signal; the current flows only in one direction. For most applications, however, this power supply is unacceptable. To make the signal more useful, the ripple in the signal can be filtered with a capacitor, as shown in Figure 5.11. Here the rectified signal charges the capacitor to the peak voltage of the source signal when the diode is conducting. When the diode is not conducting, the circuit

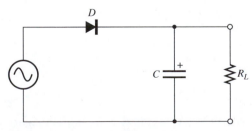

FIGURE 5.11   Filter capacitor in rectifier circuit.

is essentially shut off from the signal, forming an RC circuit with the capacitor discharging through the resistor. In general, the larger the capacitor the better the filter, but with increased capacitance comes increased cost and size. We therefore need some general rules to make a wise choice for the value of the filter capacitance.

Figure 5.12 shows the signal across the load resistor in Figure 5.11. A rough estimate of the relationship of ripple voltage ΔV, current i drawn by the load, the frequency f of the rectified signal and the filter capacitance C is given by

$$\Delta V \approx \frac{i}{Cf} \tag{5-1}$$

or

$$C \approx \frac{i}{\Delta Vf} \tag{5-2}$$

(Equation 5-1 stems from $i = C\, dV/dt$ for a capacitor.) Note that larger capacitors and higher frequencies lower the expected ripple voltage, and higher currents raise it. The *ripple factor,* a commonly quoted specification of a power supply, is defined as

$$\text{ripple factor, } r = \frac{\Delta V}{V_{DC}} \tag{5-3}$$

Although here we define the ripple factor as the peak-to-peak voltage of the AC signal component divided by the DC component, it is sometimes measured as the rms value of the AC signal component divided by the DC component or

$$r = \frac{V_{rms}}{V_{DC}}. \tag{5-3a}$$

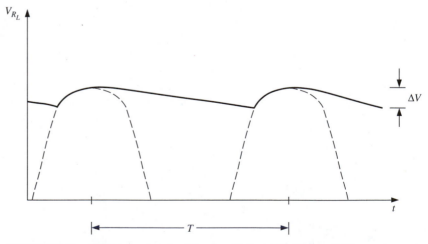

FIGURE 5.12   Voltage across load in circuit shown in Figure 5.11.

EXAMPLE 5.1

A power supply consisting of a 24 V transformer, single diode and capacitor of 10,000 μF is to supply a maximum current of 0.5 A. What is the approximate ripple voltage and the ripple factor at maximum load?

Solution: Since the rectification is half wave, the frequency of the rectified signal is 60 Hz. Then

$$\Delta V = \frac{0.5 \text{ A}}{10,000 \times 10^{-6} \text{ F} \times 60 \text{ Hz}}$$

$$= 0.8 \text{ V}$$

For a 24 V rms signal the peak voltage is 34 V. The ripple is then calculated as

$$r = \frac{0.8 \text{ V}}{34 \text{ V}}$$

$$= 0.02$$

or 2%. ■

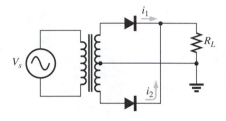

Full-Wave Rectifier

The full-wave rectifier, shown in Figure 5.13, is an improvement over the half-wave circuit. As the figure shows, this configuration includes a center-tapped (CT) transformer and two diodes. The rectified signal is doubled in frequency from that of the single wave rectifier and thus, all other parameters being the same, the ripple voltage is halved (Equation 5-1). In general, center-tapped transformers cost more than non-center-tapped ones; however, cost and space savings may be realized with a smaller capacitor.

The bridge rectifier, shown in Figure 5.14, produces the same waveform without the need for a center-tapped transformer. It requires two more diodes, but these are relatively inexpensive. Without the transformer to provide ground isolation, the circuit cannot be used with function-generator signals or other signals that have an inherent ground connection. In other words, if the transformer were not present, the junction between D_3 and D_4 would be at ground since the function generator has a ground, as would the junction between D_3 and D_1.

5-5   SPLIT POWER SUPPLIES

Often a circuit requires a power supply that provides negative voltage as well as positive voltage. We will see in later chapters that this is true for integrated circuits known as operational amplifiers. By reversing the direction of the diode and the capacitor (if it is polarized), the half-wave rectification circuits discussed above provide a negative voltage. Similarly, reversing the direction of the diodes and capacitor in the full-wave rectified supply produces a negative voltage supply.

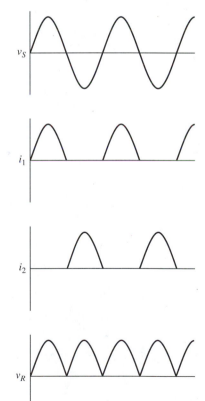

FIGURE 5.13   Full wave rectifier action.

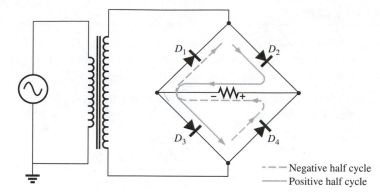

FIGURE 5.14 **Bridge rectifier action.**

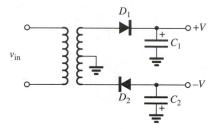

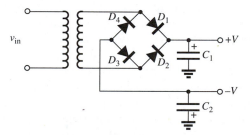

FIGURE 5.15 **Two split power supplies.**

A single center-tapped transformer can be used to make a *split power supply*—one that has positive and negative voltage. Two such circuits are shown in Figure 5.15.

5-6 VOLTAGE MULTIPLIERS

A useful modification of a rectifier circuit, the voltage doubler, is shown in Figure 5.16. During the positive half cycle, D_1 conducts and charges capacitor C_1 to the potential V_p. During the negative half cycle, D_1 is off and D_2 conducts "negatively" to charge capacitor C_2 to potential $-V_p$. The voltage across the combination of R_1 and R_2 is therefore equal to twice the peak transformer voltage. We assume here that the load does not draw a significant charge from the capacitors. Similarly, voltage triplers, quadruplers and other multipliers can be constructed as shown in Figure 5.17.

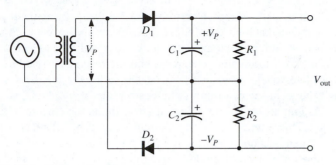

FIGURE 5.16   **Voltage doubler circuit.**

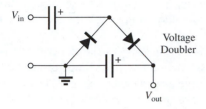

Voltage
Doubler

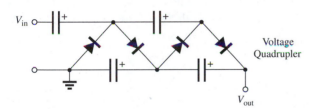

Voltage
Quadrupler

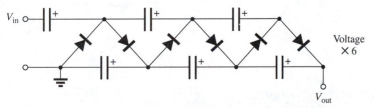

Voltage
× 6

FIGURE 5.17   **Voltage multiplying circuits. Multiplication is approximate: Diode voltage drops may have a significant effect on V_{out}.**

5-7   ZENER DIODES

Our previous discussion of the junction diode omitted an important feature of the operation of the diode: the avalanche or reverse breakdown process. Suppose that the reverse bias on a diode is greatly increased. When the voltage is sufficient, valence electrons will be freed from their positions around the nuclei that bind them. Since these electrons possess excess energy, their collisions with other atoms will knock loose additional electrons; these in turn will knock loose more electrons, and the reverse current becomes an avalanche. The voltage at which this occurs is called the *breakdown voltage.* Although the term seems to imply the destruction of the diode, such is not the case (as long as some external component limits the current to a reasonable value). This effect was first noted and utilized by Clarence Zener, for whom the phenomenon is named. A more complete version of the current-voltage behavior of a Zener diode is given in Figure 5.18. Note that for voltages more negative than the breakdown voltage, the voltage is virtually independent of the current. The breakdown, or Zener, voltage of a diode can be controlled in the manufacturing process, and these special diodes, called *Zener diodes,* are available in values from 3.9 V to several hundred volts.

It is apparent from Figure 5.18 that, as long as the Zener diode is reverse-biased, any current flow greater than a few microamperes must be accompanied by a voltage at least as large as the Zener voltage. The electronic characteristics of the Zener diode make the device suitable as a voltage regulator. Such an application is shown in Figure 5.19. The resistor R in the circuit acts as a current limiter and provides a voltage drop between the unregulated source and the load. Careful choices must be made for the resistance value and power rating of this resistor. These are dependent upon the maximum current the circuit must supply, the input voltage and the Zener diode specifications. Given the Zener diode voltage, V_Z, Zener diode power

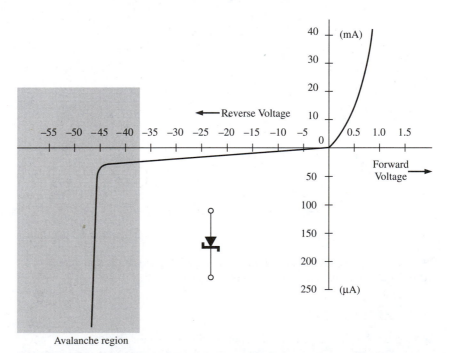

FIGURE 5.18   Zener diode behavior. Note that the units for positive current are mA and those for negative current are μA.

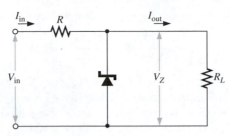

FIGURE 5.19   **Simple Zener diode voltage regulator circuit.**

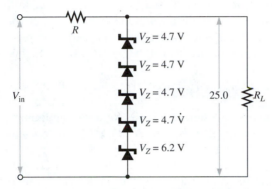

FIGURE 5.20   **Combining Zener diodes to achieve a desired output.**

rating, P_Z, maximum Zener diode current, I_{Zmax}, output current I_{out}, and input voltage V_{in}, the resistance value R can be calculated. If we design the circuit such that the current through the diode is 20% of its maximum rated value, then

$$R = \frac{(V_{in} - V_Z)}{(0.2\, I_{Zmax} + I_{out})}. \tag{5-4}$$

The power rating of this resistor must be greater than

$$P_R = I_{in}^2\, R = \frac{(V_{in} - V_Z)^2}{R} \tag{5-5}$$

Also note that the Zener diode power rating must be greater than

$$P_Z = I_Z V_Z = (I_{in} - I_{out})\, V_Z = [((V_{in} - V_Z)/R) - I_{out}]\, V_Z. \tag{5-6}$$

Zener diodes can be combined in series to produce higher voltages, as shown in Figure 5.20.

EXAMPLE 5.2

With an input voltage of 7.0 V, design a 5.1 V power source capable of supplying 10 mA.

Solution: First we choose a Zener diode which has a breakdown voltage of 5.1 V and typically a value of

$$I_{Zmax} = 50 \text{ mA (1N4733A diode)}$$

From Equation 5-4 the input current is calculated as

$$R = \frac{(7.0 - 5.1)}{(0.2)(0.5) + 0.01} = 95 \ \Omega.$$

Substituting values for $I_{in} = 0.2 \ I_{Zmax} + I_{out}$ into Equation 5-5 yields the requirement that the power rating of the resistor be greater than a mere 38 mW. Equation 5-6 states that the power rating of the diode must be greater than 5.1 mW. A 100 Ω 1/4 W resistor would be suitable. ■

In general, Zener diode regulator circuits are not very stable. Fluctuations in output current demands and the loose tolerances in the manufacturing process can lead to significant variations in the output voltage. They are, however, very useful in low power circuits not requiring precise voltage regulation. Better regulator circuits will be discussed later in this chapter.

5-8   THE DIODE CLIPPER

To protect circuitry, a design often calls for a voltage limiter. A simple solution to this design problem is a *diode clipper* (often referred to as a *diode clamp*) as shown in Figure 5.21. Since the diode will be reverse-biased, non-conducting for voltages less than $V + 0.6$ V and forward-biased, conducting for voltages greater than that value, the diode prevents the output voltage from exceeding $V + 0.6$ V. Resistor R protects the diode from excessive current.

A related diode clipper circuit is shown in Figure 5.22. It is useful in limiting the amplitude of an AC signal. Note that, in the absence of a signal, diode D_1 is reverse-biased by V_1 and D_2 is reverse-biased by V_2. During the positive half cycle of the input signal, D_1 does not conduct until V_{in} exceeds V_1. When D_1 does conduct during the positive half cycle, V_0 is limited in amplitude to V_1. The same holds true for the negative half cycle regarding D_2 and V_2. (The foregoing explanation has ignored the forward-biased voltage drop across the diode of about 0.6 V.) Also note that the batteries can be replaced with voltages from a voltage divider or directly from a power supply.

The clipper is often used in input stages of amplifier circuits. Here it acts to restrict the amplitude of signals that would otherwise drive the amplifier circuitry out of the specified operating region. The clipped signal is deformed, but any signal that has an amplitude less than the bias voltage passes by the clipper without distortion. One application of a clipper may be found in an audio amplifier which drives a speaker. To protect the speakers from overload, a clipping circuit is usually used in the amplifier.

A useful clipper circuit is formed by connecting the anodes of two Zener diodes, as in Figure 5.23. The circuit has the advantage of not needing bias voltages.

5-9   THE DIODE CLAMP

The DC-restrictive diode clamp shown in Figure 5.24 is useful for changing the reference voltage of the input signal. The circuit operation can be understood by recognizing that the diode conducts only during the negative half cycle of the input signal (when it is forward-biased). This charges the diode side of the capacitor to $+V_p$. The reference voltage is V_p, and the output

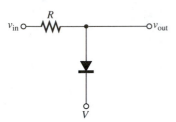

FIGURE 5.21   Diode clipper.

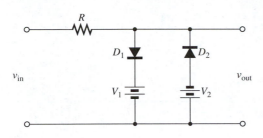

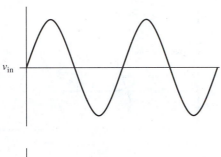

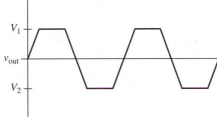

FIGURE 5.22   **Diode clipper circuit.**

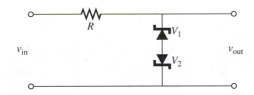

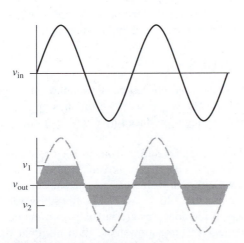

FIGURE 5.23   **Zener diode clipper circuit.**

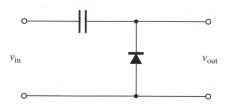

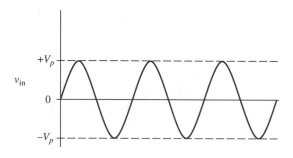

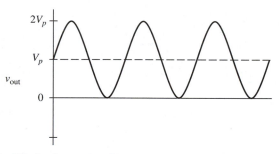

FIGURE 5.24 **Diode clamp circuit.**

waveform swings between $+2V_p$ and zero rather than between $+V_p$ and $-V_p$. By placing a voltage source in series with the diode, any desired reference voltage can be established. By reversing the diode, the output signal can be clamped below zero rather than above it. Since RC discharge of the capacitor will distort the output waveform, the following stage must have a very high input impedance for the clamp to function.

5-10 REGULATORS

Until now we have discussed filtered power supplies and low-power, Zener-diode-based supplies. Here we present two families of easy-to-use integrated circuits that regulate voltages produced by filtered supplies: fixed-voltage and adjustable-voltage regulators.

Capacitor-filtered supplies suffer from three major difficulties: the output voltage is relatively fixed at the peak voltage of the transformer used, the ripple voltage is proportional to increasing current demand, and the output voltage is dependent upon 120 V AC fluctuations. By regulating a filtered power supply, these three concerns are virtually eliminated. A voltage regulator keeps the voltage of the supply fixed with minimal ripple, even when faced with fluctuating current demands and 120 V AC fluctuations.

The LM78XX series is an industry-standard regulator for fixed positive voltages, with the related LM79XX series used for fixed negative voltages. The LM317 and LM320 series are used for adjustable positive and negative voltages, respectively. Numerous other voltage regulators are readily available, including high-voltage and precision applications. Many of the commonly used regulators and their specifications are listed in Table 5.4.

TABLE 5.4   *Continuous Voltage Regulators Selection Guide*

Adjustable Positive Voltage Regulators

Amps	Device	Output Voltage	Package
10.0	LM196K	1.25V–15V	TO-3
	LM396K	1.25V–15V	TO-3
5.0	†LM138K	1.2V–32V	TO-3
	LM338K, T	1.2V–32V	TO-3, TO-220
3.0	†LM150K	1.2V–33V	TO-3
	LM350K, T, AT	1.2V–33V	TO-3, TO-220
1.5	†LM117K	1.2V–37V	TO-3
	†LM117HVK	1.2V–57V	TO-3
	LM317K, T, AK, AT	1.2V–37V	TO-3, TO-220
	LM317HVK	1.2V–57V	TO-3
1.0	LM2941T, CT	5V–20V	TO-220
	LM78GCT**	5V–30V	TO-220 (4-Pin)
0.5	†LM117H	1.2V–37V	TO-39
	†LM117HVH	1.2V–57V	TO-39
	LM317H	1.2V–57V	TO-39
	LM317HVH	1.2V–37V	TO-39
	LM317MP	1.2V–37V	TO-202
	LM78MGCT**	5V–30V	TO-220 (4-Pin)
0.2	†LH0075G, CG	0–27V	TO-8
0.1	LM317LZ, M	1.2V–37V	TO-92, SO-8
	LM2931CT	3.0V–24V	TO-220, 5-LEAD
	†LP2951CN, J, H, M	1.24V–29V	DIP, CERDIP, HEADER, SO-8

Adjustable Negative Voltage Regulators

Amps	Device	Output Voltage	Package
3.0	LM133K	−1.2V − −32V	TO-3
	LM333K, T	−1.2V − −32V	TO-3, TO-220
1.5	†LM137K	−1.2V − −37V	TO-3
	†LM137HVK	−1.2V − −47V	TO-3
	LM337K, T	−1.2V − −37V	TO-3, TO-220
	LM337HVK	−1.2V − −47V	TO-3
1.0	LM79GCT*	−2.2V − −30V	TO-220 (4-Pin)
0.5	†LM137H	−1.2V − −37V	TO-39
	†LM137HVH	−1.2V − −47V	TO-39
	LM337H	−1.2V − −37V	TO-39
	LM337HVH	−1.2V − −47V	TO-39
	LM337MP	−1.2V − −37V	TO-202
	LM79MGCT*	−2.2V − −30V	TO-224 (4-Pin)
0.2	†LH0076G, CG	0–27V	TO-8
0.1	LM337LZ, M	−1.2V − −37V	TO-92, SO-8

TABLE 5.4   *(Continued)*

Adjustable Positive/Negative Voltage Regulator

Amps	Device	Output Voltage	Package
±0.1	LH7001	±1.2V to ±37V	DIP, TO-5

Fixed Positive Voltage Regulators

Amps	Device	Output Voltage	Package
3.0	†LM123K	5V	TO-3
	LM323K, AK	5V	TO-3
1.0	†LM109K	5V	TO-3
	†LM140AK	5V, 12V, 15V	TO-3
	†LM140K	5V, 12V, 15V	TO-3
	†LM2940T	5V, 8V, 10V, 12V	TO-220
	LM2940CT	5V, 12V, 15V	TO-220
	LM309K	5V	TO-3
	LM340AK, T	5V, 12V, 15V	TO-3, TO-220
	LM340K, T	5V, 12V, 15V	TO-3, TO-220
	LM78xxCK, T**	5V, 6V, 8V, 12V, 15V, 18V, 24V	TO-3, TO-220
0.5	LM2984CT	5V, 12V, 15V	TO-220, TO-202
	LM341T, P	5V, 12V, 15V	TO-220, TO-202
	LM78MxxCT, H**	5V, 6V, 8V, 12V, 15V, 24V	TO-220, TO-39
0.2	†LM109H	5V	TO-39
	LM309H	5V	TO-39
	LM342P	5V, 12V, 15V	TO-202
0.15	LM2930T	5V, 8V	TO-220
0.1	†LM140LAH	5V, 12V, 15V	TO-39
	LM2931Z, T	5V	TO-92, TO-220
	LM340LZ, H	5V, 12V, 15V	TO-92, TO-39
	LM78LxxACZ, H, M**	5V, 6.2V, 8.2V, 9V, 12V, 15V	TO-92, TO-39, SO-8
	LP2950CZ	5V	TO-92
0.05	LM2936Z	5V	TO-92

*The LM320 has better electrical characteristics than the LM79xx.
LM100 Series   +55°C to +150°C
LM300 Series   0°C to +125°C

Fixed Negative Voltage Regulators

Amps	Device	Output Voltage	Package
3.0	†LM145K	−5V, −5.2V	TO-3
	LM345K	−5V, −5.2V	TO-3
1.5	†LM120K	−5V, −12V, −15V	TO-3
	LM320K, T	−5V, −12V, −15V	TO-3, TO-220
	LM79xxCT, K**	−5V, −8V, −12V, −15V	TO-220, TO-3
0.5	LM320MP	−5V, −12V, −15V	TO-220
	LM79MxxCT, H**	−5V, −8V, −12V, −15V	TO-220, TO-39
0.2	†LM120H	−5V, −12V, −15V	TO-39
	LM320H	−5V, −12V, −15V	TO-39
0.1	LM320LZ	−5V, −12V, −15V	TO-92
	LM79LxxACZ, M	−5V, −12V, −15V	TO-92, SO-8

TABLE 5.4   *(Continued)*

Low Dropout Regulators

Amps	Device	Output Voltage	Package
0.050	LM2936Z	5V	TO-92
0.100	LM2931T, Z	5V, ADJ	TO-220,TO-92
	LP2950CZ	5V	TO-92
	LP2951N, J, H	ADJ	DIP, CERDIP, HEADER
0.150	LM2930T	5V, 8V	TO-220
0.500	LM2984CT	TRIPLE 5V + WATCHDOG	TO-220, 11-LEAD
0.750	LM2925T	5V WITH DELAYED RESET	TO-220, 5-LEAD
	LM2935T	DUAL 5V	TO-220, 5-LEAD
1.0	†LM2940T	5V, 8V, 10V, 12V	TO-220
	LM2940CT	5V, 12V, 15V	TO-220
	LM2941T	Adjustable (5V to 20V)	TO-220
	LM2941CT	Adjustable (5V to 20V)	TO-220

Shunt Regulators

Amps	Device	Output Voltage	Package
0.15	LM431ACZ, M	2.5V–36V	TO-92, SO-8

Building Block Regulators

Device	Title	Package
†LM104/204/304	Negative Regulator	TO-39
†LM105/205/305	Voltage Regulator (Positive)	TO-39
LM376	Voltage Regulator (Positive)	8-Pin Plastic DIP
†LM723	Voltage Regulator	14-Pin DIP, TO-39

**These products were formerly manufactured by Fairchild Corporation. The prefixes have been changed from μA to LM and may be found with the former prefix as well as the latter.

†Military qualified device. For more information, consult the Military/Aerospace Selection Guide.

Reprinted from National Semiconductor General Purpose Linear Devices Databook, 1989.

Fixed-Voltage Regulators

The fixed-voltage LM78XX regulator is easy to use and is available from a number of manufacturers. It is a three-terminal device with V_{in}, ground, and V_{out} connections. The regulators are available with current ratings from 100 mA to 1 A (with proper heat sinking). The XX indicates the output voltage of the regulator which can be 5 V, 6 V, 8 V, 9 V, 10 V, 12 V, 15 V, 20 V or 24 V. These regulators have fixed output voltages which are $+/-$ 5% of the specified value and have ripple factors of much better than 1%. In addition to superb regulation, the LM78XX series features overload protection. If the power limits of the device are exceeded, the regulator automatically shuts off until the load is removed and the circuit cools down. A typical application is shown in Figure 5.25.

In general, the input voltage specifications of the regulator mandate an input voltage of less than 30 V, but greater than 3 V above the expected output voltage. For example, a 7812 regulator will deliver 12 V if the input voltage is kept between 15 and 30 V. Note that the circuit requires three

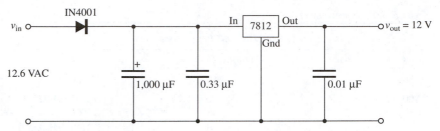

FIGURE 5.25   **Typical application of LM78XX regulator, here using the 12 V, LM7812.**

capacitors. The largest is the filter capacitor, which should be chosen such that V_p of the transformer, less V_{ripple}, is at least 3 V higher than the expected output voltage. The second capacitor, on the input, filters high frequencies which may not be shunted to ground by large electrolytic capacitors with high internal inductance. The third capacitor, on the output, is necessary for stabilization of the circuit.

EXAMPLE 5.3

Using a 12 V transformer and an LM7805 regulator, what is the smallest filter capacitor which should be used to build a 5 V power supply drawing a maximum of 1 A?

Solution: First, calculate the peak voltage from the transformer

$$V_{peak} = 2^{1/2} \times 12 \text{ V} = 17 \text{ V}$$

The regulator needs at least 5 V (its voltage rating) plus 3 V (actually, the LM7805 requires only 2 V overhead). Therefore the ripple voltage can be as high as 17 V − 8 V = 9 V. Then calculate the capacitance using Equation 5.3. Assuming half-wave rectification, the value of C is given by

$$C = \frac{i}{\Delta V f}$$

$$= \frac{1 \text{ A}}{(9 \text{ V})(60 \text{ Hz})}$$

$$= 1850 \text{ } \mu\text{F}$$

which is relatively small for a power supply capacitor. Note that large electrolytic capacitors usually have capacitance tolerances of −20%/+80%. This fact and the possible existence of line voltage fluctuations make it wise to use a larger capacitor, e.g., a 2,500 μF, 35 V electrolytic.

The negative-voltage regulators, the LM79XX series, are identical in their application, with the exception that V_{in} and V_{out} are negative.

Adjustable Regulators

The LM317 is a regulator with an output voltage that can be fixed via two external resistors. The output voltage can be set from 1.25 V to 37 V, with

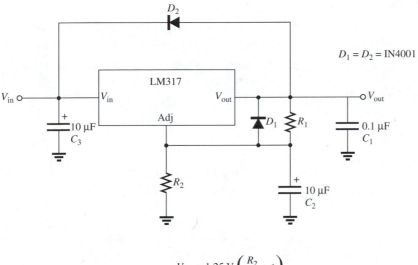

$$V_{out} = 1.25 \text{ V} \left(\frac{R_2}{R_1} + 1 \right)$$

FIGURE 5.26   LM317 regulator circuit with protection diodes (diodes not needed for $V_{out} < 25$ V and $C_1 < 25$ μF). Typical $R_1 = 220$ Ω.

varieties available that can handle up to 1.5 A. The typical circuit is shown in Figure 5.26. The output voltage is given by

$$V = 1.25 \text{ V} (1 + (R_2/R_1)). \tag{5-7}$$

R_1 should be on the order of 220 Ω while the value of R_2 is chosen to obtain the necessary voltage. Note that R_2 may be a potentiometer which makes the circuit a variable voltage power supply, with minimum voltage of 1.25 V. It is often inconvenient to have a variable power supply which does not approach zero volts. By adding three diodes in series with the output, and a resistor to ensure their conductance regardless of output voltage, the design in Figure 5.27 takes advantage of the 0.6 V drop across each diode and forces the supply to a minimum of zero volts.

Design of power supplies requires attention to many factors, such as power dissipation and heat sinking. In addition, power supplies for special applications such as constant current, high current, precision and high voltage are covered throughout the remainder of the text.

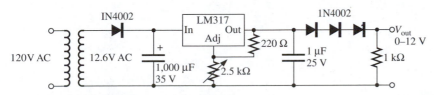

FIGURE 5.27   Variable DC voltage supply, 0–12 V.

5-11   QUESTIONS AND PROBLEMS

1.   Design a power supply using a 24 V, center-tapped transformer to supply 12 V at 500 mA, − 12 V at 100 mA and 5 V at 250 mA. What is the minimum current rating of the transformer?

2.   Design a variable voltage power supply which ranges from 5 to 15 V and is capable of supplying 250 mA.

3.   In the circuit in Figure A, what would happen if the load resistor were shorted? What would happen if the load resistor were removed? Hint: think in terms of power ratings.

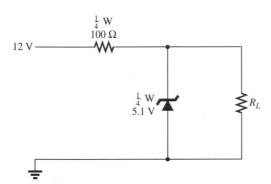

FIGURE A   Circuit for problem 3.

4.   You have two 120 V AC:24 V AC transformers, 16 high-voltage capacitors and 10 high-voltage diodes. Design, on paper, a circuit that maximizes output voltage. What is the maximum voltage of your circuit?

5.   What conditions might favor selection of a full-wave rectifier over the bridge rectifier?

6.   Design, on paper, a transformerless 300 V power supply with a ripple of 1% or less, capable of supplying up to 100 mA.

7.   Determine the necessary values of the components for the circuit of Figure B, with the requirements that the output be 15 V at 100 mV with 0.01 percent or less ripple.

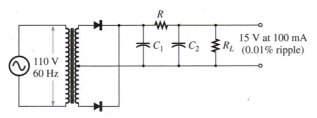

FIGURE B   Circuit for problems 7 and 8.

8.   Repeat the previous problem to achieve an output of 500 V at 10 mA with 0.1 percent or less ripple.

9.   The output and ripple expressions for the simple choke filter of Figure C are

$$V_{DC} = \frac{2V_v}{\pi} \left(\frac{R}{R + R_c} \right)$$

and

$$r = 3.1 \times 10^{-3} \, R/L$$

where R_c is the choke resistance. What is the effect upon the ripple of increasing the current?

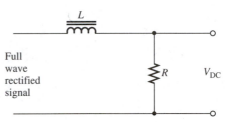

FIGURE C   Circuit for problem 9.

10.   A typical power choke, at 10 H, has a resistance of approximately 200 Ω. Using the equations supplied in the previous problem, and $R = 1000$ Ω, compute the output voltage and the ripple factor of this filter for a power supply with the transformer secondary at 200 V rms.

11.   Design a 12V, 100mA power supply using a Zener diode as a regulator.

12.   Sketch the output waveforms expected when a 100 Hz sine wave (5 V_p) is applied to each of the circuits in Figure D.

13.   Sketch the output waveforms expected when a 100 Hz square wave (10 V_{pp}) is applied to each of the circuits in Figure D.

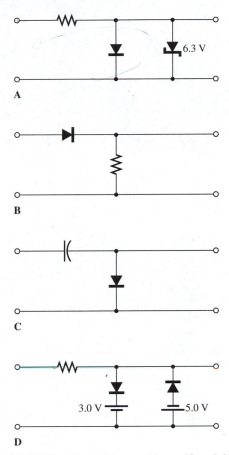

A

B

C

D

FIGURE D   **Circuit for problems 12 and 13.**

6

Test Equipment and Measurement

6-1 INTRODUCTION

Scientists and engineers are often concerned with measurement and control. Of the two, measurement has received the greatest attention, and properly so; without an adequate idea of the magnitudes of the quantities with which one is dealing, control is seldom possible. An almost infinite variety of measurement devices exists, but most of these devices operate in accordance with a few basic principles. Our attention will be directed to these basic measurement principles as we discuss the two main types of electronic-measurement devices: analog and digital.

6-2 ANALOG METERS

Most devices that use a needle or pointer to indicate the magnitude of a quantity incorporate a meter movement into the measurement. These are called analog meters because their readout is continuous, as opposed to indicating systems which use discrete integral numbers. The most widely used type of analog movement is the D'Arsonval meter, which is shown in Figure 6.1. The current passing through the meter coils induces a magnetic field in the coils. As a result, the coils, which are mounted on a pivot, turn in the static magnetic field of a permanent magnet. Because the magnitude of the induced field and the opposing tension of the suspension spring control the amount of coil movement, the pointer attached to the moving coils indicates the magnitude of the current.

Meters with D'Arsonval movements are usually rated by their full-scale sensitivity, which is the magnitude of the current which will cause the meter to read at its highest scale marking. Meter movements which have 1.0 mA full-scale sensitivity are common, while meters with sensitivities of 10 μA full scale are not uncommon. Typical accuracy (deviation of the indicated value from the true value) of the meter movement is 5%, but analog meters with better than 1% ratings are available. The D'Arsonval movement may be used to measure current, voltage and resistance when incorporated into

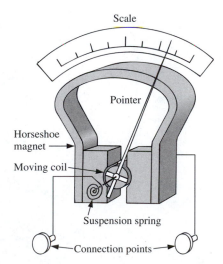

FIGURE 6.1 Principal components of the D'Arsonval meter movement.

suitable circuits as discussed below. Note here that we speak of DC signals which do not vary considerably with time.

Ammeter

The D'Arsonval movement fundamentally responds to current, and is thus inherently an ammeter. The range of current over which the basic movement can be used is limited by the full-scale sensitivity. With additions, however, the movement can be converted to measure an almost limitless range of currents. We can draw the equivalent circuit of the meter movement as shown in Figure 6.2. Here the real meter movement has been separated into an ideal meter (with zero resistance) represented by the letter A and the actual resistance of the meter movement, R_m. The range of the meter can be increased readily by causing most of the current to bypass the meter through a shunt resistor, R_s, as shown in Figure 6.3. The current to be measured, I, is split into that current which goes through the shunt, I_s, and that which goes through the meter, I_m. Thus,

$$I = I_m + I_s. \tag{6-1}$$

Since $I_s R_s = I_m R_m$ (no voltage drop across the ideal meter), we obtain

$$I = I_m + \left(\frac{I_m R_m}{R_s}\right) = I_m \left(1 + \left(\frac{R_m}{R_s}\right)\right). \tag{6-2}$$

Since I_m and R_m are known, the value of the shunt resistance, R_s, needed for a given current range I may be readily calculated.

EXAMPLE 6.1

What shunt resistor should be added to a 1 mA movement which has an internal resistance of 2300 Ω in order to read 10 A full scale?

Solution: In this situation we have $R_m = 2300\ \Omega$, $I_m = 1$ mA and $I = 10$ A. From Equation 6-2, the calculated value for the shunt resistance is $R_s = 0.230\ \Omega$. (Note that the voltage drop across the meter with the added shunt resistor is 2.3 V.) ∎

Voltmeter

A voltmeter is easily constructed from a meter movement by placing a resistance R_v, in series with the meter, as shown in Figure 6.4.

FIGURE 6.2   **Equivalent circuit of a meter.**

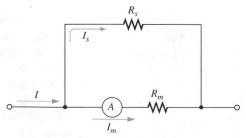

FIGURE 6.3   **Ammeter shunting circuit.**

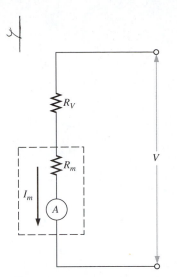

FIGURE 6.4   Voltmeter circuit.

From inspection, $V = V_m$, thus

$$V = I_m(R_v + R_m). \qquad (6\text{-}3)$$

Using the values I_m and R_m of the previous example, when there is no series resistor ($R_v = 0$), the full-scale deflection voltage, V_{fs}, is

$$V_{fs} = 1 \times 10^{-3}\text{A} \times 2300\ \Omega = 2.3\ \text{V}.$$

If the meter is to be used to measure a full-scale reading other than this, then the added series resistor value can be calculated using Equation 6-3.

EXAMPLE 6.2

What series resistor should be added to a 1 mA movement which has an internal resistance of 2300 Ω in order to read 10 V full scale?

Solution: From Equation 6-3, $V = I_m(R_v + R_m)$, R_m 2300 Ω, $I_m = 1$ mA, $V_{fs} = 10$ V, the calculated value for the series resistor is $R_v = 7700\ \Omega$. ■

It is common practice to specify the sensitivity of an analog meter in terms of ohms per volt. This is exactly equivalent to a full-scale current sensitivity, since a 1.0 mA scale current is rated as

$$1/(1 \times 10^{-3}\text{A}) = 1000\ \Omega/\text{V}.$$

The use of the Ω/V designation is seen by considering the following example. Using a 20,000 Ω/V voltmeter (a commonly available sensitivity) on its 50-volt scale, the meter's resistance is calculated as

$$50\ \text{V} \times 20,000\ \Omega/\text{V} = 1 \times 10^6\ \Omega = 1\ \text{M}\Omega.$$

The full-scale current sensitivity of the meter movement is given by

$$50\ \text{V}/1\ \text{M}\ \Omega = 50 \times 10^{-6}\text{A} = 50\ \mu\text{A}.$$

Ohmmeter

A simple analog ohmmeter (resistance-measuring meter) incorporating the D'Arsonval movement is shown in Figure 6.5. In the operation of this circuit, it is understood that the scale of the meter is calibrated initially by shorting the leads ($R_x = 0$) and adjusting R_A to achieve a full-scale deflection. With a meter movement of I_{fs} and a battery of voltage V we can write

$$V = I_{fs}(R_m + R_A)$$

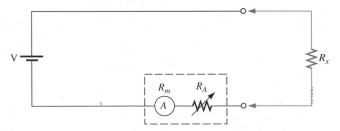

FIGURE 6.5   Simple ohmmeter circuit.

or

$$(R_m + R_A) = \frac{V}{I_{fs}}. \tag{6-4}$$

If we now introduce R_x into the circuit, the current, I, will be reduced and we can write

$$V = I(R_m + R_A + R_x).$$

Solving for R_x and substituting in Equation 6-4 we have

$$R_x = V(1/I - 1/I_{fs}). \tag{6-5}$$

Note that the relationship between R_x, the resistance being measured, and I, the current being measured by the meter, is not linear. Analog meters which use this type of measurement must have carefully calibrated, non-linear scales.

Analog Multimeter

With the proper switching arrangements, the D'Arsonval movement can be used in a so-called *multimeter* which is capable of measuring voltage, current and resistance. Such a device is usually called a volt-ohm-milliammeter, or VOM. Although VOMs and analog meters are rapidly being replaced by their digital counterparts, they are still quite useful in experimental apparatus, particularly in applications where the value of a time varying parameter, such as the speed of an automobile, must be seen at a glance of the readout.

6-3   INPUT IMPEDANCE

A perfect meter would not significantly change the phenomena being measured or controlled. A good scientist or engineer will spend a great deal of time designing a test procedure which does not change the phenomena being measured or controlled in a significant way. Unfortunately, a measurement rarely has no effect on the experiment, but we can minimize this effect, or at least account for it, with careful consideration toward instrumentation design.

Voltmeter

Consider the simple DC circuit in Figure 6.6A consisting of two resistors in series with a 10 V battery. To measure the voltage across resistor R_2 one would connect a voltmeter as shown in Figure 6.6B. In the figure the resistances of the meter and the series resistor, R_m and R_v, respectively, have been replaced by one resistor, R_{in} in order to simplify the following calculations. This resistance is called the *input resistance,* or *input impedance,* of the meter. The voltage across R_2, or between points A and B, in Figure 6.6A is given by

$$V_{R2} = \frac{(V R_2)}{(R_1 + R_2)}$$

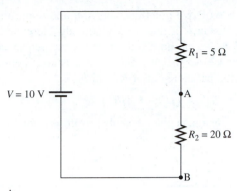

A

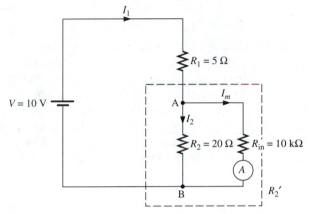

B

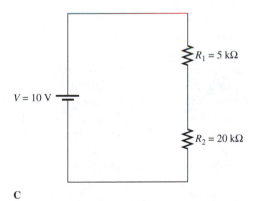

C

FIGURE 6.6   **Resistor network showing loading effect of voltage measurement across R$_2$.**

or

$$V_{R2} = \frac{(10 \times 20)}{(5 + 20)} = 8.0 \text{ V.} \qquad \textbf{(6-6)}$$

For the circuit in Figure 6.6B we must consider the additional input resistance of the meter (the resistance of the meter is already accounted for in the input resistance; therefore, as pictured, the ammeter element itself behaves ideally, having no resistance) so that the effective resistance between points A and B now becomes the parallel sum of R_2 and R_{in} or

$$R_2' = (1/R_2 + 1/R_{in})^{-1} \qquad \textbf{(6-7)}$$

and the actual measured voltage is given by substituting R_2' for R_2 into Equation 6-6.

EXAMPLE 6.3

Calculate the effect that presence of the meter has on the voltage measurement across R_2 in the circuit shown in Figure 6.6C using the meter from Example 6.2.

Solution: From Example 6.2, the meter has a 1 mA full-scale movement and internal resistance $R_m = 2300 \; \Omega$, and with the added series resistor of $R_v = 7700 \; \Omega$ the meter can read up to 10.0 V. The input resistance is therefore 10,000 Ω. From Equation 6-7 we see that the effective resistance is 19.96 kΩ, as opposed to 20 kΩ, and the voltage drop across the resistor using Equation 6-6 is 7.9968 V. Keeping in mind significant figures and the fact that an analog meter cannot be read with extreme accuracy, the voltage across points A and B would be the same as that without the meter, namely 8.0 V. ∎

Now if we consider a similar circuit with higher resistance values as shown in Figure 6.6C, an interesting measurement occurs. Equations 6-6 and 6-7 still accurately describe the phenomenon, but instead of having a resistance of 20 kΩ across points A and B we now have 6.67 kΩ, and the associated voltage is 5.7 V, which is decidedly different from the expected 8.0 V.

For this reason, voltmeters with high input impedances are desirable. However, very high input impedance voltmeters have inherent problems with noise and stray signals which make them more difficult to use properly. It is worth noting that a voltmeter with an input resistance 10 times the value of the resistor across which it is placed causes only a 5% decrease in the voltage measured. If you are concerned with only rough voltage measurements, this may not interfere with your work. On the other hand, if greater accuracy is needed, careful consideration should be given to the input impedance of the meter.

Ammeter

In the case of an ammeter, the input resistance is kept as low as possible. This can be understood by analyzing the simple circuit of Figure 6.7, where a battery of voltage V is connected across a resistor of value R. To measure current, an ammeter is placed in series with the circuit. Note that the current in the circuit, without the ammeter, is simply $I = V/R$; however, the introduction of the ammeter reduces the current to $I' = V/(R + R_{in})$. If the resistance of the meter were zero, then I' would be, of course, equal to I. Ideally, the value of R_{in} should therefore be kept as small as possible.

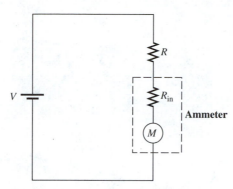

FIGURE 6.7   **Simple ammeter circuit.**

EXAMPLE 6.4

Using the meter from Example 6.2, calculate the current indicated by the ammeter in Figure 6.7 if $V = 10$ V and $R = 10$ kΩ.. Also calculate the current flowing in the circuit without the ammeter.

Solution: Here we use the 1 mA meter movement from the previous examples without the added series resistor. $R_{in} = 2.3$ kΩ, the same as R_m, and we find that the current indicated by the ammeter is $I' = V/(R + R_{in}) = 10$ V/ $(10$ kΩ $+ 2.3$ kΩ$) = 0.81$ mA, whereas the current in the circuit without the meter is given by $I = V/R = 10$ V/10 kΩ $= 1.0$ mA, a difference of about 20%. ∎

At this point you might wonder what purpose a meter with a relatively large (2.3 kΩ) resistance could possibly have as an ammeter. One such use is in a high-voltage circuit with V on the order of thousands of volts and a circuit resistance greater than a few hundred kilohms. In that case it can easily be shown that the introduction of this ammeter would have little effect on the circuit.

6-4   DIGITAL METERS

Digital meters differ from their analog counterparts in a number of ways. The most apparent difference is that the readout is numeric as shown in Figure 6.8. The *digital voltmeter* (DVM) reduces reading error and increases reading speed, since the digits display the voltage, current or resistance directly. The D'Arsonval meter is a *current* measuring device which can be used to measure voltage and resistance as well. The DVM measures *voltage,* but with some additional circuitry, it can be used to measure current and resistance. Instruments that incorporate the DVM to measure voltage, current and resistance are called *digital multimeters,* or DMMs.

There are several methods by which DVMs operate; these are discussed in detail in Chapter 14, but a brief overview is in order. The quantity being measured is an analog one (that is, continuous, rather than discrete) while the readout is digital (discrete, rather than continuous). Thus, the use of a DVM (or DMM) necessarily involves an analog-to-digital conversion (ADC). Single integrated circuits which accomplish this conversion, modify the

A

B

FIGURE 6.8   **Typical bench-top DMMs: a. 3-½ digit meter, b. 7 digit meter. Courtesy Keithley Instruments.**

incoming signal, and drive the auto-polarity (" + " or " − ") numeric display are readily available at modest cost from a number of manufacturers. These and other circuits have been integrated into a wide range of commercially available instruments. Inexpensive digital multimeters with very good accuracies are readily available. More expensive models may feature a wider range of readings, higher resolution, true rms AC measurement, better accuracy, auto-ranging (the meter displays the most appropriate range rather than the user choosing it), frequency counter, a transistor tester, a diode tester, capacitance measurement and sometimes even a built-in function generator.

Since the DMM is constructed around a DVM, a voltage meter, the measurement of current and resistance is fairly straightforward. To read current, the DMM is placed in the circuit so that current flows through the meter circuitry. When current measurement is selected, the current passes through a precision resistor causing a voltage drop equal to the value of the current multiplied by the value of the resistance. This voltage is connected internally across the input of the analog-to-digital converter which reads the voltage. Since the resistance is known and the voltage is measured, the voltmeter can display the current directly. For a resistance measurement, the meter circuitry passes a known current through the resistor whose value is to be measured. A voltage drop equal to the product of this current and the value of the unknown resistance is produced across the meter's terminals and is measured by the analog-to-digital converter. Similar to the current measurement, the resistance measurement is displayed directly on the face of the meter.

There are five important parameters which describe the digital multi-meter: number of digits, range, resolution, accuracy and input impedance.

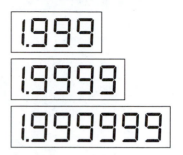

FIGURE 6.9   **Display of 3-½, 4-½ and 6-½ digit DVMs.**

Number of Digits: Figure 6.9 shows a 3-1/2 digit, a 4-1/2 digit and a 6-1/2 digit display. Most inexpensive meters have 3-1/2 digit displays, which are adequate for many testing purposes. As the number of digits in the display increases, so does the cost of the instrument, often·rather dramatically. The digit to the left (the most significant digit) is considered a "half" digit since it can only display the value of 0 or 1. The remaining digits can display any value from 0 to 9.

Range: Because the display can indicate 0 to 1999 for a 3-1/2 digit meter, the range of readings is indicated by the maximum value plus one least signifi-cant digit, which for voltage is usually 200 mV, 2 V, 20 V, 200 V and 2000 V. Resistance ranges are normally 200 Ω, 2 kΩ, 20 kΩ, 200 kΩ, 2 MΩ 'and 20 MΩ. These voltage and resistance ranges are fairly common even for the expensive 6-1/2 digit meters. Usually the maximum current measured by a DMM is 2.0 A, but 10.0 A models are available (with decreased resolution). The lower limit of current measurement varies widely among meters, with 200 mA being a common minimum range, but meters which can measure in the 200 μA range can be easily found.

Resolution: Resolution is the value of the least significant digit (the one most to the right). For example, a 4-1/2 digit meter which has a lowest voltage range of 200 mV is expected to have a best DC voltage resolution of 0.01 mV (199.99 mV maximum reading on this range) or 10 μV, whereas a 3-1/2 digit meter would have a best DC voltage resolution of 100 μV (199.9 mV maxi-mum). Note that a 6-1/2 digit meter would have a resolution of 0.1 μV, a resolution 1000 times finer than the much less expensive 3-1/2 digit meter.

Accuracy: Accuracy specifications are usually better than 1% for most digital meters, even the inexpensive 3-1/2 digit meter. The accuracy of a digital me-ter is usually presented as an addition of two percentages, namely % of read-ing + % full scale. A bottom-of-the-line, hand-held 3-1/2 digit meter might have an accuracy of ± (0.8% of reading + 0.2% of full scale), whereas a top-of-the-line, bench-top 4-1/2 digit meter could have a rating of ±(0.6% of reading + 0.06% full scale). Sometimes the accuracy of the meter is given as a percentage of the reading and a number of least significant digits. An example of this is a 6-1/2 digit meter which claims an accuracy of ±0.5% of reading ±20 in the last two digits. Because of the quantization of the signal being measured, even a perfect digital meter can have an accuracy no better than ±1/2 of the value of the last digit.

Input Impedance: Unlike the analog meter, the digital meter usually has a fixed-input impedance which is the same for all voltage ranges (sometimes, however, there is an aberration in the very low voltage range which is worth noting). Common input impedances are 1 MΩ and 10 MΩ. A simple scheme sometimes used to accomplish this is shown in Figure 6.10. Note that for

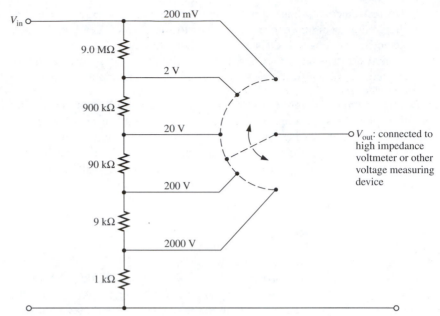

FIGURE 6.10   Voltage divider scheme for constant input impedance over 5 voltage ranges in a DVM.

each range the input impedance is the series sum of the five resistors, 10 MΩ (assuming that the circuitry that follows the switch has a very large impedance with respect to 10 MΩ) and that the A/D converter circuitry only needs to measure a small voltage range, normally from 0 to 200 mV, for any of the ranges.

According to most available literature, the parameter which describes the input impedance of the meter during a current measurement is usually the voltage drop at full scale, but sometimes the actual value of the input impedance is given. For the best meters this voltage drop is about 200 mV, and for the least expensive it is about 350 mV. If the introduction of a voltage drop of this magnitude is significant to your measurement, then careful analysis which includes the induced voltage drop is in order, or a less detrimental measurement technique must be found.

6-5   AC MEASUREMENT

Until now, we have not discussed the measurement of AC currents or voltages. It should be clear that alternating currents and voltages cannot be measured directly with the D'Arsonval movement or the digital meter, since the positive and negative halves of each cycle would cancel each other, resulting in a net reading of zero. However, if the alternating signal can be made unidirectional (rectified), the AC measurement becomes straightforward. Figure 6.11 depicts a simple AC Voltmeter using a D'Arsonval movement and an additional diode and capacitor. The diode rectifies the AC signal while the capacitor filters the rectified signal. The capacitor charge drains through R_{in} and is chosen such that the product $R_{in}C > T$, the period of the signal, thus allowing the meter to respond to variations in the peak amplitude of the sig-

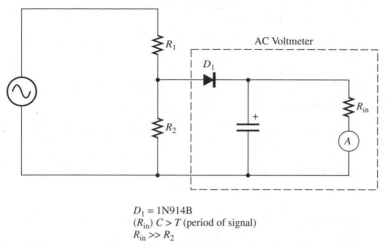

D_1 = 1N914B
$(R_{in}) C > T$ (period of signal)
$R_{in} \gg R_2$

FIGURE 6.11 **Simple AC circuit showing a possible way to measure AC voltages using an analog meter.**

nal, but not to the alternating variation. This arrangement measures the peak voltage of any AC signal which is centered at zero volts. Note that a DC offset would cause a discrepancy in the measurement.

Digital meters are by far superior to their analog counterparts in accuracy, frequency range and in the fact that many of them measure the true rms value of voltage or current. Those that do not, read rms values for sine waves and full rectified sine waves only, but the literature that accompanies the meter usually gives multiplier factors that allow for the calculation of peak-to-peak, peak, rms and average voltage or current values for simple signals such as the square wave, triangle wave, and pulse waveform. DMMs are available that will measure true rms voltages and currents from 20 Hz to 50 kHz, but a more common frequency range is 45 Hz to 1 kHz.

6-6   THE OSCILLOSCOPE

The measuring devices described thus far have limitations. Chief among these is the inability to correctly measure the parameters of a complex (non-sinusoidal) waveform. The oscilloscope, although more costly, is ideal for the display, and hence measurement, of complex waveforms. The cost of an oscilloscope is directly proportional to its quality, which ranges from about $300 to over $5,000 per unit. The circuitry within an oscilloscope is much more complicated than the measuring devices already described. Most of the basic circuits in an oscilloscope are, in fact, fully discussed in later chapters, so that only the principles of the circuit operations will be described here. The several functional building blocks of the oscilloscope are given in Figure 6.12. Each is discussed in some detail by examining how the waveform signal is processed by each part of the circuit.

At the heart of the device is the cathode ray tube (CRT), upon which the waveform is displayed. In simple terms, the CRT is a vacuum tube in which electrons are produced and directed to a phosphorescent coating on the tube face. The electrons striking this coating produce visible light pulses. It is critical that this beam of electrons truly represents the waveform. Several

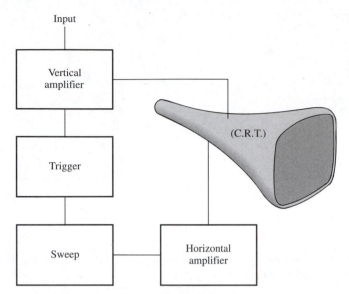

FIGURE 6.12   Block diagram of cathode ray oscilloscope.

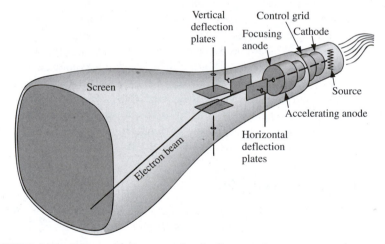

FIGURE 6.13   Schematic diagram of cathode ray tube.

parts of the CRT permit the proper display of the signal; they are diagrammed in Figure 6.13.

The source of electrons is the cathode, which is heated by the filament (or heater). The electrons are "boiled" off the filament and are directed toward the screen at the opposite end of the tube. The collimation and focusing of these electrons is necessary so that a point, rather than a diffuse spot, is obtained at the screen. The beam is electrostatically focused by succeeding stages. The properly collimated and focused beam is then acted upon by the deflection plates. The deflection plates reflect the qualities of the input waveform and perturb the electron beam so that its movement, in the horizontal and vertical directions, represents the waveform. The accelerating section of the CRT is necessary to impart sufficient energy to the beam so that the electrons will produce phosphorescence when they strike the screen. The magnitude of the voltage in the accelerating section is of the order of kilovolts, and it must be uniform in space, since a non-uniformity would undo the rather elaborate efforts to produce a point-like beam.

The vertical amplifier attenuates or amplifies the input waveform signal so that it is of sufficient magnitude to drive the vertical deflection plates of the CRT. In addition, the vertical amplifier also supplies a signal, related to the input signal, which begins the triggering process eventually leading to the horizontal deflection plates. This latter feature is critical, since the displayed waveform would be skewed in an unpredictable manner unless the horizontal and vertical deflection plates were synchronized. Among the desirable features of a vertical amplifier are a wide range of frequencies that can be amplified correctly, the ability to handle large magnitude signals and a low amount of noise introduced by the amplifier. Often the input capability allows either DC or AC input. The AC input finds use when the alternating portion of the signal is small relative to a fixed or DC portion of the signal and the former is the signal of interest. A simple *RC* attenuator, such as the one shown in Figure 6.14, produces the desired result. It should be recognized that the *RC* circuit may alter the signal character, as discussed in Chapter 2.

The sweep generator is simply a ramp generator whose output may be initiated by the input signal to the trigger circuit. This "sawtooth" signal, shown in Figure 6.15, drives the horizontal deflection plates and thereby creates a time axis. The linearly increasing voltage moves the electron beam at a constant rate from left to right on the screen. The rapidly descending portion of the sawtooth is used to bring the beam back to the left side again.

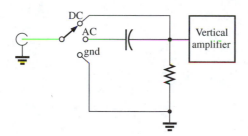

FIGURE 6.14   **Input selection circuit for a CRT.**

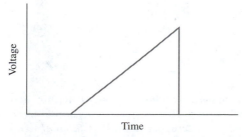

FIGURE 6.15   **Sawtooth, or ramp, waveform used to drive the CRT horizontal amplifier.**

Generally, the beam is shut off during the time the beam is returning; the "fly back" signal would cause confusion if it were visible, so it is "blanked" by making the cathode of the CRT positive. The more sophisticated oscilloscopes incorporate several features in the sweep generator section which permit independent control of the sweep rate.

The trigger section is included to make the input signal compatible with the sweep generator. It is quite possible that the input signal is such that it is not capable of initiating the sweep. The trigger section modifies the input signal to ensure that the sweep is properly initiated. The trigger section permits selection of the signal level used to begin the sweep. For instance, a noisy signal would probably initiate the sweep prematurely, so the triggering level is set above the noise level to ensure that the sweep will begin when the signal (rather than the noise) reaches the prescribed level. The trigger section also allows the operator to select either the rising or the descending portion of the input waveform to initiate the sweep. This feature is important when phase relationships are being examined.

In addition to the time base signal ($y - t$) mode, oscilloscopes generally offer $x - y$ operation. In this mode, one signal (y) is fed to the vertical amplifier and the other (x) is fed directly to the horizontal amplifier. The triggering section is disconnected from the vertical amplifier in this operation. The qualities of the horizontal amplifier must be exactly the same as those of the vertical amplifier. Frequently, the $x - y$ mode is used to measure the frequency of the signal applied to the vertical amplifier, with a signal source of controllable frequency connected to the horizontal amplifier. The resulting waveforms, some of which are shown in Figure 6.16, are called *Lissajous figures*. After counting the number of nodes in the two directions, the unknown frequency is determined by multiplying the known frequency by the ratio of vertical nodes to horizontal nodes.

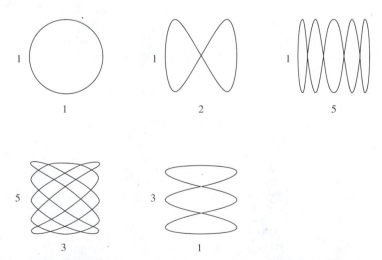

FIGURE 6.16 Lissajous figures for frequency determination. Shape of figure is dependent on ratio (F) of vertical to horizontal frequency. Upper left, $F = 1$; upper center, $F = 2$; upper right, $F = 5$; lower left, $F = 0.6$; lower center, $F = 0.33$.

When the two signals are of the same frequency, the phase angle between them can be determined, as illustrated in Figure 6.17. Note that at phase angles of 0° and 180°, a straight line is obtained. Naturally the slope of this line is related to the relative gains of the two amplifiers. You may recognize that this situation suggests using the oscilloscope as a high speed $x - y$ recorder. Photographing the signal on the oscilloscope face produces a permanent record. For many rapid transients, this is the only satisfactory method of observing and recording the phenomena.

6-7   THE SIGNAL GENERATOR AND FREQUENCY COUNTER

Two related instruments which often prove invaluable in the laboratory are the *function generator* and the *frequency counter*. The name frequency counter implies the function of this device: to measure the frequency of a waveform signal. The operation of the instrument is fairly straightforward. The frequency counter counts, for a given time, the number of transitions from negative to positive that the applied signal makes across the zero voltage point in a given time and displays this information directly. In effect, the frequency counter is actually a pulse counter and indeed most of these devices can be used directly as pulse counters. Most frequency counters can measure the period of a signal as well as the frequency.

The signal generator, or function generator, is used to produce waveforms of various types. In general, signal generators which produce waveforms with frequencies of less than 1 Mhz are known as audio generators (although the audio range is 20 Hz to 20 kHz), and those above are known as radio frequency generators (often producing signals with frequencies of over 100 Mhz). Signal generators are rarely capable of spanning both the audio and radio frequencies. Also, note that most generators have a ground connection which is not floating; that is, the ground is connected to the ground of the 120 V AC line voltage as described in an earlier chapter.

The signal amplitude, frequency and type of waveform can be controlled using the knobs and buttons on the front of the generator; this is universal to all devices of this sort. The most common signals are the sine, square and triangular waves, but many generators are capable of producing sawtooth (ramp) waves and pulses as well (see Figure 6.18). Additional features may include a DC offset which allows the imposition of a DC voltage along with the AC signal and a duty cycle adjustment which allows the user to manipulate the shape of the waveform as shown in Figure 6.19 for a pulse wave. Some generators are capable of signal modulation, both AM, amplitude modulation, and FM, frequency modulation, as shown in Figure 6.20. Many units are available with built-in frequency counters to ensure accurate frequency settings.

When selecting a function generator, compare parameters such as the frequency range, signal amplitude, DC offset range, flatness of signal (how well the amplitude remains constant when the frequency is changed), signal distortion (how well the signal approximates a perfect signal) and output impedance which is normally 50 Ω or 600 Ω. The output impedance of a device such as the function generator is most readily explained by means of

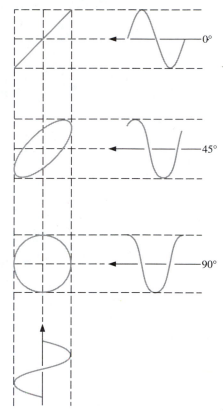

FIGURE 6.17   **Scope figures used to determine the phase angle between two signals of same frequency.**

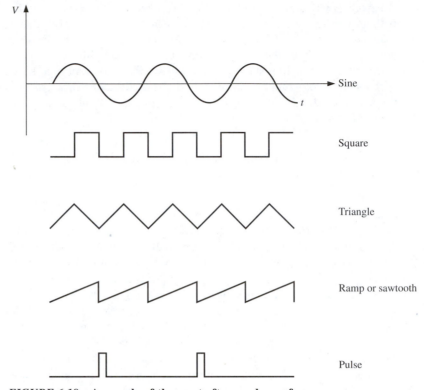

FIGURE 6.18   A sample of the most often used waveforms.

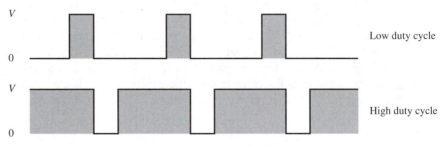

FIGURE 6.19   Pulse train with low duty cycle (above) and high duty cycle (below).

a diagram. Figure 6.21 depicts an ideal function generator in series with a resistor, R_{out}. R_{out} is the output impedance of the function generator. Most signal generators do not produce signals with peak amplitudes of more than 25 V. If the terminals of a 50 Ω output impedance generator are shorted, the function generator will produce 500 mA, whereas a generator with a 600 Ω output impedance will produce just over 40 mA. Keep in mind that these values represent the maximum current that can be drawn from the generator. For general usage, avoid the purchase of high-impedance signal generators because of this limitation and since the output voltage will drop when connected to a low-impedance load.

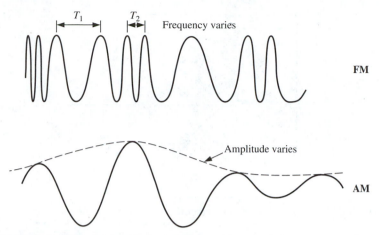

FIGURE 6.20   Waveform undergoing frequency modulation (constant amplitude, varying frequency) and amplitude modulation (constant frequency, varying amplitude).

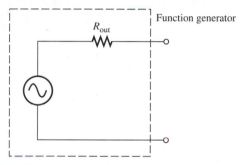

FIGURE 6.21   Function generator showing series output resistance.

6-8   WHEATSTONE BRIDGE

The precise method of resistance measurement is the Wheatstone bridge method, discussed in Chapter 1 and shown in Figure 6.22. Balance is achieved by adjusting R_3 so that no current flows through the detector. When this occurs,

$$\frac{R_1}{R_2} = \frac{R_3}{R_x} \tag{6-8}$$

If $R_1 = R_2$, then $R_x = R_3$. A range control is built into the bridge, since the ratio of R_3 to R_x is equal to that of R_1 to R_2. The accuracy of evaluating R_x is limited only by the accuracies of the other arms. To a degree, the sensitivity of the detector may be increased by increasing V. This is of limited value, however, since the power dissipation of precision resistors is generally low. If they are overheated, the precise values of resistance are changed. At balance, when the detector indicates no current flow, significant current is flowing in the $R_1 + R_3$ and $R_2 + R_x$ branches. For this reason, care should be exercised in measuring the resistance of an element with a low power capacity.

It is of interest to consider the possibility of continuously monitoring a changing resistance. This is not readily done by balancing a Wheatstone

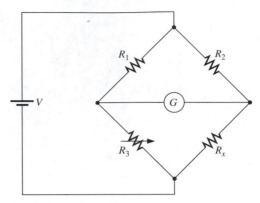

FIGURE 6.22 **Wheatstone bridge circuit.**

bridge; however, the "out of balance" signal taken across the detector, under the proper conditions, affords an excellent method of continuous monitoring. Consider the Wheatstone bridge in Figure 6.23A. The voltages and currents developed across the detector for various values of R_x are given in Table 6.1 and plotted in Figure 6.23B. The "out of balance" signal is clearly linear

TABLE 6.1 Voltage and Current Measurements across Wheatstone Bridge Circuit in Figure 6.23A

R_x(ohms)	V (mV)	I (μA)
90	−26.3	−245
92	−20.8	−198
94	−15.5	−143
96	−10.2	−94
98	−5.1	−46
99	−2.5	−23
100	0.0	0
101	2.5	23
102	5.0	45
104	9.8	88
106	14.5	131
108	19.2	172
110	23.8	212

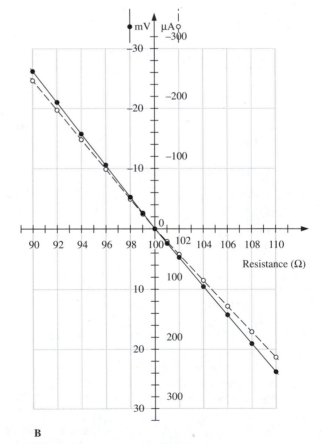

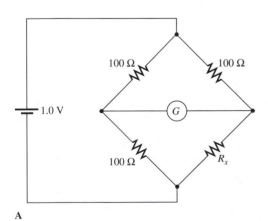

A

FIGURE 6.23A **Wheatstone bridge circuit operated in "out of balance" conditions.**

B

FIGURE 6.23B **Current and voltage *versus* out of balance resistance.**

in both voltage and current. It can therefore be used to convert an otherwise static measurement into a continuous monitoring process, as there are many recording devices capable of responding well to a signal in this voltage range.

6-9   AC BRIDGES

A number of input transducers converting a non-electrical signal into an electrical one produce a change in either a capacitance or an inductance. Heretofore, discussions have centered on the DC components of voltage, current and resistance. The measurement of inductance and capacitance is now presented.

The general AC bridge is equivalent to the Wheatstone bridge except that the arms may be impedance elements other than resistance. The battery is replaced by an AC signal source, and the detector must be capable of detecting AC signals. An oscilloscope is usually used for this purpose, although certain less critical applications permit use of an AC meter. The generalized AC bridge circuit is given in Figure 6.24. The condition of balance is

$$Z_x = \frac{Z_2 Z_3}{Z_1} \tag{6-9}$$

Examples of AC bridges abound, most bearing the name of the developer of that particular bridge. One of the most widely known is the Wien bridge, the circuit shown in Figure 6.25. In this bridge, a parallel *RC* arm is used to balance a series *RC* arm. The conditions at balance for this bridge, or any other AC bridge, may be derived by inserting the appropriate terms into Equation 6-9 and rearranging. For the Wien bridge,

$$Z_1 = R_1 \tag{6-10}$$

$$Z_2 = R_2 \tag{6-11}$$

$$Z_3 = \frac{R_3}{1 + j\omega R_3 C_3} \tag{6-12}$$

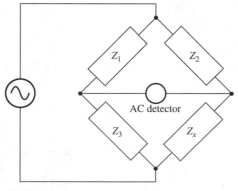

FIGURE 6.24   Generalized AC bridge.

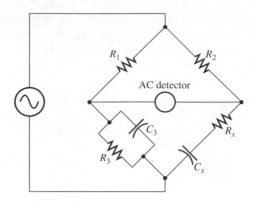

FIGURE 6.25   The Wien bridge.

$$Z_x = R_x + \frac{1}{j\omega C_x} = \frac{1 + j\omega R_x C_x}{j\omega C_x} \tag{6-13}$$

When these values are used in Equation 6-9, we obtain

$$\frac{1 + j\omega R_x C_x}{j\omega C_x} = \frac{R_2}{R_1}\left(\frac{R_3}{1 + j\omega C_3 R_3}\right) \tag{6-14}$$

Rearrangement yields

$$R_1 + j\omega R_1 R_3 C_3 + j\omega R_1 R_x C_x - \omega^2 R_1 R_3 R_x C_3 C_x = j\omega R_2 R_3 C_x \tag{6-15}$$

The balance conditions are found by separating the real and imaginary terms:

$$\omega^2 R_1 R_3 R_x C_3 C_x = R_1 \tag{6-16}$$

$$j\omega R_2 R_3 C_x = j\omega(R_1 R_3 C_3 + R_1 R_x C_x) \tag{6-17}$$

or

$$R_2 R_3 C_x = R_1 R_3 C_3 + R_1 R_x C_x \tag{6-18}$$

These may be rewritten in the more usual forms: from Equation 6-16,

$$\omega^2 R_3 R_x C_3 C_x = 1 \tag{6-19}$$

and from Equation 6-18,

$$\frac{R_x}{R_3} = \frac{R_2}{R_1} - \frac{C_3}{C_x} \tag{6-20}$$

If these equations are cast in terms of the known values of R and C, the results are

$$R_x = \frac{R_2 R_3}{R_1 + \omega^2 C_3^2 R_1 R_3^2} \tag{6-21}$$

and

$$C_x = \frac{R_1}{R_2}\left(\frac{1 + \omega^2 C_3^2 R_3^2}{\omega^2 R_3^2 C_3}\right) \tag{6-22}$$

Note that the balance conditions contain ω. This means that the balance is frequency dependent. In other words, in addition to the values of R_3 and C_3 required for balance, there is one frequency at which balance occurs. This requires a stable, accurate AC voltage source or oscillator. Alternatively, when the values of the arms of the bridge are known, the Wien bridge may be used to measure the frequency.

Frequently, the assessment of a component's quality is as important as measurement of its magnitude. The AC bridge is often used to evaluate both features. The quality, Q, of an inductor is defined as the ratio of the inductive reactance to the resistance. A similar definition for the Q of a capacitor is

$$Q = \frac{|X_c|}{R} = \frac{1}{\omega CR} \qquad (6\text{-}23)$$

More often than not, a capacitor's "lossiness" or dissipation factor, D, is an apt description. These two terms are reciprocally related,

$$D = \frac{1}{Q} = \omega CR \qquad (6\text{-}24)$$

This definition implies an RC series model for the capacitor, in which a perfect or ideal capacitor has $R = 0$. Thus, D is zero for an ideal capacitor. In practice, a dissipation factor for a polystyrene capacitor is low (0.0001), while an electrolytic capacitor's value may reach 0.1.

The series capacitance bridge, illustrated in Figure 6.26A, is the most common capacitance bridge. In practice, R_x represents the leakage resistance of the capacitor being tested, and C_x represents its pure capacitance. At balance,

$$C_x = C_3 R_1 / R_2 \qquad (6\text{-}25)$$

$$R_x = R_2 R_3 / R_1 \qquad (6\text{-}26)$$

and

$$D = \omega C_3 R_3 \qquad (6\text{-}27)$$

As shown in the figure, R_1 and R_3 are varied to achieve balance. When one is dealing with lossy capacitors (D is large), the bridge balance may require inordinately large values of R_3. The series capacitance bridge can be replaced by the parallel capacitance bridge (Figure 6.26B) for these situations. In effect, changing the bridge configuration means changing the model of the capacitor from a series to a parallel one. For the parallel bridge, the balance conditions are the same as for the series one, except that the definition of Q for the parallel model is $Q = R/|X_c|$, so

$$D = \frac{1}{\omega C_3 R_3} \qquad (6\text{-}28)$$

This permits the balancing of the bridge with smaller values for R_3. Note that while the *expressions* for C_x and R_x are the same for both bridges, the *values* of the variable components are different; on the other hand, although the *expressions* for D are different, the *value* of D is a property of the capacitor being tested and is the same for both measurements. If it is necessary, the

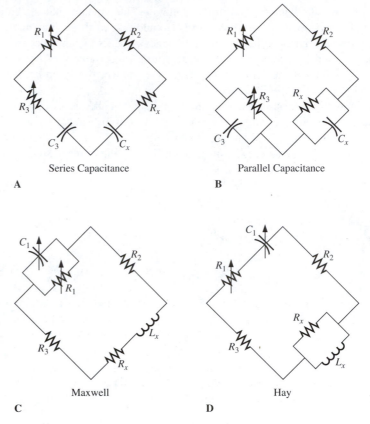

Series Capacitance

A

Parallel Capacitance

B

Maxwell

C

Hay

D

FIGURE 6.26 Various AC bridges.

series model values can be calculated from the equivalent parallel model values:

$$C_s = (1 + D^2)C_p \qquad \text{(6-29)}$$

and

$$R_s = \frac{D^2}{1 + D^2} R_p \qquad \text{(6-30)}$$

where the subscript s signifies the series value and p signifies the parallel value.

The Maxwell bridge (Figure 6.26C) is often used to evaluate inductance. Notice that the balance is achieved with a parallel RC arm. This configuration eliminates the need for standard inductors, which are rare and costly. Again, R_x represents the resistance of the physical coil, and L_x represents its pure inductance. The conditions of balance are

$$L_x = C_1 R_2 R_3 \qquad \text{(6-31)}$$

$$R_x = \frac{R_2 R_3}{R_1} \qquad \text{(6-32)}$$

and

$$D = \omega C_1 R_1 \qquad (6\text{-}33)$$

As shown, balance is achieved by adjusting C_1 and R_1. For very high Q inductors, R_3 needs to be very large to achieve balance. The Hay bridge, shown in Figure 6.26D, is used when this situation occurs. Note that the LR arm is now parallel, while the RC arm is in series. By analogy to the series and parallel capacitance bridges, one can properly deduce that the balance conditions for the Hay bridge are the same as those of the Maxwell bridge except that Q is expressed as the inverse,

$$Q = \frac{1}{\omega C_1 R_1} \qquad (6\text{-}34)$$

Using the same subscripting as in the previous conversion example,

$$L_s = \frac{Q^2}{1 + Q^2} L_p \qquad (6\text{-}35)$$

and

$$R_s = \frac{1}{1 + Q^2} R_p. \qquad (6\text{-}36)$$

The reader should verify the balance conditions given for each of the bridges discussed.

Several commercial "general purpose" or "universal" bridges are available. Such instruments contain a built-in detector and oscillator and, most importantly, a capability for switching from one bridge configuration to another so that different components can be measured with relative ease. Generally, the bridge configurations available in a commercial instrument are those discussed in the preceding paragraphs, including the Wheatstone bridge. As a general purpose laboratory instrument, it has the advantage of being versatile, self-contained, and fully calibrated. Available accuracies are in the 0.1% to 1.0% range.

6-10   HIGH-VOLTAGE MEASUREMENT

Very few digital multimeters and oscilloscopes can measure voltages greater than about 1000 V, yet potentials beyond this value are often found in the laboratory. Extreme caution must be exercised when using such possibly lethal high-voltage sources. Although current measurement can be done in the normal manner, voltage measurements require a different strategy. Consider the schematic representation of a 5000 V DC power supply in Figure 6.27A which has a series resistor R_L to limit the current which may be drawn from the supply. (Such current limiting resistors are usually found in high-voltage, low-current supplies for safety reasons.) Now consider the voltage divider circuit connected to a meter as shown in Figure 6.27B which essentially divides the voltage across the input terminals by a factor of 1000, thus lowering the measurement to a range within the limits of the meter. But, if R_L is 10 MΩ, the voltage indicated by the meter would be 1/2 of the expected

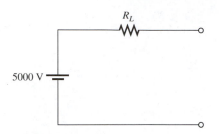

FIGURE 6.27A **High voltage supply with series current-limiting resistor.**

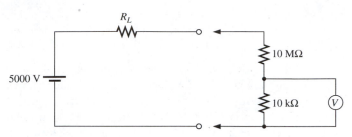

FIGURE 6.27B **Two-resistor circuit addition to measure high voltage with DVM.**

5000 V. To avoid this situation, the resistance values of the voltage divider circuit should be much higher than the output impedance of the power supply. Also note that many resistors are rated to be used with voltages of not more than a few hundred volts and that resistors with values greater than 22 MΩ are quite rare. Also, a low-wattage resistor's power-dissipation rating can easily be exceeded in a high-voltage measurement circuit. To overcome these shortcomings, a number of resistors are connected in series in the voltage divider circuit.

6-11   FOUR-WIRE RESISTANCE MEASUREMENT

One way to measure the resistance of a device is to pass a known current through the resistor, measure the voltage drop and calculate the resistance using Ohm's law, $V = IR$. Consider the circuit shown in Figure 6.28A in which the current source shown on the right produces a known current that passes through resistor R. If the voltage between points A and B was mea-

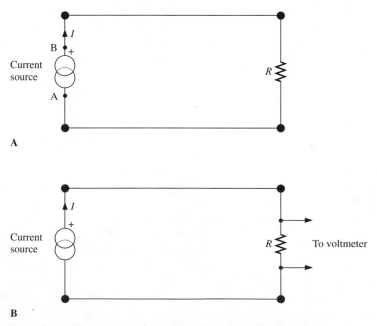

FIGURE 6.28 **Four-wire resistance measurement circuit.**

sured, the reading would include the voltage drop across the wires which connect the current source to the resistor and the voltage drops caused by the resistances of the wire connections to the source and resistor. Under many circumstances, the resistances mentioned are minimal compared to the resistance of the resistor, and thus the method works well. Essentially, this is the technique a DMM uses to measure resistance.

Sometimes a resistance must be measured with better precision than that available using the above technique, or the resistance value is on the order of the resistance of the leads and contacts. In these cases, the four-wire resistance measurement shown in Figure 6.28B is appropriate. Here, a voltmeter reads the voltage across the resistor. Ideally no current flows into the voltmeter and thus, there are no voltage drops over the leads or at the contact points between the leads and resistor. The four-wire resistance measurement technique is common in many data-collecting circuits.

6-12 QUESTIONS AND PROBLEMS

1. Suggest a circuit which can be used to evaluate the resistance of a meter.

2. The circuit of Figure A is used to obtain the data given beside the figure. What is the meter resistance?

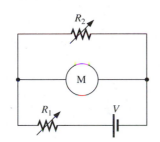

$R_1\ \Omega$	$R_2\ \Omega$	V Volts	1 mA
100	∞	1.5	1.36 (full scale)
100	1000	1.5	1.25

FIGURE A   Circuit for problem 2.

3. A meter movement, having $R = 80\ \Omega$ with full-scale deflection current of 5 mA is to be converted into a 25 mA full-scale meter. What is the value of the shunt resistance needed?

4. Using the meter with the specifications of problem 3, it was found that the actual meter resistance was 78.8 Ω. What error is involved in a meter reading of 20.0 mA?

5. Develop a circuit using a meter movement of 10 mA full scale ($R_M = 150\ \Omega$) for an ammeter having ranges of 0.1, 1.0, and 10 A.

6. What is the sensitivity of the meter in problem 5?

7. What is the full-scale voltage that can be measured with the meter of problem 5?

8. Develop an ohmmeter circuit having ranges of 100 Ω, 1 kΩ and 10 kΩ, using the meter of problem 5.

9. Using Kirchhoff's equations, derive the balance conditions for the potentiometer and the Wheatstone bridge.

10. Certain experimental situations require that an alternating signal be used as the voltage source for a Wheatstone bridge. In that case, an oscilloscope is used as the detector. Describe the waveforms at balance and on either side of balance.

11. The Wien bridge is often used as the basis of a variable frequency oscillator. Compute the bridge arm values for such an oscillator where the frequency range is from 1.5 to 15 kHz, using only one arm of the bridge to achieve the variation in frequency.

12. Measuring very large or very small resistances is difficult. What precautions need to be taken in each case?

13. Devise a circuit which would measure wattage directly.

14. If a quality capacitor is to be evaluated, which AC bridge should be used and why? If an electrolytic capacitor (non-quality) is to be evaluated, which AC bridge should be used and why?

15. Radio circuits use small (μH) air-core inductors. Which AC bridge should be used to evaluate the inductance? What other considerations are needed in this evaluation?

16. Comment on the advantages and disadvantages of using a DVM (set to measure AC volts) as a detector for an AC bridge, as opposed to using an oscilloscope.

Transducers

7-1 INTRODUCTION

Most phenomena that are of interest to scientists and engineers are not inherently electrical, yet most are measured by using electronic instrumentation. One reason for this is the great ease with which electrical signals can be used, studied and recorded. A nuclear physicist may be interested in measuring the energies of certain decay processes; a chemist may wish to evaluate the acidity of a solution. A civil engineer evaluating the stress likely to be imposed on a steel "I" beam used in constructing a bridge will almost inevitably turn to electronic instrumentation to make these measurements.

Instruments developed to measure or control a given process may or may not be adequate for some newer situation. In order to develop instrument packages or systems to meet ever-arising needs, one must be able to recognize several functions of a given instrument. The transducer concept is useful for clarifying the various functions of an instrument. An *input transducer* is a device which converts a non-electrical signal into an electrical signal. A modifier adjusts or alters the incoming electrical signal, and an *output transducer* converts the modified electrical signal into a non-electrical signal which can be read.

An example will illustrate this concept. Consider an audio system composed of a tape player, amplifier and a speaker. The head, or pick-up, of the tape player is the input transducer, in that it converts the magnetically encoded sound on the tape into a weak electrical signal. The amplifier modifies the signal in terms of tone and balance and raises the power level of the information so that it can drive the coils of the speaker. The speaker is an output transducer since it converts the electrical signal into non-electrical (sound) signals. This is diagramed in Figure 7.1.

Another example of the transducer concept is the determination of the force with which a human's eyelid closes. A very small electrical signal is generated when the eyelid closes; hence the body itself is considered to be the input transducer. The remaining functions of the "instrument" are block diagramed in Figure 7.2. The electrical signal generated by the closing of the eyelid is proportional to the distance the eyelid travels. Differentiation of the

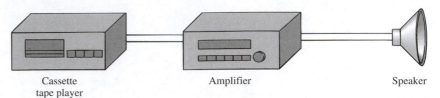

FIGURE 7.1   **Audio system viewed with the transducer concept.**

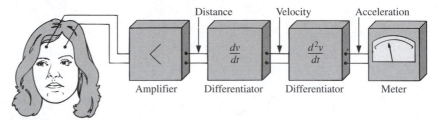

FIGURE 7.2   **Transducer system in block diagram form.**

signal with respect to time (after amplification) results in a signal proportional to the velocity with which the eyelid closes. Further, a second differentiation leads to a signal proportional to the acceleration of the eyelid. Since the acceleration is directly proportional to the force, the output transducer (the meter) reads in units of force. The calibration of the meter would involve evaluation of the proportionality constant, which is the mass of the eyelid.

The input transducer is the electrical component that becomes part of the experimental apparatus. At best, it does not disturb the experiment when it conveys critical information pertaining to the phenomena being investigated. All transducers have limitations. These are described in detail with the literature that accompanies them or is available from the manufacturer, but here we have noted some generalities which apply to the individual components.

The kinds of measurements a civil engineer makes are markedly different from those made by a chemist. The differences in instrumentation are, however, largely in the type of input transducer used. Once an electrical signal is obtained to represent the physical quantity being measured, it is processed by amplifiers and recorders that are common to all fields. A general overview of input transducers is presented here to provide the reader with a basic introduction. The sections that follow are meant to be illustrative, not encyclopedic. The reader is referred to the references cited at the end of the text for further information on this subject.

7-2   TEMPERATURE

All scientific and engineering disciplines are concerned with temperature measurement in some form or another. Whether it is the astronomer observing the temperature of a star, a low-temperature physicist working with millidegree temperatures, an engineer certifying the integrity of a sealing ring or the chemist monitoring the change in temperature of a reaction, accurate temperature measurements are critical in their experimentation. We present four foundational transducers used in temperature measurement in the laboratory: the thermistor, platinum resistance temperature detector, thermocouple and solid-state temperature transducer. The first two devices are resistance-based, the thermocouple is voltage-based and the solid-state transducer is current-based.

Thermistor

The *thermistor* is a device constructed of a transition metal-oxide or of a semiconducting material with a resistance which changes dramatically with temperature. In these devices, resistances remarkably decrease with increasing temperature (4 to 7% per °C). This has led to the wide acceptance of the thermistor as a routine substitute for other temperature-measuring systems where accuracies of ±0.005°C are sought. The resistance-vs.-temperature response for a typical thermistor is given in Figure 7.3. The response is decidedly nonlinear, but over a small range of temperatures a linear approximation is valid. Although the plot appears to be of the form

$$R = R_o e^{-\alpha T} \tag{7-1}$$

it is actually not, as can be seen from the absence of straight-line behavior on the logarithmic plot of resistance vs. temperature shown in Figure 7.4. The resistance of an individual thermistor can be approximated fairly well by the Steinhart-Hart equation

$$1/T = A + B\ln(R) + C(\ln(R))^3 \tag{7-2}$$

where T is the temperature in °K, R the resistance of the thermistor and A, B and C are constants determined by fitting three calibrated values to the equation. With such a fit, Equation 7-2 gives the temperature T to better than ±0.02°C if the temperature range is kept under 100°C. A simpler fit can be made using the following

$$T = [B/\ln (R A)] - C \tag{7-3}$$

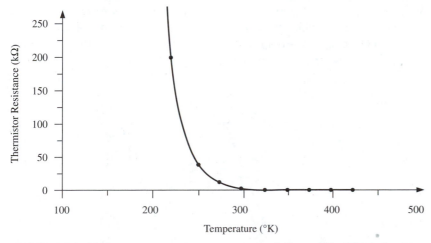

FIGURE 7.3 **Resistance-*versus*-temperature graph for a typical thermistor.**

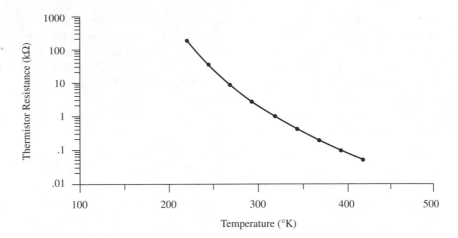

FIGURE 7.4 Log plot of resistance versus temperature for the thermistor of Figure 7.3.

where *A, B* and *C* are curve-fitting constants. Over a narrower range of temperature than the Steinhart-Hart equation, this fit results in similar accuracies.

The temperatures over which the thermistor can be used are limited to the range from approximately −75°C to 150°C. In addition to its high temperature coefficient, the thermistor's small size (1–2mm) and its relatively low price give it a decided advantage over its competitors. The allied circuitry consists only of a Wheatstone bridge, or 4-wire circuit and a voltmeter. Care must be taken to minimize i^2R heating in the device itself caused by excessive currents. This problem can be severe because of the thermistor's highly negative temperature coefficient. If i^2R heating begins, the resistance decreases and the current increases, resulting in catastrophic failure unless the circuit includes some current-controlling electronics.

Thermistors are manufactured in many sizes and shapes. They may be encapsulated in glass for use in corrosive environments which would otherwise destroy them. When a glass covering is used, the time for response of the thermistor to sudden temperature changes increases significantly, from milliseconds or less to perhaps one second. Several types of thermistors are illustrated in Figure 7.5. When high accuracy is demanded, a matched pair

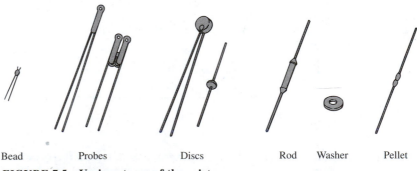

Bead   Probes   Discs   Rod   Washer   Pellet

FIGURE 7.5 **Various types of thermistors.**

of thermistors can be used, with one established as the reference. Since the matching occurs over a temperature range rather than at a single temperature, temperatures can be readily measured to a resolution of $\pm 0.0005°C$ when the usual precautions are observed.

A number of interesting applications of thermistors can be found in the literature. One such application uses a thermistor probe in a plastic holder, which is inserted in a person's nostril. The difference in temperature between the inhaled and exhaled air constitutes a means of monitoring the patient's breathing rate.

Platinum Resistance Temperature Detector (PRTD)

The value of any resistor is temperature dependent. The equation for this dependence is usually of the form

$$R_{T_2} = R_{T_1}[1 + \alpha(T_2 - T_1)] \tag{7-4}$$

where T_1 and T_2 are temperatures. The proportionality factor α is called the temperature coefficient of the resistive material, and varies between $2 \times 10^{-2}/°C$ and $2 \times 10^{-5}/°C$ for various materials. In contrast to the thermistor, the resistance of the PRTD increases with increasing temperature. Note that the temperature coefficient itself changes over large ranges of temperature, making accurate temperature measurements over a wide range very difficult. Table 7.1 lists values of α for some common materials.

The PRTD is considered the standard for temperature measurement over the range of $-190°C$ to $+660°C$. Platinum is chosen for its relatively linear dependency (α fairly constant over this range) despite its rather modest temperature coefficient. Generally, the PRTD is made of thin platinum wire, bifilar wound on a glass or ceramic bobbin, and then encased in glass. At 0°C the length and area of the wire are usually chosen such that the resistance of the PRTD is 100.0 Ω.

Naturally, care must be taken to avoid i^2R heating, which becomes more significant as the current becomes larger. The 4-wire resistance-measurement circuit can be used to measure temperatures to $\pm 0.00001°C$ at midrange. The accuracy falls off to $\pm 0.001°C$ as the upper limit of the range (about 600°C) is approached. Commercial PRTDs are available with $\alpha = 0.00385/°C$ (European Curve) and with $\alpha = 0.00392/°C$ (American Curve). The temperature coefficient is determined by the purity of the platinum. Figure 7.6 shows the resistance-vs.-temperature characteristics for a PRTD with $\alpha = 0.00392/°C$. Although Equation 7-4 is useful in estimating the temperature of a PRTD given its resistance, errors as great as 5% can be obtained through its use at high temperatures. One method of avoiding this error is the use of resistance-temperature tables often supplied with the device. An approximation which is more accurate than that given by Equation 7-4 is the Callendar–Van Dusen equation:

$$R_T = R_o + R_o\alpha[T - \delta((T/100)-1)(T/100)$$
$$- \beta((T/100) - 1)(T^3/100)] \tag{7-5}$$

where R_T is the resistance of the PRTD at temperature T and R_o is its resistance at $T = 0°C$. The temperature coefficient is $\alpha \, (=0.00392/°C)$, $\delta = 1.49$ typically for $\alpha = 0.00392/°C$, $\beta = 0$ for $T > 0°C$ and $\beta = 0.11$ typically

TABLE 7.1   Temperature Coefficients of Various Materials

Material	Temperature Coefficient $(1/°C)(at\ 0°C)$
Nickel	6.7×10^{-3}
Copper	4.3×10^{-3}
Silver	4.1×10^{-3}
Iron	4.0×10^{-3}
Platinum	3.9×10^{-3}
Mercury	9.9×10^{-4}
Carbon	-7.0×10^{-4}

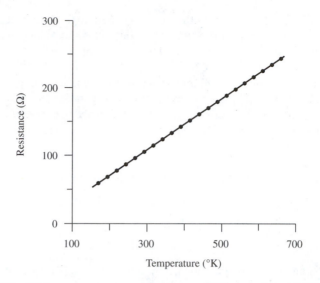

FIGURE 7.6 **Resistance-*versus*-temperature graph for platinum RTD (American curve).**

for $T < 0°C$. Precise values for α, δ and β are obtained by using four calibrated resistance temperature measurements and solving the simultaneous equations for these parameters.

Thermocouple

Whenever two dissimilar metals are in contact, a potential is developed that is proportional to the temperature of the junction. This is called the thermoelectric effect, and it can be approximated over small temperature changes by

$$V = s\Delta T \tag{7-6}$$

where V is the potential developed, ΔT is the temperature difference between the junction temperature and some reference temperature and s is the Seebeck coefficient (temperature coefficient) of the particular thermocouple used. A schematic diagram of a thermocouple is shown in Figure 7.7. In Figure 7.8, the response curves for several different thermocouples are given. Note that s is the slope of the curve, and it is not constant over a large temperature range. It is also obvious that the voltage developed is quite small, on the order of $3\mu V/°C$. This immediately suggests that a high-precision mea-

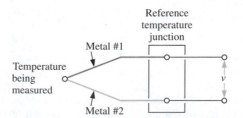

FIGURE 7.7 **Schematic representation of a thermocouple system.**

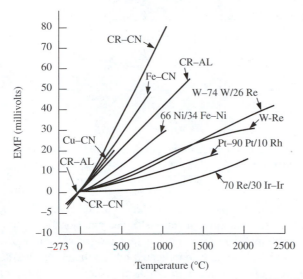

FIGURE 7.8   **Response curves for various thermocouple combinations.**

surement of the voltage developed is necessary if the temperature is to be accurately determined. Standard thermocouples are designated with a letter type. Some of these are shown in Table 7.2.

Thermocouple voltage measurement must be done with care, as we shall soon see. Consider the circuit shown in Figure 7.9, in which a thermocouple is connected to an accurate digital voltmeter. At first glance the circuit seems sound, until we realize that there are in fact three thermocouple junctions,

TABLE 7.2   *Standard Thermocouple Types and Useful Temperature Range*

Letter Designation	Metals	Approximate Temperature Range (degrees Celsius)
Type K	Chromel/Alumel	−200 to 1250
Type J	Iron/Constantan	0 to 750
Type T	Copper/Constantan	−200 to 350
Type E	Chromel/Constantan	−200 to 900
Type S	Platinum/Platinum 10% Rhodium	0 to 1450
Type R	Platinum/Platinum 13% Rhodium	0 to 1450

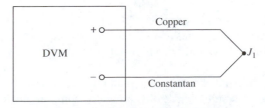

FIGURE 7.9   **Thermocouple connected directly to DVM.**

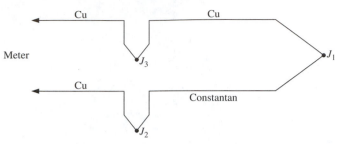

FIGURE 7.10 **Three thermocouple junctions in the DVM-thermocouple circuit of Figure 7.9.**

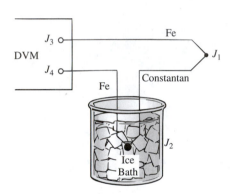

FIGURE 7.11 **Two-thermocouple, ice-bath reference measurement.**

the original thermocouple and one at each meter connection. Figure 7.10 schematically shows the three junctions. Since junction J_3 is a copper–copper junction, it exhibits no thermoelectric effect and the circuit reduces to simply junctions J_1 and J_2. The voltage indicated by the meter will be that produced by J_1 at temperature T_1 less the emf of J_2 produced at T_2 (subtracted since J_2 is connected opposite that of J_1). To calculate T_1, we must therefore know T_2, the temperature of the meter connection itself.

Not all thermocouples have one copper lead, and often meters do not have copper connections. In any case, measuring the temperature of the meter connections would be difficult since the temperature may change considerably over the course of the experiment. To avoid these complications, another identical thermocouple, held at a known temperature, is used to provide an external reference junction. An ice bath at $T = 0°C$ is generally the most convenient source of a known temperature. This arrangement is shown in Figure 7.11 for two iron–constantan thermocouples. Now we have four thermal junctions: J_3 and J_4 are identical but reversed. Since they are at the same temperature, their voltages cancel and we are left with J_1 and J_2, with J_2 at 0°C. The emf measured by the meter is then given by

$$V = V_{J1} - V_{J2} \approx s(T_{J1} - T_{J2}) \tag{7-7}$$

and if T is in °C then Equation 7-7 reduces to

$$V = sT_{J1} \tag{7-8}$$

The National Bureau of Standards references their thermocouple voltage tables to an ice point reference. By using the arrangement of Figure 7.11, we can simply measure a voltage and look up the corresponding temperature in a table. A good approximation can be made by fitting the data to a third-order polynomial, namely

$$V = A + BT + CT^2 + DT^3 \tag{7-9}$$

where A, B, C and D are constants which need to be determined. This can be done by picking four points and solving the four equations which represent these points for A, B, C and D or by using a third-order polynomial fit done by computer.

Recently, electronic component manufacturers have begun producing single integrated circuits which electronically compensate for the ice-bath junction and provide an amplification of the thermocouple voltage. A typical

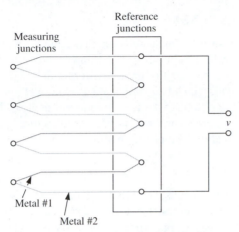

FIGURE 7.12   **Schematic representation of a thermopile.**

integrated circuit of this type, the AD 595, is manufactured by Analog Devices.

By properly combining several thermocouples, as in Figure 7.12, increased sensitivity is gained—in direct proportion to the number of thermocouples used. This assembly is called a *thermopile*. Commercially produced thermopiles are available with as many as twenty junctions.

Solid-State Temperature Transducers

Two-terminal integrated-circuit temperature transducers allow currents to flow that are linearly proportional to the temperature of the devices. A typical unit of this sort is the AD590 produced by Analog Devices. The temperature range of the AD590 is from $-55°C$ to $+150°C$ with a linearity of $\pm 0.3\%$ over the entire range. The unit is powered by a voltage from 4 to 30 V and produces a *current* of 1.0 μA/K. A typical circuit that exemplifies the simplicity of the AD590's use is shown in Figure 7.13. Because V_{out} is equal to 1 mV per °K, a simple DVM can display the temperature directly. The 100 Ω resistor is used to initially calibrate the transducer, because it can have an output current offset error of ± 0.5 μA (although not correcting for this offset would cause an error of only $\pm 0.5°C$).

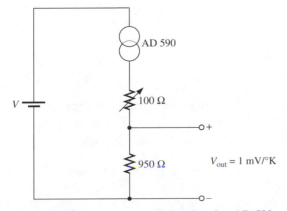

FIGURE 7.13   **Temperature measurement circuit using AD 590.**

7-3   LIGHT INTENSITY

The measurement of light intensity is another laboratory technique that cuts across many disciplines. A number of transducers are available for such measurements. Here we present the photoresistor, which is a resistance device, the photovoltaic cell, which is a voltage device, and the phototransistor, photodiode and photomultiplier, which are all current devices. Infrared radiation detection, which crosses the boundary of temperature and light measurement, is discussed at the end of the section.

Photoresistor

Photoresistors, or photocells, are made of cadmium sulfide, a compound which exhibits a change in resistance when exposed to different intensities of light. An equation which approximates the resistance of the photocell as a function of light intensity is

$$R = (R_o/I_o)\, I^{-K} \qquad \text{(7-10)}$$

where R_o/I_o is a constant, I the intensity of light and K a constant which is less than 1. Figure 7.14 shows a log–log graph of resistance vs. light intensity for a photocell which has $K = 0.75$ and $R_o = 2000$. Like thermistors, these devices have large variations in resistance over a small range, making them extremely sensitive devices, although they are generally not used for exacting measurements. A typical parameter listed in the literature is the resistance of the cell in the dark, which can range into the megohm region. Specific resistances are sometimes given at various light intensities, which are usually listed in footcandles (ftc). The footcandle is a standard unit used for intensity

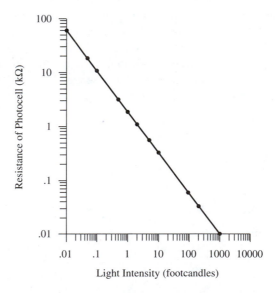

FIGURE 7.14   Log plot of resistance *versus* light intensity for typical photoresistor.

and is equal to 10.7 lumens/m^2; 1 lumen/m^2 is equal to 1 lux, so 1 ftc is approximately equal to 10 lux. To give you an appreciation for these units, the indoor lighting of a building varies in the range of about 25 lux to 500 lux, depending on the function of the room.

Photovoltaic Cell

The photovoltaic cell, or solar cell, is a solid-state device composed of a *p-n* junction. When photons strike the cell, electrons move from the p to the n region and holes move in the opposite direction, thus producing a potential difference. (In fact, this behavior is exhibited in common signal diodes which are nothing more than *p-n* junctions.) If the solar cell is connected as a voltage source in a circuit, a current will flow as long as there is ample light striking the cell. The maximum voltage a solar cell can produce with a fair load (very little current produced), regardless of the light intensity striking its surface, is about 0.6 V. The voltage-vs.-intensity characteristics of the device are far from linear; however under heavy load (high current demand) the current-vs.-intensity behavior is, to a good approximation, linear. Thus the solar cell can, under some circumstances, be used as a linear light-intensity-to-current transducer. On the other hand, since the maximum voltage peaks at 0.6 V, a few cells in series can be used as a voltage source in low current circuits such as calculators. Commercial applications use large numbers of solar cells connected in parallel to produce large currents, cells connected in series to produce higher potentials and, of course, combinations of the two.

Photodiode

A photodiode is similar to a solar cell except that it is manufactured to detect light, not to power devices. Photodiodes are much smaller, sometimes as small as the head of a pin, and have very fast response times, often on the order of nanoseconds. These are usually used in circuits where the diode is reverse-biased, a negative voltage applied to the anode and a positive voltage (respectively) to the cathode. In this mode, the current which flows through the junction is linearly proportional to the intensity of the light which strikes the diode. A simple circuit using a photodiode as a light-intensity meter is shown in Figure 7.15.

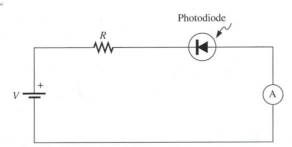

FIGURE 7.15   **Simple light-intensity meter using photodiode.**

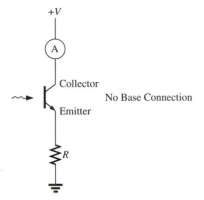

FIGURE 7.16 **Simple light-intensity meter using phototransistor.**

Phototransistor

Phototransistors are two-terminal devices, like those described above, which allow current to flow proportionally to the intensity of light striking the surface of the unit. They are more sensitive than photodiodes, but are also slower. A simple circuit which measures the current as a function of light intensity is shown in Figure 7.16. More practical circuits are discussed later in Chapter Nine.

Photomultiplier

For extremely low light situations and fast response, the photomultiplier (PM) tube is difficult to replace. Photomultiplier tubes are very bulky and require high-voltage power sources, along with special electronics. They are used to detect very low levels of light in astronomical and biological applications, as well as indirect sources of light such as the burst of photons produced by a particle in a scintillator. The schematic drawing of the tube, given in Figure 7.17, is helpful in understanding its operation. The photon, striking the photocathode, releases an electron. The electron is directed to the first anode (dynode) by a voltage difference. The first electron striking the dynode causes, for example, two electrons to be released. These two electrons are directed to the second dynode, where, in this example, four electrons are released. This process continues through each of the dynode stages to the last one. Typically, nine or more dynodes are incorporated in a PM tube. Each dynode is held at a voltage more positive than the preceding one, as shown in Figure 7.18, so that the electrons will be accelerated from one to the next. Quite often, the physical placement of the dynodes and their shape are designed to control the gain within the PM tube. The supply voltages necessary to power the tube vary from 600 to 3000 V.

The current that flows in the PM tube when no light impinges on it is called the dark current. The dark current is a result of thermal excitation of electrons in the photocathode, and is included in the photocurrent at all higher light levels. Since this current, like the photocurrent, is amplified to a non-negligible amount, we must pay attention to it. When the light levels are such that the photocurrent and the dark current are approximately equal, no quantitative significance can be attributed to the measured current, because

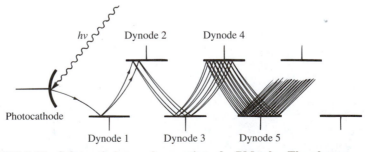

FIGURE 7.17 **Schematic internal operation of a PM tube. The photon striking the cathode releases a single electron. The cascading effect through the dynodes is illustrated.**

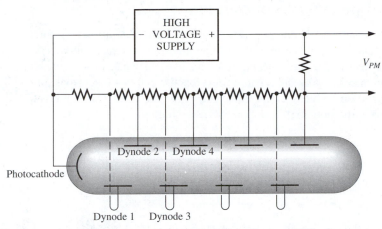

FIGURE 7.18 **Schematic of power supply connection to a PM tube to maintain the dynodes at the appropriate voltages.**

the dark current is dependent on temperature and tube geometry and may fluctuate.

Infrared Radiation Detection

Some of the radiation transducers already discussed in this chapter show a response only to the very near infrared region and shorter wavelengths as illustrated in Figure 7.19. They fail to respond to longer wavelength infrared radiation, however, because its photon energy is insufficient to initiate the processes that operate the transducers. Infrared radiation does cause thermal effects, and temperature measurements can detect these. The temperature transducers already discussed can thus be used to measure infrared radiation.

A *bolometer* is nothing more than a resistance thermometer which has been blackened to increase the efficiency of interaction between the radiant energy and the thermometer. The temperature changes produce changes in resistance. Likewise, the thermocouple and thermopile are temperature transducers whose voltages change with temperature and can be blackened to increase their efficiency to detect infrared radiation.

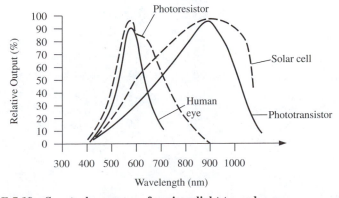

FIGURE 7.19 **Spectral response of various light transducers.**

7-4 pH AND CONDUCTIVITY

The broad field of potentiometric measurement of solution concentrations offers another application of the voltage transducer. The chemical composition of a solution, or the concentration of a particular pair of ions in solutions, is often measured by the potential developed when two electrodes (one indicating and one reference) are immersed in the solution. The composition is related to the potential, V, by the Nernst equation,

$$V = V_o - \frac{0.059}{n} \log \frac{\text{Concentration of } M^{+m}}{\text{Concentration of } M^{+m-n}} \tag{7-11}$$

for the electrode reaction

$$M^{+m} + ne^- \rightleftharpoons M^{+m-n} \tag{7-12}$$

where the species of interest are M^{+m} and M^{+m-n}, n is the number of electrons involved in the reaction, and V_o is a constant related to the particular species being investigated. The choice of the electrode material depends upon those ions in solution which are to be monitored. A number of different electrode pairs have been used extensively, and the reader is referred to one of the several excellent monographs in this field for more information.

The most extensive application of this phenomenon is in the measurement of solution acidity (or alkalinity). For those unfamiliar with the commonly used terms of this area, a very brief review is included here. The quantitative measure of acidity is called the pH. The pH unit is defined as

$$\text{pH} = -\log_{10} (\text{Concentration of } H^+) \tag{7-13}$$

where H^+ is the hydrogen or hydronium ion. If the pH is less than 7, the solution is acidic. (The lower the pH, the greater the acidity.) A neutral solution has a pH of 7, and alkaline (basic) solutions have a pH greater than 7. Because of the widespread interest in pH, many potentiometric approaches to its measurement have been made. By far the most successful has been the use of the "glass" electrode, as shown in Figure 7.20. A particular combi-

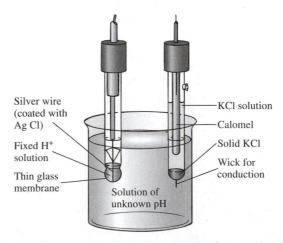

Silver wire (coated with Ag Cl)

Fixed H^+ solution

Thin glass membrane

Solution of unknown pH

KCl solution

Calomel

Solid KCl

Wick for conduction

FIGURE 7.20 **Details of a glass electrode/calomel reference electrode system.**

nation of materials in the glass produces a surface which is sensitive to pH. The interior of the glass membrane is filled with a solution of fixed pH. A surface potential develops which is proportional to the pH,

$$V = R + 0.059 \text{ pH} \qquad \qquad \textbf{(7-14)}$$

where R is a constant which depends upon the choice of a second (reference) electrode. Note that equation 7-14 is equivalent to the general Nernst Equation 7-11. Because glass is an insulator, extremely high resistances (10^{10} to 10^{14} Ω) are encountered when a glass-electrode pH measurement is attempted. The normal type of null potentiometric measurement will not suffice because of the very high resistances. This feature is made clear by reference to the accuracy with which the measurement is made. Consider that an accuracy of ± 0.01 pH unit is desired. This means, using Equation 7-14, that the maximum allowable voltage error is 0.6 mV. The voltage measurement must draw very little current because the current flow produces an IR drop across the glass. The very high resistance, 10^{10} Ω, indicates that the maximum current flow is

$$\frac{6 \times 10^{-4} \text{ V}}{1 \times 10^{10} \text{ } \Omega} = 6 \times 10^{-14} \text{ A}$$

If the current flow exceeds this value, then the measurement error is in excess of ± 0.01 pH unit. Only selected, specialized circuits can meet these stringent conditions.

Besides H^+ specific electrodes, many other ion-selective electrodes have been developed, and potentiometric measurements are experiencing a great deal of activity. These developments have markedly enhanced the number of concentration-voltage transducers. The concentrations of the following ions, among others, are presently being measured in this manner: sodium (Na^+), potassium (K^+), calcium (Ca^{++}), fluoride (F^-), sulfate (SO_4^-), sulfide $S^=$), nitrate (NO_3^-) and perchlorate (ClO_4^-).

Electrolytic Conductance

A large number of resistive input transducers operate by variation in length or thickness (l), area (A), or resistivity (ρ). The relationship given in Chapter 1 is used:

$$R = \rho l / A$$

It is important that only one of the parameters be variable in the device so that proper interpretation can be made. For instance, the concentration of an electrolyte solution is determined by measuring ρ (or its reciprocal, conductivity) when l and A are kept constant. Figure 7.21 shows two typical conductance cells. A calibration solution of known conductivity is used to evaluate l/A (known as the *cell constant*) for the cell. Thereafter, the resistance measurement of the electrolyte solution is directly related to its specific conductivity. A potential flaw in this measurement is that the electrodes are exposed to the solution whose resistance is being determined. This may lead to fouling of the electrodes and to erroneous results.

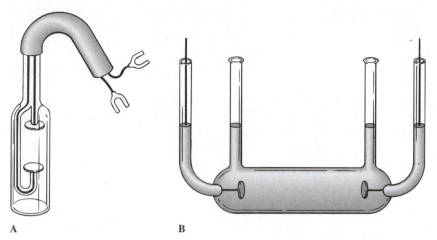

FIGURE 7.21   Conductance cells for measurement of electrolyte concentration.

7-5   FORCE

Strain Gauge

The strain gauge is an outstanding example of a resistive input transducer whose change in resistance is related to changes in length. Small changes are accurately measured in this manner. These devices, one of which is shown in Figure 7.22, are the size of a postage stamp or smaller. The backing of the gauge is coated with adhesive for mounting on the element to be tested. Changes in the length of the long axis, as the test member is stretched or compressed, produce changes in resistance, according to the equation

$$\frac{\Delta R}{R} = G \epsilon \tag{7-15}$$

where G is the gauge factor, determined under standard conditions, and ϵ is the strain, or change in length, $\Delta l/l,$ of the member. The change in resistance is measured by a Wheatstone bridge. Because the resistance of the gauge depends on its temperature, a second strain gauge (of properties similar to the first) held at the same temperature, but not attached to the test member, is used in another arm of the Wheatstone bridge to compensate for any tem-

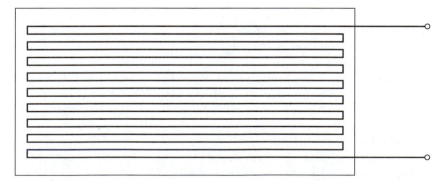

FIGURE 7.22   Typical strain gauge.

perature variation during the measurement. The resistance variation observed is generally less than 1.0% of the total resistance. The bridge voltage generally should be less than 10 V to avoid I^2R heating. Displacements of 0.05 mm are easily measured when a temperature-compensated bridge is used, which means that the out-of-balance bridge signal is 10 to 50 mV, and the bridge output is conveniently recorded. The deviation from linearity is less than 1.0% for a properly constructed gauge.

Piezoelectric

Certain crystalline materials (Rochelle salt, quartz) and ceramics (barium titanate) generate a voltage when deformed. This is known as the piezoelectric effect. Thus, force or pressure on a piezoelectric material will produce a voltage which is directly proportional in sign and magnitude to the applied stress, as shown in Figure 7.23. In addition to force and pressure, acceleration can be transduced through the use of an inertial mass attached to the piezoelectric crystal. The force required to accelerate the mass is applied through the crystal and a voltage develops which is proportional to the force, and therefore the acceleration. A voltage applied to a piezoelectric transducer will deform the transducer, producing a means of changing, and therefore controlling, position.

7-6   POSITION

Potentiometer

The simplest form of position transducer is the potentiometer. Linear and logarithmic potentiometers are available which can rotate from 3/4 of a turn to 10 turns, with the output voltage precisely related to angle (typically with 0.1% repeatability and 0.5% accuracy). Angular measurement is accomplished quite simply as shown in Figure 7.24. Linear measurement can be

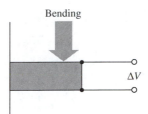

FIGURE 7.23   Schematic representation of use of piezoelectric transducers.

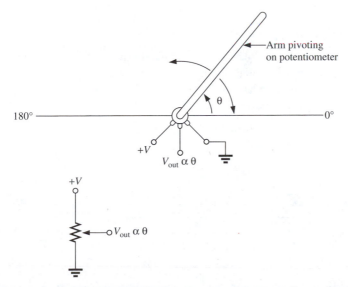

FIGURE 7.24   Mechanical arm pivoting on potentiometer and associated electrical connections to measure angle.

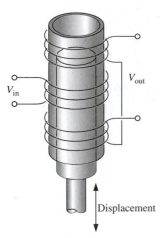

FIGURE 7.25 **Linear Variable Differential Transformer.**

accomplished by mechanically attaching the potentiometer to the object whose position is being monitored.

Linear Variable Differential Transformer

The Linear Variable Differential Transformer (LVDT) is a device which consists of a set of coils through which a movable metal core is placed. The first coil is driven by an AC voltage. The position of the core, how far it is in or out of the coils, is measured by monitoring the voltage out of the second (center-tapped) coil. This voltage depends on the degree of mutual induction between the coils, which in turn depends on the degree of coupling through the core. An example of an LVDT is illustrated in Figure 7.25. LVDTs are available from under 0.1 inch to up to 24 inches of travel and with linearities as good as 0.1%. Sensitivity is typically 0.001 inch, which translates to a few mV. Angular versions are available as well.

Ultrasonic Measurement

A scheme for measuring larger displacements consists of producing a high frequency burst of ultrasonic sound waves while at the same time starting a timer of a suitable type (a digital clock or a charging capacitor). The sound waves propagate from the source, strike the target and travel back to the detector. When they arrive back at the detector, which is often the same transducer which produced the sound, the timer is stopped. Knowing this time and the speed of sound in the medium, the position of the object can be calculated. Such transducers are commonly used in auto-ranging cameras. Here we describe a measuring scheme rather than an individual transducer, but a piezoelectric sender/receiver and associated drive and clock circuitry is available at modest cost from Texas Instruments, which produces a linear signal proportional to position. These units are generally used in ranges of less than 10 m and can be used for sensitivities as good as 1mm under certain conditions.

Capacitive Input Transducer

Displacements may be measured capacitively by attaching a plate to the object which is moving, and another plate to a fixed point. The direction of movement of the first plate can be either parallel or perpendicular to the fixed plate, as shown in Figure 7.26. The empirical relationship is

$$C(\text{pF}) = 0.0885\ \epsilon_r A/d \tag{7-16}$$

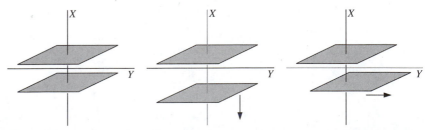

FIGURE 7.26 **Representation showing how displacements affect a capacitive input transducer.**

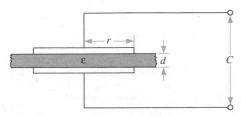

FIGURE 7.27   Schematic representation of a general capacitive transducer.

where A is the area of the plate, d the distance that separates them and ϵ_r the dielectric constant of the material between them, as is illustrated in Figure 7.27. Note that Equation 7-16 is only approximate, in that fringe effects (edge effects) are neglected; that is, it is assumed that a is much greater than d. Appropriate equations can be written when this is not the case.

An interesting modification of this basic application uses a movable plate between two fixed plates. The movement or displacement in the x direction is measured. This arrangement is called a differential capacitor and is presented in Figure 7.28A. In this situation the two capacitances are

$$C_1 = \frac{\epsilon A}{d+x} \tag{7-17}$$

and

$$C_2 = \frac{\epsilon A}{d-x} \tag{7-18}$$

The output signals may be obtained by measuring either the difference in voltages or the ratio of capacitances. In the former method,

$$V_1 = \frac{VC_2}{C_1+C_2} = V\frac{d+x}{2d} \tag{7-19}$$

and

$$V_2 = \frac{VC_1}{C_1+C_2} = V\frac{d-x}{2d} \tag{7-20}$$

Hence, the difference, ΔV, is

$$\Delta V = V\frac{x}{d} \tag{7-21}$$

Note the linear relationship between the displacement, x, and the difference in voltage. The sensitivity of the method, $\Delta V/\Delta x$, is $1/d$. The method has general applicability, since neither the electrode area nor the dielectric constant affects the measurement. Generally, an alternating voltage is applied to the fixed plates and the voltage difference is measured between the movable plate and one of the fixed plates.

The measurement of the actual capacitance is accomplished by a bridge. In this method,

$$\frac{C_2}{C_1} = \frac{d+x}{d-x} \tag{7-22}$$

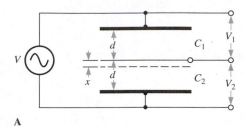

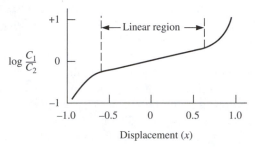

A

B

FIGURE 7.28 **Differential capacitive input transducer, (A) Schematic, (B) response curve.**

For small displacements ($d \gg x$),

$$\frac{C_2}{C_1} \cong 1 + \frac{2x}{d} \tag{7-23}$$

These conclusions lead to a calibration curve similar to that of Figure 7.28B.

The thickness of a material, such as a plastic sheet in a continuous manufacturing process, can be monitored conveniently with the capacitance input transducer shown in Figure 7.29. The material passes between the plates of a transducer in which the movable plate is spring-loaded to press lightly on the material. The output is fed back, with suitable modification, to the rollers that control the thickness of the material being produced.

This principle can be extended to the determination of moisture content, as in the paper industry (where the moisture of the newsprint is an important manufacturing variable). The output from the transducer is fed into an AC bridge. While the paper thickness is relatively constant, the variation in moisture content produces a change in the phase angle of the signal. The "moisture meter" is, in reality, a "phase angle meter."

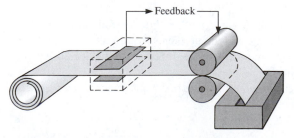

FIGURE 7.29 **Capacitive thickness transducer system.**

7-7   PRESSURE, VACUUM AND SOUND

Pressure

Gas pressure may be measured electronically in a number of ways. Two schemes that are commonly used are a strain gauge and a piezoelectric diaphragm element. Both utilize a small vacuum chamber on one side of a diaphragm, with the other side exposed to the gas whose pressure is being measured. In the case of the strain-gauge device, the output is resistive, whereas the piezoelectric device gives a voltage output which is linearly proportional to the pressure difference. One such device is manufactured by Sensym, Inc. Also, pressure transducers made of diffused semiconductor material are available for higher pressure applications (up to 350 atmospheres).

Vacuum

Low pressures are readily determined via a resistive input transducer called a Pirani gauge. At atmospheric pressures, the ability of a gas to conduct away heat is nearly independent of pressure. At low pressures, however, relatively few gas molecules are present to collide with a hot wire and carry some of the heat away. The pressure of the gas can therefore be measured by the rate at which heat in a resistance wire is dissipated. If the resistance wire is heated with a constant current, the measured resistance depends on pressure. Since gases possess specific thermal conductivities, the composition of the gas must be known in order to properly interpret the transducer's output. Commonly, a bridge circuit similar to that of Figure 7.30 is used to determine the resistance, and a calibration curve such as that in Figure 7.31 gives the corresponding pressure. Pressures in the range from 10^{-1} to 10^{-4} torr (mm of mercury) are measurable by this technique. The accuracy is about 2%, and resistance changes are seldom more than 10%.

To measure pressures from 10^{-3} to 10^{-11} torr, the ionization gauge is generally used. The basic elements of such a gauge are illustrated in Figure 7.32. Electrons are released by the electrically heated cathode and accelerated toward a grid, which is maintained at a positive potential. Since the grid is actually a coiled wire, most of the accelerated electrons pass through the grid and collide with the gas molecules whose pressure is being measured.

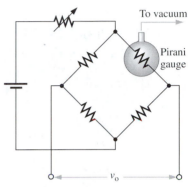

FIGURE 7.30   **Bridge assembly for a Pirani gauge.**

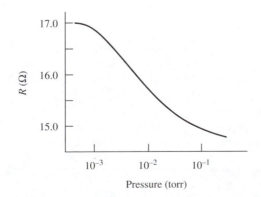

FIGURE 7.31   **Resistance-*versus*-pressure response curve of a Pirani gauge.**

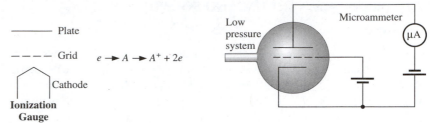

$e \rightarrow A \rightarrow A^+ + 2e$

Ionization Gauge

FIGURE 7.32   **Ionization gauge system.**

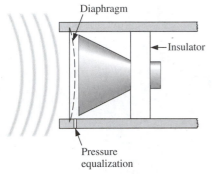

FIGURE 7.33   **Microphone as a capacitive input transducer.**

These collisions cause ionization of the molecules into positive ions and more electrons. The negative plate attracts the positive ions and repels the electrons. The current measured is directly proportional to the number of gas molecules ionized, and thus to the pressure. The current that results is in the μA range or less; it can be amplified if necessary.

One of the earliest applications of capacitive transduction, and one that is still in general use today, is in the area of sound measurement, or, more properly, sound pressure. The usual microphone is a capacitive transducer, in which sound pressure causes a diaphragm to move in and out relative to a fixed plate, as shown in Figure 7.33. Because the output from such a transducer is generally a voltage, the microphone's response is not usually recognized as capacitive. Usually, a large DC voltage is applied to the two capacitor plates. This results in a constant charge on the electrodes. The vibrations of the diaphragm result in an AC voltage proportional to the sound pressure and at the same frequency as the vibrations. The output must be fed into a high-impedance stage to eliminate the possibility of leaking the constant charge off the electrodes or plates. This, in fact, is accomplished in the body of the microphone by an emitter follower. Further amplification is then carried out in a straightforward manner. The features of interest in a microphone are the frequency response and the directional character. Microphones with a frequency response equaling or exceeding that of the human ear (20 Hz to 20,000 Hz) are generally available. The directional character of a microphone changes considerably with frequency. Typically, there is no measurable directional character at a few kHz, while at the upper frequency limit of the human ear there is an extreme directional character.

Sound

The speaker is a common output transducer (although it can be used as an input transducer to pick up sound) which is used to produce a variation of air pressure, or sound. Most large speakers consist of a diaphragm which is moved in and out. This is accomplished by attaching a coil of wire to a diaphragm through which a current, the sound signal, is passed. The coil is surrounded by a permanent magnet. The coil's magnetic field therefore interacts with the fixed field of the magnet, causing the coil and diaphragm to move in and out. Smaller speakers which are used at higher frequencies and lower amplitudes often use the piezoelectric effect to move the speaker element.

7-8   HUMIDITY

A commercially available device manufactured by the N. V. Philips Company allows the electronic measurement of relative humidity. It consists of a 1-cm piece of plastic that has a thin gold coating on both sides. Since the dielectric constant of the film changes with the relative humidity, the capacitance changes and thus the relative humidity can be measured. The capacitance changes over a range of about 115 pF to 160 pF and is fairly linear over the 0 to 100% relative-humidity range.

7-9   MAGNETIC FIELD

When a current-carrying conductor is placed in a magnetic field, charge is displaced from its path and collects on opposite sides of the conductor. The separation of charge creates a voltage potential across the two charged sides of the material. The produced voltage is proportional to the strength of the magnetic field and the amount of current flowing through the conductor. This phenomenon is called the *Hall effect*.

The Hall effect can be used to measure the strength of a magnetic field. Transducers that make use of this effect are usually made of semiconductor material and typically have four leads, two for passing current through the semiconductor region and two for measuring the Hall-effect voltage across the device. F. W. Bell is a major manufacturer of these devices, which have common output voltages in the range of mV/kilogauss.

7-10   PARTICLE DETECTORS

Geiger-Müller Tube and Proportional Counter

The Geiger-Müller (GM) tube is used to detect ionizing radiation. Ionizing radiation is the very high energy radiation associated with nuclear disintegrations. Materials transparent to intense radiation and high operating voltages must be used in a GM tube. The window is made of mica, since mica absorbs little radiation of this energy, and it can withstand evacuation prior to filling the tube with a gas. The construction of a GM tube is shown in Figure 7.34. The metal container has an insulated wire at the center of the cylinder. The

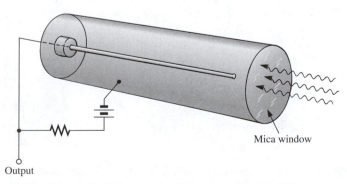

FIGURE 7.34   Schematic diagram of a GM tube and its circuitry.

GM Output

After Differentiation

FIGURE 7.35 **Output of a GM tube, with and without a stage of differentiation (filtering).**

anode is the central wire, and the container is the cathode. The container is not evacuated but contains a gas at a low pressure. As the radiation enters the GM tube, it causes ionization of the gas particles into electrons and positive ions. The high positive voltage of the anode attracts electrons to it, while the positive ions are drawn to the cathode. A single particle causes complete ionization within the chamber. The current pulses in the external circuitry are a measure of the number of particles entering. The movement of the electrons to the anode is much faster than the movement of the positive ions to the walls; this means that each pulse has a considerable decay time associated with it, as shown in Figure 7.35. The current flow associated with particles of different energies is the same because of the very high internal amplification of the GM tube. The rate at which particles can be counted is relatively slow, approximately 10^3 counts per second, because the tube must *quench* (cease ionization) between particles if they are to be counted individually.

The GM tube is operated under a wide range of applied voltages. The response of the tube to radiation under these conditions is illustrated in Figure 7.36. As the voltage is first increased, a plateau is reached which corresponds to ion collection before any recombination takes place. This is the region where the device functions as an ionization chamber. The currents are quite small because no secondary ionization exists. Raising the voltage causes acceleration of the ions to the electrodes; soon an energy is reached at which additional ionization occurs. The current is then proportional to the number of initial ions; hence the device functions as a proportional counter. Because some secondary ionization has occurred (some internal amplification) the current is significantly higher than for the same situation in the ionization chamber region. Clearly, the energy of the particle affects the

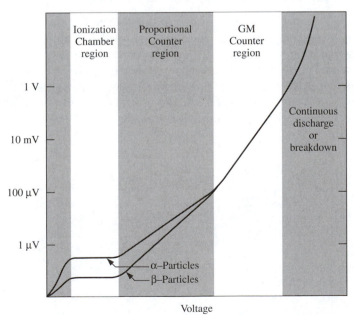

Voltage

FIGURE 7.36 **The transfer curve for a GM tube, showing the several regions of normal operation.**

magnitude of the output. At high enough voltages, there is so much ion acceleration that an avalanche effect occurs. As soon as an alpha or beta particle forms one ion, that ion is accelerated; it strikes a second gas molecule and ionizes it, and the avalanche quickly ionizes all of the gas in the tube. Thus any particle capable of producing one ion will cause an output pulse of the same magnitude as that for any other ionizing particle. The voltage region in which this occurs is the counting region, where the tube is normally operated. If the voltage becomes too high, however, spontaneous ionization will occur and will destroy the device.

The methods of discriminating among various particle energies after they have been detected will be dealt with in some detail in the following chapter.

Scintillation Counters

It has been known since the 1880s that particles emitted in radioactive decay can cause certain materials to fluoresce. Initially, scientists relied on tedious visual counting methods. The development of the photomultiplier tube during the 1940s permitted instrumental counting methods to be developed. A convenient arrangement is indicated in Figure 7.37, where the fluorescent material, called the scintillator, is attached directly to the face of the PM tube. The sample to be counted can be placed in a well drilled in the scintillation crystal, providing a very highly reproducible geometry. A variety of organic compounds have proven to be excellent materials for use as scintillators. Terphenyl is but one example.

Since fluorescence is generally completed in 10^{-8} seconds, the recovery time of the scintillation counter is excellent. In order to overcome the dark current contribution from the PM tube, the entire assembly is often housed in a refrigerator. As in a proportional counter, the scintillation counter signals are proportional to the energy of the particles.

Solid-State Detectors

A relatively recent application of semiconductors has been for the detection of ionizing radiation. Consider a *p-n* junction with a very thin *n* region. If the device is reverse biased, depletion takes place in the region adjacent to the junction. Of course, depletion represents the relative absence of charge carriers. If ionizing radiation enters through the *n* region and crosses the junction into the *p* region, electron hole pairs are created, particularly in the latter region. The electrons drift toward the junction and the holes move away from the junction. The current resulting from this event is monitored, generally as a voltage drop across a resistor.

Two solid-state devices for the detection of ionizing radiation are the *lithium drifted germanium detector* (Ge(Li)) and its associated silicon counterpoint the (Si(Li)). Their names describe the method of manufacture; in the Ge(Li) detector lithium atoms are thermally drifted into a bar of germanium, making it a heavily doped *n*-type material. The device must be operated *and* stored at liquid nitrogen temperatures. In operation, a very large voltage, 1500 V, is applied to it. This voltage completely depletes the donor level of the conductor, leaving the lithium ions in the valence band. At room

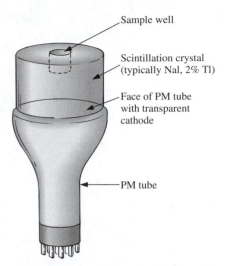

FIGURE 7.37   **Schematic diagram of the PM tube for scintillation counting.**

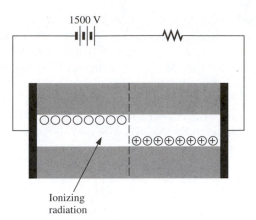

FIGURE 7.38 **Schematic diagram of the operation of a lithium drifted germanium detector.**

temperatures, this stripping of the donor level would be followed instantaneously by destruction of the device. At the lower temperatures, the process stops at depletion. This situation is depicted in Figure 7.38.

High-energy particles ionize the substrate of the device, producing a voltage pulse across R. The pulse height is proportioned to the number of electrons produced by the high-energy particles. Since the path length of the particle penetration is proportional to its energy and the number of electrons produced are proportional to the path length, the pulse voltage is related directly to the energy of the particle. The principle is simple, and the cost, while still relatively high, can be expected to decrease.

7-11   SUMMARY AND CONCLUSIONS

This has been a brief survey of some of the standard methods of input transduction. By now, you may have conceived of alternative and imaginative methods. This chapter is intended to show only what may be done in this area of instrumentation, and not to cover the field in any rigorous manner. As the conditions of measurement become clearly defined, it can be assumed that the transducer manufacturers will hasten to meet those needs. Such features as the applicable range of transducer operation, sensitivity, accuracy and speed of response all become important. Little attempt has been made here to discuss these important considerations. For those who are faced with this type of problem, there are a number of excellent references to be found, in addition to the manufacturers' literature.

7-12   QUESTIONS AND PROBLEMS

1. Suggest what kind of transducer might be used to measure each of the following. Indicate the electrical property which results.

    (a) temperature
    (b) viscosity
    (c) density
    (d) wind speed

2. Brackish water generally means water which has been contaminated with salt. Suggest a transducer which will monitor the salt concentration.

3.   A measurement of interest in the summer is the THI—the temperature-humidity index, consisting of the sum of the temperature and the percentage relative humidity. How would you measure this?

4.   A measurement of interest in the winter is the wind-chill factor. The factor is calculated from wind speed and temperature. How would you measure this?

5.   Suggest methods for measurement of flow velocity in conducting fluids and in non-conducting fluids.

6.   Suggest transducers for measurement of pressures:
(a) which are essentially atmospheric.
(b) which are well above 1 atmosphere.
(c) which are 1 torr or less.

7.   Describe the advantages and disadvantages of using an AC voltage (as compared with a DC voltage) for strain-gauge measurements.

8.   Some sports cars have engine speed tachometers, as well as speedometers, on their dashboards. What differences, if any, are there between the two in terms of transducers?

CHAPTER

8

Transistors

8-1 INTRODUCTION

The first transistor was developed over forty years ago. After a relatively slow start, transistors succeeded in edging out vacuum tubes almost completely. The past two decades have seen the extensive development of new solid-state devices which often replace the transistor. The most impressive development has been the integrated circuit (IC). ICs are arrays of diodes, transistors, resistors and sometimes capacitors appropriately connected together on a single substrate. The techniques of manufacturing the discrete transistor have been extended to a state where it is no more difficult to produce a single "chip" containing an entire circuit than it is to produce the naked transistor. As a result, the prices of many ICs are no more than those of single transistors. There may be a day in the not too distant future when a discrete transistor will have become as much of a curiosity to some as the vacuum tube has become.

In order to utilize ICs most effectively, we must thoroughly understand transistors. In a real sense, the transition in comprehension from transistors to ICs is a very small one, while the prerequisite step, that of understanding and using transistors, is more difficult. This chapter introduces transistor nomenclature, operation and methodology.

The transistor represents one large class of semiconductor devices based on the *p-n* junction described in Chapter 5. We shall encounter a number of other classes in this and later chapters. No attempt will be made to cover all of the devices available today. In fact, as you read this, new devices for particular applications will become commercially available, some of which were not even known when this was written. If you want the very latest developments, you must continually peruse the manufacturers' literature. What you learn from this chapter should be of value in understanding and using these new devices.

8-2 THE BIPOLAR JUNCTION TRANSISTOR

A junction transistor is made when two *p-n* junctions are formed on the same substrate crystal. This device is frequently called a *bipolar junction* transistor. There are two types: *pnp* and *npn*. The usual circuit connections and terminal designations are shown in Figures 8.1 and 8.2. Regardless of the transistor type, the emitter-base junction is normally forward biased as shown, and the collector-base junction is normally reverse biased. The majority carriers in the *pnp* device are holes, while electrons are the majority carriers in the *npn* transistor.

In the typical manufacturing process, the base region is kept physically quite thin. Moreover, the base region is lightly doped. The emitter region is richly doped to reduce its resistance. The collector region is lightly doped to reduce the junction capacitance of the collector-base junction.

Transistor Operation

The mode of operation of the *npn* transistor shown in Figure 8.2 can be understood by considering that electrons (I_E) from V_{EE} are injected into the emitter region (*n* material). The forward biased base-emitter junction, acting like a diode, does not interfere with the drift of the electrons into the base

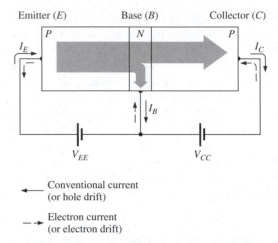

FIGURE 8.1   **Symbolic representation of *pnp* transistor action.**

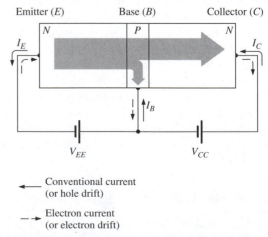

FIGURE 8.2   **Symbolic representation of *npn* transistor action.**

region (p material). A small amount of electron-hole recombination occurs in the base region, and appears in the external circuit as the base current , I_B. Since the base region is quite thin, however, the vast majority of the electrons reach the collector-base junction and cross it under the influence of the positive polarity of the collector. The collector is held positive, relative to the base and the emitter, by V_{CC}.

For the effective operation of a transistor, more than 95% of the majority carriers must pass through the base region from the emitter to the collector without recombination. This suggests that the collector current, I_C, is only slightly less than the emitter current, I_E. The exact ratio is called α, defined by

$$\alpha = \frac{I_C}{I_E} \tag{8-1}$$

Typical values of α for commercial transistors are between 0.95 and 0.99. The remainder of the emitter current, diverted by recombination in the base region, appears at the external base terminal, and is given by Kirchhoff's Law, $I_B = I_E - I_C$. The *current gain, β,* is defined as the ratio of the collector current to the base current; using the Kirchhoff relationship, this can be written as

$$\beta = \frac{I_C}{I_B} = \frac{\alpha}{1 - \alpha} = h_{FE} \tag{8-2}$$

Note that a transistor with α = 0.99 has a β of 99. It should be emphasized that this discussion has concentrated on current because the transistor is inherently a current amplifying device.

It is important to recognize that a change in V_{EE} will change the collector current, I_C. Recall the diagram of diode operation (Figure 5.8). Increasing V_{EE} is equivalent to increasing the forward bias on the emitter-base junction. Because of the steep slope of the voltage-current curve, a small increase in the voltage will cause a large increase in the current crossing the emitter-base junction. Rather than leaving the junction through the base, however, most of the current appears at the collector as I_C. Thus, the transistor is a control device in which a small change in the base voltage can produce a relatively large change in the collector current.

The *pnp* transistor operates in an analogous manner, except that holes are the majority current carriers. The base, being *n* material, provides electrons to recombine with the holes. The reader should test his or her understanding of transistor action by using the discussion of the *npn* transistor as a guide and applying it to the *pnp* transistor. The standard circuit symbols for both types are given in Figure 8.3. In both cases, the arrow points in the direction of conventional current flow.

External Circuit Configurations

The junction transistor can be used as a circuit device in three distinct configurations. The descriptions of these configurations refer to the lead common to both the input and output. Thus, in the common-base (CB) configuration the input signal is applied to the emitter and base leads while the output is taken across the collector and base leads. The three configura-

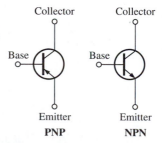

FIGURE 8.3   Transistor configurations.

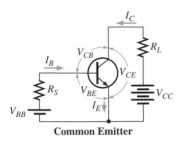

Common Emitter

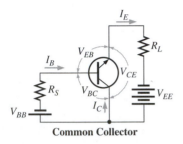

Common Collector

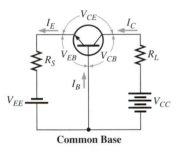

Common Base

FIGURE 8.4 **Electronic symbols for bipolar junction transistors.**

tions, common emitter (CE), common collector (CC) and CB are given in Figure 8.4. The important circuit properties of each configuration, such as the input and output resistances and the voltage and current gains, are given in Table 8.1.* Each of the configurations possesses certain desirable circuit characteristics. The choice of the configuration to be used for a given application is based upon the most important particular characteristics. For example, the CE configuration is most frequently used because both current and voltage gains greater than unity are achieved. The weaknesses of this configuration are the relatively low input resistance and the relatively high output resistance. If a high input resistance is needed, the CC configuration should be selected even though a voltage gain of less than unity is achieved. By having a CC input stage followed by a CE amplification stage, both high input impedance and voltage gain for the overall circuit are obtained.

Characteristic Curves

A more quantitative insight into the operation of a transistor can be obtained by examination of the *characteristic curves* of the device. These are frequently supplied by the manufacturer. When not supplied, they can be readily obtained by some straightforward explanation. For the CE configuration, curves of I_C versus V_{CE} at several values of I_B will be like those shown in Figure 8.5B. One simple circuit used to obtain these curves is given in Figure 8.5A. (A number of automatic "curve tracers" are commercially available which produce these curves on the face of an oscilloscope.) Notice that small changes in I_B produce large changes in I_C. This is the essence of transistor operation in the CE configuration.

In the CE configuration, the input signal is inserted in series with V_{BB} to produce a change in I_B. If the negative portion of an AC signal were large enough to overcome V_{BB}, the base-emitter junction would become reverse biased and the transistor would be turned off, effectively clipping the corresponding portion of the amplified output signal. In order to avoid this, V_{BB} is set large enough to maintain I_B greater than zero for all expected signals. This

*TABLE 8.1   Transistor Circuit Properties**

	Configuration		
	CB	**CE**	**CC**
Power gain $A_p > 1$?	yes	yes	yes
Voltage gain $A_v > 1$?	yes	yes	no
Current gain $A_i > 1$?	no	yes	yes
Input resistance (typical)	30 Ω	3.5 kΩ	580 kΩ
Output resistance (typical)	3.1 MΩ	200 kΩ	35 Ω
Voltage phase change?	no	yes	no

*The values depend upon the particular transistor and other circuit components used. To obtain the values in this table, a 2N3904 transistor was used with R_L = 5000 Ω and R_s = 500 Ω (abbreviations as in Figure 8.4).

Gain in general is defined as the ratio of an output quantity to the corresponding input quantity. For example, voltage gain, A_v, is the ratio v_{out}/v_{in}.

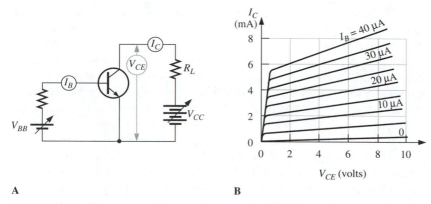

FIGURE 8.5   Characteristic curves for an *npn* transistor in the common-emitter configuration.

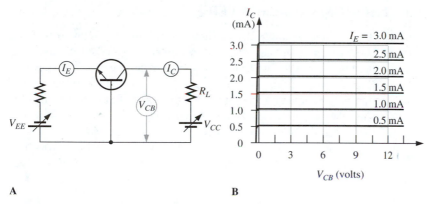

FIGURE 8.6   Characteristic curves for an *npn* transistor in the common-base configuration.

fixed base current is called the *quiescent point,* and its proper selection ensures that the transistor will stay on for all portions of the applied signal.

The characteristic curves for the CB configuration are shown in Figure 8.6B, and the circuit used to obtain them is given in Figure 8.6A. In this situation, the important feature to note is that I_C, for a given value of I_E, is virtually independent of V_{CB}. In other words, the CB configuration represents a kind of constant current generator.

AMPLIFICATION

Everyone is qualitatively familiar with amplification, but it is important to understand the principles clearly and quantitatively. The next section deals with the mathematical modeling that results in the quantitative evaluation of transistor amplification. Before beginning this discussion, however, it is well to recall the qualitative nature of the process.

Amplification is the process by which the magnitude of a signal is increased, whether it be voltage, current, or power. This is accomplished by using the input signal to control the amount of power being supplied to the output from an independent reserve. This reserve, as well as the input signal, is essential to the process. The control device is, of course, the transistor. For most situations, an additional quality is sought: the character of the input signal should be faithfully reproduced in the output signal. In other words, the information content of the input should not be degraded by the amplifier. This feature will be studied in more detail in later sections.

From time to time, the novice loses sight of the essential requirement of the power reserve for the amplification process. Of course, this power reserve is the DC power supply. For all practical purposes, it is considered here to have an infinite capacity to supply DC power; but the designer must bear in mind the considerations of regulation versus load current that were discussed in Chapter 5. Always keep in mind that amplification is *not* something for nothing. The power reserve must always be there.

8-3   THE COMMON-EMITTER AMPLIFIER

Consider the common-emitter circuit shown in Figure 8.7. Notice that the batteries shown in previous illustrations have been replaced by a single DC power source, V_{CC}, but the collector voltage remains greater than the base voltage, which is positive, and both are greater than the voltage at the emitter. If we capacitively couple an incoming AC signal, Δv_B, and measure the output AC signal Δv_C through a capacitor (to block the DC signal) as shown in Figure 8.8, we find that the circuit is a common-emitter amplifier.

First, we must realize that the base-emitter junction acts as a forward-biased diode and thus has an associated voltage drop of about 0.6 V. In other words,

$$V_E = V_B - 0.6 \text{ V}. \tag{8-3}$$

A varying signal at the base will thus cause the identical signal to appear at the emitter (a time derivative of Equation 8-3 will make this fact readily apparent), or

$$\Delta v_E = \Delta v_B. \tag{8-4}$$

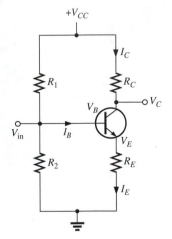

FIGURE 8.7   The common-emitter amplifier.

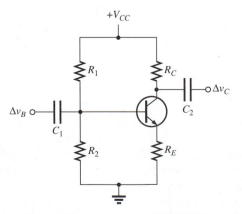

FIGURE 8.8   Capacitor-coupled common-emitter amplifier.

Dividing both sides of the equation by R_E results in

$$\Delta v_E/R_E = \Delta v_B/R_E$$

or

$$\Delta i_E = \Delta v_B/R_E. \qquad \text{(8-5)}$$

Recall that

$$i_C = \beta i_B,$$

and

$$i_E = i_C + i_B \qquad \text{(8-6)}$$

so for large values of β, Equation 8-6 becomes

$$i_E \approx i_C$$

and then

$$\Delta i_E \approx \Delta i_C. \qquad \text{(8-7)}$$

Substituting Equation 8-5 for Δi_E and realizing that the voltage drop $\Delta v_C = -\Delta i_C R_C$ we find

$$\Delta v_B/R_E = -\Delta v_C/R_C$$

or

$$\text{voltage gain} = \Delta v_C/\Delta v_B = -R_C/R_E. \qquad \text{(8-8)}$$

For values of $R_C > R_E$ the common-emitter circuit is a voltage amplifier. The negative sign indicates a phase shift of 180°.

For stability in high-gain AC common-emitter amplifiers, a relatively large ($\approx 10\mu F$) bypass capacitor, C_E, is connected in parallel across R_E. When using the amplifier in a high-gain configuration, V_E is generally low enough that fluctuations in the forward-bias voltage drop across the base-emitter junction caused by temperature changes have a significant effect when using AC signals. By adding the bypass capacitor, DC voltages remain unchanged, and thus the bias voltage on the base is unchanged, but AC signals see the capacitor as the main source of emitter impedance. Thus R_E is now replaced by

$$Z_E = R_E/(1 + R_E j2\pi f C_E) \qquad \text{(8-8a)}$$

where f is the frequency of the signal. If we choose $Z_C > 10\ R_E$ for a minimum signal frequency f, then C_E is chosen to be

$$C_E = 10/(2\pi f R_E) \qquad \text{(8-8b)}$$

For $R_E = 270\ \Omega$ and the lowest frequency that the amplifier is to handle $f = 1$ kHz, $C_E = 5.9\ \mu F$. By using a capacitor of 10 μF, the circuit in question would be stable. At low DC frequencies the resistor is the main source of impedance, whereas at high frequencies the capacitor takes over.

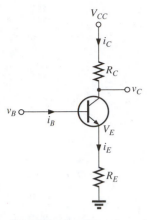

FIGURE 8.9 **Circuit used to calculate input and output impedances of common-emitter amplifier.**

Input and Output Impedance

Consider the circuit shown in Figure 8.9. To calculate the input impedance into the base of the transistor we write

$$R_{in} = v_B/i_B \qquad (8\text{-}9)$$

Neglecting the 0.6 V drop across the junction we can write this as

$$R_{in} = v_E/i_B$$

or

$$R_{in} = i_E R_E/i_B. \qquad (8\text{-}10)$$

For large values of β we can write

$$R_{in} = \beta R_E \qquad (8\text{-}11)$$

which is the input impedance of the circuit pictured.

The input impedance of the common-emitter amplifier shown in Figure 8.8 (neglecting the capacitor), is R_1 and R_2 in parallel with βR_E, or

$$Z_{in} = (1/R_1 + 1/R_2 + 1/(\beta R_E))^{-1}. \qquad (8\text{-}12)$$

The capacitor passes high frequencies and blocks DC and low frequency signals. The choice of the capacitance value depends on the expected frequency range of the input signal (recall the material of Chapter 3).

The output impedance is simply R_C. As with the input capacitor, inclusion of the output capacitor into the analysis yields a high-pass filter. The output capacitor should be chosen accordingly.

Biasing the Base

Two rules should be followed when choosing R_1 and R_2, the resistors that make up the voltage divider providing the bias voltage at the base. The first is that the current flowing through R_1 and R_2 should be significantly greater than that flowing into the base, i_B. To accomplish this, the resistor network should have a significantly lower impedance to the signal than to the transistor, or the parallel combination of R_1 and R_2 should be significantly less than the transistor current gain β times R_E. In other words

$$\beta R_E >> (1/R_1 + 1/R_2)^{-1}. \qquad (8\text{-}13)$$

The second rule states that the voltage swing, $V_C \pm \Delta v_C$, must be within the DC power supply limits, where V_C is the collector voltage when no input signal is applied. It is called the *quiescent collector voltage*. For maximum swing, R_1 and R_2 are chosen such that the quiescent collector voltage is 1/2 the supply voltage V_{CC}. A common rule is to choose R_1 to be 10 to 20 times as great as R_E.

EXAMPLE 8.1

Design a common-emitter amplifier with an output impedance of 4.7 kΩ and a gain of -10 using a transistor with a $\beta = 200$ powered by a 12 V DC power supply.

Solution: For the maximum swing in output voltage, we must choose a quiescent voltage of $1/2\ V_{CC}$, or in our case 6.0 V. Since the output impedance is to be 4.7 kΩ we choose R_C to be this value. Then the collector current is given by

$$i_C = (12 - V_C)/R_C = 6.0\ \text{V}/4.7\ \text{k}\Omega = 1.28\ \text{mA}$$

which must be checked against the maximum allowable collector current for the transistor being used. However, nearly every transistor manufactured can handle this small a current! Next, given a gain of -10, we can choose R_E using Equation 8-8. We find

$$\text{gain} = -10 = R_C/R_E \text{ or } R_E = 470\ \Omega.$$

Now to determine the base voltage, V_B, we first calculate V_E

$$V_E = i_E R_E \approx i_C R_E = (1.28\ \text{mA})(470\ \Omega) = 0.60\ \text{V}$$

and to determine V_B we add the voltage drop across the junction which is another 0.6 V for a silicon transistor so

$$V_B = V_E + 0.6\ \text{V} = 0.60\ \text{V} + 0.6\ \text{V} = 1.2\ \text{V}.$$

Choosing $R_2 = 10\ R_E$ we have $R_2 = 4.7\ \text{k}\Omega$ and must now determine the value of R_1. The voltage drop across R_1 is due to the current through R_1 which is made up of the current through R_2 and the current into the base i_B which are calculated as

$$i_2 = V_B/R_2 = 1.2\ \text{V}/4.7\ \text{k}\Omega = 0.26\ \text{mA}$$

and

$$i_B = i_C/\beta = 1.28\ \text{mA}/200 \approx 0.01\ \text{mA}$$

so that

$$i_1 = i_2 + i_B = 0.27\ \text{mA}.$$

The voltage drop across R_1 must be $V_{CC} - V_B$ which implies that R_1 should be

$$R_1 = (V_{CC} - V_B)/i_1 = (12\ \text{V} - 1.2\ \text{V})/0.27\ \text{mA} = 40\ \text{k}\Omega$$

but can be chosen to be the closest standard resistor size, 39 kΩ, which satisfies the condition imposed by Equation 8-13. ∎

Designing a transistor amplifier is somewhat more of an art than a science. Always keep in mind the limits of the transistor used and the fact that many of the parameters are not very accurately rated. The transistor gain, β, is a function of so many other quantities such as current demand, operating voltages and temperature of the transistor, that manufacturers give ranges over which the value of β may lie. In fact two identical transistors in identical circuits will rarely have the same value of β because of inconsistencies in the manufacturing process. A circuit that depends heavily on this quantity being a certain value is doomed to fail.

8-4 COMMON COLLECTOR–EMITTER FOLLOWER

Quite often, a low-output impedance is desired in the terminating stage of an amplifier to drive some circuitry such as a meter. On the other hand, a high-input impedance into the driver will ensure that the signal source is not distorted by the power drain of the amplifier. These rather common requirements can be met by the use of the common collector transistor configuration. Another name for the common collector is the *emitter follower*.

Consider the circuit shown in Figure 8.10. Note that the output voltage at the emitter, V_E, is equal to the input voltage at the base, V_B, less the 0.6 V junction drop, or

$$V_E \approx V_B - 0.6 \text{ V}. \tag{8-14}$$

At first glance, you might ask why this circuit is useful. The answer is seen when the input and output impedances of the emitter follower are analyzed. In essence, the emitter follower is a current amplifier which does not modify the voltage of a signal.

Calculation of the input impedance is fairly straightforward. For ease of clarification, we present DC voltages and currents and neglect the 0.6 V drop across the base-emitter junction. Given

$$R_{\text{in}} = V_B / I_B \tag{8-15}$$

and

$$V_B = V_E \tag{8-16}$$

Dividing both sides of the equation by R, we write

$$V_B / R = V_E / R = I_E. \tag{8-17}$$

We also know from previous discussion that $I_E = I_B + I_C$ which can also be written as

$$I_E = \beta I_B + I_B = I_B(1 + \beta). \tag{8-18}$$

Combining Equations 8-17 and 8-18 we have

$$I_B(1 + \beta) = V_B / R \tag{8-19}$$

and rearranging using Equation 8-15 we find

$$R_{\text{in}} = V_B / I_B = R(1 + \beta) \approx R\beta. \tag{8-20}$$

Because β is assumed to be large, the input impedance can easily be in the megohm range.

The output impedance of the emitter follower is given by

$$R_{\text{out}} \approx R_{\text{source}} / \beta \tag{8-21}$$

where R_{source} is the impedance of the source driving the base of the transistor. The value of R_{out} is usually very small since β is chosen to be relatively large.

Note that the input signal in the circuit above was assumed to be somewhere between V_{CC} and 0.6 V. Any signal centered around zero volts would only appear at the emitter when the signal is greater than 0.6 V. Biasing, as previously described for the common emitter circuit, is often required for the emitter follower to bring the input signal within the limits implied.

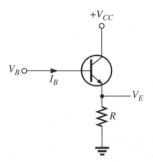

FIGURE 8.10 **Emitter follower circuit.**

8-5 THE DARLINGTON CONFIGURATION

The Darlington configuration is a name given to a particular transistor pair in which the emitter of one transistor is coupled directly into the base of the second, as shown in Figure 8.11. Note that the emitter follower configuration is used, in which the input impedance of the second transistor acts as the emitter resistor of the first transistor. All of the emitter current of Q_1 flows into the base of Q_2. The current gain of the two stages is essentially the product of the gains of the individual stages,

$$\beta(\text{overall}) = \beta_1\beta_2 \qquad (8\text{-}22)$$

When a single stage gain is insufficient, the use of the Darlington configuration offers a satisfactory solution. To a good approximation, the input resistance is

$$R_{\text{in}} = (\beta)^2 R_E. \qquad (8\text{-}23)$$

This means that an input resistance of greater than a megohm is common. The output resistance is approximately given by

$$R_{\text{out}} = \frac{R_{\text{source}}}{\beta^2} \qquad (8\text{-}24)$$

Hence, the output resistance is quite small.

The Darlington configuration can be thought of as a single transistor stage having the relationships developed above: high current gain and input impedance with low output impedance. Further, a minimum of biasing resistors are needed. Darlington transistors, or high β transistors, are single three-terminal devices which internally use this type of connection.

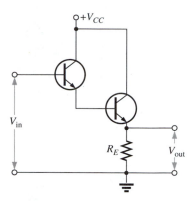

FIGURE 8.11 **Darlington configuration amplifier.**

8-6 THE TRANSISTOR SWITCH

Although transistor voltage amplifiers and bipolar transistor emitter followers are rapidly being replaced by simpler integrated circuits, the transistor still has a solid foundation in its use as a switch. Figure 8.12 shows a standard transistor switch. When the mechanical switch is open, no current flows into the base, and therefore no collector current flows. Since no current flows into the collector, no current flows through the load resistor, and the voltage across R_L is zero, or $V_{\text{out}} = V_{CC}$.

On the other hand, if the switch is closed, the current into the base will be given by

$$i_B = V_{CC}/R_B \qquad (8\text{-}25)$$

and the current flowing through the load resistor, R_L into the collector will be

$$i_C = \beta i_B = \beta V_{CC}/R_B. \qquad (8\text{-}26)$$

The voltage drop across R_L is thus given by

$$V_{RL} = i_C R_L = (\beta V_{CC}/R_B)R_L$$

which can also be stated as

$$V_{RL}/V_{CC} = \beta(R_L/R_B). \qquad (8\text{-}27)$$

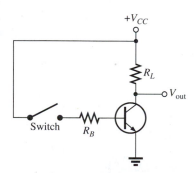

FIGURE 8.12 **Standard transistor switch. $V_{\text{out}} \cong V_{cc}$ when switch is open, $V_{\text{out}} \cong \text{OV}$ when switch is closed.**

Clearly, the voltage across the load resistor cannot exceed the power supply voltage! When this ratio is greater than 1, β changes accordingly and the transistor is considered to be in *saturation*. This is the equivalent of the transistor being a short circuit since V_C is essentially at zero volts (actually it is a fraction of a volt above ground).

In summary, when the switch is closed, providing current to the base, the transistor acts as a short circuit (this state is often referred to as the transistor being ON) and when the switch is open, no current flows and the transistor acts like an open circuit (referred to as the transistor being OFF). The resistor R_B is an absolutely necessary component in the circuit. Since $V_E \approx V_B - 0.6$ V, and given that V_E is at ground, V_B must be about 0.6 V. Hence, R_B serves to drop the difference between the power supply voltage and the base voltage, $V_{CC} - 0.6$ V. Without R_B, or with an improperly selected value of this resistor (the specified transistor base current limit must not be exceeded), the transistor will certainly not last very long!

Transistor switches are very useful when driving loads from high impedance sources, such as microcomputers. The mechanical switch in Figure 8.12 and its connection to V_{CC} can easily be replaced with a voltage/current output from another device. Keep in mind that the device must be able to provide enough current to drive the transistor into saturation. The load resistor can certainly be replaced by any DC load, such as a lamp or a relay. It is important to note that when a transistor switch is controlling an inductive load, such as a coil-driven relay, it must be accompanied by a diode to protect the transistor against the inductive voltage spikes that may occur when the coil is turned off. The diode is oriented so that it is reverse biased in normal operation and can shunt the voltage spike to the power supply when the relay is turned off.

In summary, using the transistor switch allows a small current to drive a device which requires greater current to operate, or more generally, a small power to control a larger power; the voltage source which provides the base current does not necessarily have to equal V_{CC}, but they must share a common ground connection.

In the transistor switch discussed above, the voltage at the collector is V_{CC} when the switch is open, or no signal is present at the base, and essentially zero volts when the switch is closed. The signal is *inverted,* that is the output is high when the input is low and vice-versa. Often a *non-inverting* switch is required. Figure 8.13 shows a non-inverting transistor switch which is essentially an emitter follower circuit driven into saturation. The difference between this configuration and that of the previous circuit is that when the switch is closed, the voltage at the emitter is V_{CC}, and it is zero volts when the switch is open.

8-7 JUNCTION FIELD EFFECT TRANSISTORS

The bipolar junction transistor is a splendid device that revolutionized the electronics industry. One of its weaknesses is low input impedance, which is caused by the forward-biased base-emitter junction. In order to overcome this deficiency a common collector stage must be used before a common emitter stage, so that the latter supplies the gain while the former provides a high input impedance. Another way to obtain high input impedance is to reverse

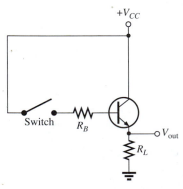

FIGURE 8.13 **Saturated emitter follower as a switch. $V_{\text{out}} \cong V_{cc}$ when switch is closed, $V_{\text{out}} \cong$ OV when switch is open.**

bias the input "diode" junction, as is the case for the junction field effect transistor (JFET or FET).

The *n*-channel FET is shown as a physical entity in Figure 8.14. Both portions of the gate (the shaded areas) are internally connected. The bar of *n*-type material acts as a simple resistance between the drain (*D*) and the source (*S*). The gate (*G*) is *p*-type material. This construction results in only majority carrier conduction, which accounts for the name "unipolar" sometimes given to the FET. The absence of minority carrier conduction and the fact that the device is operated with reverse bias between the gate and source yield a high input resistance. As the gate is reverse biased with respect to the source, depletion of carriers occurs in the channel. The degree of depletion depends on the magnitude of the reverse bias, as shown in Figure 8.15. For a given value of V_{GS} (voltage between the gate and the source), the drain current, i_D, increases linearly with increasing V_{DS} (voltage between the drain and the source) until the channel is depleted. At this point, further increases in V_{DS} essentially do not increase i_D. The characteristic curves (i_D vs. V_{DS} graphs) for a particular FET are given in Figure 8.16 for a family of V_{GS} values. The voltage at which further increases in V_{DS} do not produce increases in i_D is termed the pinch voltage, V_p. Clearly, V_p is a function of i_D; it is shown as a dashed line in Figure 8.16. The relationship between V_p and i_D is

$$i_D = i_{DSS}(1 - (V_{GS}/V_p))^2 \qquad \textbf{(8-28)}$$

where i_{DSS} is the drain-source current for $V_{GS} = 0$ V (shorted), a parameter which is specified in the manufacturer's literature.

The region to the left of the dashed line representing V_p is called the ohmic region, since the current-voltage behavior of the FET is linear, similar

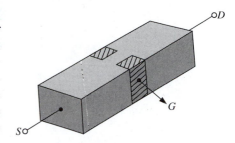

FIGURE 8.14   Physical representation of a junction field effect transistor (JFET).

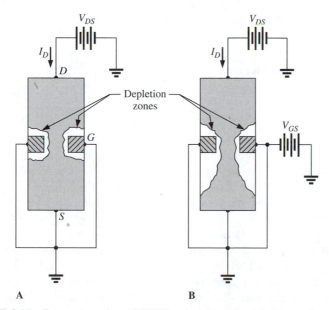

FIGURE 8.15   Representation of JFET operation (a) with $V_{GS} = 0$, and (b) $V_{GS} < 0$.

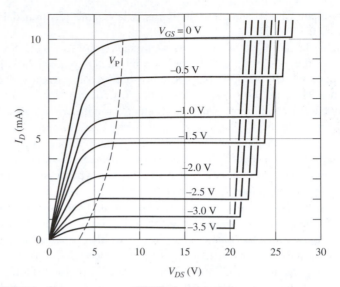

FIGURE 8.16   Common source JFET characteristic curves.

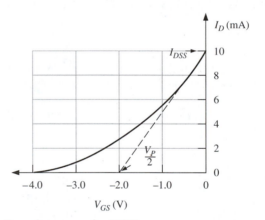

FIGURE 8.17   Transfer curve of a JFET.

to that of a resistor. There are occasions when this effect is useful, and it cannot be obtained with a bipolar transistor. One such use is that of a voltage-controlled resistor. In this region, the FET is just that.

To the right of the dashed line, the current-voltage graphs become flat. This region is called the active, constant-current, pinch-off or saturation region. Here the FET acts like a voltage-controlled current source whose value is controlled by the gate voltage, V_{GS}.

The transfer curve for an FET is a plot of i_D versus V_{GS} at a fixed value of V_{DS} (Figure 8.17). The transfer curve can be derived from the family of characteristic curves. The slope of the transfer curve is a direct measurement of the device's ability to amplify. Notice that the curve becomes non-linear for increasingly negative values of V_{GS}. If the linear portion of the curve (near $i_D = i_{DSS}$) is used for extrapolation, the intercept can be shown to have a value of $V_p/2$, as marked on the graph.

The standard electronic symbols for *n*-channel and *p*-channel FETs are given in Figure 8.18. The *p*-channel device is related to the *n*-channel in the same way that *pnp* and *npn* bipolar transistors are related. Note that the arrow is representative of the direction current that will flow if the junction is forward biased. When V_{GS} is greater than zero for an *n*-channel FET, the junction simply behaves as a diode.

8-8   THE FET AMPLIFIER

The most frequently encountered configuration for an FET is the common-source circuit; that is, the source is common to both the input and the output. The usual circuit is shown in Figure 8.19. One can obtain the correct value of V_{GS} in a number of ways. One way is that the manufacturer's specifications may state the value of V_{GS} used to assess the FET's performance. Another way is by using Equation 8-28 since all other parameters in that equation will be available.

Because of the construction of the FET, the gate current is extremely small. This fact leads to two useful conclusions: first, the current at the FET's source is essentially the same as the drain current, i_D; and second, $V_S = -V_{GS}$. Saying that the gate is negative (reverse biased) with respect to the source is exactly the same as saying that the source is positive with respect to the gate; it is a question only of which one is taken as the reference potential. This approach, in which the drain current passing through R_S maintains the bias voltage according to

$$V_S = i_D R_S = -V_{GS} \qquad \text{(8-29)}$$

is called self-biasing. It is useful in the analysis of the FET because the gate current is so small; but it would fail completely with a bipolar transistor, where the base current is appreciable (and the polarity is wrong). In designing the FET circuit, Equation 8-29 is used with the known values of i_D and V_{GS} to find R_S.

Now the value of R_D can be calculated, using

$$R_D = (V_{DD} - V_{DS})/i_D - R_S. \qquad \text{(8-30)}$$

Generally, the value of R_G is not critical. The only criterion is

$$i_G R_G << V_S = i_D R_S. \qquad \text{(8-31)}$$

Since i_G for the reverse-biased junction is typically measured in nanoamperes, a value of R_G in the range of 0.1 to 10 MΩ is appropriate. Note that this value is effectively the input resistance of the amplifier.

The function of C_S is to provide a low impedance path to ground for the high-frequency components of V_{DS}, so that AC signals will not cause swings in the bias voltage. This is similar to the bypass capacitor across R_E in the common emitter amplifier. Typically, the impedance of the capacitor is chosen to be large with respect to that of the resistor. If we choose $Z_{CS} = 10 R_S$, the resulting equation for choosing the value of C_S is given by

$$C_S = 10/(2\pi f R_S) \qquad \text{(8-32)}$$

where *f* is the minimum frequency at which the amplifier is expected to operate.

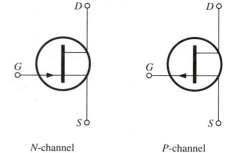

N-channel   *P*-channel
   JFET   JFET

FIGURE 8.18   Electronic symbols for JFETs.

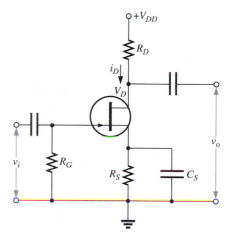

FIGURE 8.19   A typical common-source JFET amplifier.

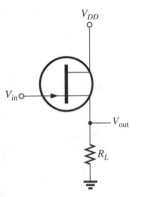

FIGURE 8.20   FET follower circuit.

The transconductance, g_m, of the FET is defined as the ratio of the change in output current to the change in input voltage (over a small range). It is similar to the current gain, β, for the bipolar transistor, but since the FET is a voltage controlled device, not a current controlled device, reference to the control voltage is made. In terms previously defined, g_m is written as

$$g_m = \Delta i_D/\Delta V_{GS} \tag{8-33}$$

and the voltage gain, or amplification, can be found as

$$A = \Delta V_D/\Delta V_{GS} = R_D g_m. \tag{8-34}$$

Unfortunately the transconductance, g_m, is a function of the drain current and will thus vary for different values of i_D giving a non-linear gain over a relatively large range.

8-9   FET FOLLOWER

Since the FET has an enormously high input impedance, it is a natural component to be used as a high-impedance voltage follower. The circuit shown in Figure 8.20 accomplishes this task. Working backwards, we first write the change in V_{out} in terms of a change in the drain current, i_D.

$$\Delta V_{out} = R_L \Delta i_D \tag{8-35}$$

Rewriting Equation 8-31 gives us

$$\Delta i_D = g_m \Delta V_{GS} = g_m \Delta(V_{in} - V_{out}). \tag{8-36}$$

Solving for ΔV_{out} gives

$$\Delta V_{out} = \Delta V_{in}[R_L g_m/(1 + R_L g_m)]. \tag{8-37}$$

If the product $R_L g_m$ is chosen to be much greater than 1, then Equation 8-37 reduces to

$$\Delta V_{out} = \Delta V_{in} \tag{8-38}$$

which states that the circuit shown in Figure 8.20 is a voltage follower. Many uses can be found for this circuit, particularly those that require little current drain from a device whose voltage is being measured, such as the voltage of a small, charged capacitor, or of a capacitive element like a microphone.

8-10   THE INSULATED GATE FIELD EFFECT TRANSISTOR

The insulated gate field effect transistor or IGFET, usually referred to as the metal oxide semiconductor field effect transistor or MOSFET, is a logical outgrowth of the desire to increase the input impedance of an amplifier. Typical MOSFETs can make available input impedances in the range from 10^9 to 10^{14} Ω.

The operation of the MOSFET is best understood in terms of its structure. To achieve an insulated gate, a layer of silicon dioxide is coated over the basic FET, as shown in Figure 8.21. A protective layer of silicon nitride is

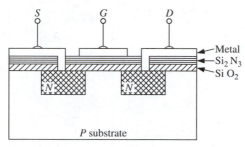

FIGURE 8.21   Physical representation of a MOSFET.

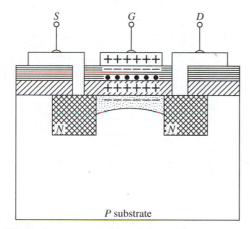

FIGURE 8.22   Physical representation of the enhancement mode MOSFET.

coated over that, producing an input impedance on the order of 10^{15} Ω. Holes are etched through the layers for the drain and source contacts to the *n*-type material.

There are several kinds of MOSFET, each of which operates either in the *enhancement mode* or in the *depletion mode,* or, in certain instances, in a combination of both. The enhancement mode is illustrated in Figure 8.22. A positive charge applied to the gate induces a negative charge in the channel between the source and drain, allowing the drain current to flow. As the positive charge on the gate increases, the current flow in the channel is enhanced. In other words, the channel resistance is controlled by the gate voltage. The transfer characteristic curve (I_D versus V_{GS}) is given in Figure 8.23, and the characteristic curves for the device (I_D versus V_{DS}) are shown in part B. The circuit symbols for the *n*-channel and *p*-channel MOSFETs are shown in Figure 8.24. Note that a substrate or active bulk (*B*) connection is available. It is usually connected externally to the source, which is the common (ground) point for the external circuit, but it supplies a degree of freedom in certain applications where the inner electrode capacitance may be a problem.

Operation of the MOSFET in the depletion mode begins with a different structure, as shown in Figure 8.25. The drain and source regions are heavily doped material (designated as N_+), while the channel contains diffused *n*-

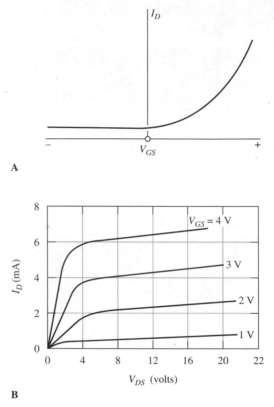

A

B

FIGURE 8.23 **Characteristic curves and transfer curve for the enhancement mode MOSFET.**

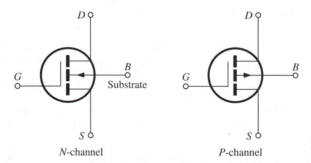

N-channel   *P*-channel

FIGURE 8.24 **Electronic symbols for enhancement mode MOSFETs.**

type material. All of this is on a *p*-type substrate. When the gate is made negative, the induced positive charge (carrier depletion) in the channel impedes the source-drain current flow. Increasing the negative charge at the gate decreases the drain current. The transfer curve and characteristic curves are given in Figure 8.26. Note that the transfer curve shows that this type of MOSFET design permits both positive and negative gate signals. The mechanism for enhancement is the same as that in the MOSFET designed only for enhancement mode operation. The decisive difference between the two designs is the allowable current flow at zero gate voltage. When virtually zero

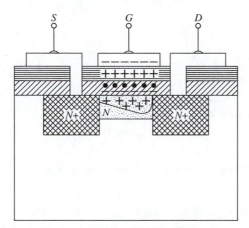

FIGURE 8.25   **Physical representation of a depletion mode MOSFET.**

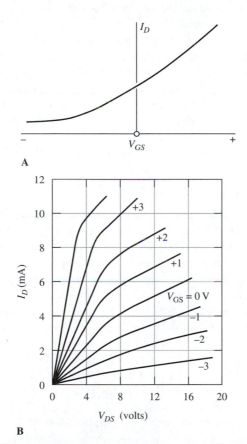

FIGURE 8.26   **Characteristic curves and transfer curve for the depletion mode MOSFET.**

current at zero gate voltage is required, the enhancement mode device must be used. Otherwise, the combined enhancement and depletion mode of operation offers a wider variation in acceptable gate signals. The circuit symbols for the depletion mode devices are given in Figure 8.27.

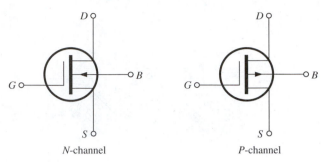

N-channel P-channel

FIGURE 8.27 **Electronic symbols for depletion mode MOSFETs.**

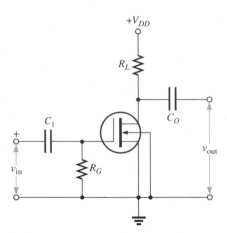

FIGURE 8.28 **A MOSFET amplifier circuit.**

A typical example of a MOSFET amplifier is presented in Figure 8.28. MOSFETs are used as amplifiers in precisely the same manner as are JFETs. The depletion mode device requires that the gate be appropriately biased. The enhancement-mode device may be operated at zero gate voltage without clipping an AC input signal.

Because of the extremely high resistance of the MOSFET gate, one must exercise great care in using these devices. The normal handling and soldering techniques might place a charge on the gate which would lead to puncture of the gate. The ordinary electrostatic charge associated with a dry air environment will build up on the gate unless precautions are taken, such as using jumper leads while soldering with a hot, but unplugged, iron. This extreme condition is not usually encountered in other devices because lower resistances permit enough leakage to release any electrostatic build-up.

8-11   MOSFET ANALOG SWITCHES

FETs, or more precisely, JFETs, operate only in the depletion mode; that is, the gate voltage must be made negative for current to flow through the drain and source. MOSFETs can operate in either depletion mode or in enhancement mode. Enhancement mode MOSFETs are by far the most common type of this transistor family available. In the n-channel, enhancement-mode MOSFET, the device is conducting when the gate voltage (with respect to the voltage of the source) is positive and non-conducting when the gate voltage is negative or zero volts. Power MOSFETs of this variety are available with ON resistances (the resistance between the drain and source when the device is conducting) of a few tenths of an ohm and OFF resistance in the range of 10^3 MΩ. Units are available that can handle hundreds of volts and pass hundreds of amperes of current. This behavior makes them quite useful as switches, both digital and analog.

Figure 8.29 shows an example of a digital MOSFET switch. In this application you might ask why the transistor was used; why not simply connect the switch in series with the load resistor? This type of application is common when small currents are used to control larger current-consuming devices. The switch might be a small relay capable of handling only a few milliamperes and the load may be a 100 W heater element. Figure 8.30 illustrates a MOSFET analog switch. With the switch set to V_{DD} we can write

$$V_S = V_{in} = V_{out} \tag{8-39}$$

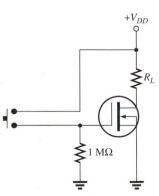

FIGURE 8.29 **MOSFET switch: ON when switch is closed, OFF when switch is open.**

and in our case

$$V_G = +V_{DD}, \qquad \text{(8-40)}$$

then the gate source control voltage is given by

$$V_{GS} = V_{DD} - V_{in}. \qquad \text{(8-41)}$$

Attention must be given that V_{GS} does not become negative ($V_{in} > V_{DD}$) when the transistor should be conducting or that it become positive if the transistor should be non-conducting; otherwise the MOSFET will change states unexpectedly!

8-12   THE SCR AND TRIAC

The *Silicon Controlled Rectifier* (SCR) and *Triode AC Switch* (TRIAC) are two power-switching devices which often prove useful in controlling AC currents. The SCR controls currents in one direction only while the TRIAC controls currents in both directions.

SCR

Figure 8.31 illustrates a bipolar *npn* transistor connected to a *pnp* bipolar transistor, with labels G for gate, A for anode, and C for cathode. Consider the transistors as switches. When the switch is open, the *npn* transistor is in the non-conducting mode and no current flows between the anode and the cathode. When the switch is closed, current flows and continues even after the switch is open. This is the equivalent operation of an SCR. Figure 8.32 shows the structure of the SCR and its schematic representation.

Since the device remains in the conducting state even after the switch is closed, one might think that the device is not very useful, other than for a one-time latch (a latch is a device which holds its state after the controlling signal is removed). In fact, the device is often found in many control applications, which stands to reason since the SCR returns to the non-conducting (OFF) state once the source currents are removed. Such a device is commonly used in the control of AC loads such as lamps or motors and is even found in automotive engine circuitry. One such application, an AC motor control circuit, is shown in Figure 8.33. Neon lamps are devices that have a high input impedance (essentially an open circuit) for voltages less than the

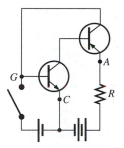

FIGURE 8.30   MOSFET analog switch.

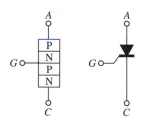

FIGURE 8.31   *npn-pnp* latch.

FIGURE 8.32   *p-n* junction representation and schematic symbol for SCR.

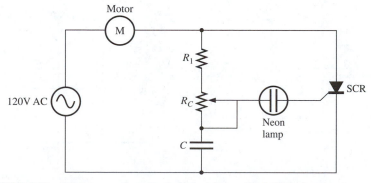

FIGURE 8.33   Motor speed control using AC rectified SCR circuit.

light's *discharge voltage*. For voltages higher than the discharge voltage, the neon lamp lights and conducts allowing current to flow through it. The capacitor charges until the discharge voltage of the neon lamp is reached whereupon breakdown in the lamp occurs and current is dumped from the capacitor into the SCR's gate. The SCR conducts until the AC supply goes negative and the cycle repeats itself once the supply goes positive. Note that the charging rate, and thus the time for which the motor is powered, is controlled by $R_1 + R_C$.

TRIAC

The use of the SCR is limited because it can pass current only in one direction. If back-to-back SCRs are connected as in Figure 8.34A this problem is alleviated. Figure 8.34B shows the schematic representation of the TRIAC. The terminals are labeled MT1 and MT2 since the anode-cathode scheme is no longer sensible. TRIACs are available in a single package and can be switched with either positive or negative currents. Specifications list the various gate voltages and currents which trigger these devices. Typically the values are a few milliamperes at a couple of volts, and can control currents in the tens of amperes. A typical application of a TRIAC in a lamp dimmer circuit is shown in Figure 8.35. Triggering the gate voltage can be accomplished by a number of techniques such as an RC, neon lamp circuit.

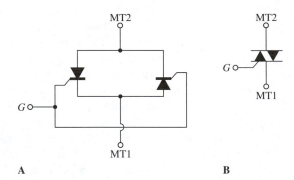

A   B

FIGURE 8.34   **Dual SCR circuit and TRIAC symbol.**

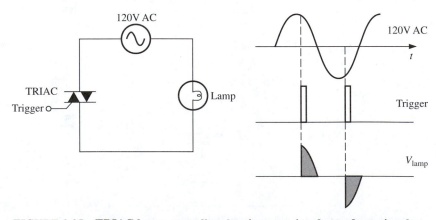

FIGURE 8.35   **TRIAC lamp controller showing associated waveform signals.**

8-13 FEEDBACK

Feedback has been occurring for a long time; nature employs it with great success. The field of electronics is far better because of feedback. Most circuit operations improve dramatically when it is incorporated. In order to employ the ideas of feedback successfully, the general principles must be developed and understood.

The inherent voltage gain, A_v, of an amplifier, as represented in Figure 8.36, is the ratio of the output, v_0, to the input, v_s. If the circuit is modified, as in Figure 8.37, so that a fraction, β, of the output is fed back into the input, then the "new" output, v_0', is

$$v_0' = v_s A_v. \tag{8-42}$$

Since $v_s = v_i + \beta v_0'$, we have

$$v_0' = A_v(v_i + \beta v_0'). \tag{8-43}$$

Upon rearrangement, we can define A_v', the gain of the amplifier with feedback:

$$A_v' = \frac{v_0'}{v_i} = \frac{A_v}{1 - \beta A_v} \tag{8-44}$$

By substituting values into this equation, one can see that A_v' goes to infinity as βA_v approaches unity. This situation is designated *positive feedback,* since β is positive. In actual practice, the infinite gain of the positive feedback amplifier results in oscillation. This is the basis of a family of oscillators which is self-sustained and draws power from a supply in an oscillating manner. The criteria for oscillation, called the Barkhausen criteria, are that βA_v be positive and equal to unity.

If, on the other hand, β is made negative, the phenomenon known as *negative feedback* occurs. Notice that A_v' is always less than A_v under this condition, because βA_v is less than zero and the denominator is greater than unity. As A_v is made large, A_v' approaches $-1/\beta$ as a limit and is independent

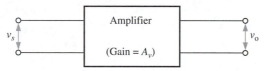

FIGURE 8.36 Schematic of an amplifier with a gain of A_r.

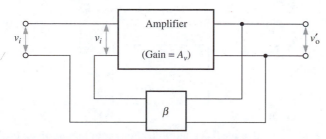

FIGURE 8.37 Schematic of an amplifier with feedback.

of A_v. Thus, any variation in A_v that may occur because of temperature, aging effects and the like will not affect $A_v{'}$.

Remember that β is a fraction. A simple voltage divider suffices to fix the value of β if the divider is not loaded. Consider an amplifier with $A_v = 35,000$. Now, say that β is fixed at -0.010 in a feedback loop. The amplifier's gain with feedback is 100. The amplifier's components may age or the operating temperature may change, but the gain of the amplifier will stay at 100. In other words, a 10.0 mV input to that amplifier will always produce a 1.00 V output, regardless of the temperature or state of aging of the amplifier.

Negative Feedback

The properties of an amplifier with negative feedback can be significantly better than those of the equivalent amplifier without feedback. The effect of negative feedback, to reduce, but not eliminate, variations in gain with component variation, is just one of the advantages of this approach. Consider, for example, the problem of distortion.

Assume a distortion-free signal at the input to an amplifier without negative feedback, as in Figure 8.38. The distortion, D, appears at the output. If we now incorporate feedback, as in part B of the figure, then a fraction of the output equal to $\beta(v_0 + D)$ is fed back into the input, and the "new" distortion at the output is

$$D' = D + Av\beta D' \tag{8-45}$$

or

$$D' = \frac{D}{1 - \beta A_v} \tag{8-46}$$

When the gain of the amplifier is large, the distortion is greatly reduced.

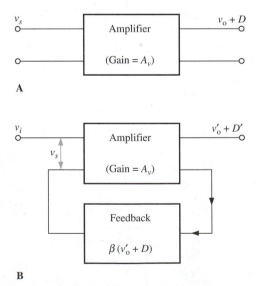

A

B

FIGURE 8.38 **Negative feedback amplifier for reduction of distortion (and noise).**

Although the issue was not addressed previously, you should recognize that an amplifier can (and always will) add noise to the amplified signal. The methods of reducing noise are discussed fully in Chapter 15. At this point, however, we simply assume that a noisy output is extant. Thus the unfeedback output has a noise component, which is equivalent in every way to the distortion just discussed. Therefore, we can write by analogy

$$N' = \frac{N}{1 - \beta A_v} \qquad (8\text{-}47)$$

where N' is the noise under negative feedback conditions.

We should also consider the effects of feedback on the input and output impedances of the amplifier. The input impedance without feedback is

$$Z_{\text{in}} = \frac{v_s}{i_{\text{in}}} \qquad (8\text{-}48)$$

and

$$v_s = v_i + \beta v_0. \qquad (8\text{-}49)$$

This may be written as

$$v_i = v_s - \beta A_v v_s. \qquad (8\text{-}50)$$

The input impedance with feedback, Z_{in}', is then

$$Z_{\text{in}}' = \frac{v_i}{i_{\text{in}}} = Z_{\text{in}} (1 - \beta A_v). \qquad (8\text{-}51)$$

Thus, since βA_v is less than zero, the input impedance is increased. In a similar manner, we can show that the output impedance is reduced from the value without feedback, Z_0, to

$$Z_0' = \frac{Z_0}{1 - \beta A_v}. \qquad (8\text{-}52)$$

8-14   SOME EXAMPLE CIRCUITS

Below are some circuits which can be constructed using bipolar transistors.

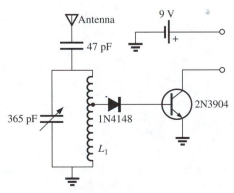

A   L1: Antenna Coil
        Antenna: Long (> 3m) wire

FIGURE 8.39a   **AM radio.**

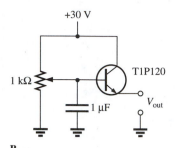

B

FIGURE 8.39b   **0-25V adjustable power supply.**

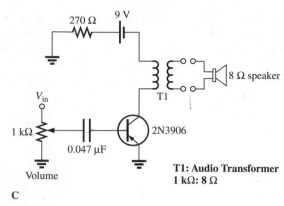

C

FIGURE 8.39c   **Audio amplifier.**

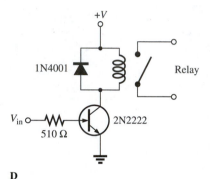

D

FIGURE 8.39d   **Relay driver. When V_{in} is high, relay is energized. Diode protects transistor from back EMF when coil is switched off.**

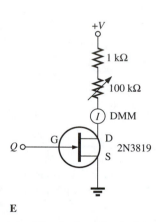

E

FIGURE 8.39e   **Electrometer. Short wire attached to open gate connection is used to detect static charge.**

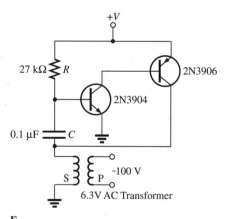

F

FIGURE 8.39f   **Oscillator/Hi-voltage supply. Frequency can be changed by changing the values of R and/or C. Transformer can be replaced with speaker for metronome.**

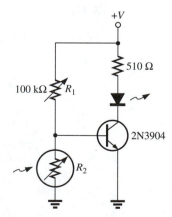

FIGURE 8.39g   **Light Activated Switch. R_2 is a photoresistor. When light falls on R_2, its resistance goes low, causing the voltage at the base to be low and the transistor to switch off. The transistor turns on when R_2 is in the dark. The light emitting diode and 510 Ω resistor can be replaced with a relay (and protective diode) to drive other loads.**

8-15   QUESTIONS AND PROBLEMS

1.   Typically, the absolute maximum voltage rating for the base-emitter voltage, V_{BE}, is the lowest of all values. Because of this low value, the base-emitter junction must be protected against too high a reverse voltage. A diode is used to protect it, as shown in Figure 8.A for an *npn* transistor. Explain its function and state the conditions under which protection is given. Draw the equivalent circuit for a *pnp* transistor.

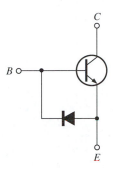

FIGURE 8.A   **Circuit for problem 1.**

2.   Design a *CE* amplifier with gain of 10 using a 2N3906 transistor and a 20 V power supply. Calculate the required resistor and capacitor values for the circuit, assuming a minimum frequency of 10 Hz.

3.   Evaluate the voltage, current and power gains for the circuit of problem 2, assuming 100 Ω for the signal resistance.

4.   Design a *CC* amplifier with gain of 10 using a 2N3906 transistor and a 15 V power supply. Calculate the required resistor and capacitor values for the circuit, assuming a minimum frequency of 10 Hz.

5.   Evaluate the voltage, current and power gains for the circuit of problem 4, assuming 100 Ω for the signal resistance.

6.   Compare the input and output impedances of the *CE* and *CC* circuits of problems 2 and 4.

7.   Design an amplifier using a JFET and a 20 V power supply. Calculate the required resistor and capacitor values for the circuit, assuming a minimum frequency of 10 Hz.

8.   Evaluate the voltage, current and power gains for the circuit of problem 7, assuming 100 Ω for the signal resistance.

9.   Calculate the input and output impedance of the FET circuit of problem 7 and compare them to those of the *CE* bipolar transistor circuit of problem 2.

10.   In selecting a power transistor, which particularly important features are found in the absolute maximum ratings of the spec sheet?

11.   Describe how the circuit shown in Figure 8.39a works.

12.   Since some potentiometers do not go to zero ohms, the circuit shown in Figure 8.39b sometimes fails to produce zero volts. Suggest an appropriate modification to the circuit that would eliminate this problem.

13.   The circuit shown in Figure 8.39c has an inherent distortion problem. Explain why the distortion problem exists and suggest a modification to the circuit to eliminate such distortion.

14.   Explain why the diode is necessary in parallel to the relay in the circuit shown in Figure 8.39d.

15.   Explain why the circuit shown in Figure 8.39d works.

16.   Design a buffer circuit to measure the voltage on a small capacitor using a bipolar transistor and a DVM. Design a similar circuit using an FET transistor. Which is better and why?

17.   Design an amplifier which has a gain of 100 using a 2N3904 transistor. What is its input impedance?

18.   Design an amplifier which has a gain of 100 using a JFET transistor. What is its input impedance?

19.   Design a circuit which will drive a 100 Ω relay with 40 mA using a 2N3904 and a 5 V signal capable of sourcing 10 mA.

9

Operational Amplifiers

9-1 INTRODUCTION

When you begin to design circuitry toward a particular end, you will often have the uncomfortable feeling that the circuit on paper will not function properly when transferred into practice. Unfortunately, these fears are sometimes justified, and it's "back to the drawing board." A few cycles of this operation can discourage all but the most persistent from attempting another design.

As the medicine-show pitchman heralded the magic elixir as the panacea for "whatever ails you," the operational amplifier is prescribed for the novice designer's ills. Just as the elixir wasn't good for every illness, the operational amplifier has its limitations; fortunately, however, these limitations are generally insignificant, particularly when compared with the device's advantages. The operational amplifier offers a direct and immediate entry into design with a high assurance of success.

The use of the operational amplifier evolved from the field of analog computation. These *amplifiers* were specifically designed to perform certain mathematical *operations* such as addition, subtraction, multiplication, differentiation and integration. They were designed, furthermore, to perform these operations with a high degree of accuracy and reliability. The result of this has been to make convenient, functional building blocks available to the instrument designer. The task before us is to use operational amplifiers effectively, being cognizant of the few simple "rules" of their use.

This chapter begins with an introduction to the ideal operational amplifier and its circuit properties. Then the practical operational amplifier will be described. At this point, the limitations of the real device will be clear. A number of useful circuits which employ operational amplifiers will be discussed so that some appreciation of the breadth of their application will be gained. Finally, IC operational amplifiers will be described. These are the latest additions to this field, and have contributed significantly by offering much value for very little expenditure of time and money. Indeed, they are "where the action is."

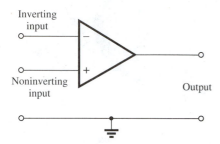

FIGURE 9.1   **Electronic symbol and lead designations for an operational amplifier.**

The Ideal Operational Amplifier

If you were given the task of designing an ideal or perfect amplifier, what properties would you seek? Referring to the earlier chapter on transistor amplifiers, we see that the gain, the input impedance and the output impedance are the most important parameters. The perfect amplifier should have infinite input impedance so that any signal could be supplied to it without loading problems. The output impedance should be zero so that the power supplied by the amplifier is not limited. Finally, for ease of operation we would want the gain to be infinite. These are precisely the features of the ideal operational amplifier. Thus for an ideal operational amplifier $Z_i = \infty$, $Z_o = 0$, and $A = \infty$. The usual circuit symbols for the operational amplifier are given in Figure 9.1; the *inverting input* notation signifies only that the phase angle of the output will be 180° different from that of the input applied at this terminal. Often the ground line is omitted from circuit diagrams for simplicity; *it must be understood to be present at all times,* however.

The key to operational amplifier use is negative feedback, which gives the operational amplifier the advantages and versatility suggested above. The DC voltage gain is between 10^4 and 10^9. Since the closed loop performance depends on the gain, the negative feedback configurations improve gain stability, reduce output impedance and improve linearity. With the very high gain of the operational amplifier, the circuit performance is usually determined *entirely by the feedback components* and is independent of the amplifier's internal characteristics. Another important feature of the operational amplifier is that it has very little DC offset (that is, when the input is zero, the output is also zero or very nearly so). When a feedback connection is made between the inverting input and the output (with the non-inverting input grounded), the input is maintained at *virtual ground* and the output is fed back to maintain this condition, as shown in Figure 9.2. In more general terms, we can think of this as both the positive and negative inputs being maintained at the same potential. The final essential feature of the operational amplifier is that very little current is drawn by the input. Saying that the input circuit is zero is the same as stating that the input impedance is infinite, since, for a finite voltage, the impedance must be infinite for a cur-

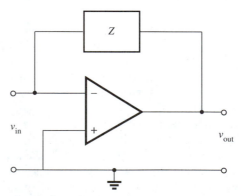

FIGURE 9.2   **Operational amplifier with a feedback loop between the output and inverting (−) input.**

rent of zero. This feature ensures that the signal inputs to the operational amplifier will not degrade (or load) the signal source.

Four rules that describe the behavior of an ideal operational amplifier simplify op-amp circuit design tremendously. These are:

1. The circuitry of an operational amplifier with a closed, negative-feedback loop will adjust its output in any way possible to make the inverting input (IN_-) and non-inverting input (IN_+) terminals of the device equal in voltage.
2. The inputs draw no current (the input impedance is infinite).
3. The gain, or voltage amplification, is infinite.
4. The output impedance is zero.

The operational amplifier is usually available as an integrated circuit in an 8-pin dual, in-line package (DIP) configuration, although other package types are available. Some operational amplifier integrated circuits feature two or more op-amps on one chip. All operational amplifiers have a non-inverting input, an inverting input, an output, and V_+ and V_- power supply connections. The op-amp is rarely connected to ground, except in such circumstances where the V_- connection is zero volts. The three other connections (on an 8-pin single operational amplifier IC) usually serve to nullify manufacturing imperfections, or have no connection, as we shall see later in the chapter.

Please note that in this chapter we have chosen to label voltage and current in capital letters which previously have represented DC signals. Here the analysis holds for both AC and DC signals—the choice is simply a matter of preference. The analysis below assumes the rules above to be true and is considered by the authors to be the most useful in terms of design and understanding. For the purist, a more formal analysis can be found in Appendix A.

9-2 VOLTAGE FOLLOWER

In chapter 8 we examined the emitter-follower circuit using a bipolar transistor. The operational amplifier provides the basis of a circuit which supersedes the emitter-follower in performance and simplicity of design. Consider the circuit in Figure 9.3. Since the op-amp operates to make the IN_+ and IN_- terminals equal (Rule 1 from above), the only way for it to accomplish this

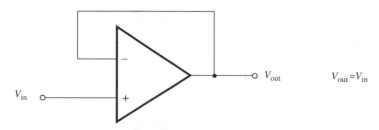

FIGURE 9.3   **Operational amplifier voltage follower.**

is to make V_{out}, which is connected to IN_-, equal to IN_+. For the circuit shown in Figure 9.3 the relationship between V_{out} and V_{in} is simply

$$V_{in} = V_{out} \qquad \textbf{(9-1)}$$

Since the ideal op-amp has infinite input impedance, it draws no current, and thus has no effect on the incoming signal V_{in}. Additionally the op-amp has a low output impedance, which is ideal for a voltage follower, or buffer.

9-3   INVERTING AMPLIFIER

Consider the circuit in Figure 9.4. At first glance it may be difficult to determine how V_{in} will be related to V_{out}. A simple analysis will shed some light on the situation. Following Rule 2 above, we can re-draw the circuit branch on the top of the diagram as shown in Figure 9.5. Since no current flows into the IN_- input, the two resistor circuit is unchanged by the connection to IN_-. However, the operational amplifier can affect V_{out}. We must also assume that the op-amp will behave in such a way that Rule 1 is obeyed—namely that it will work to make the voltage at point A be the same as the voltage at point B—zero volts. Following these rules we can solve for V_{out} in terms of R_1, R_2 and V_{in}. Since point A is at zero volts, but is not connected to ground, it is often referred to as a *virtual ground*. Using Kirchoff's law for voltage drops, we can write by inspection

$$V_{in} - IR_1 = 0 \qquad \textbf{(9-2)}$$

and

$$0 - IR_2 = V_{out}. \qquad \textbf{(9-3)}$$

Solving Equation 9-2 for I and substituting into Equation 9-3 yields

$$V_{out} = -V_{in} R_2/R_1$$

or

$$V_{out}/V_{in} = -R_2/R_1. \qquad \textbf{(9-4)}$$

The circuit shown in Figure 9.4 is an inverting amplifier; that is, the signal is amplified and switched in polarity. Notice that the design of the amplifier

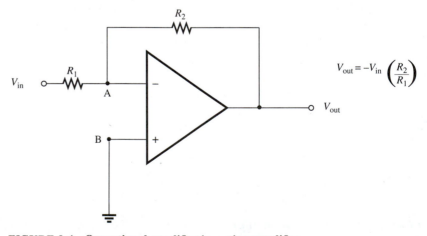

FIGURE 9.4   Operational amplifier inverting amplifier.

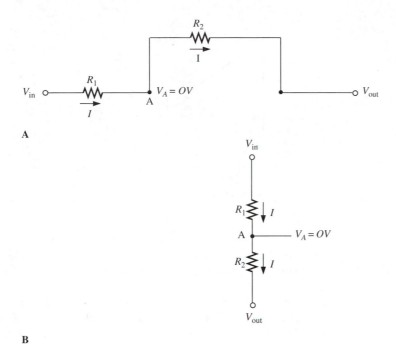

FIGURE 9.5   Feedback circuit for inverting amplifier. Note that *A* and *B* are identical.

consists of the choice of two resistors—certainly a much easier chore than designing a bipolar transistor amplifier. The minus sign implies that if the incoming voltage is positive, the voltage out will be negative, and vice-versa. An AC signal will thus have a 180° phase shift. The input impedance of the amplifier is simply given by

$$Z_{in} = R_1 \tag{9-5}$$

since point A is at virtual ground.

Notice that if $R_1 = R_2$ then the amplifier has unity gain (gain $= 1$) and inverts the incoming signal. This can often be useful if, for example, you have a negative voltage from a transducer that must be read by an instrument which only responds to positive voltages. Note also that R_2 may be replaced by a variable resistor which makes the inverting amplifier a variable-gain, inverting amplifier. It is a wiser choice to vary R_2 rather than R_1 since the input impedance, equal to R_1, should be fixed for a well designed amplifier. Making R_1 very small, which would be possible with a variable resistor, would make the input impedance very small, and would certainly disturb the incoming signal. Since R_1 could equal zero ohms with a variable resistor, the input impedance could equal zero and the theoretical gain would go to infinity.

9-4   NON-INVERTING AMPLIFIER

Now let's analyze the circuit pictured in Figure 9.6. As in the previous example, we separate the feedback circuit from the op-amp, which is possible since the *IN₋* input draws no current and thus has no effect on the circuit. Rule 1 mandates that the voltage at point A be equal to the voltage at point

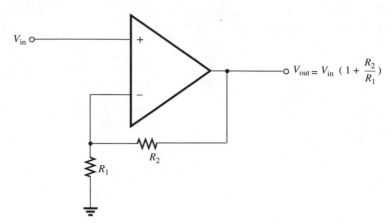

FIGURE 9.6 **Non-inverting operational amplifier circuit. Note that inverting and non-inverting inputs are switched from previous diagrams.**

B, which is equal to the input voltage, V_{in}. This is indicated in the feedback circuit shown in Figure 9.7. The analysis is nearly identical to that of the inverting amplifier. By inspection we write

$$V_{out} - V_{in} = IR_2 \qquad\qquad (9\text{-}6)$$

and

$$V_{in} - 0 = IR_1. \qquad\qquad (9\text{-}7)$$

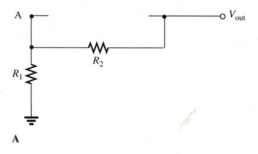

A

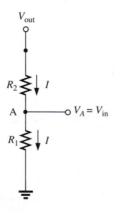

B

FIGURE 9.7 **Feedback circuit for non-inverting amplifier. Note that A and B are identical.**

Solving for I and substituting the result into Equation 9-6 yields

$$V_{out} = V_{in}(1 + R_2/R_1)$$

or

$$V_{out} / V_{in} = (1 + R_2/R_1). \tag{9-8}$$

Thus the circuit under analysis is a non-inverting amplifier. Again, its amplification is chosen by two external resistors, R_2 and R_1. Here, however, the minimum amplification is unity. If R_2 equals zero ohms, or a short circuit, the result is an output voltage equal to the input voltage. A variable-gain amplifier can be made by using a variable resistor for R_2. The input impedance in this circuit is infinite in the case of an ideal operational amplifier.

9-5   DIFFERENCE AMPLIFIERS

The circuits above are very useful in modern electronic design; however, instrumentation circuits often call for reading the voltage difference between two points, neither of which is ground. Differential operation can also be obtained with an ideal operational amplifier. Such a circuit is shown in Figure 9.8. Using Kirchoff's law and separating the top branch from the op-amp as we did in the previous analysis we can refer to Figure 9.9 and write

$$I_1 = (V_1 - V_A) / R_1 \tag{9-9}$$

and

$$I_1 = (V_A - V_{out}) / R_f. \tag{9-10}$$

Combining these two equations and solving for V_A yields

$$V_A = (V_1 R_f + V_{out} R_1) / (R_1 + R_f). \tag{9-11}$$

For the lower branch of the circuit, shown in Figure 9.9 a similar analysis yields

$$V_B = V_2 R_3 / (R_2 + R_3). \tag{9-12}$$

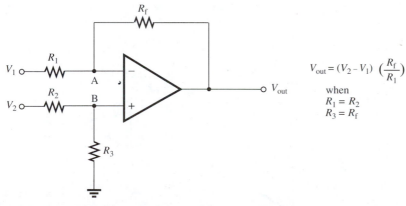

$$V_{out} = (V_2 - V_1)\left(\frac{R_f}{R_1}\right)$$

when
$R_1 = R_2$
$R_3 = R_f$

FIGURE 9.8   **Operational amplifier difference amplifier.**

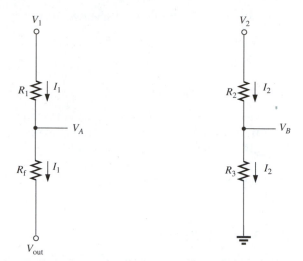

FIGURE 9.9 **Feedback circuits of difference amplifier.**

Since Rule 1 requires that $V_A = V_B$ we can write

$$(V_1 R_f + V_{out} R_1) / (R_1 + R_f) = V_2 R_3 / (R_2 + R_3) \tag{9-13}$$

which when solved for V_{out} gives us

$$V_{out} = V_2 (R_3 / R_1) [(R_1 + R_f) / (R_2 + R_3)] - V_1 R_f / R_1. \tag{9-14}$$

This equation may seem unwieldy at first, but if $R_1 = R_2$ and $R_3 = R_f$, then

$$V_{out} = (V_2 - V_1) R_f / R_1 \tag{9-15}$$

which is quite straightforward. It is very important to keep in mind that the resistor values must be precisely matched for Equation 9-15 to hold true.

The difference amplifier suffers from a number of problems, one being its relatively low input impedance. An improved three-op-amp model is shown in Figure 9.10. This configuration of op-amps is the classical configuration for an instrumentation amplifier. The analysis of the gain for the instrumentation amplifier is somewhat cumbersome, but worth the effort. We begin by realizing the circuit consists of two voltage followers that provide a high input impedance. The outputs of the voltage followers are fed into a difference amplifier identical to the one just discussed. Note that the voltage

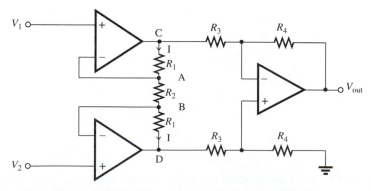

FIGURE 9.10 **Instrumentation amplifier circuit using op-amps.**

at points A and B of the circuit must equal the input voltages V_1 and V_2 in accordance with Rule 1. The current, I, flowing through the three series resistors is given by

$$I = (V_A - V_B) / R_2 = (V_1 - V_2) / R_2 \qquad \textbf{(9-16)}$$

where the voltages at points A and B are already defined. The voltage at point A is equal to the voltage at point C less the voltage drop IR_1. Similarly, the voltage at point B is the voltage at point B plus the voltage drop IR_1. The voltages into the difference amplifier section of the circuit, V_C and V_D, can therefore be written as

$$V_C = V_1 + IR_1 \qquad \textbf{(9-17)}$$

and

$$V_D = V_2 - IR_1. \qquad \textbf{(9-18)}$$

From Equation 9-15, the voltage out of the difference amplifier can be written, after substituting V_C and V_D for V_1 and V_2 of Equation 9-15 and the appropriate resistor values, as

$$V_{out} = (V_D - V_C) (R_4 / R_3) \qquad \textbf{(9-19)}$$

Substituting Equations 9-16, 9-17 and 9-18 into Equation 9-19 gives us

$$V_{out} = R_4 / R_3 [V_2 - V_1 - 2 (V_1 - V_2) R_1 / R_2]. \qquad \textbf{(9-20)}$$

Choosing $R_2 = 2 R_1$ we can rewrite this as

$$V_{out} = (2 R_4 / R_3) (V_2 - V_1). \qquad \textbf{(9-21)}$$

This states that the amplifier produces a voltage with respect to ground which is proportional to the difference between the two input voltages. The gain is set by the ratio of R_4 and R_3.

Here again the value of R_2 must be very accurately equal to $2 R_1$ for this simple relationship to hold. Other performance parameters degrade as well when this relationship is not held, even for an ideal op-amp. Compounded by the problems of realistic components, this circuit, which must meet stringent demands if it is to be used for data collection, is not very simple to design. Fortunately, instrumentation amplifiers with closely matched feedback resistors and excellent performance parameters are available commercially in one integrated circuit package; there is little need to construct one from scratch using precision resistors and three op-amps. Many units feature a gain which can be chosen simply by connecting a single resistor, or by connecting different pins of the IC to one another. The internal-component block diagram for the instrumentation amplifier AD524 manufactured by Analog Devices is shown in Figure 9.11, along with a typical configuration.

9-6   MATHEMATICAL FUNCTIONS: SUMMATION, DIFFERENTIATION, INTEGRATION

Summation

As was mentioned in the introduction to this chapter, a number of mathematical operations can be performed with operational amplifiers. One of the most straightforward is addition. The circuit in Figure 9.12 shows such a

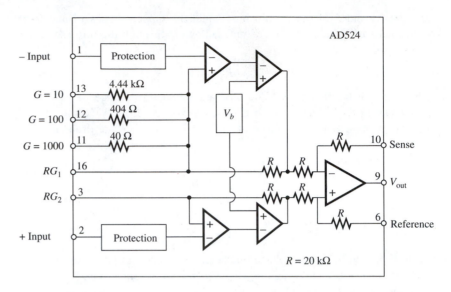

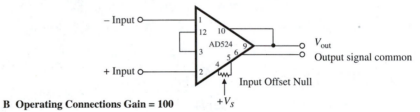

A Block Diagram

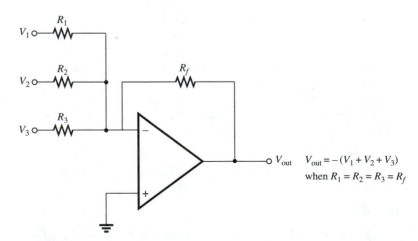

B Operating Connections Gain = 100

FIGURE 9.11 **Single chip instrumentation amplifier. (A) Block diagram showing some pin connections. (B) Typical operating connections for Gain = 100. Power supply connections $+V_s$ and $-V_s$ not shown.**
Figure courtesy of Analog Devices, Inc.

$V_{out} = -(V_1 + V_2 + V_3)$
when $R_1 = R_2 = R_3 = R_f$

FIGURE 9.12 **Voltage adder circuit.**

circuit. Analysis of the circuit following the methods of previous examples (the summing-point voltage at IN_- is equal to the voltage at IN_+, which is ground) yields

$$V_{out} = -(V_1 / R_1 + V_2 / R_2 + V_3 / R_3) R_f. \qquad \textbf{(9-22)}$$

Quite naturally, this is called a *voltage summer*. When $R_1 = R_2 = R_3 = R_f$, we obtain

$$V_{out} = -(V_1 + V_2 + V_3). \qquad \textbf{(9-23)}$$

Note that the input voltages can be either positive or negative or even AC voltages, and that any number of voltages can be summed. This example used three input voltages.

Integration and Differentiation

Other mathematical operations can be performed by using non-resistive components. Consider the inverting amplifier configurations shown in Figure 9.13. The configuration in Figure 9.13a, with a capacitor C in the feedback element, is called an *integrator*. By now we can write the feedback equations (using Kirchoff's equations) by inspection,

$$I = V_{in} / R \qquad \textbf{(9-24)}$$

$$I = -C \, dV_{out} / dt \qquad \textbf{(9-25)}$$

or

$$V_{in} / R = -C \, dV_{out} / dt \qquad \textbf{(9-26)}$$

or

$$V_{out} = -(1 / RC) \int V_{in} \, dt. \qquad \textbf{(9-27)}$$

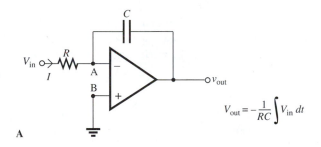

A

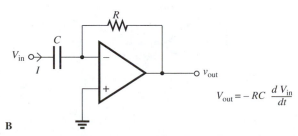

B

FIGURE 9.13 **(A) Integrator op-amp circuit. (B) Differentiator op-amp circuit.**

The output of the circuit is thus the integral of the input, multiplied by $-1/RC$. Interchanging R and C in the circuit, as in Figure 9.13b, yields a *differentiator*. For this circuit,

$$I = C\, dV_{in}/dt \tag{9-28}$$

and

$$I = -V_{out}/R. \tag{9-29}$$

Therefore,

$$V_{out} = -RC\, dV_{in}/dt. \tag{9-30}$$

Now the output is the derivative of the input, multiplied by $-RC$.

9-7   CURRENT AMPLIFIERS

Current to Voltage Conversion

We have seen that many transducers output a current, whereas all the amplifier circuits we have discussed amplify voltage. Oftentimes very small currents must also be measured, a task which is not easily done. The operational amplifier, used as a current-to-voltage converter as shown in Figure 9.14, is ideal for many of these applications. The reference voltage, V_R, is usually held at ground, but need not be. The output voltage is given by

$$V_{out} = V_R - I_{in}R_f. \tag{9-31}$$

If we set $V_R = 0$, this circuit is quite similar to the inverting amplifier of Figure 9.4 with $R_1 = 0$. But the inverting amplifier has gain $= -(R_2/R_1)$, which indicates an infinite voltage gain, not that of Equation 9-31. In fact, there is no contradiction since the current into the op-amp is fixed. This can be easily seen by example. If we set $R_2 = 10$ MΩ and $I_{in} = 1$ μA, then Equation 9-31 predicts $V_{out} = -10$ V (assuming $V_R = 0$). We can produce the input current by connecting one end of a 1 MΩ resistor to the inverting input of the operational amplifier and the other end to a 1 V supply. In doing this we form the inverting amplifier with gain $= -(R_2/R_1) = -10$ and with $V_{out} = -10$V.

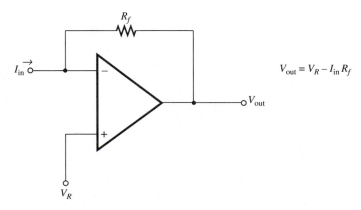

FIGURE 9.14   Current-to-voltage converter.

Current-to-Current Amplifier

A true current amplifier, sometimes called a current-to-current amplifier, is illustrated in Figure 9.15. The relationship of I_{out} to I_{in} is given by

$$I_{out}/I_{in} = 1 + R_f/R_s \qquad (9\text{-}32)$$

and is independent of R_L. Analysis of this circuit follows a route similar to that of the inverting and non-inverting amplifiers.

9-8   FILTERS

The ideas inherent in the approach outlined above are extremely useful in designing band pass amplifiers, in which either a given frequency band is amplified to the exclusion of all others, or a given frequency band is rejected while all others are amplified. Reexamination of the circuit of Figure 9.16 suggests that the integrator-differentiator configuration is useful for this purpose.

Twin-*T* Filter

A still better network is called the twin-*T* filter network (Figure 9.17A). When the values of the components are selected as indicated in the figure, the center-band frequency can be computed as

$$f_o = \frac{1}{2\pi RC} \qquad (9\text{-}33)$$

The frequency response is given in Figure 9.17B. Note that the filter is really a combination of a high pass filter (*C, C, R/2*) and a low pass filter (*R, R, 2C*). A point of clarification is in order for the twin-*T:* the impedance of this circuit is a maximum at f_o. Therefore, when this configuration is used in the feedback loop of an operational amplifier, the gain at f_o is maximum, and decreases on either side of f_o because the impedance decreases. On the other

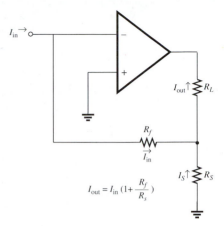

FIGURE 9.15   Current-to-current amplifier.

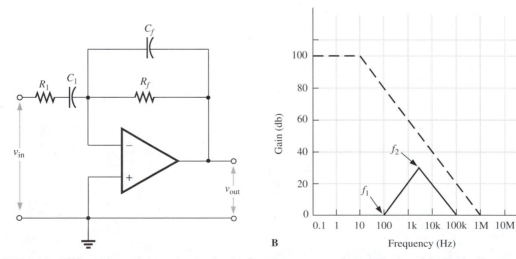

A   B

FIGURE 9.16   Differentiator–integrator (to fix the frequency range of operation) and its Bode diagram.

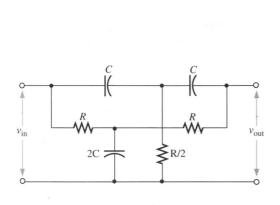

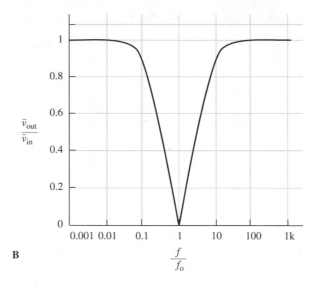

A

B

FIGURE 9.17 **Twin-*T* filter and its transfer curve.**

hand, when the twin-*T* is used in place of R_i, the output of the operational amplifier goes virtually to zero at f_o because the feedback to input impedance ratio is minimized. These features will be clarified by examination of Figures 9.18 and 9.19, which give typical circuits and their Bode plots. Hence, either a band-pass or a band-reject amplifier can be made, using the twin-*T* at the appropriate place. The sharpness of the band is directly related to the degree of component matching. In practice, we begin with selection of the capacitors to achieve as close a match as possible for the frequency of interest. In all probability, the value of the matched capacitors is not quite that desired, so trimming resistors are used to achieve both a resistor match and the desired frequency. Care and effort at this stage give high returns in terms of a sharply "notched" response from the twin-*T* network.

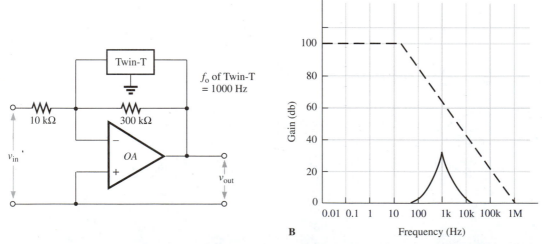

A

B

FIGURE 9.18 **Twin-*T* filter in the feedback loop of an operational amplifier and the resulting Bode diagram.**

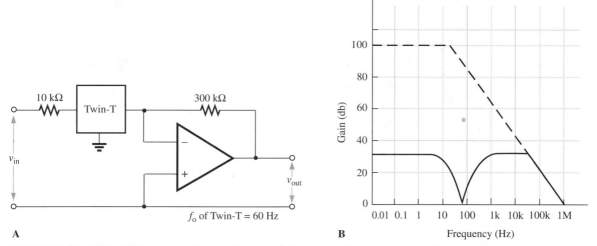

10 kΩ   Twin-T   300 kΩ

v_{in}

v_{out}

f_o of Twin-T = 60 Hz

A

B

FIGURE 9.19   **Twin-*T* filter at the input of an operational amplifier and the resulting Bode diagram.**

9-9   COMPARATOR

Consider the circuit in Figure 9.20A. The reference voltage, V_R, may be set at ground or any other voltage. When V_{in} is greater than V_R, V_{out} will saturate to the voltage of the positive supply since the gain of the amplifier is assumed infinite. When V_{in} is less than V_R, V_{out} will swing to the voltage value of the negative supply. This is made readily apparent by realizing that the op-amp multiplies the difference in voltage between its inputs by a large factor, usually on the order of 10^5. With a difference of even 1 mV, the output voltage is theoretically 100 V, but is limited to that of the power supply. In practice the output voltage is often a few volts less than the supply voltage. The sign of the output voltage is dictated by which voltage is greater, that on the inverting input or that on the non-inverting input. If $V_{inverting} > V_{non-inverting}$, then the output will be less than zero and vice-versa, as shown in Figure 9.20B.

Since the operational amplifier compares two signals and produces a voltage which relies on this comparison, it can be used as a *comparator*. A comparator is a circuit that does just that; it has two inputs—if one is greater than the other, the output is one value and if the other is greater, the output is a different value. Figure 9.21 shows a simple use of such a device.

Consider the signal with noise in Figure 9.22 which represents the output of a thermocouple. When the voltage produced by the temperature transducer reaches a certain point, shown here as V_R (R = reference) we wish to turn on a light emitting diode (LED). At any time before t_o, the signal into the inverting input of the transducer is smaller than V_R and therefore the op-amp output is swung to the positive voltage of the power supply. When the input voltage reaches V_R the output swings negative, and the LED is turned on. The input voltage, however, goes below V_R because of the noise on the signal, and the LED goes off again before returning to the ON state. This type of bouncy behavior is very common near the transition point of a comparator.

To rectify this problem, a *Schmitt trigger* is used. A Schmitt trigger is an addition to a circuit that causes the reference voltage, V_R, to change when

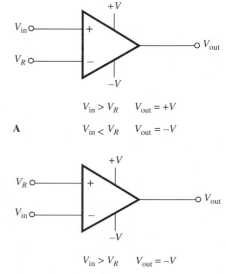

$V_{in} > V_R$ $V_{out} = +V$

A $V_{in} < V_R$ $V_{out} = -V$

$V_{in} > V_R$ $V_{out} = -V$

B $V_{in} < V_R$ $V_{out} = +V$

FIGURE 9.20   **Comparator circuits.**

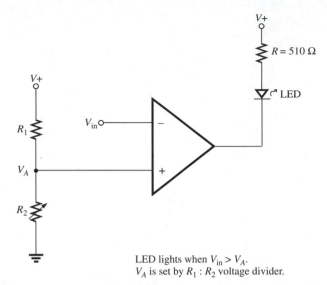

FIGURE 9.21   Simple comparator circuit.

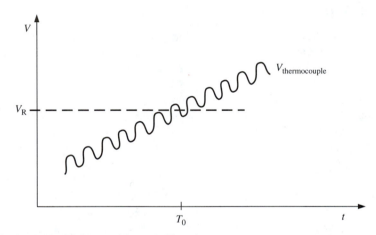

FIGURE 9.22   Voltage with periodic noise.

a transition occurs. Such a modification is shown in Figure 9.23. Although the values of the resistors are chosen here to change V_R dramatically, usually they are chosen for a much milder change, depending on the nature of the input voltage. When the output voltage is negative, V_R is zero. When the output voltage is positive, $V_R = 2/3\ V_+$. For our circuit, the reference voltage is initially $2/3\ V_+$. When the input voltage first reaches this value, the output goes negative, and thus the reference voltage becomes zero. To reset the comparator, the input voltage would need to go below this point.

Op-amp comparator circuits are relatively slow in response to an incoming signal. Special, dedicated, integrated comparator circuits are manufactured with much faster response times than their op-amp equivalents. One of the most widely used comparators is the LF311. Its output is an open collector transistor, as shown in Figure 9.24, which calls for a dropping resistor to be connected externally as shown if V_{out} is to behave as expected.

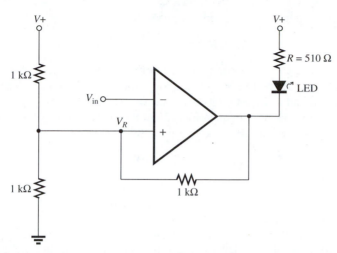

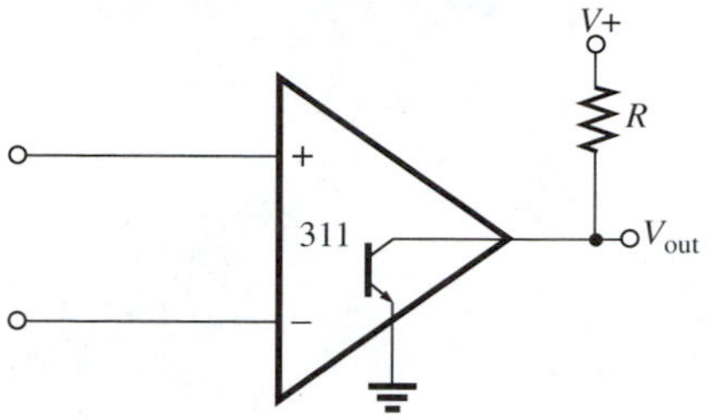

FIGURE 9.24   LF311 comparator.

FIGURE 9.23   Comparator circuit similar to Figure 9.21, but with Schmitt trigger feedback.

9-10   REAL OPERATIONAL AMPLIFIERS

Gain

Just how much is lost in going from the ideal to a real operational amplifier? This is a fair question. Consider first the effect of finite gain (as opposed to the infinite gain postulated for the ideal amplifier). Referring to Figure 9.25 and introducing the voltage at the inverting input terminal as V_{IN-} (not assuming that the previous rules hold true, except that the inputs draw or source no current), we have

$$(V_{in} - V_{IN-}) / R_1 + (V_{out} - V_{IN-}) / R_2 = 0 \qquad (9\text{-}34)$$

The open loop gain (no feedback resistors), A, of the amplifier circuit is given by the ratio of the output voltage to the difference between the input voltages,

$$A = V_{out} / (V_{IN+} - V_{IN-}) \qquad (9\text{-}35)$$

which, for $V_{IN+} = 0$, yields

$$A = -(V_{out} / V_{IN-}). \qquad (9\text{-}36)$$

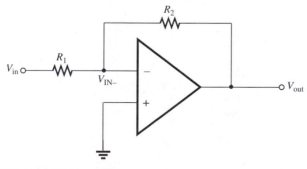

FIGURE 9.25   Inverting amplifier.

Substituting Equation 9-36 into 9-34 gives us

$$-V_{out} / V_{in} = R_2 / R_1 / [1 + (1 + R_2 / R_1)/A]. \qquad (9-37)$$

The denominator of the above relationship reduces to unity as the open loop gain gets large. Table 9.1 shows the error involved in assuming a large open loop gain instead of Equation 9-37 for various values of A and R_2 / R_1. Note that the magnitudes of both quantities affect the amount of error. When 0.1% accuracy or better is desired, the following relationship must be obeyed:

$$A R_1 / R_2 \geq 10^3. \qquad (9-38)$$

The limitation is not severe, since most commercially available operation amplifiers have open loop gains well in excess of 10^4. When the criterion of the above relationship is met, the assumption of ideal behavior is excellent. Thus, in this instance, there is no significant difference between ideal and real behavior. However, as we shall soon see, the open loop gain is a function of frequency, and falls off dramatically at higher frequencies.

Output Impedance

The output impedance is an important feature of operational amplifiers. Recall that Z_{out} was assumed to be zero for the ideal operational amplifier. A real amplifier has a finite, non-zero value of Z_{out}. Consider the circuit of Figure 9.26, in which the output impedance is assumed to all be lumped at the output of an ideal voltage follower. A change in output voltage, ΔV_{out}, results from a change in current, ΔI_{out}, since

$$\Delta V_{out} = \Delta I_{out} Z_{out}. \qquad (9-39)$$

This, in turn, produces a change in the voltage between the input terminals, ΔV_s, of the amplifier, because the summing point is connected directly to the output. This change is written as

$$\Delta V_s = -\Delta V_{out} / A. \qquad (9-40)$$

TABLE 9.1 *Error in Assuming a Large Open Loop Gain*

A	Ideal	Real	% Error
		$R_2 / R_1 = 100$	
10^1	100	9.0	91.0
10^2	100	49.8	50.2
10^3	100	90.8	9.2
10^4	100	99.0	1.0
10^5	100	99.9	0.1
10^6	100	99.99	0.01
		$R_2 / R_1 = 10$	
10^1	10.0	4.76	53.4
10^2	10.0	9.01	9.9
10^3	10.0	9.89	1.1
10^4	10.0	9.99	0.1
10^5	10.0	9.999	0.01
10^6	10.0	9.9999	0.001

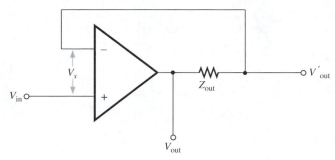

FIGURE 9.26   Circuit used to analyze output impedance of voltage follower.

The voltage observed at the output, V'_{out}, must therefore change by ΔV_s. The closed loop output impedance, Z_{co}, is thus

$$Z_{co} = \Delta V'_{out} / \Delta I_{out} = \Delta V_s / \Delta I_{out}. \qquad \textbf{(9-41)}$$

This is the quantity of interest, although manufacturers' specifications list Z_{out} values. Using Equations 9-39 and 9-40, Z_{co} can be put in terms of Z_{out}:

$$Z_{co} = (\Delta V_{out} / A) (Z_{out} / \Delta V_{out}) = Z_{out} / A. \qquad \textbf{(9-42)}$$

The output impedance, Z_{co}, takes on remarkably small values when the amplifier is used in a closed-loop configuration. A typical value of Z_{out} is 5 kΩ. With an open loop gain, A, of 10^5, the output impedance for closed-loop operation is only 0.05 Ω. Certainly, for all but the most demanding applications we may ignore this small impedance. Once again the difference between real and ideal is small. Note, however, that operational amplifiers are not high current sources or sinks (output or input)—they are mainly used merely for signal modification. Typical current limitations are on the order of 20 mA. Larger currents can be obtained by placing a current amplifier, such as a transistor, on the output of the operational amplifier.

Input Impedance

The input impedance of a real operational amplifier can be developed with reference to the circuit in Figure 9.27, which is also a voltage follower. A

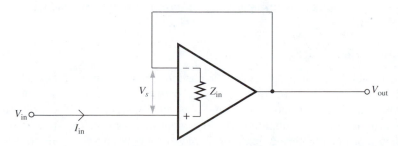

FIGURE 9.27   Circuit used to analyze input impedance of voltage follower.

change at the output, ΔV_{out}, results in a change at the summing point, ΔV_s. Since it is in the voltage follower configuration, the relationship is

$$\Delta V_s = \Delta V_{\text{out}} / A = \Delta V_{\text{in}} / A. \qquad (9\text{-}43)$$

This voltage change forces a change in current, ΔI_{in}, which is related to the input impedance, Z_{in}:

$$\Delta I_{\text{in}} = \Delta V_s / Z_{\text{in}} = \Delta V_{\text{in}} / A\, Z_{\text{in}} \qquad (9\text{-}44)$$

where the definition of open loop gain, A, is used. The closed loop impedance, Z_{ci}, is defined as

$$Z_{\text{ci}} = \Delta V_{\text{in}} / \Delta I_{\text{in}}. \qquad (9\text{-}45)$$

Using Equation 9-42, the closed loop input impedance is related to the impedance by

$$Z_{\text{ci}} = \Delta V_{\text{in}} A\, Z_{\text{in}} / \Delta V_{\text{in}} = A\, Z_{\text{in}}. \qquad (9\text{-}46)$$

This means that the effective impedance at the input under closed-loop conditions is quite large. A typical value of Z_{in} for an inexpensive bipolar operational amplifier is 50 MΩ, so that a gain of $A = 10^5$ produces a value of Z_{ci} of about 10^{12} Ω. Common input impedances of 10^{12} Ω for FET operational amplifiers result in astronomical theoretical values of Z_{ci}.

In all fairness, it should be pointed out that the voltage follower provides the highest input impedance of the several configurations described. The impedance of the inverter, on the other hand, is simply the input resistance, R_1, because the summing point is at virtual ground.

Input Bias and Offset Current

Since the input impedance is so large, more attention is generally given to the input bias current, i_B, and input difference current (or input offset current), I_{offset}. For the ideal amplifier circuit, both of these parameters are equal to zero; however, in real applications, this is not true. The input bias current is defined as the maximum current required at either input terminal in order that the output be equal to zero. Often, I_B is given as the average of the current required at the inverting terminal and that required at the non-inverting terminal. Typical values range from nA down to a few pA. The input difference current is the current difference between I_B for each terminal. Typical values are 10% of I_B. Both of these parameters must be taken into consideration for low-current, low-voltage applications, generally by choosing an op-amp with input bias and offset currents low enough to be neglected.

Input Offset Voltage

The input offset voltage, V_{os}, is defined as the voltage required at the input to drive the output to zero. Another way of stating this in terms related to the circuits previously discussed is that V_{os} is the DC voltage at the summing point of an inverting amplifier with the input to R_1 grounded. It is present because of manufacturing limitations, and many operational amplifiers have connections which allow external components to compensate for V_{os}. This voltage can be significant even for modest circuits. Typical values are in the

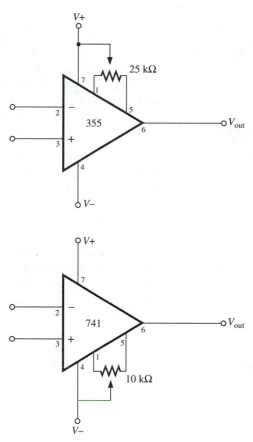

FIGURE 9.28   **Input offset nulling connection for LF355 and LM741 operational amplifiers.**

mV range. Figure 9.28 shows a typical offset-voltage trim circuit for the LF355 and LM741 operational amplifiers.

Common Mode Rejection Ratio (CMRR)

The Common Mode Voltage is defined as the output voltage when the input terminals of an operational amplifier are at the same voltage. Ideally, this condition would result in an output voltage of zero. In a real amplifier, this is not true for several reasons, one being that the gain of the inverting input may differ slightly from the gain of the non-inverting input. The ability of the operational amplifier to respond to only the difference of the signals at its inputs is expressed in terms of the so-called Common Mode Rejection (CMR). Usually the gain is unity for such a common-mode input, which is relatively good since the differential gain that we have been analyzing in detail is generally of the order of 10^5. The ratio of the gain for the differential mode, to the gain of the common mode is the Common Mode Rejection Ratio (CMRR). In general, the higher the value of CMRR, the higher the quality of the operational amplifier. The CMR is usually given in dB, where CMR = $20 \log_{10}$ (CMRR).

Slew Rate

The slew rate is a measure of how fast the output of an operational amplifier can respond to changes in the input voltage. Typically it is of the order of a volt per μs. While this is an impressive figure, certain operations can be degraded with slew rates of this magnitude. Most manufacturers offer special, high-slew-rate devices for use in critical, high-speed applications.

Frequency Response

The slew rate gives rise to the frequency dependence of the gain of an operational amplifier. The frequency response of an operational amplifier is one of its important aspects. The term *frequency response* refers to the range of frequencies in which the voltage gain remains constant; an amplifier whose gain changes with changes in the input frequency tends to distort a signal. Frequency response is best described by a Bode plot (see Appendix A for an explanation of the basis for this type of diagram). The Bode plot of Figure 9.29 describes an operational amplifier with a voltage gain of 10^5 (that is, 100 dB) and a unity gain (or crossover) frequency of 1 MHz. Since the gain of most operational amplifiers falls off ("rolls off") at higher frequencies at a uniform rate of 20 dB per decade (or 6 dB per octave; an octave is the interval between any given frequency f_0 and its first harmonic, $f_1 = 2f_0$), the Bode plot is easily constructed. This representation is approximate, as there is no sharp break at 10 Hz (the actual behavior is represented by the dashed curve), but this is a minor deficiency. An inverting amplifier with a gain of 100 ($100 = R_2 / R_1$) is represented by the plot in Figure 9.30. It is apparent that gain must be sacrificed to obtain a greater frequency response range; the

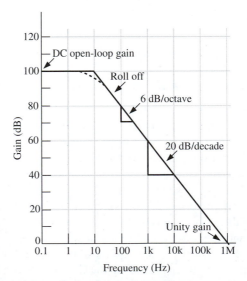

FIGURE 9.29   Bode diagram for operational amplifier performance showing an open-loop gain of 10^6, a unity gain frequency of 1 MHz and a 6 dB/octave (20 dB/decade) rolloff.

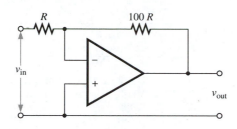

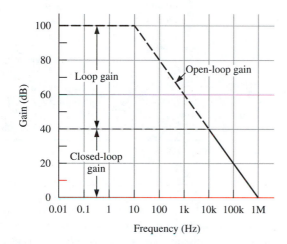

FIGURE 9.30   **Bode diagram for an inverting amplifier with a gain of 100.**

high frequency cut-off of the 100 dB amplifier is 10 Hz, while that of the 40 dB amplifier extends to 10 kHz.

Drift

Although drift has so far not been mentioned, it can cause problems. Either current or voltage drift can occur, both being primarily due to temperature changes. Amplifiers generally have drift rates on the order of 1nA / °C or less for current and 1μV / °C or less for voltage. Low-drift operational amplifiers are commercially available.

Operational Amplifier Comparison

Table 9.2 lists some common operational amplifiers and their characteristics. This list is by no means complete—there are hundreds of amplifiers available.

9-11   CONCLUSION

While operational amplifiers have been shown to perform a wide variety of functions with extreme reliability and predictability, not all operational amplifiers will meet all needs. Several parameters have been discussed which bring the ideal operational amplifier and its analysis into the real world. The

TABLE 9.2 Operational Amplifier Characteristics

Op-Amp	Bandwidth (MHz)	Slew Rate (V / μs)	I_B (nA)	V_{os} (mV)	CMRR (dB)	Comments
General Purpose						
LM741	1	0.5	200	6	70	classic general purpose
AD548	1	1.8	0.02	2	90	low power consumption
LF356	4.5	12	0.05	1	85	
LF411	3	15	0.2	2.0	70	
Low Input Current						
AD549	1	NA	0.25pA	20	62	
High Speed						
AD844	900	2000	450	0.05	NA	very fast
LF351	4	13	0.05	10	70	
Precision						
AD707	0.9	0.3	2.5	100	140	
LF347	4	13	0.05	2	70	low total harmonic distortion
Multiple Package						
AD713	4	20	150	20	NA	quad package
LM324	1	NA	45	2	65	quad package, single supply
Low Input Offset Voltage						
OP-07	0.6	0.3	1.0	.025	126	ultra-low offset voltage

LM, LF: National Semiconductor   AD: Analog Devices   OP: Precision Monolithics

NA: not directly available in data sheet

user must be aware of potential problems which may arise, particularly with low-voltage, low-current applications. For most applications, the ideal operational amplifier design procedures work rather well. When critical applications are called for, detailed analysis of the real operational amplifier and its limitations is appropriate. The references given at the end of the text provide an excellent source of information along these lines.

9-12 SOME EXAMPLE CIRCUITS

Figures 9.31a through 9.31k illustrate some circuits which can be constructed using operational amplifiers.

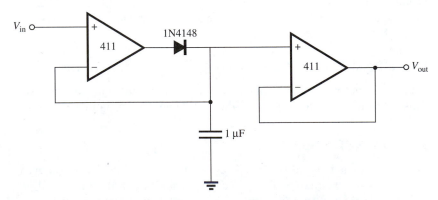

FIGURE 9.31a Peak voltage detector. V_{out} will equal the maximum voltage that V_{in} obtains. To reset, or zero, add a pushbutton or n-channel JFET across the capacitor.

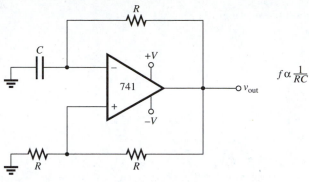

$$f \alpha \frac{1}{RC}$$

FIGURE 9.31b Square wave oscillator. When $V_{in-} > V_{in+}$, $V_{out} = -V$ and the capacitor discharges through R until $V_{in-} < V_{in+}$ at which point $V_{out} = +V$ and the capacitor charges through R, etc.

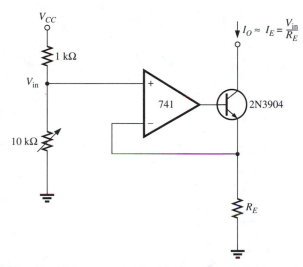

FIGURE 9.31c Variable current sink. 1 kΩ resistor limits V_{in} to 0.91 V_{cc} (upper limit).

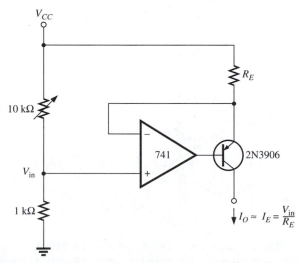

FIGURE 9.31d Variable current source. 1 kΩ resistor limits V_{in} to 0.09 V_{cc} (lower limit).

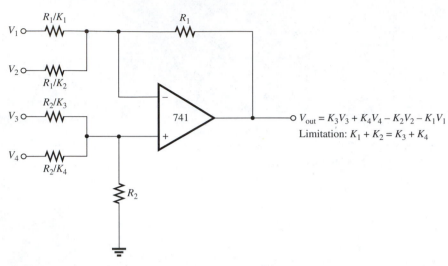

$$V_{out} = K_3 V_3 + K_4 V_4 - K_2 V_2 - K_1 V_1$$

Limitation: $K_1 + K_2 = K_3 + K_4$

FIGURE 9.31e **Generic summer. Extra inputs may be added.**

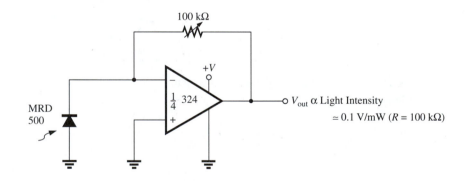

$V_{out} \propto$ Light Intensity

$\simeq 0.1$ V/mW ($R = 100$ kΩ)

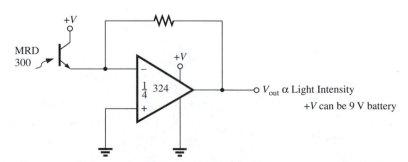

$V_{out} \propto$ Light Intensity

+V can be 9 V battery

FIGURE 9.31f and g **Single power supply photometers. Photodiode and phototransistor pass currents proportional to light intensity. Op-amp is used as current to voltage converter.**

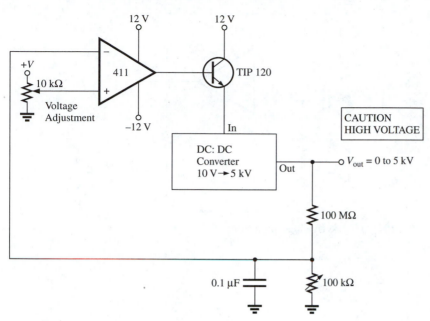

FIGURE 9.31h   **High voltage power supply.**

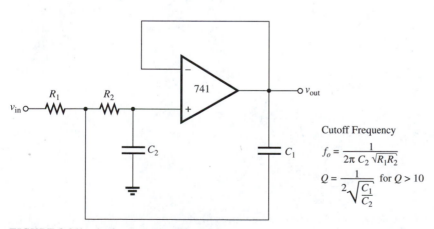

Cutoff Frequency

$$f_o = \frac{1}{2\pi C_2 \sqrt{R_1 R_2}}$$

$$Q = \frac{1}{2\sqrt{\dfrac{C_1}{C_2}}} \quad \text{for } Q > 10$$

FIGURE 9.31i   **Active low pass filter.**

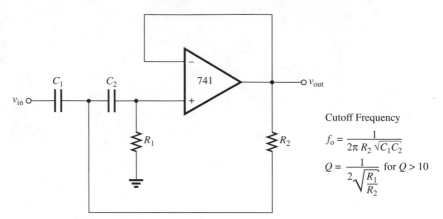

Cutoff Frequency

$$f_o = \frac{1}{2\pi R_2 \sqrt{C_1 C_2}}$$

$$Q = \frac{1}{2\sqrt{\dfrac{R_1}{R_2}}} \quad \text{for } Q > 10$$

FIGURE 9.31j   **Active high pass filter.**

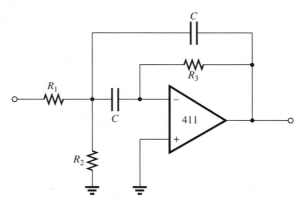

Choose Gain (G), center frequency (f_o) and Q, then

$$R_1 = \frac{Q}{2\pi f_o GC}$$

$$R_2 = \frac{Q}{(2Q^2 - G)\, 2\pi f_o C}$$

$$R_3 = \frac{2Q}{2\pi f_o C}$$

$$G \geq \sqrt{Q}$$

For audio frequencies choose $C \simeq 0.1\ \mu F$

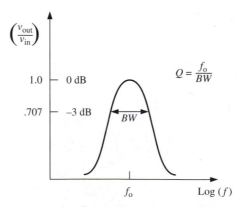

$$Q = \frac{f_o}{BW}$$

FIGURE 9.31k   **Generalized bandpass filter.**

9-13   QUESTIONS AND PROBLEMS

1.   Solve Equation 9-27 for constant V_{in}. What type of signal is generated by the integrator under these conditions?

2.   In order to control the slope of a ramp produced by an integrator, the values of V_{in}, R and C may be varied. Assuming a fixed value for C of 10 μF, would it be better to have a relative large V_{in} with a correspondingly large value of R, or a relatively small V_{in} with a relatively small corresponding value of R? Explain your choice in terms of the current offset at the summing point.

3.   For the circuit of Figure 9A, determine V_{out} as a function of V_{in}. Also evaluate the input impedance of the circuit.

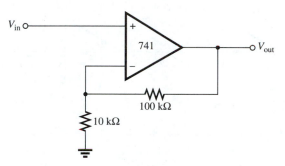

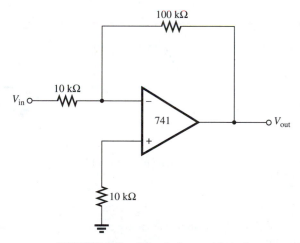

FIGURE 9.A   **Circuit for problem 3.**

4.   For the circuit of Figure 9B, determine V_{out} as a function of V_{in}. Also evaluate the input impedance of the circuit.

FIGURE 9.B   **Circuit for problem 4.**

5.   Write the expression for the output voltage in terms of the input voltages for the circuit shown in Figure 9C.

6.   Sketch a graph of V_{out} vs. V_{in} for the circuit shown in Figure 9D.

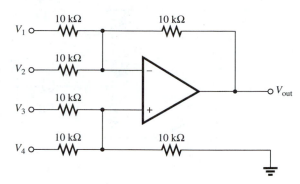

FIGURE 9.C   **Circuit for problem 5.**

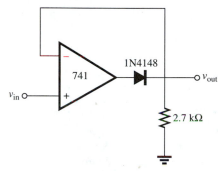

FIGURE 9.D   **Circuit for problem 6.**

7.   Sketch a graph of V_{out+} vs. V_{in} and V_{out-} vs. V_{in} for the circuit shown in Figure 9E.

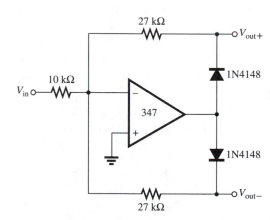

FIGURE 9.E   **Circuit for problem 7.**

8.   What is the function of the circuit of Figure 9F? Determine the output-input relationship for the circuit for an AC signal and a DC voltage input.

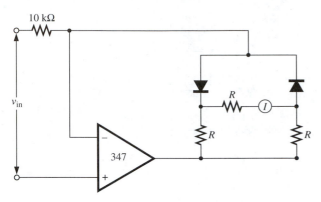

FIGURE 9.F   Circuit for problem 8.

9. What is the function of the circuit shown in Figure 9G? Could it function if the diode were reversed? Explain.

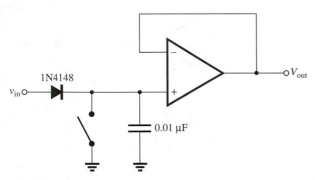

FIGURE 9.G   Circuit for problem 9.

10. Design a circuit, using operational amplifiers, which can multiply and divide. Note that the circuit should be capable of handling both negative and positive values.

11. Without elaborate circuitry, suggest how the input difference current (or input offset current), I_{offset}, may be evaluated.

12. Figure 9H is a square-wave generator. Derive the relationship between frequency and the values of R and C.

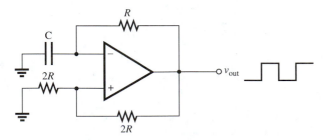

FIGURE 9.H   Circuit for problem 12.

13. Design operational amplifier circuits which will generate the waveforms shown in Figure 9I.

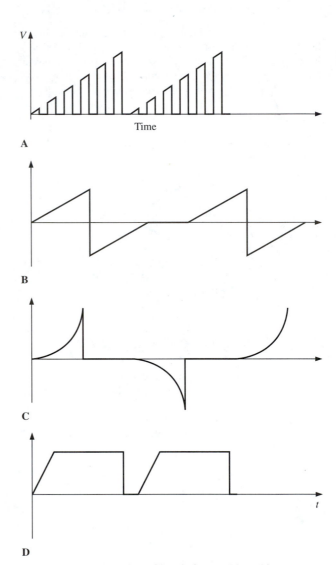

FIGURE 9.I   Circuit for problem 13.

14. Set up an operational amplifier circuit to solve the equation

$$\frac{d^2x}{dt^2} + 5\frac{dx}{dt} + 7x + 3 = 0.$$

15. The circuit shown in Figure 9J is a difference amplifier. What is the V_{out} in terms of the input voltages V_a and V_b?

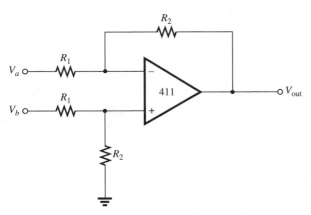

FIGURE 9.J   **Difference amplifier circuit for problem 15.**

16.   The circuit shown in Figure 9K is a difference amplifier, or more specifically, an instrumentation amplifier. Derive the transfer equation (V_{out} in terms of V_{in} and other
components). What is the advantage of this circuit over that
in problem 15?

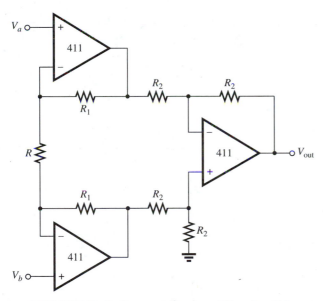

FIGURE 9.K   **Instrumentation amplifier circuit for
problem 16.**

17.   Discuss why operational amplifiers should not be used
to directly drive loads greater than 20 mA (such as a lamp or
relay).

18.   Design a circuit which, at a rate of 0.2 Hz, turns on
and off a relay that in turn drives a 5 V, 1 A lamp using
components up to and including those discussed in this
chapter.

19.   Is it possible to design a single operational amplifier
circuit with gain of exactly 96.000 using simple components?
Explain.

20.   A non-inverting operational amplifier circuit is constructed to have a gain of 100 with an input signal of 1 kHz.
Would you expect the circuit to have the same gain with an
input signal of 1 MHz? Explain.

21.   You need to measure the voltage of a 100 pF very-
low-leakage capacitor which has one end connected to
ground. The voltage is known to be about 1 V. Using a digital voltmeter with 10 MΩ input impedance and an operational amplifier (choose one from Table 9.2), design a circuit
which can measure this voltage before considerably discharging the capacitor.

CHAPTER

Waveform Generators

10

10-1 INTRODUCTION

We examined AC signals in some detail in the early part of this text. The generation of these waveforms, with the exception of commercial-function generators and AC line voltage, has not been covered. Many applications call for a periodic signal of some sort, be it for the turning on and off of an LED or the timing of a microcomputer. This chapter relies heavily on an integrated-circuit approach to waveform generation and the associated waveform shaping, but some discrete component oscillators are discussed as well. Digital waveforms—signals that are on or off, voltage levels of high or low—are fundamental in the study of digital circuitry.

10-2 WAVEFORM SHAPING

Comparator

In the previous chapter we introduced the comparator circuit as a means of detecting when an input signal crossed a certain threshold, or reference voltage. Here we analyze the comparator with additional Schmitt trigger circuitry as a wave-shaping device. This type of comparator is often used to modify an incoming signal and produce a pulse wave. Consider the comparator circuit shown in Figure 10.1. If V_{in} is less than the reference voltage, V_R, on the positive input, then the output of the comparator is V_+, or a high signal. "High" indicates a voltage equal to that of the positive power supply. When the output of the comparator is high, V_R is given by

$$V_R = V_+ R_1/(R_1 + R_2). \tag{10-1}$$

For our example the reference voltage is thus equal to 1.8 V. Once the input signal exceeds this voltage, the output goes low (equal to the value of the negative power supply V_-) and the reference voltage V_R changes to

$$V_R = V_- R_1/(R_1 + R_2) \tag{10-2}$$

which in our case is -1.8 V. Now, for the output to change from low to high, the input signal must fall below this *new* reference voltage. The *hyster-*

215

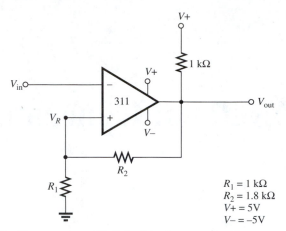

FIGURE 10.1 **Comparator with Schmitt trigger connection.**

esis voltage is defined as the voltage difference between these two reference voltages, or

$$\text{hysteresis voltage} = (V_+ - V_-)R_1/(R_1 + R_2). \quad \textbf{(10-3)}$$

Hysteresis is important since it allows the circuit designer to nullify the effects of noise present on any signal, particularly on those which vary slowly as a function of time. Note that V_- may be any voltage, including zero. The relationships of these voltages as they pertain to the circuit behavior is shown in Figure 10.2.

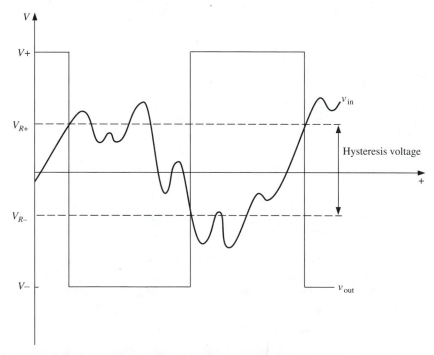

FIGURE 10.2 **Input and corresponding output signals of comparator with Schmitt trigger.**

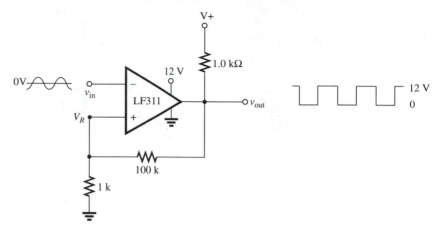

FIGURE 10.3   Sine-wave to square-wave converter.

The comparator circuit with hysteresis is invaluable as a waveform shaper. An example is shown in Figure 10.3 where an incoming sine wave is converted to a clean (very little noise) square wave. If a Schmitt trigger were not used, the output might oscillate rapidly between high and low at the transition points. Note that the reference voltage, V_R, changes from zero to about 100 mV when the output changes from zero to 12 V, causing the produced square wave to be asymmetric.

Differentiation and Integration

Resistor and capacitor circuits, integrators and differentiators, can also be used to shape waveforms, as we have seen in earlier chapters. Active RC circuits, using operational amplifiers discussed in the previous chapter, play an important role in the modification as well as amplification of repetitive signals. One common use of an integrator circuit is to convert a square wave to a triangular wave. The differentiator is used to convert a square wave to a pulse wave. In conjunction with a comparator, the differentiator circuit can produce a high-quality pulse train, or if the input is a single pulse, the output pulse width can be modified with such a circuit.

10-3   RESONANT OSCILLATORS

Recall that a parallel LC circuit is used to store energy, as diagrammed in Figure 10.4. As the charge stored on the capacitor begins to flow through the coil, the induced magnetic field expands, as indicated in frames t_1 and t_2. When the capacitor is completely discharged, the magnetic field begins to collapse, resulting in a charge of the opposite polarity on the capacitor (t_3 and t_4). The process then reverses itself (t_5 through t_8), until the circuit returns to its original state. The whole process repeats itself at a resonance frequency, f_0, determined by

$$f_0 = 1/(2\pi\sqrt{LC}). \tag{10-4}$$

Because the components of the resonant circuit are real rather than ideal, this oscillation deteriorates because of power losses (notably i^2R losses in the

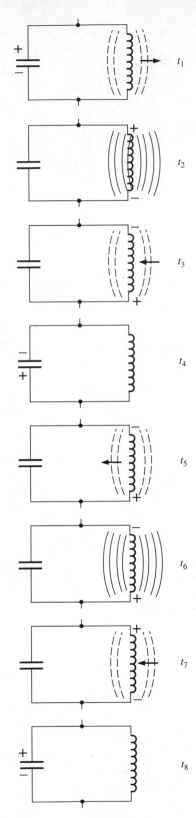

FIGURE 10.4 Schematic representation of the action of a resonant circuit in a time sequence.

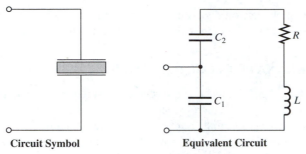

Circuit Symbol   **Equivalent Circuit**

FIGURE 10.5   **Electronic symbol for a crystal and equivalent circuit of a crystal.**

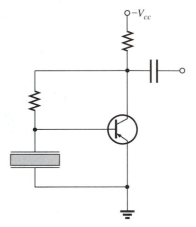

FIGURE 10.6A   **Pierce crystal oscillator circuit.**

coil). This deterioration is overcome, however, when the inductor-capacitor combination is placed in a circuit capable of resupplying energy at the appropriate times. That is, voltage of the same polarity as that on the capacitor must be applied at time t_1, t_4 and t_8. To accomplish this, the resonant circuit is placed in the feedback loop of an amplifier.

An excellent frequency standard involves the use of a piezoelectric crystal oscillator. The equivalent circuit of a quartz crystal is given in Figure 10.5. A crystal meant to operate in the 0.5 MHz region would have the following values (referred to in Figure 10.5); C_1 = 6 pF, C_2 = 0.04 pF, L = 3 H and R = 4 kΩ; however, crystals are usually specified in terms of their oscillation frequency directly. A transistor–crystal oscillator is shown in Figure 10.6A. With care, the frequency of the output can be maintained with an accuracy of 0.001% or better. With extreme care, accuracy of one part in 10^7 is possible. This circuit, whose output is a sine wave, is called the *Pierce crystal oscillator*. Another version of the Pierce crystal oscillator involves an inverter circuit as shown in Figure 10.6B. Although the inverter circuit (the triangular shaped component with one input and one output connection) is discussed in

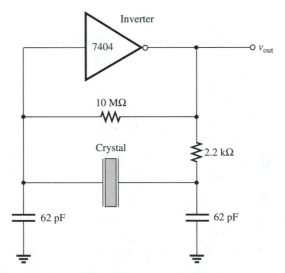

FIGURE 10.6B   **Crystal oscillator square wave circuit.**

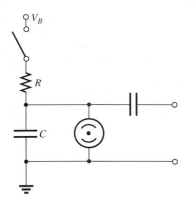

FIGURE 10.7 **Neon lamp relaxation oscillator circuit.**

the following chapter on digital electronics, suffice it to say that the inverter does just that: when the input is low, the output is high and vice-versa.

10-4 RELAXATION OSCILLATORS

The waveforms generated by the circuits presented thus far have been the square and sine wave. Numerous occasions arise when another waveform is needed, particularly in timing applications where the feature of interest is the waveform's periodicity rather than its shape. The relaxation oscillator offers a straightforward route to a controllable, repetitive waveform.

Neon Lamp Oscillator

The simplest relaxation oscillator uses an RC charging circuit in which a neon lamp shunts the capacitor. This circuit is given in Figure 10.7. When the switch is closed, current flows to charge C. The charge rate, you will recall, is related to the RC time constant. Because the lamp is shunting the capacitor, a further consideration is involved. A neon lamp will "fire" when a potential (V_f) of approximately 60 V is applied to it. It will go out when the voltage (V_x) on it drops below approximately 40 V. Thus, when the lamp is not on, its resistance is quite large, in excess of 20 MΩ. When the lamp is on, its resistance is quite small, less than 1 kΩ. As long as V_B is greater than V_f ($=60$ V), the circuit operates as an oscillator. The capacitor is charged exponentially until the voltage across it is sufficient to fire the lamp. At that point, the resistance of the lamp drops suddenly, and the capacitor discharges through it at a rate determined by the RC constant of the lamp (at low resistance) and the capacitor. Because the resistance of the lamp is so low, the discharge is very rapid. When the voltage across the capacitor drops below the voltage necessary to sustain the lamp, the lamp goes off and the charging begins again. This process is illustrated in Figure 10.8. The parameters which control the period of the waveform are V_B, R, C, V_f and V_x. Consideration of

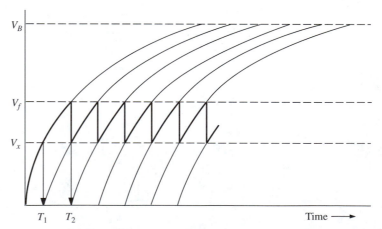

FIGURE 10.8 **RC charging and discharging processes within a neon lamp oscillator.**

the *RC* charging and discharging equations of Chapter 2 leads to the following expressions for T_1 and T_2 (as defined in Figure 10.8):

$$T_1 = -RC \ln \frac{V_B - V_x}{V_B} \qquad (10\text{-}5)$$

and

$$T_2 = -RC \ln \frac{V_B - V_f}{V_B} \qquad (10\text{-}6)$$

Since the frequency is the reciprocal of the period and the period is the difference between T_1 and T_2, the following expression results:

$$f_0 = \frac{1}{T_2 - T_1} = \frac{1}{RC \ln \dfrac{V_B - V_x}{V_B - V_f}} \qquad (10\text{-}7)$$

Attempts to significantly increase the frequency result in a deterioration of the waveform, as shown in Figure 10.9. The cause of this is the fact that the method of increasing the frequency, decreasing the value of *RC*, eventually decreases the charging time to the point where it is roughly equal to the discharge time.

Low Frequency

High Frequency

FIGURE 10.9   Waveforms produced by a neon lamp oscillator. Upper: low-frequency waveform; lower: high-frequency waveforem.

The Unijunction Transistor (UJT)

The unijunction transistor is a device whose special characteristics are attributable to careful, controlled fabrication. A bar of *n*-type silicon with a small *p*-type emitter probe implanted in it represents both the qualities of operation and the physical configuration of the device, as shown in Figure 10.10. The emitter is not equidistant from the ends of the silicon bar; rather, the emitter is deliberately placed closer to one end than to the other.

When a positive voltage, V_{BB}, is applied to B_2, a continuous current flows through the bar to B_1. Therefore, the applied emitter voltage, V_E, can be related to a fraction of V_{BB}, called η, the uniform voltage drop across the silicon bar (B_2 to B_1). When $V_E < \eta V_{BB}$, the emitter-B_1 junction is reverse biased and no current flows into the emitter (except the minority carrier current, I_{EO}). As V_E is increased to ηV_{BB}, the emitter current I_E flows to B_1. As I_E increases, the conduction increases and the voltage drop V_{EB_1} *decreases*. This is the negative resistance behavior shown in the V_E–I_E characteristic curve of Figure 10.10. Note that the points of particular interest are the peak and valley points which define the negative-resistance region. The emitter diode is reverse biased in the cutoff region and forward biased when $V_E > \eta V_{BB}$. In the saturated region the current is limited by the recombination rate of holes and electrons.

It should be clear that appropriate use of the UJT requires that the peak point be properly stabilized against temperature-induced variations. The peak voltage, V_p, is equal to the fraction of the supply voltage at the emitter, ηV_{BB}, plus the forward voltage drop across the emitter diode, V_{EB};

$$V_p = \eta V_{BB} + V_{EB}. \qquad (10\text{-}8)$$

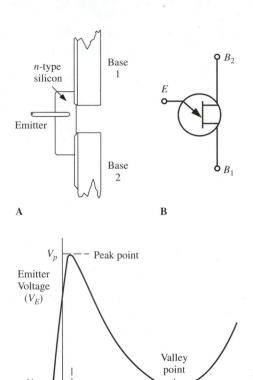

A   **B**

C

FIGURE 10.10   The unijunction transistor. (A) The morphology; (B) the electronic symbol; and (C) the current-voltage characteristics of the base-emitter.

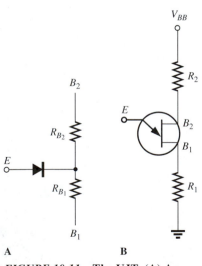

A   **B**

FIGURE 10.11   The UJT. (A) An equivalent circuit representation; (B) a stabilized circuit.

Since the UJT is silicon, $V_{EB} \cong 0.7$ V. This consideration leads to the equivalent circuit, shown in Figure 10.11A, in which

$$\eta = \frac{R_{B_1}}{R_{B_1} + R_{B_2}} \qquad \textbf{(10-9)}$$

Using this representation, stabilization is achieved by using resistors in both base leads, R_1 and R_2, as shown in Figure 10.11B. As the current increases (due to temperature increase), the voltage drop across R_2 increases. This increases the voltage drop across the UJT, which maintains the peak voltage essentially constant. The appropriate value of R_2 is given approximately by

$$R_2 \cong \frac{\eta(R_{B_1} + R_{B_2})V_{BB}}{0.7} \qquad \textbf{(10-10)}$$

which assumes a voltage divider network for the supply voltage of $R_1 + R_{B_1} + R_{B_2} + R_2$. The peak voltage can be readily computed from Equation 10-8.

A straightforward use of the UJT is in the relaxation oscillator circuit shown in Figure 10.12. The UJT may be thought of as replacing the neon

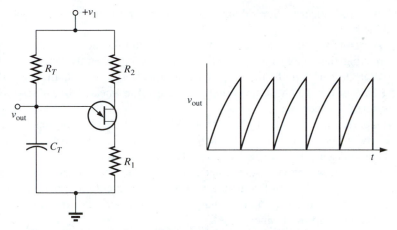

FIGURE 10.12 UJT relaxation oscillator and its waveform.

lamp of the earlier circuit, but the principle is the same. The capacitor, C_T, is charged toward V_1 through R_T. When the voltage on the capacitor exceeds the peak voltage, V_p, of the UJT, the device conducts strongly, discharging C_T quickly. When the discharge is such that $V_E < V_p$, the charging process begins again. Since the capacitor is grounded, the waveform generated returns to ground at the beginning of each period. This may be an advantage over the neon lamp variety of oscillator.

Several conditions of operation with the UJT must be considered. The value of R_T must be small enough to permit sufficient current to flow in the emitter of the UJT. This ensures that the negative resistance side of the current-voltage curve is reached. This criterion is stated as

$$\frac{V_1 - V_p}{R_T} > I_p \tag{10-11}$$

Although sufficient current is needed to reach the proper region of the operating curve, too much current will move the operation into the saturation region. If this occurs, the UJT will not turn off. In order to avoid this condition, we must have

$$\frac{V_1 - V_p}{R_T} < I_v \tag{10-12}$$

The lowest useful value of C_T in this circuit is approximately 0.005 μF. This means that the upper frequency limit for the oscillator is about 100 kHz. This assumes a value of $R_T = 3$ kΩ and $\eta = 0.5$ in the expression for the frequency:

$$f_0 = \frac{1}{R_T C_T \ln\left(\dfrac{1}{1 - \eta}\right)} \tag{10-13}$$

which is valid when $R_T > R_2, R_1$.

The linearity of the sawtooth waveform is important in certain situations. The circuit just discussed is somewhat sensitive to load and does not

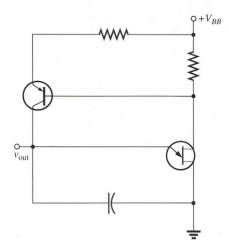

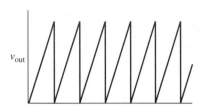

FIGURE 10.13   UJT with a constant-current source to produce a linearized sawtooth waveform.

exhibit good linearity. Several approaches to this problem exist; two will be illustrated.

Separate supplies may be used for the UJT and for the charging process. By making the supply for the charging process significantly larger (100 to 200 V), the fraction of full charge on the capacitor is very small. Since the first 10% of the exponential charging curve is quite linear, a linear sawtooth waveform is generated.

Good frequency stability may be sacrificed with this last approach, since small changes in the large voltage supply will produce noticeably different voltage–time responses for a fixed V_E. A much better approach is shown in Figure 10.13, where a constant current source feeds the capacitor. The constant current source is, of course, a *pnp* transistor in common base configuration. Because the capacitor is charged at a constant rate, the linearity is significantly improved, since

$$\frac{dV}{dt} = \frac{i}{C} = \text{constant}$$

when C is fixed and i is made constant. This produces the waveform shown in Figure 10.13; the improvement is seen when it is compared with Figure 10.12.

The circuits utilizing the UJT in sawtooth generation are also pulse-generating circuits. This feature, not mentioned earlier, becomes obvious when the remainder of the UJT is examined. When the emitter fires or conducts, a current pulse flows in the B_1 and B_2 portions as well. Therefore, the signal taken at R_1 has a pulse character. Naturally, the frequency of the pulse is the same as the sawtooth frequency. The pulse is positive at R_1 and negative at R_2.

Other Relaxation Oscillators

A simple, not very accurate, square-wave generator can be made from an operational amplifier, or comparator, using the circuit shown in Figure 10.14.

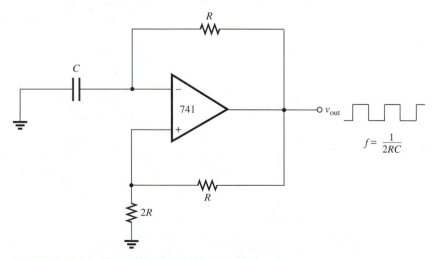

FIGURE 10.14   Operational amplifier oscillator.

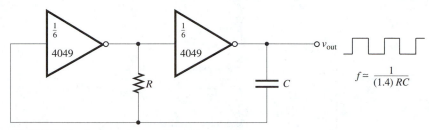

FIGURE 10.15   **Inverter oscillator.**

The output signal has a frequency equal to about 1/(2RC). Inverters can be used to form a similar square-wave generator as shown in Figure 10.15. Here the oscillation frequency is equal to about 1/(1.4 RC). A large variety of square wave oscillators can be made from digital circuits, which are discussed later in the text.

10-5   THE 555 TIMER

For generating oscillations and accurate time delays, the 555 (pronounced "five-fifty-five") timer circuit is probably the most versatile and popular integrated circuit used in electronics. It can be operated in two different modes: the *astable* mode in which it is an oscillator and the *monostable* mode in which it behaves as a single-pulse generator. As a monostable device, the necessary timing is accurately controlled by just two external components, a capacitor and a resistor. In the oscillator mode, the frequency of oscillation is established by one capacitor and two resistors. The operation of the circuit can also be triggered and reset by an external waveform. The output current can supply or sink up to 200 mA, a usefully large current. Although the 555 timer chip consists of numerous resistors and transistors, its operation can be understood by analyzing the block diagram shown in Figure 10.16. Here the pin connections for an 8-pin DIP (dual-inline) package are given as well. The 556 timer is a 14-pin DIP circuit which contains two 555 timers, while the 7555 is a low-power version of the 555.

Astable Operation

Figure 10.17 shows the 555 connected as a square-wave generator. When the output is low, or zero, the capacitor discharges via the internal transistor connected to pin 7 through resistor R_2. When the voltage across the capacitor reaches $V_+/3$, the output goes high, or V_+, and the transistor is turned off. The capacitor thus begins to charge through resistors R_1 and R_2. When the capacitor voltage reaches $2 V_+/3$, the output goes low and the cycle is repeated. The voltage across the timing capacitor as it relates to the output voltage is shown in Figure 10.18. Charging and discharging times are therefore given by (recall the RC circuit exponential charge and discharge characteristics in Chapter 2)

$$t_1 = 0.693(R_1 + R_2)C \quad \text{charging, output high} \qquad \textbf{(10-14)}$$

$$t_2 = 0.693\, R_2 C \qquad\qquad \text{discharging, output low.} \qquad \textbf{(10-15)}$$

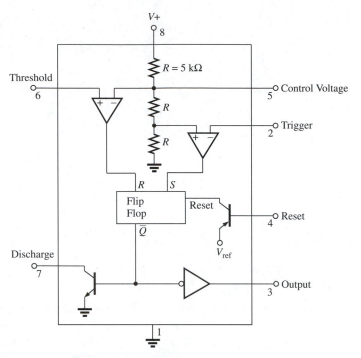

FIGURE 10.16 **555 timer equivalent circuit.**

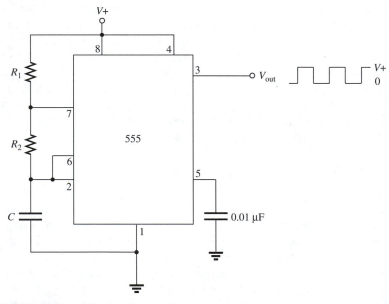

FIGURE 10.17 **555 timer in astable operator mode.**

The frequency of oscillation is thus given by $f = 1/T$ where $T = t_1 + t_2$ or in terms of R and C

$$f = 1.44/(R_1 + 2R_2)C \qquad \textbf{(10-16)}$$

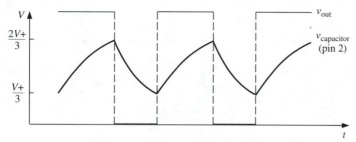

FIGURE 10.18   Voltage relationships in astable 555 oscillator.

and the *duty cycle, D.C.,* the ratio of the time the output is low to the total period is given by

$$D.C. = t_2/(t_1 + t_2) = R_2/R_1 + 2R_2. \qquad \textbf{(10-17)}$$

Usually the duty cycle is defined as the ratio of the high-output time to the total period; however, here we maintain the manufacturer's definition of the 555 duty cycle.

A symmetrical square wave has a duty cycle of 0.5. The voltage is high for the same amount of time that it is low; however, Equation 10-17 suggests that it is impossible to get *D.C.* exactly equal to 0.5. Of course, *D.C.* can be made very close to 0.5 if the value of R_1 is much smaller than R_2.

Since it varies the reference voltage of the comparators, pin 5 of the 555 timer can be connected in a number of ways to create some rather interesting circuits. Most data books give a variety of such applications which need not be discussed here.

Monostable Operation

The circuit shown in Figure 10.19 is similar to the astable multi-vibrator circuit, with a few exceptions. The external capacitor is initially held dis-

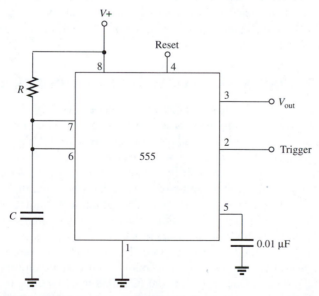

FIGURE 10.19   555 timer in monostable mode.

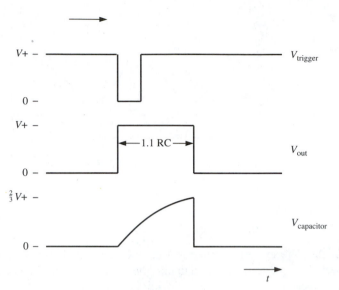

FIGURE 10.20   Voltage relationships in monostable 555 circuit.

charged and the output is low. When a negative-going trigger pulse of voltage less than $V_+/3$ is applied to the trigger input, the output goes high. The capacitor charges for a period equal to 1.1 RC, after which the voltage across the capacitor is $2V_+/3$, and the output goes low. If another trigger pulse is applied during the charging period, nothing is changed. A similar pulse to the reset input will, however, terminate the charging and return the monostable to its original state. If the reset is not used, it should be connected to V_+ to ensure stable operation. Figure 10.20 shows the timing diagram for this mode of operation. Since the output pulse is of a chosen duration, regardless of the trigger pulse duration, and since the circuit produces only one pulse per trigger, it is often called a "one-shot" circuit.

Loads can be connected between the V_{out} terminal and ground, for current flow when the signal is high, or between V_{out} and V_+ for current flow when the signal is low.

10-6   PRECISION WAVEFORM GENERATOR: THE ICL8038

With today's family of integrated circuits, a square-wave signal may be produced relatively easy. Through a variety of techniques such as filtering, duty-cycle variance, differentiating and integrating, other useful signals such as the triangle, ramp, pulse and sine waves can be produced from the square wave. Circuits which accomplish the production of these waveforms often require multiple chips and a number of design decisions, and then are rarely operable over a broad range of frequencies. Often such limitations are acceptable, but sometimes an application calls for an adjustable frequency and duty cycle, high-quality sine or triangle signal. The design of a circuit to accomplish this is far from trivial. Luckily, some manufacturers have designed single-chip circuits which produce low distortion waveforms using little external circuitry. One such IC is the ICL8038 manufactured by Intersil.

The ICL8038 is a precision waveform generator which will produce high-quality square, triangle and sine waves (all at the same time) over a frequency range of 0.001 Hz to over 300 kHz. The signal can be swept over a given range, or frequency modulated, by an externally applied control voltage. Figure 10.21 shows the pin-out arrangement of this 14-pin, DIP chip, while Figure 10.22 shows a typical application. The frequency of the output signal is simply given by $f = 0.15/(RC)$ for the circuit shown. Data sheets

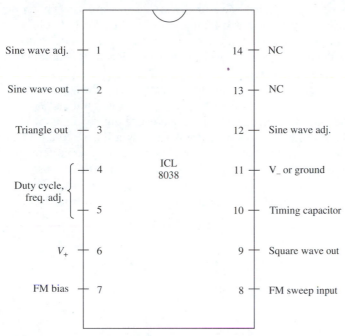

FIGURE 10.21   Pin arrangement of Intersil ICL 8038 function generator IC.

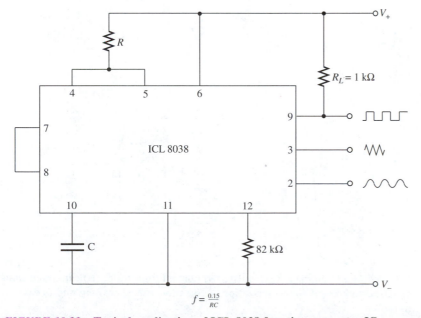

$$f = \frac{0.15}{RC}$$

FIGURE 10.22   Typical application of ICL 8038 function generator IC.

are included in the appendix which give other applications as well as specifications for the integrated circuit.

10-7   CHARACTERIZATION OF WAVEFORMS

Before proceeding, it is advantageous to spend some time defining real waveform parameters. The preceding text has assumed that a sine-wave oscillator produces a "pure" sine wave or that a square wave generator's output has extremely sharp leading and trailing edges and perfectly flat horizontal portions. In practice, this is not always the case. In several circuits developed earlier, the quality of the waveform was mentioned. It is appropriate to discuss waveforms quantitatively.

   Although we are generally more familiar with the sine wave, characterization of the square wave is both more straightforward and more useful, in that the latter is used extensively in pulse and digital circuitry. Consider the real single-cycle square-wave pulse illustrated in Figure 10.23. The so-called square-wave output from a real oscillator is clearly imperfect; the leading edge is rounded. The principal features of these imperfections are labeled in the figure: rise time, overshoot and tilt, or sag. The rise time is the time needed for the leading edge of the square wave to go from 10 to 90% of its

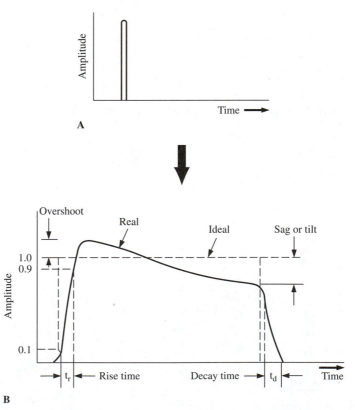

FIGURE 10.23   Square wave pulse: (A) normal time scale; (B) expanded time scale showing usual characterizations—rise time, overshoot, sag and decay time.

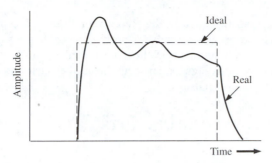

FIGURE 10.24   **Ringing in a square wave pulse.**

full amplitude. The overshoot is the amount of initial rise above the nominal amplitude and the sag is the amount of decay below the nominal amplitude at the end of the half cycle. In cases of severe overshoot, "ringing" is observed; this is shown in Figure 10.24. In most low-frequency situations, these imperfections are not readily apparent. As the frequency of the signal increases, they become much more obvious. This is understandable, because the circuit reactances increasingly degrade the wave shape at higher frequencies.

The product of the rise time, t_r, and the upper frequency cutoff, f_2, is a constant,

$$t_r f_2 = 0.37 \qquad\qquad (10\text{-}18)$$

where the upper frequency cutoff is that frequency at which the voltage gain drops to $1/\sqrt{2}$ of this midrange value, the -3dB point. This result is completely consistent with the constancy of the gain–bandwidth product associated with amplifiers. It can be seen from this equation that an amplifier that is to handle pulses with very short rise times must have a very good high-frequency response.

Waveforms can be, and often are, distorted as they are processed through a circuit. The amount of distortion is often predictable for a quality waveform; however, a similar waveform already having some distortion will be processed in a manner that produces unpredictable extraneous shapes. This can be of critical importance when the information contained in the waveform depends on its shape.

The idea that waveforms contain information is fundamental to their proper use in electronic circuitry. The most familiar example of this is in radio broadcasting, where the shape and the frequency are both important. In analog instrumentation, the information may not be as clearly understood. The amplitude of the waveform is often related to a physical quantity, while the frequency is often selected simply for the convenience of the operator. The phase, and sometimes the shape of the wave as well as the amplitude, are the waveform features generally useful in the analog field. In most of these situations, however, the frequency remains relatively unimportant in terms of useful information. On the other hand, the field of digital circuitry makes extensive use of frequency information. Because the speed of information gathering and processing is so important, high frequencies are very common. In the basic sense, the amplitude is unimportant in the digital information area; most frequently, the presence or absence of a pulse is all that must be measured. Because a pulse is nothing more than a square wave, its qualities demand our attention. When the digital signal is processed, the appearance of spurious signals can complicate the retrieval of information carried by the series of pulses.

10-8   SOME EXAMPLE CIRCUITS

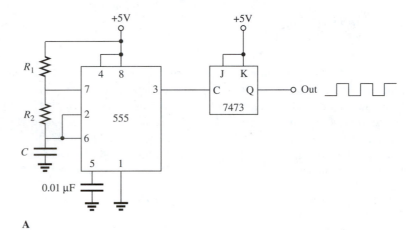

A

FIGURE 10.25A   **555 Astable Multivibrator with 50% duty cycle. 7473 is a clocked flip-flop connected as a binary divider.**

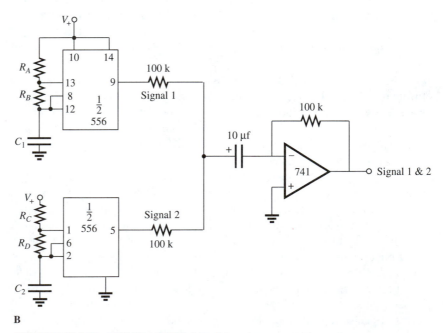

B

FIGURE 10.25B   **556 (Dual 555) based two-tone generator.**

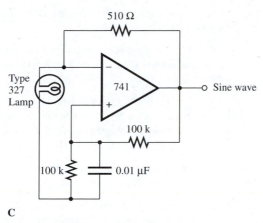

C

FIGURE 10.25C Wein bridge based sine wave generator using positive temperature coefficient lamp for stability.

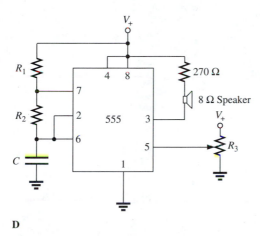

D

FIGURE 10.25D Voltage-Controlled Oscillator (VCO) made with 555. An increase in voltage at pin 5 caused by potentiometer voltage divider R_3 increase the frequency of the signal. Thus, R_1, R_2 and C control the signal frequency as well as the voltage on pin 5.

10-9 QUESTIONS AND PROBLEMS

1. Design a circuit that produces 5 V pulses with 1 ms of duration every second.

2. Design a 555 circuit which generates pulses of $1\,\mu s$ with a duty cycle of 30%.

3. Sketch the waveforms at v_a, v_b and v_{out} for the comparator relaxation timer in Figure 10A. What is the relationship between the frequency of the output signal, the connected external components and the DC input voltage V_{in}?

4. Design a safe circuit which converts 60 Hz, 120 V AC line voltage to a 60 Hz square wave with amplitude of $\pm$ 5 V.

5. Using a UJT, design a sawtooth oscillator which produces a signal of 1 kHz.

6. Design a circuit which converts a 1 kHz square wave to a triangle wave.

7. Design a circuit which converts a 1 kHz triangle wave to a square wave.

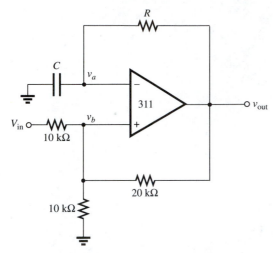

FIGURE 10A **Circuit for problem 3.**

8. Design a one-shot circuit which produces a 1 ms pulse when a 1 s pulse is applied.

9. Discuss what is meant by duty cycle for square, pulse, sine and triangle waves. Sketch waveforms of these types which have 30% duty cycles.

10. Design a voltage-controlled oscillator circuit; that is, one that produces a signal that varies in frequency when a variable DC input voltage is applied. Choose the frequency range to be centered about 10 kHz.

11. Is it possible to produce a square wave using a 555 timer that has exactly a 50% duty cycle? Explain.

12. A pulse signal is needed to drive an accurate clock at 100 kHz. Discuss why a 555 timer would be an unwise choice for such a circuit design.

13. Discuss how a square wave might be converted to a sine wave using filters.

Digital Basics

11-1 INTRODUCTION

Very few areas of human endeavor have undergone a more explosive growth than the field of digital electronics. From a technological standpoint, integrated circuit digital electronics is one of the most impressive accomplishments of the United States' space program of the sixties. The spin-offs from it still continue, from calculators and microprocessors to digital watches and video games. As we settle into the 1990s, no other nation yet has the computer development and capability which we presently enjoy here. The field of digital electronics is so much a part of our scientific and technological base that we must constantly inquire into its unrealized full potential.

The transition from analog to digital is not an easy one. The change is largely one of perspective. The components are similar. The circuits are not very different, either. Yet the meanings given the circuit responses can be significantly different. The rapid development of the digital field, to which both computer mathematicians and electronic engineers contributed, has created a "language" which is a curious admixture from both areas. New and unusual words and symbols are applied freely to operations and devices. Yet, the words and symbols have specific meanings which must be understood and followed. This chapter and the ones which follow will present this new language.

This chapter develops the basic ideas and components of digital circuitry: gates and flip-flops. The particular uses are described with some emphasis upon Boolean algebra. In order to give the reader some appreciation for past developments, a brief survey is included, but the emphasis in this and the following chapters is on those devices encountered in current use and practice. For those in particular, sufficient detail is provided that the reader can actually use them.

The understanding of digital electronics is rooted in the concept of having *two states;* no more and no less. How one refers to the two states is a matter of both convention and personal preference. One of the easiest perspectives to use regarding two-state logic involves the concepts of ON and OFF. A device is either on or off; it is never partially on or 72% off. More-

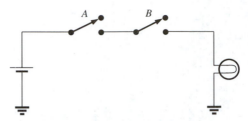

FIGURE 11.1   **An AND gate developed from switches.**

over, if it is not on, it must be off. Of course, the opposite condition is also true. There are other two-state designations which have garnered some acceptance and use. These include true/false, high/low, and one/zero. In fact, the last of these is used extensively because of its symbolic utility. The matter of assignment of the quality of an electrical signal to one of the two states will be dealt with later, once the elements of this digital domain are firmly in hand. In general the states correspond electronically to the power-supply positive voltage (high, true, 1) and ground (low, false, 0), but this is not always the case, as we shall see.

11-2   GATES

Gates are the fundamental building blocks of digital circuitry. The devices function by "opening" or "closing" to admit or reject signals. Only a handful of different kinds of gates are available; through combinations of these, a vast array of gating circuits can be assembled.

AND Gate

Consider the circuit shown in Figure 11.1. The switches, *A* and *B,* are either open or closed and the lamp is either on or off as a result of the switch combinations. The only set of switch combinations which will turn the lamp on is *A* closed *and* B *closed.* Because both switches *A* and *B* must be closed for the lamp to be on, the circuit is called an AND gate. In a digital AND gate the inputs must both be high to cause the output to be high; all other combinations yield a low output. Table 11.1 shows the possible combinations and the common shorthand convention for the designations of the states, where a "1" and "0" symbolize the two states, high and low. Such a tabular designation is called a *truth table.*

The accepted electronic symbol for an AND gate is illustrated in Figure 11.2. This particular one would be described as a two-input AND gate, because it has two inputs. AND gates with three, four and even eight inputs are also readily available. The logic of the AND gate holds true for gates with more than two inputs; that is, when all inputs are high, the output is high; otherwise the output is low.

OR Gate

The parallel combination of the switches used in the previous circuit is shown in Figure 11.3. For this circuit, the lamp is on when one or both of

TABLE 11.1   Truth Table for AND Gate

Switch A	Switch B	Lamp
open	open	off
closed	open	off
open	closed	off
closed	closed	on

A	B	Out
0	0	0
1	0	0
0	1	0
1	1	1

when "1" = closed/on
     "0" = open/off

FIGURE 11.2   **Electronic symbol for an AND gate.**

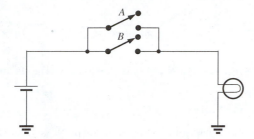

FIGURE 11.3   An OR gate developed from switches.

TABLE 11.2   Truth Table for OR Gate

A	B	Lamp
open	open	off
open	closed	on
closed	open	on
closed	closed	on

A	B	Out
0	0	0
0	1	1
1	0	1
1	1	1

when "1" = closed/on
"0" = open/off

the switches is closed. We designate this an OR circuit since the description of its function is when *A or B* is closed, the light is on. Note that when *A and B* are closed, the light is also on. The truth table, Table 11.2, illustrates this operation. The electronic symbol for an OR gate is given in Figure 11.4. In this instance a two-input OR gate is shown. OR gates are also available with three, four and eight inputs. With more than two inputs, the logic is the same; when any of the inputs is high, the output is high and only when all inputs are low is the output low.

FIGURE 11.4   Electronic symbol for an OR gate.

Inverter

Consider the situation depicted in Figure 11.5. When V_{in} is low, or at ground, there is no voltage on the base of the transistor. The transistor does not conduct and V_{out} is equal to V_{cc}, or, as we have defined in logic language, high. When V_{in} is high, the transistor is conducting and since the collector-emitter voltage drop is quite small, V_{out} is low. These conditions are summarized in Table 11.3, using the previous definitions of 1 and 0 for high and low, respectively. Since this circuit inverts the input signal, it is called an inverter, negator or NOT gate. Figure 11.6 shows the electronic symbol for an inverter. The indication of signal inversion is the circle on the output lead just after the triangle. A triangle symbol of this type without the circle indicates a buffer (driver) gate which passes a digital signal unchanged, or non-inverted. Buffer gates, both inverting and non-inverting, are sometimes useful in isolating gates or allowing them to drive higher loads.

TABLE 11.3   Truth Table for Inverter

In	Out
0	1
1	0

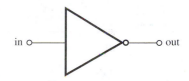

FIGURE 11.6   Electronic symbol for an inverter or negator.

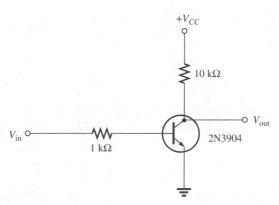

FIGURE 11.5   Transistor inverter.

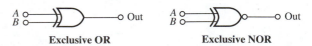

FIGURE 11.7 Synthesis of a NAND gate from an AND gate and an inverter.

NOR and NAND Gates

OR gates and AND gates can be negated to form NOR (Not OR) and NAND (Not AND) gates. The synthesis of a NAND gate from an AND gate and inverter is shown in Figure 11.7 along with the signal symbol used for a NAND gate. Here again, note the circle on the output of the gate indicating the negated feature. Figure 11.8 shows the symbol for a NOR gate. The truth tables for the NAND and NOR gates are given in Table 11.4.

FIGURE 11.8 Electronic symbol for a NOR gate.

TABLE 11.4   Truth Table for AND, NAND, OR and NOR Gates

			Out		Out
A	B	AND	NAND	OR	NOR
0	0	0	1	0	1
1	0	0	1	1	0
0	1	0	1	1	0
1	1	1	0	1	0

TABLE 11.5   Truth Tables for Exclusive OR and Exclusive NOR Gates

Exclusive OR			Exclusive NOR		
A	B	Out	A	B	Out
0	0	0	0	0	1
0	1	1	0	1	0
1	0	1	1	0	0
1	1	0	1	1	1

Exclusive OR and Exclusive NOR Gates

To complete our survey of simple gates, we introduce the Exclusive OR and Exclusive NOR gates often referred to as XOR and XNOR. These have truth tables similar to those for the OR and NOR gates previously mentioned, but differ in a way depicted in Table 11.5. For the XOR gate, the output is high when only one input is high and the other is low, not when both are high as with the OR gate. The XNOR gate yields a high output when both inputs are the same, whether they be high or low, and a low output when the inputs are different. The schematic representation of these gates is shown in Figure 11.9.

Schmitt Trigger

The characteristics of the Schmitt Trigger have been covered in Chapter 9, but as a reminder, the scheme is a voltage-level-sensing circuit with a hysteresis effect. This is seen in Figure 11.10, where a sine-wave input is applied to the inputs of the 4093B CMOS NAND gate shown in the diagram. With its two inputs connected together, a NAND gate functions as an inverter (check the NAND gate truth table to verify this). The difference between V_{ON} and V_{OFF} is the hysteresis voltage and is sometimes designated as the dead zone.

A Schmitt Trigger is extremely useful in cleaning up a noisy pulse signal, such as that from a mechanical switch or transducer. By passing the noisy pulse through a Schmitt Trigger, a virtually noise-free pulse is created. This is illustrated in Figure 11.11 which uses two CMOS Schmitt Trigger inverters of a 40106B chip. The hysteresis for this particular IC is 4 V when powered by a 10 V supply.

One gate of the 40106B Schmitt Trigger inverter can be used along with a resistor and capacitor to form a simple, low-power oscillator circuit. The

Exclusive OR   Exclusive NOR

FIGURE 11.9   Symbol for Exclusive OR and Exclusive NOR gates.

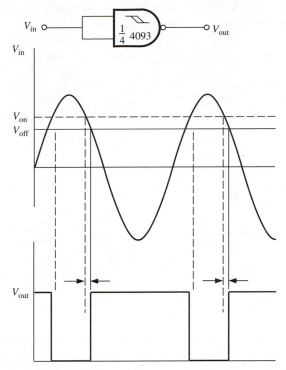

FIGURE 11.10   Schmitt Trigger performance. (Notice the symbol inside the NAND gate schematic which indicates that the device is a Schmitt Trigger type.)

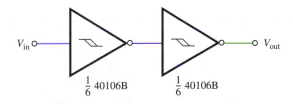

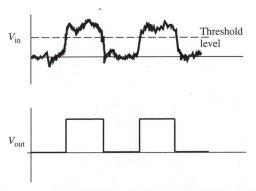

FIGURE 11.11   Use of Schmitt Triggers to "clean up" a noisy square wave.

circuit shown in Figure 11.12 is such an oscillator circuit. Note that the timing of the circuit transitions is dependent upon the threshold voltages V_{T+} and V_{T-} and the supply voltage V_{DD}. Although the predictable accuracy of such a circuit is low, its simplicity makes it quite useful as an oscillator circuit.

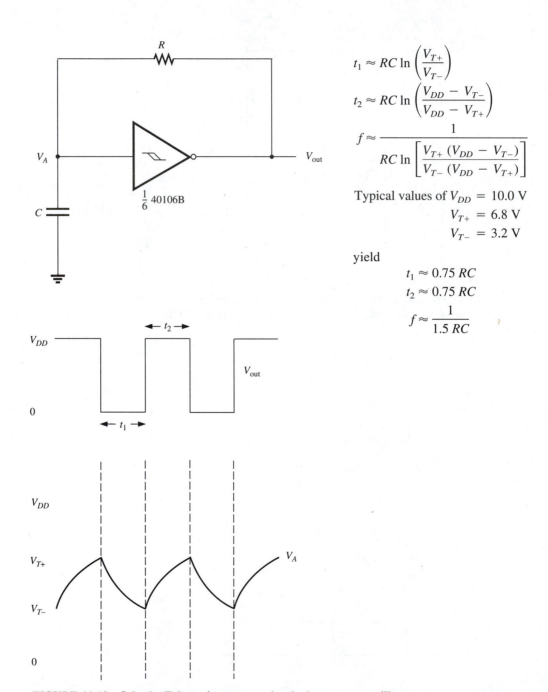

$$t_1 \approx RC \ln \left(\frac{V_{T+}}{V_{T-}} \right)$$

$$t_2 \approx RC \ln \left(\frac{V_{DD} - V_{T-}}{V_{DD} - V_{T+}} \right)$$

$$f \approx \frac{1}{RC \ln \left[\dfrac{V_{T+} (V_{DD} - V_{T-})}{V_{T-} (V_{DD} - V_{T+})} \right]}$$

Typical values of $V_{DD} = 10.0$ V
$$V_{T+} = 6.8 \text{ V}$$
$$V_{T-} = 3.2 \text{ V}$$

yield

$$t_1 \approx 0.75 \, RC$$
$$t_2 \approx 0.75 \, RC$$
$$f \approx \frac{1}{1.5 \, RC}$$

FIGURE 11.12 **Schmitt Trigger inverter as simple, low-power oscillator.**

11-3   SURVEY OF IC GATES

The realization of the logic operations just discussed began with vacuum tubes and relay circuits. These were completely superseded by discrete transistor circuits, which, in turn, have been completely replaced with integrated circuit devices. Several IC logic product "lines," or groups of compatible devices, are available. Two of these are CMOS (Complementary Metal Oxide Semiconductor) and TTL (Transistor-Transistor Logic). Within each group, there may be a number of different families. Families within the TTL line include such numeric designations as 74, 74L, 74LS, and 74ALS. Several lines with their associated families will be described here. Before we do so, some definition and understanding of the more important specifications attached to the product lines must be developed. The specifications of most general interest are fanout, threshold levels, propagation delay and power dissipation.

The term *fanout* refers to the number of other devices of that same logic line which can be successfully attached to the device under study before unexpected behavior occurs. Typical fanout for CMOS and TTL circuits in a given family is approximately 10. When mixing family types, special care must be taken to ensure that the current out of the device is ample to drive the inputs of the connected circuits.

Threshold levels are the voltages at which the circuits realize a low or high input or output signal. Ideally, a high output would be of a given precise voltage and a low output another voltage, but these design limitations are quite impractical. The range over what is acceptable as a signal and what is not varies from line to line. Typical specifications for TTL circuits are shown in Figure 11.13.

The *propagation delay, T_d,* is the time that a signal takes to get through a device. Specifically, it is defined as the difference between the time at which a signal change is initiated at the input and the time at which it is observed at the output. Obviously, this specification is of intense interest to those who design high-speed instruments, where speed is of prime consideration.

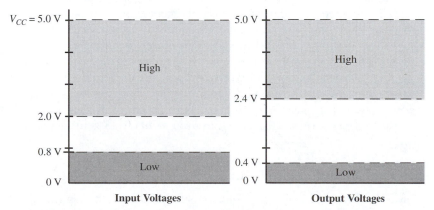

FIGURE 11.13   **Voltage ranges and threshold voltages for typical TTL component.**

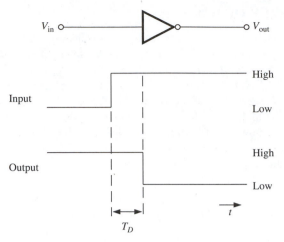

FIGURE 11.14 Propagation delay for inverter circuit.

Figure 11.14 illustrates this parameter, which is on the order of 10 ns for a TTL gate.

The *power dissipation, P_d,* is a measure of the power requirements of a logic device. Because a large number of devices is likely to be used, the total power requirement defines the power-supply demands of the system.

One final point is quite worthy of mentioning before we continue. The logic families shown here, comprised of individual, or discrete, components are given for explanatory reasons. Actual use of logic circuits is quite simple and does not involve the complex design calculations and decisions encountered in previous discrete component circuits.

Transistor-Transistor Logic (TTL)

The logic line which is currently in widest use is transistor-transistor logic (TTL). Most major semiconductor manufacturers offer a wide range of devices within this line. The basic two-input NAND gate is given in Figure 11.15. The operation of this circuit may be understood by recognizing that grounding either input (or both) prevents current from reaching the base of Q_2, thereby shutting it off. When Q_2 is shut off, Q_3 is turned on and Q_4 is shut off. The output, taken at the collector of Q_4, is approximately V_{CC} less V_{BE} of Q_3 and less the forward voltage drop of the diode, or 3.6 V. When voltages in excess of 0.7 V are applied to the inputs, the emitters of Q_1 are reverse biased. This means that Q_2 is on. With Q_2 on, Q_3 is shut off and Q_4 is turned on. This causes the output to drop virtually to zero (actually to $V_{CE(sat.)}$).

Emitter-Coupled Logic (ECL)

The emitter-coupled logic (ECL) line represents one of the highest speed logic lines commercially available. (It is also one of the most expensive lines on a per-gate basis.) The speed is achieved in ECL by nonsaturating operation. That is, the transistors are controlled so that neither "full on" nor

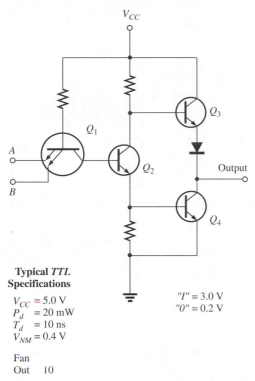

V_{CC}

Q_1

A

B

Q_2

Q_3

Output

Q_4

**Typical *TTL*
Specifications**

V_{CC} = 5.0 V
P_d = 20 mW
T_d = 10 ns
V_{NM} = 0.4 V

Fan
Out   10

"I" = 3.0 V
"0" = 0.2 V

***FIGURE 11.15*   Transistor-Transistor-Logic (TTL) gate with its
specifications. Note Q_1 is a multiple emitter *npn* transistor.**

"full off" is ever reached. As a result, the junction capacitance is reduced
significantly.

A two-input NOR gate in ECL is illustrated in Figure 11.16. Notice that
the ECL line requires two power-supply voltages, $V_{EE} = -5.2$ V and V_{BB}
$= -1.15$ V (which is midway between the "1" and "0" voltages, -0.75 V
and -1.55 V, respectively). With no signal applied to either A or B, the input
transistors Q_1 and Q_2 are off. Notice that the emitters of Q_1, Q_2 and Q_3 are
common. Under the off condition, the emitters are at approximately -1.90
V ($= V_{BE} = V_{BB}$). If a voltage (-1.55 V) equivalent to a "0" is applied to
either or both inputs, nothing happens since V_{BE} is not overcome. The current
in the common emitter resistor is proportional to V_{BB} since the transistors are
operated under nonsaturating conditions. Moreover, $I_2 > I_1$; hence V_1 is high
and V_2 is low. If a "1" (-0.75 V) is applied to A, then $V_A > V_{BB}$. In this
situation, $I_1 > I_2$ and V_1 is low while V_2 is high. Since the output is taken at
the emitter of Q_5, it is low under this condition. When $V_{BB} > V_A$ or V_B, the
output is high. Since nonsaturation conditions are maintained throughout the
device's operation, a large number of similar devices can be driven by the
output. In the case of ECL, more input transistors can be added in parallel
with Q_1 and Q_2 to obtain a large fan-in as well.

Although ECL logic is extremely fast, it does not have the general use
one might expect in today's common instrumentation applications. The devices themselves suffer from no major engineering difficulties, but the

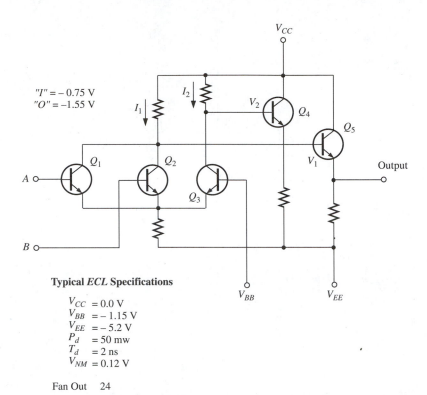

"I" = – 0.75 V
"O" = –1.55 V

Typical *ECL* Specifications

V_{CC} = 0.0 V
V_{BB} = – 1.15 V
V_{EE} = – 5.2 V
P_d = 50 mw
T_d = 2 ns
V_{NM} = 0.12 V

Fan Out   24

FIGURE 11.16   Emitter-Coupled-Logic (ECL) gate with its specifications.

sub-nanosecond speeds at which they excel bring about high-speed circuitry design problems which are not easily addressed by the novice designer.

Complementary Metal Oxide Semiconductor (CMOS) Logic

The use of FETs as logic devices is to be expected, since they offer certain advantages over bipolar transistors, particularly with their very high input resistances. The high gate resistance means power dissipation of less than 1 mW/gate. When a *p*-channel device is used in a complementary manner with an *n*-channel device, only a single power supply is needed; hence the name complementary metal oxide semiconductors (CMOS). The logic devices currently employing IGFETs are manufactured for enhancement mode operation. The transfer characteristics for *p*-channel and *n*-channel enhancement mode devices are given in Figure 11.17 A and B. Notice that with $V_{GS} = 0$, neither device conducts. An example of this complementary arrangement is given in Figure 11.17C. Notice that the positive power supply is connected to the *p*-channel source lead, while the *n*-channel source lead is grounded. The drain leads are connected together and form the output. When the input is zero, the *n*-channel device is off. On the other hand, the *p*-channel device is on since $V_{SG} = V_{DD}$ or $-V_{GS} > 0$. With the *p*-channel-on resistance of a few hundred ohms and the *n*-channel-off resistance of a few megohms, the output is essentially V_{DD}. If V_{IN} is made sufficiently positive to turn off the *p*-channel device ($V_{IN} \approx V_{DD}$), the *n*-channel device is turned on by V_{IN} ($V_{IN} > V_{GS}$). The output becomes virtually zero. Thus, this configuration acts as an inverter.

n Channel Enhancement Model *Mosfet*

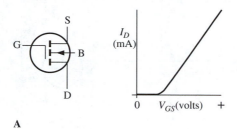

A

p Channel Enhancement Model *Mosfet*

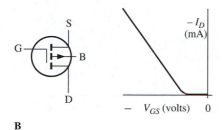

B

FIGURE 11.17A and B   **Enhancement-mode MOSFET operation.**

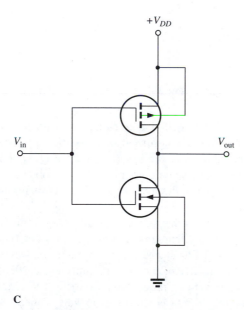

C

FIGURE 11.17C   **CMOS inverter circuit.**

A CMOS two-input NOR gate is depicted in Figure 11.18. Notice that each input resembles the configuration of the previous figure, except that the p-channel MOSFETs are in series, while the n-channel MOSFETs are in parallel. The operation of the circuit can be understood by building on the previous circuit. When the input to A is zero, Q_1 is on and Q_4 is off. If the input to B is also zero, Q_2 is on and Q_3 is off. Under these conditions the output is

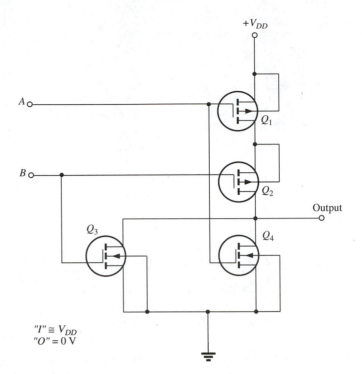

Typical *CMOS* Specifications

V_{DD} = 3 to 15 V
P_d = 0.3 mW
T_d = 200 ns
V_{NM} = 0.4 V_{DD}

Fan Out   50

FIGURE 11.18 **A CMOS NOR gate.**

essentially at V_{DD}. However, if A is zero and B is positive and sufficiently large to turn on Q_3, then Q_2 is turned off. The output drops to a very low value (approaching zero) since the low resistance of Q_3 is in parallel with the high resistance of Q_4. The same output condition is obtained when B is zero and A is positive. Likewise, the low output condition is produced when both A and B are positive.

CMOS logic offers some substantial advantages, including very low power dissipation and excellent noise margin; at the same time, however, it is generally slower than TTL. It has a further, unique advantage. That is, its operation is quite insensitive to the power supply used (4000B series CMOS logic ICs can be powered in a range from 3.0 to 18.0 V). Moreover, its noise margin may be increased by the simple expedient of increasing the value of V_{DD}.

CMOS logic is quite susceptible to damage by electrostatic energy. An electrostatic discharge into a CMOS chip can easily destroy the thin gate oxide insulator in the device. Early in the development of CMOS logic, such damage was more common. Today's devices provide some protection against electrostatic discharge. In fact, National Semiconductor's CD4000 line can withstand a 400 V discharge from a 100 pF capacitor through a 1.5 kΩ resistor. Special care must still be taken when working with this logic line, since the act of walking across a tiled floor or even the sliding of chips in a plastic storage box can generate a potential of thousands of volts. In general, soldering iron tips should be grounded and the table top of your workbench should be conductive and grounded. CMOS chips should be stored in special shipping tubes which minimize electrostatic discharge or inserted into conducting foam. In many laboratories, grounded wristbands are worn to minimize electrostatic build-up which occurs simply by sliding movements of the body.

Summary

This survey of logic lines is not thorough, but is intended to be representative of the types of IC logic devices available. One may properly deduce that a constant relationship exists between speed, power and ease of use. ECL is the highest speed line, but consumes the most power. At the other end of the spectrum, we find CMOS. It appears that TTL represents a compromise between the two extremes.

Within each group of logic, there are further breakdowns into families. For example, we already mentioned the TTL line has families such as 74, 74L, 74S, 74LS, 74F, 74 AS and the 74ALS. Each of these families has different characteristics, mainly in the areas of speed and power consumption. The CMOS line has the families 4000B, 74HC, 74HCT and 74AC. (To confuse matters further, there is also a similar line with the same lettering but with a 54 designation instead of the 74 designation. The 54 simply indicates that the device will perform adequately over a wider temperature range than its counterpart.) Here the differences are again in speed and power consumption, with an additional characteristic that the 74-series CMOS chips have the same pin-outs and power supply requirements as the 74-series TTL chips. This was done to make the two series compatible, to some extent. The 4000B series is quite different in power supply requirements, numbering, speed and pin-outs. For example, a 74LS14 is a TTL Hex Inverter (six inverters on one chip) with Schmitt Trigger, the same as the 74HCT14 which is a CMOS version. The pin-outs are identical and both run from a 5.0 V supply. In the 4000B CMOS series, a 4014B is an 8-stage static shift register and can be powered from 3 to 18 V—an entirely different animal!

Table 11.6 gives a brief summary of the various CMOS and TTL families. This table is meant as a general reference; individual chip specifications may vary. For example, buffers are designed to produce high output currents and will therefore exceed the output current rating listed in the table. Not listed are the older 74 and 74L series of TTL circuits, among others, that are still respectable for many applications and can still be found, often at very low prices. Note that the power requirements given in the table are typical for a simple gate circuit such as a NAND gate. More complex chips, such as multiplexers, will undoubtedly require more power, but the table is a good comparison tool between families. Power consumption is rated at 1 Mhz operation. In this range CMOS still has an advantage over TTL, and at slower speeds, its power consumption decreases. In the static state, CMOS inputs draw virtually no current; only at a transition point is current required.

TABLE 11.6   CMOS and TTL Family Characteristics

	T_d (ns)	P_d(mW/Gate) @ 1 Mhz	V_{cc} (Volts) Min	Max	I_{out} (mA)
CMOS					
4000B	50	0.5	3	18	± 3
74HC	15	0.5	2	6	± 25
74HCT	15	0.5	4.5	5.5	± 25
74AC	5	0.1	2	6	± 24
74ACT	5	0.1	4.5	5.5	± 24
TTL					
74LS	10	2	4.75	5.25	+ 24/− 25
74S	3	20	4.75	5.25	+ 20/− 1.0
74ALS	4	1	4.5	5.5	+ 8/− 0.4
74AS	1.5	7	4.5	5.5	+ 20/− 2.0

General use is often dictated by what is readily available in the widest selection of flavors. When low power consumption and extended power-supply voltage ranges are required, such as in battery operation, the 4000B CMOS line is a wise choice. For low-speed applications, CMOS is often the best choice as well, since it draws virtually no current in the static state. When existing TTL circuits need to be compatible with new designs, the 74 HCT is a good choice since the line is readily available, but staying with the TTL line is acceptable as well.

11-4   BOOLEAN ALGEBRA

The two-state logic system or algebra was developed well over 100 years ago by an Englishman, George Boole. It remained virtually unused for almost a century until 1938, when Claude Shannon showed how the concepts of Boolean algebra were applicable to switching theory. The last quarter century has seen wide dissemination of Boole's concepts. The concepts of Boolean algebra are very useful in simplifying or reducing arrays of gates.

Boolean algebra includes just two mathematical operations: OR and AND. The OR operation is reminiscent of addition and is symbolized by a + sign. The statement, "If A is 1 or B is 1 or C is 1, then the output is 1" is written in equation form as

$$A + B + C = \text{out} \tag{11-1}$$

The AND operation is reminiscent of multiplication and is symbolized by a "·" sign. The Boolean equation,

$$A \cdot B \cdot C = \text{out} \tag{11-2}$$

is read as, "If A is 1 and B is 1 and C is 1, the output is 1." Sometimes, the operational symbol is omitted, as in $ABC = \text{out}$. This condition, of course, is frequently found in normal algebra as well. A third symbol is of special use in Boolean algebra, one of negation. It is symbolized by a bar over the letter, as $\overline{A}$. Since Boolean algebra has only two states, a negation (not . . .) statement has definite and precise meaning. For example, if A is not 1, it must be 0.

Some of the commonly encountered theorems of Boolean algebra are listed in Table 11.7. Many are self-evident, while others may require some study to prove the identity. Several approaches can be taken to establish these and other identities. We can construct truth tables and show, by complete comparison, that the results are the same. The eleventh theorem can be proved readily in this way; see Table 11.8. Another way to verify a theorem is to "derive" it using other theorems. For example, theorem 12 can be proved in this manner, by expanding the left side:

$$A\,(\overline{A} + B) = A \cdot \overline{A} + A \cdot B \qquad (11\text{-}3)$$

Since $A \cdot \overline{A} = 0$ from the eighth theorem,

$$0 + A \cdot B = A \cdot B \qquad (11\text{-}4)$$

Applying theorem 5, $A + 0 = A$, the proof is obtained,

$$A \cdot B = A \cdot B \qquad (11\text{-}5)$$

This derivation approach is useful in securing simpler solutions to logic statements. Consider $(A + C)(A + B)$. This may be expanded:

$$(A + C)(A + B) = A \cdot A + A \cdot C + A \cdot B + B \cdot C \qquad (11\text{-}6)$$

The terms containing A may be factored:

$$A \cdot A + A \cdot C + A \cdot B + B \cdot C = A(A + B + C) + BC \qquad (11\text{-}7)$$

Theorem 11 may be used to simplify it:

$$A(A + B + C) + BC = A + BC \qquad (11\text{-}8)$$

Thus,

$$(A + C)(A + B) = A + BC \qquad (11\text{-}9)$$

The last two entries in Table 11.7 are called *De Morgan's theorems* and are extremely useful in gate simplification. It is important to carefully observe the "mechanics" of applying De Morgan's theorems. It may be visualized as slicing through the bar over both letters and changing the symbol between them. (The theorem states that $\overline{A \cdot B} \neq \overline{A} + \overline{B}$.)

With De Morgan's theorems it is easy to develop AND, OR and NOR gates from NAND gates. When the inputs of a two-input NAND gate are connected together, it becomes an inverter or negator. Therefore, an AND gate is made with two NAND gates as shown in Figure 11.19A. In Figure 11.19B, the synthesis of an OR gate from three NAND gates is shown. When the output of an OR gate is negated, as in Figure 11.19C, a NOR gate is obtained. Notice that De Morgan's theorem was used to verify the OR gate. As an exercise, the reader should attempt to show how NOR gates may be used to synthesize the other gates.

One other important gate circuit is very useful, the Exclusive OR gate. Figure 11.20 indicates how it is generated from NAND gates. The outputs of each gate may be written down to derive the Exclusive OR conditions. The output of gate 1 is $\overline{AB}$. By De Morgan's theorem this becomes

$$\overline{AB} = \overline{A} + \overline{B} \qquad (11\text{-}10)$$

TABLE 11.7   Boolean Theorems

1. $A + A = A$
2. $A \cdot A = A$
3. $A + 1 = 1$
4. $A \cdot 1 = A$
5. $A + 0 = A$
6. $A \cdot 0 = 0$
7. $A + \overline{A} = 1$
8. $A \cdot \overline{A} = 0$
9. $\overline{\overline{A}} = A$
10. $A + A \cdot B = A$
11. $A(A + B) = A$
12. $A(\overline{A} + B) = A \cdot B$
13. $A + \overline{A} \cdot B = A + B$
14. $\overline{A} + A \cdot B = \overline{A} + B$
15. $\overline{A} + A \cdot \overline{B} = \overline{A} + \overline{B}$
16. $\overline{A \cdot B} = \overline{A} + \overline{B}$
17. $\overline{A + B} = \overline{A} \cdot \overline{B}$

TABLE 11.8   Proof of Theorem 11

A	B	A + B	A(A + B)
0	0	0	0
1	0	1	1
0	1	1	0
1	1	1	1

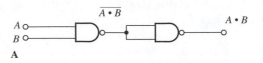

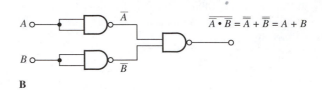

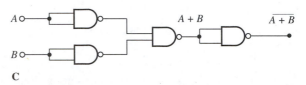

FIGURE 11.19 NAND gate implementation of Boolean statements. (A) AND, (B) OR, (C) NOR.

This output is NANDed with A by gate 2 to give

$$\overline{A(\overline{A} + \overline{B})} = \overline{A} + AB \qquad (11\text{-}11)$$

In an analogous manner, the output of gate 3 is

$$\overline{B(\overline{A} + \overline{B})} = \overline{B} + AB \qquad (11\text{-}12)$$

These latter outputs are applied to NAND gate 4, whose output is

$$\overline{(\overline{A} + AB)(\overline{B} + AB)} = (\overline{A} + AB) + (\overline{B} + AB) \qquad (11\text{-}13)$$

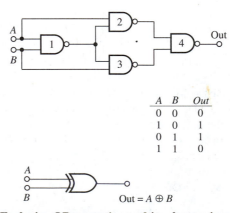

A	B	Out
0	0	0
1	0	1
0	1	1
1	1	0

Out = $A \oplus B$

FIGURE 11.20 Exclusive OR operation and its electronic symbol.

This reduces to

$$\overline{\overline{A}}\,(\overline{AB}) + \overline{\overline{B}}(\overline{AB}) = A(\overline{AB}) + B(\overline{AB}) \qquad \textbf{(11-14)}$$

Factoring out the $\overline{AB}$ term yields

$$\overline{AB}(A + B) = (\overline{A} + \overline{B})(A + B) \qquad \textbf{(11-15)}$$

This latter term can be expanded to

$$A\overline{A} + \overline{A}B + A\overline{B} + \overline{B}B \qquad \textbf{(11-16)}$$

Since $A\overline{A} = \overline{B}B = 0$, the appropriate result for the output is

$$A\overline{B} + \overline{A}B = \text{output} \qquad \textbf{(11-17)}$$

Figure 11.20 presents the truth table for the Exclusive OR. Because of its frequent use, the special symbol $\oplus$ is used to designate it, as shown in the figure. An Exclusive NOR gate is obtained when the output of the Exclusive OR is negated. This gating operation is symbolized by $\odot$.

11-5 NUMBERING SYSTEMS

Our earliest contact with the world around us forcibly introduces us to the decimal system. Because we are so familiar with this system, we do not readily appreciate its limitations or accept other, equally good counting systems. The binary system offers significant advantages over the decimal system in digital electronics. Any electronic computation system based on decimal counting would be unbelievably complex. On the other hand, the binary system has only two states and offers a straightforward approach to electronic counting and computation.

Any counting system, be it decimal or binary or anything else, is used to describe quantities. In the familiar decimal system, the digits 0 through 9 are used, while in the binary system only 0 and 1 are used. In either system, when the quantity being described exceeds the number of symbols (digits) available, the counting continues by use of two or more of the digits together. This is performed in the decimal system as follows. The number 389 uses three digits, and can be analyzed as consisting of $3 \times 10^2 + 8 \times 10^1 + 9 \times 10^0$. This is, of course, $300 + 80 + 9 = 389$.

The counting procedure in the binary system is completely analogous, except that there are just the two symbols: 0 signifies a "nothing quantity" and 1 signifies a "single" quantity. Representation of binary numbers is made in a manner analogous to decimal numbers, except that the numbers are powers of two. Consider the binary number 101011. To find its decimal equivalent, write each column as a *base two* number: $1 \times 2^5 + 0 \times 2^4 + 1 \times 2^3 + 0 \times 2^2 + 1 \times 2^1 + 1 \times 2^0$. This is $32 + 0 + 8 + 0 + 2 + 1 = 43$ in decimal notation. Two quantities, in binary notation, are symbolized as 10, while three quantities are symbolized as 11. The available symbols have been used up in the two columns. The next higher quantity to be counted, four, requires three columns for its representation: 100. Table 11.9 gives the representations of both decimal and binary numbers for the first twenty-five quantities.

Obviously, the operation of digital instruments can require very long binary numbers. For ease of manipulation, the binary numbers are broken up

TABLE 11.9 Equivalent Decimal and Binary Representations

Decimal	Binary
0	0
1	1
2	10
3	11
4	100
5	101
6	110
7	111
8	1000
9	1001
10	1010
11	1011
12	1100
13	1101
14	1110
15	1111
16	10000
17	10001
18	10010
19	10011
20	10100
21	10101
22	10110
23	10111
24	11000
25	11001

TABLE 11.10 Equivalent Decimal and Binary Coded Decimal Representations

Decimal	BCD
0	0000
1	0001
2	0010
3	0011
4	0100
5	0101
6	0110
7	0111
8	1000
9	1001

TABLE 11.11 Equivalent Decimal and Octal Representations

Decimal	Octal
0	0
1	1
2	2
3	3
4	4
5	5
6	6
7	7
8	10
9	11
10	12
11	13
12	14
13	15
14	16
15	17
16	20

into groups. Each group is called a *character,* and the number of binary digits in it is specified. A binary digit is called a *bit.* A *byte* consists of 8 binary digits, or bits.

Several *codes* have been developed which facilitate conversion from decimal to binary and back to decimal. The simplest code to understand is the *binary-coded decimal (BCD)* system. This system uses a 4-bit character to represent each of the nine decimal values. Examples of BCD codes are given in Table 11.10. The decimal number 389 becomes 0011 1000 1001 in BCD code. More extensive codes have been developed to meet particular needs. The *American Standard Code for Information Interchange (ASCII),* for example, uses a 7-bit character code to handle the alphabet, symbols, and punctuation marks, as well as numbers. Other codes will be introduced later.

Other systems besides binary may be encountered. The *octal* system is based on the use of the digits 0 through 7. These are applied in exactly the same manner as is the binary system, as can be seen by examination of Table 11.11.

The *hexadecimal* system uses characters and numbers to count in base sixteen. The "new" symbols for the hexadecimal system are given in Table 11.12. Compare the octal system with the hexadecimal system. Both counting systems maximize the efficiency of the bits per character. Table 11.13 is provided to allow comparisons among the several numbering systems discussed here. It is not a difficult matter to represent numbers in either the octal or hexadecimal system. The octal is a *base eight* system. Thus, the quantity in decimal units of 389 is represented in octal as 605_8; that is, $6 \times 8^2 + 0 \times 8^1 + 5 \times 8^0 = 384 + 0 + 5$. The representation of this octal number in binary form is 110 000 101. In a completely analogous manner, the hexadecimal representation of the same quantity is $1 \times 16^2 + 8 \times 16^1 + 5 \times 16^0$ or 185_{16}. Its binary form is 0001 1001 0101.

The reason for presenting the ideas of coding and of alternative numbering systems may not be clear at this point. Frequently, the input and the output signals in a computer or instrument system will be in decimal, while the manipulation of the signal will be in binary. A judiciously selected code will permit easy decoding back into the decimal output. The octal and hexadecimal systems are often used in computers because of their more efficient bit use. They are equivalent to a shorthand system for the computer. These modifications, and others, have been made to decrease computation time and to increase ease of use.

11-6 CODES

There are several ways in which a 4-bit character may be used to represent a decimal digit. The most straightforward way of doing this is, as previously mentioned, binary-coded decimal or BCD. It is sometimes called the 8 4 2 1 code because the most significant bit (MSB) determines whether the decimal value will be eight or less. The least significant bit (LSB) is the last one, the 1. The BCD code is nothing more than the 4-bit character representation of the decimal digit in pure binary. The *Excess 3* code is obtained by adding a binary 3 to the BCD representation of the decimal. This code offers the advantage over BCD of easier error detection. For example, the appearance of a 0000, a 0001 or a 0010 signals an error. The *Gray* code finds use in analog-to-digital conversions. The principle of the Gray code is that the representation of a decimal quantity is only 1 bit different from its predecessor and from its successor. The comparison of the three codes can be made through examination of Table 11.14. Naturally, other codes may be developed. One such example in use today is the *Excess 3 Gray*. In this case, the Excess 3 code is converted into the Gray code, rather than BCD being converted into the Gray. As would be expected, arrays of gates are used to convert from one code to another.

TABLE 11.12   Equivalent Decimal, Binary and Hexadecimal Representations

Decimal Number	Read-Out Display	Hexadecimal Number	Binary
0	0	0000	0
1	1	0001	1
2	2	0010	2
3	3	0011	3
4	4	0100	4
5	5	0101	5
6	6	0110	6
7	7	0111	7
8	8	1000	8
9	9	1001	9
10		1010	A
11		1011	B
12		1100	C
13		1101	D
14		1110	E
15	Blank	1111	F

TABLE 11.13   Comparison of Numbering Systems

Decimal	Binary	BCD		Octal	Hexadecimal
0	0		0000	0	0
1	1		0001	1	1
2	10		0010	2	2
3	11		0011	3	3
4	100		0100	4	4
5	101		0101	5	5
6	110		0110	6	6
7	111		0111	7	7
8	1000		1000	10	8
9	1001		1001	11	9
10	1010	0001	0000	12	A
11	1011	0001	0001	13	B
12	1100	0001	0010	14	C
13	1101	0001	0011	15	D
14	1110	0001	0100	16	E
15	1111	0001	0101	17	F
16	10000	0001	0110	20	10
17	10001	0001	0111	21	11
18	10010	0001	1000	22	12
19	10011	0001	1001	23	13
20	10100	0010	0000	24	14
21	10101	0010	0001	25	15
22	10110	0010	0010	26	16
23	10111	0010	0011	27	17
24	11000	0010	0100	30	18
25	11001	0010	0101	31	19
99	110011	1001	1001	143	63

TABLE 11.14   Common Coding Systems

Decimal	BCD	Excess 3	Gray	Excess 3 Gray
0	0000	0011	0000	0011
1	0001	0100	0001	0100
2	0010	0101	0011	0110
3	0011	0110	0010	0101
4	0100	0111	0110	0100
5	0101	1000	0111	1100
6	0110	1001	0101	1101
7	0111	1010	0100	1111
8	1000	1011	1100	1110
9	1001	1100	1101	1010

11-7   OPEN-COLLECTOR AND THREE-STATE LOGIC

Consider the circuit in Figure 11.21 which consists of two NAND gates of a single 7400 TTL chip. At first glance, the circuit could be a solution to the need for high output when any of the four inputs is low. If all inputs A through D are high, then the outputs of the individual gates V_1 and V_2 are equal and low, so no problem exists. If A is made low, the output of the top gate, V_1, goes high, but this conflicts with the output of the lower gate, V_2, which is still low. Here V_{out} may or may not become the state we predicted.

Open-Collector Logic

The first solution to this problem was accomplished by manufacturing gates which have open collectors at their outputs (Figure 11.15 without Q_3 and the diode attached to its emitter). You may think of the outputs as being two *npn* transistor–inverter switches (Figure 8.13) in parallel sharing the same load resistor. Figure 11.22 shows the circuit in question using an open-collector NAND gate chip. Note that a pull-up resistor, R_p, must be connected between the output and the positive power supply in order to connect the open-collector transistor to a power source. The value of R_p is not very critical; 1 kΩ is generally acceptable, but specifications are given in the data sheets for the particular chip.

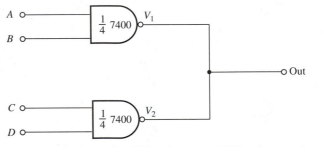

FIGURE 11.21   Poor logic configuration for output high when any input is low.

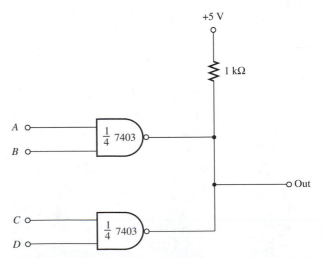

FIGURE 11.22   Open-collector gates yielding high output when any input is low.

Open-collector logic circuits are often useful in driving loads, since more current can be sourced through the pull-up resistor than through the gate. Here the load is connected to the positive power supply and not ground. Many basic gates are available in this configuration.

Three-State Logic

A second solution to the problem of connecting multiple parallel circuits is three-state logic. In addition to the usual binary output conditions, 1 or 0, three-state logic elements are capable of assuming a third—a very high output impedance. This latter state, when activated, overrides the other two possible states, essentially disconnecting the output of the circuit from the line. When the chip is enabled, the third state is deactivated, and the circuit behaves as it normally would. These types of devices allow multiple parallel circuits to be connected to one line, or bus, and to transmit a signal upon selection of the particular device. Three-state logic is often called Tri-State which is a trade name registered by National Semiconductor.

Figure 11.23 shows the truth table and schematic for a three-state buffer circuit. Notice that this particular buffer is enabled with a low signal, indicated on the schematic by the circle on the input of the enable line. Many computer logic elements, which are connected to common lines, or buses, are three-state. Other circuits such as digital-to-analog converters, which are designed to be interfaced to computers, also have enable controls.

11-8   FAMILY INTERFACING

Sometimes a design will call for a mixing, or more likely an interfacing, of different families of logic devices. One such example is an add-on computer board which uses 4000B CMOS chips which must connect to a TTL-based computer. We already mentioned some aspects of compatibility when we spoke of the relative ease of combining TTL with the HCT and ACT CMOS families. With other families and lines, the combination may be more difficult. TTL, HCT and ACT components share the same power-supply requirements, have compatible logic-level specifications and have ample output drive capabilities (TTL circuits of the same family may have fanouts of 10 or more, but mixing TTL families indiscriminately can often yield a fanout of 1 or 2). When mixing families, these three parameters demand special attention.

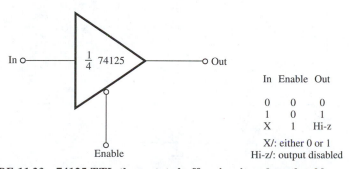

In	Enable	Out
0	0	0
1	0	1
X	1	Hi-z

X/: either 0 or 1
Hi-z/: output disabled

FIGURE 11.23   **74125 TTL three-state buffer circuit and truth table.**

Connection of CMOS to TTL is often best accomplished with the use of a buffer such as the 4049 inverting or the 4050 non-inverting buffer. By placing a buffer between the CMOS and TTL components, the buffer can drive up to two TTL loads. Such an arrangement is shown in Figure 11.24. Note that the buffer chip is powered from the same supply which powers the TTL component, which may not necessarily be the same as the CMOS supply. The supply grounds, however, must be made common.

Connection of TTL to 74HC, 74AC, 74C and 4000B (the latter powered by 5 V) can be accomplished by adding a pull-up resistor to V_{cc}, the 5 V power supply. Such a configuration is shown in Figure 11.25.

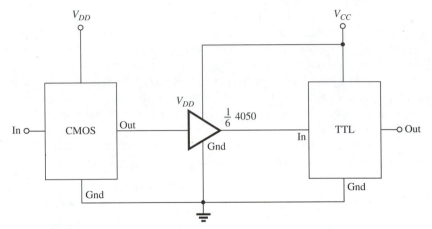

FIGURE 11.24 CMOS to TTL connection using CMOS buffer.

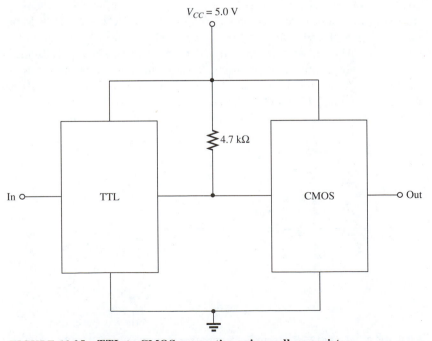

FIGURE 11.25 TTL to CMOS connection using pull-up resistor.

11-9   SOME EXAMPLE CIRCUITS

Figure 11.26 illustrates examples of the circuits discussed in this chapter.

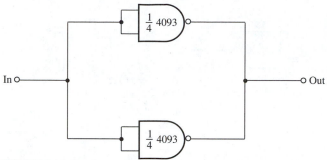

FIGURE 11.26A   Increased output current for NAND gate.

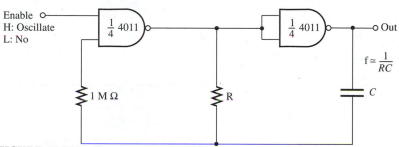

FIGURE 11.26B   Gated oscillator, CMOS.

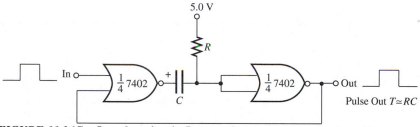

**FIGURE 11.26C   One-shot circuit. Input pulse causes output pulse with
period approximately equal to *RC*.**

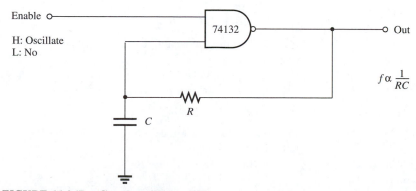

FIGURE 11.26D   Gated oscillator, TTL.

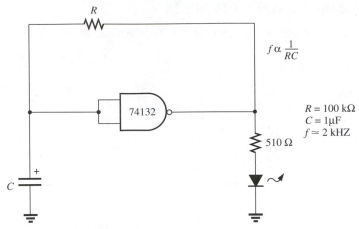

FIGURE 11.26E **LED flasher.**

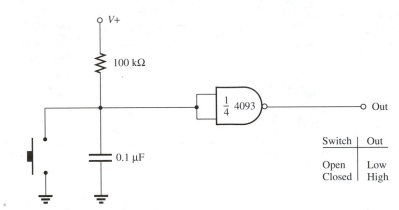

FIGURE 11.26F **Bounceless switch.**

11-10 QUESTIONS AND PROBLEMS

1. The open-collector option for TTL NAND gates (e.g., the 7401 shown in Figure 11A, which is similar to Figure 11.15 but without Q_3 and the diode) is sometimes used when two gate outputs are to be tied together or when an output is to drive a number of inputs. In those situations, the gate output is supplied with V_{cc} through a load resistor. Why might this be preferable to the usual NAND gate configuration?

2. Develop a truth table for the TTL gate circuit shown in Figure 11B.

3. Interfacing one logic line with another can produce problems. Consider a TTL gate driving a CMOS gate directly, with no special consideration given to interfacing the two chips. If both chips are powered by 5 V, do you anticipate any problems? Explain.

4. Consider the reverse of problem 3; that is, a CMOS gate driving a TTL gate. What problems do you expect?

5. Prove the following Boolean algebraic equalities.
(a) $\overline{\overline{A}} = A$
(b) $A + BC = (A + B)(A + C)$
(c) $A + \overline{A}B = A + B$
(d) $AB + \overline{A}C = AB + BC + \overline{A}C$
(e) $AB + A\overline{B} + \overline{A}B = A + B$

6. Reduce the following Boolean expressions to their simplest forms.
(a) $ABCD + A\overline{B}CD$
(b) $AB + \overline{AC} + A\overline{B}C(AB + C)$
(c) $A + B(C + \overline{DE})$

7. Using NAND gates, develop circuits to solve the following expressions.
(a) $AB + \overline{A}B + A\overline{B} + \overline{AB}$
(b) $(AB + C)(AB + D)$

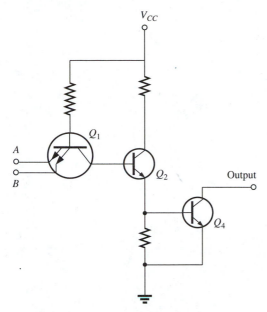

FIGURE 11A   **Circuit for problem 1.**

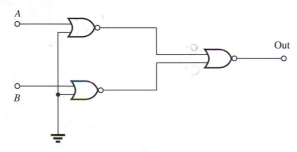

FIGURE 11B   **Circuit for problem 2.**

(c) $(\overline{AB})(AB) + AB$
(d) $(1 + B)(ABC)$

8.   For the output $= A\overline{BC} + AB,$ under what input conditions will the output $= 1$?

9.   Using NOR gates, show how AND, OR and NAND gates are made.

10.   Develop an Exclusive OR gate using only NOR gates.

11.   The And-Or-Invert (AOI) gate is sometimes encountered. It is symbolized in Figure 11C. Develop its truth table.

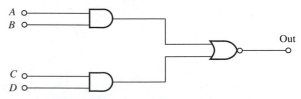

FIGURE 11C   **Circuit for problem 11.**

12.   Using only AOI gates show how NAND, AND, NOR, OR and Exclusive OR gates can be assembled.

13.   Complete the following table:

Decimal	Binary	Octal	Hexadecimal
92			
	1000010		
	1111000		
		778	
37		777	
			FF

14.   Complete the following table:

Decimal	BCD	Binary
	1000 1000 1000	
41		
1993		
		10000001
	0001 0110 0100	
		101010

15.   Discuss why battery powered logic circuits usually use CMOS ICs.

16.   Using logic gates, design an alarm system which lights an LED when any of 5 doors is open.

17.   Design an LED flasher circuit which lights two LEDs alternately at a frequency of about 2 Hz.

18.   A device which indicates the status of a gate output is called a logic probe. Design a logic probe with two LEDs. One should light when the output is high and the other when the level is low.

19.   Design a 1 kHz oscillator using NAND gates, a resistor and capacitor.

CHAPTER

12

Digital Circuitry

12-1 INTRODUCTION

This chapter builds upon the basic concepts and attitudes of digital electronics developed in Chapter 11. The material of this chapter has a certain stand-alone character. That is, several useful circuits and concepts are developed here. Beyond this aspect, however, this chapter is a necessary prerequisite to Chapter 13 on microprocessors. The latter chapter contains very few new circuits or ideas. Rather, it rests upon the already developed concepts which are packaged in particular configurations that justify the terms microcomputer or microprocessor.

The main topics of this chapter are counters, registers and control circuits. In each of these sections, we shall see that the basic circuits of earlier sections are combined to accomplish a particular objective. As in Chapter 11, there is also the need to explain or define words which have unique or special meanings for digital electronics. As much as is practical, the emphasis will be placed upon the use of medium and large scale integrated circuits (MSI and LSI), which are widely available.

12-2 FLIP-FLOPS

The basic counting unit for electronic circuitry is called the flip-flop (FF). The number of counts accumulated can be ascertained by examining the contents of the flip-flop at any given time. Although flip-flops are the most widely used digital building blocks, they were initially developed for counting purposes in analog circuits. This section begins with a description of a digital flip-flop, using gates, then continues with several types of IC digital counters which are in wide and varied use today, mainly employing TTL circuitry.

Basic RS Flip-Flops

Flip-flops can be synthesized from gates. One such example is shown in Figure 12.1. This one is a very basic FF using NAND gates. Its truth table is

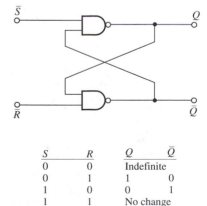

S	R	Q	$\bar{Q}$
0	0	Indefinite	
0	1	1	0
1	0	0	1
1	1	No change	

FIGURE 12.1 A flip-flop (FF) synthesized from NAND gates.

included in the figure. When $\overline{R}$ is made 0, Q goes to 0 and $\overline{Q}$, the complement, goes to 1. When $\overline{S}$ is set to 0, Q goes to 1 (and $\overline{Q}$ to 0). If both $\overline{S}$ and $\overline{R}$ are supplied with 1's, no change in the output occurs, i.e., it remains at the values it had before 1's were applied to $\overline{S}$ and $\overline{R}$. When $\overline{S}$ and $\overline{R}$ are both supplied with 0's, both outputs go to 1's. The first input which is then returned to 1 causes its output to go to zero. Several problems are associated with this primitive form of a digital FF. First, a 1 at both inputs produces no output change. Zeros at both inputs may or may not produce a change. Finally, the state of the FF before counting begins is not known.

An improved FF, still using NAND gates, is illustrated in Figure 12.2. It is called an RSFF since R stands for reset and S for set. These terms correctly imply that this FF can be "set" to begin counting or "reset" after counting is concluded. The C (clock) input controls whether or not the FF is active. When the C input is 1, the FF is enabled or made active. When the C input is 0, the FF is inactive or disabled. In digital terminology, the outputs are said to be *latched* when the system is disabled. This terminology leads to the circuit being designated a latch. Latches are used extensively in digital circuits because they can "hold" a signal upon command.

The truth table for the RSFF is given in Figure 12.2. Comparison of this truth table with the previous one, Figure 12.1, shows that the outputs are the same for the same S and R inputs. (Note that the inputs of Figure 12.1 were designated as inverted values.) The behavior of the RSFF can be described as "set to 1 and reset to 0." Obviously, setting a FF means $Q = 1$ (and $\overline{Q} = 0$), while resetting means $Q = 0$ (and $\overline{Q} = 1$).

This circuit represents an example of a different kind of control of FFs. It is called clock control and is realized through the use of the C input. In

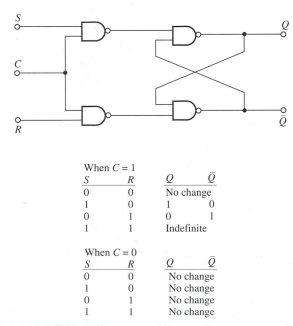

When $C = 1$

S	R	Q	$\overline{Q}$
0	0	No change	
1	0	1	0
0	1	0	1
1	1	Indefinite	

When $C = 0$

S	R	Q	$\overline{Q}$
0	0	No change	
1	0	No change	
0	1	No change	
1	1	No change	

FIGURE 12.2 **An RSFF and its truth table.**

the circuit of Figure 12.2, the FF is enabled or active only when $C = 1$. The FF counts only when this condition is obtained, regardless of the states of S and R. This clocking approach offers the important advantage of controlling the rate at which the FF operates so that it may be synchronous with other devices in the circuit. Moreover, when the clock disables or inhibits the FF, extraneous signals are excluded. (Even when the clock enables, these extraneous signals are counted only at the clock rate.)

Master-Slave Flip-Flop

The master-slave FF configuration is illustrated in Figure 12.3. The important improvement gained here is control of the signal through the circuit. It is under clock control. The "master" portion of the circuit is a clocked RSFF. The "slave" portion is also a clocked RSFF, except that it is clocked in a complementary fashion. That is, when the master is enabled and acquires information from S and R, the slave portion is inhibited. When the clock changes state, the master portion is inhibited while the slave portion is enabled. Since the inputs of the slave portion are the outputs of the master portion, the circuit outputs become those prescribed by the circuit inputs (S and R) after the clock has gone through one cycle, $0 \rightarrow 1 \rightarrow 0$. One important advantage of the master-slave arrangement is that it overcomes *racing,* a condition in which input signals catch up with signals already in the FF, thereby producing erratic results.

Data Flip-Flop

The data flip-flop, or DFF, adds another dimension to the clocked RSFF. It has two inputs, one for data and the other for the clock, and two outputs Q and $\overline{Q}$. The circuit, composed of discrete gates, for a DFF is shown in Figure 12.4 along with its truth table. Note that the outputs are valid as indicated in the truth table when the clock is high. When the clock goes low, changing the data input does not change the output state. In this sense, the FF stores the last value of the data input when the clock goes low. ICs are available which contain these gates with as many as 8 DFF circuits on one chip.

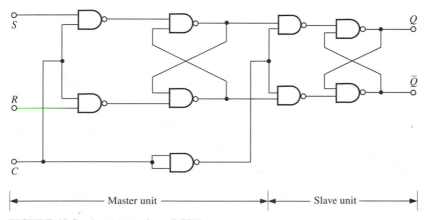

FIGURE 12.3   **A master-slave RSFF.**

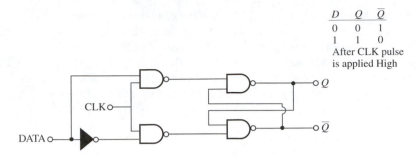

D	Q	$\overline{Q}$
0	0	1
1	1	0

After CLK pulse
is applied High

FIGURE 12.4   Data flip-flop using 4 NAND gates.

Toggle Flip-Flop

A toggle flip-flop (TFF) is a true counter or binary circuit. It is a simple extension of the master-slave FF in which the outputs are fed back into the inputs. The clock input becomes the counts input. These modifications are illustrated in Figure 12.5, where Q is fed back to R and $\overline{Q}$ is fed back to S. This guarantees that $S \neq R$. Therefore, the output states will "toggle" between 1 and 0 with no indefinite conditions. Of course, this FF is not clocked synchronously with other devices because the clock control has been reassigned as the input. The input-output performance of the TFF is exactly that of the discrete FF which began this section. Naturally, the output of a TFF may be fed into the C input of a second TFF for continued counting purposes. When a symmetrical square wave is applied to the input of a TFF, the output square wave frequency is exactly one-half of that of the input; hence the frequently used name, "Divide by two" circuit. TFF circuits made from other FFs are shown in Figure 12.6.

JK Flip-Flop

The TFF represents the realization of a useful counter. It is not, however, the ultimate counter. Not all of the input possibilities are utilized, and it cannot be set or reset. The JK flip-flop overcomes these limitations. The JKFF is widely available, generally as two such devices in a single IC package. One example of this is the 7476, a TTL-based dual JK flip-flop in a 16-pin dual

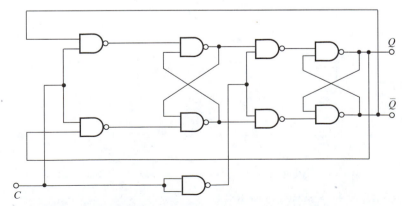

FIGURE 12.5   A toggle flip-flop (TFF).

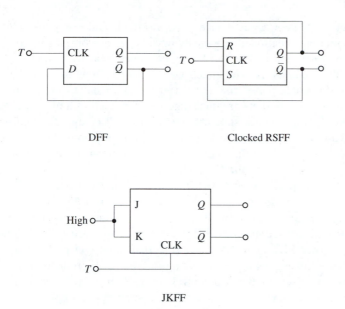

DFF   Clocked RSFF

JKFF

FIGURE 12.6   **Toggle flip-flop circuits from other flip-flops.**

in-line package (DIP). The circuit for one of the two JKFFs on the chip is given in Figure 12.7, together with the standard symbol for FFs. The truth table for the JKFF is also included in the figure. The truth table shows that every possible pair of J and K inputs produces a specific result when the FF

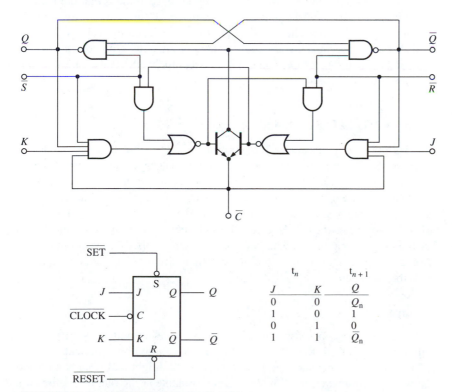

	t_n		t_{n+1}
J		K	Q
0		0	Q_n
1		0	1
0		1	0
1		1	Q_n

FIGURE 12.7   **A JKFF and its truth table.**

is clocked. Moreover, the device is both settable and resettable. Thus, all of the limitations mentioned previously have been overcome with the JKFF.

The operation of the JKFF can be understood by examining its truth table. The clocking is symbolized by the designations of t_n and t_{n+1} (the "before the clock pulse" and the "after the clock pulse" conditions). When $J = K = 0$, the outputs are unchanged as the result of a clock pulse. When $J = K = 1$, the output state changes as a result of the clock pulse. When $J = 0$ and $K = 1$, Q goes to 0 (regardless of its previous state) as a result of the clock pulse. When the opposite input conditions are used, $J = 1$ and $K = 0$, Q goes to 1 when the FF is clocked.

With the specific type of JKFF used in this illustration, $\overline{\text{CLOCK}}$ signifies that the negative edge of the clock pulse, the return from 1 to 0, is the active portion. The $\overline{\text{SET}}$ connection is used to place a known condition at the outputs. Recall that the earlier RSFF used $S = 1$ to place $Q = 1$. The bar in this case over SET indicates that a 0 to that input will produce the $Q = 1$ state. In an analogous manner, placing $\overline{\text{RESET}} = 0$ produces a $Q = 0$ output. Note that these latter two controls override the clock. We may correctly deduce that the normal conditions of FF operation require that $\overline{\text{SET}} = 1$ and $\overline{\text{RESET}} = 1$ in order to enable the clocking.

The JKFF is a versatile counting unit. It can be controlled by the J and K inputs while the signal to be counted is supplied to the clock input. For example, if $J = K = 1$ and the signal is fed to the clock input, the counter becomes a TFF. Alternatively, if an inverter is added to the inputs, as illustrated in Figure 12.8, it becomes a data flip-flop or DFF. In this configuration only one input for control of the FF is needed; it guarantees that the state of the J input will be opposite that of the K input at all times. The truth table in Figure 12.7 shows that the value of Q at t_{n+1} will be the value of J at t_n. This versatility permits the assembly of a counting circuit to perform specific functions which, otherwise, could not be accomplished readily. These counters are integral parts of computers.

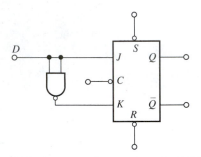

FIGURE 12.8   Conversion of a JKFF into a DFF.

12-3   DIGITAL READ-OUTS

To this point, our discussion has ignored how the state of a gate or FF is ascertained. We will now turn our attention to that important aspect of digital circuitry. The development of a circuit to perform a certain function is not a success until the operator has some means of assessing the output. One measures the voltages developed at the output and deduces the state of the circuit. Using a voltmeter for this is acceptable but unnecessary. A better indication of the output condition is whether a lamp is on or off. Because digital electronics is a two-state system, the on or off condition quickly and completely specifies the state.

Lamps

One obvious way to sense the digital signal level is with a lamp. If TTL circuitry is involved, then the 1 level ($\cong 3.0$ V) should be sufficient to cause a lamp to go on. Incandescent lamps, such as flashlight bulbs, can be turned on with 3 V; however, the limited current available at the output requires the

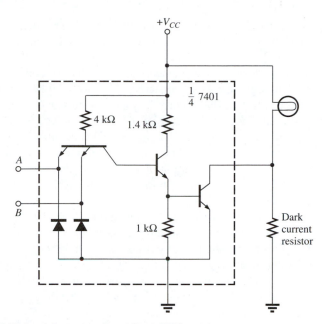

FIGURE 12.9   **A lamp read-out for a TTL gate.**

judicious selection of a particular lamp. The open-collector option of a TTL NAND gate, given in Figure 12.9, provides a useful route to accomplishing this. The output transistor, when on, can pass up to 16 mA with a 0.4 V drop across it. The dark current resistor functions as a by-pass when the output transistor is off. While this read-out solution is acceptable, it requires relatively high currents, with the attendant risk of burning out the output transistors or at least decreasing the fan-out capability of the gate or FF. Moreover, the response time of incandescent lamps is quite slow.

Neon lamps offer some improvements over incandescent lamps. They are cheaper and require less current. However, they require larger voltages. This places a different set of requirements on the output or control transistor, as illustrated in Figure 12.10. In order to turn the lamp on, the control transistor is grounded (turned on). When the control transistor is off, it must withstand the lamp's starting voltage, 55 V, without appreciable (<0.2 mA) leakage current. The circuit shown in Figure 12.10 is a good example of how to interface high-voltage devices with logic ICs.

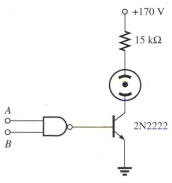

FIGURE 12.10   **A neon lamp read-out for a NAND gate.**

Light Emitting Diodes

Light emitting diodes (LEDs) offer an attractive alternative to lamps as read-out devices. Under conditions of forward bias, certain specially compounded semiconductor materials will emit light. The usual material used is gallium arsenide doped with phosphorus. The wavelength of light emitted is 660 nm (red) for this material. By far, the red LED is the least expensive and most common. By increasing the amount of diode material, more light can be produced. Green and yellow LEDs are also available, at somewhat greater cost. LEDs which emit one color when forward biased and another color when reverse biased (alternating voltages yield another color and these are therefore

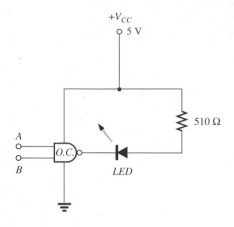

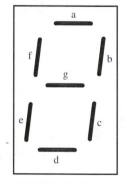

A	B	NAND	LED
0	0	1	OFF
1	0	1	OFF
0	1	1	OFF
1	1	0	ON

FIGURE 12.11 **An LED read-out using an open collector (OC) NAND gate with its truth table.**

FIGURE 12.12 **A seven-segment read-out schematic.**

sometimes called tricolor) are also readily available, as are infrared diodes. Blue LEDs can be obtained at the time of this printing, but their cost is significantly higher than the other units.

The current and voltage requirements of the LED are compatible with TTL circuitry. The light intensity is linearly related to the forward current of the diode, once conduction has occurred. Approximately 1.7 V produces an easily detected light signal with a current of about 10 mA. Figure 12.11 illustrates a typical readout, using an LED, for a TTL open collector NAND gate. Notice from the truth table, which is included in the figure, that the gate behavior, as indicated by the condition of the LED, is that of an AND gate. If NAND gate indication is needed, an inverter between the gate output and the LED is required.

These relatively primitive on-off indicators have been somewhat superseded by devices which give an alphanumeric indication of the output condition. This is accomplished first by decoding the number of counts to be displayed and then by appropriately activating a multi-element read-out device. These operations are frequently performed on a single chip, called a decoder driver. Several different types are available, depending upon the kind of read-out device to be used.

LED, LCD and incandescent displays are available with segments as shown in Figure 12.12. The seven-segment read-out devices all use the principle that an alphanumeric character may be developed by powering the appropriate (up to seven) segments. The segment identification, using lower case letters, is standard. For instance, when b, c, f and g are powered, a 4 appears. The 7447 is an example of the BCD-to-seven-segment decoder driver used with the incandescent display. Its specification sheet is given in Figure 12.13. The *blanking* feature permits operation of the display device in an AC mode. This mode materially reduces the current requirements. Notice, from the figure, that all sixteen combinations of the BCD input produce unique outputs. This is an important feature of the seven-segment devices, since it permits full symbolic representation for the hexadecimal numbering system.

Liquid Crystal Displays

Liquid crystal displays, or LCDs, have made a major impact on the world of electronics in the last few years. They require much less power than their equivalent in LEDs. LCDs do not emit light, rather they scatter it. These units must be powered by AC voltages, but drivers are readily available to provide the necessary frequency and power requirements. These devices are often packaged as panels rather than single-character units. They are useful in high illumination areas where an LED display would be washed out by the ambient light conditions. In addition, LCD displays are available in a number of more complex variations than that of the readout scheme shown in Figure 12.12. These generally have built-in drivers to compensate for the added complexity of their use.

Figure 12.14 shows the connections between a CMOS 4543 BCD-to-seven-segment LCD driver and a generic LCD display. Notice that the display driver has four inputs which control seven outputs in accordance with the BCD encoding given for displays in Table 11.13. The driver has three other inputs for reversing the truth table: phase (Ph—a high at this pin reverses the

BCD-TO-SEVEN SEGMENT
DECODER/DRIVERS

MC5446L • MC7446L,P*
MC5447L • MC7447L,P*

Compatible with MC5400/7400 Series devices.

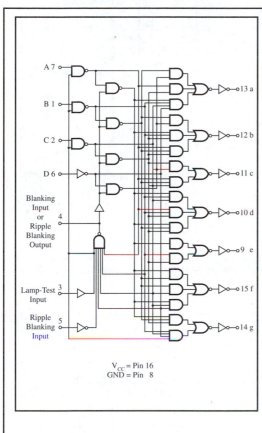

V_{CC} = Pin 16
GND = Pin 8

These devices decode 4-bit binary coded decimal data, dependent on the state of auxiliary inputs, and provide direct driving of incandescent, seven-segment, display indicators.

Ripple blanking inputs provide capability for suppression of non-significant zeros in a system. The blanking input can be used to control lamp intensity.

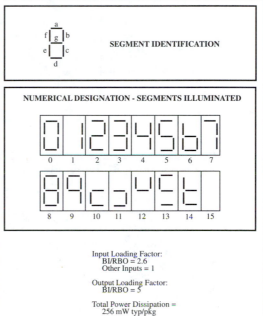

SEGMENT IDENTIFICATION

NUMERICAL DESIGNATION - SEGMENTS ILLUMINATED

Input Loading Factor:
BI/RBO = 2.6
Other Inputs = 1

Output Loading Factor:
BI/RBO = 5

Total Power Dissipation =
256 mW typ/pkg

TRUTH TABLE

DIGIT OR FUNCTION	INPUT							OUTPUT						
	LT Pin 3	RBI Pin 5	D Pin 6	C Pin 2	B Pin 1	A Pin 7	BI/RBO Pin 4	a Pin 13	b Pin 12	c Pin 11	d Pin 10	e Pin 9	f Pin 15	g Pin 14
0	1	1	0	0	0	0	1	0	0	0	0	0	0	1
1	1	X	0	0	0	1	1	1	0	0	1	1	1	1
2	1	X	0	0	1	0	1	0	0	1	0	0	1	0
3	1	X	0	0	1	1	1	0	0	0	0	1	1	0
4	1	X	0	1	0	0	1	1	0	0	1	1	0	0
5	1	X	0	1	0	1	1	0	1	0	0	1	0	0
6	1	X	0	1	1	0	1	1	1	0	0	0	0	0
7	1	X	0	1	1	1	1	0	0	0	1	1	1	1
8	1	X	1	0	0	0	1	0	0	0	0	0	0	0
9	1	X	1	0	0	1	1	0	0	0	1	1	0	0
10	1	X	1	0	1	0	1	1	1	1	0	0	1	0
11	1	X	1	0	1	1	1	1	1	0	0	1	1	0
12	1	X	1	1	0	0	1	0	1	1	1	1	0	0
13	1	X	1	1	0	1	1	0	1	1	0	1	0	0
14	1	X	1	1	1	0	1	1	1	1	0	0	0	0
15	1	X	1	1	1	1	1	1	1	1	1	1	1	1
BI	X	X	X	X	X	X	0	1	1	1	1	1	1	1
RBI	1	0	0	0	0	0	0	1	1	1	1	1	1	1
LT	0	X	X	X	X	X	1	0	0	0	0	0	0	0

X = Don't care

*L suffix = 16-pin dual in-line ceramic package (Case 620).
P suffix = 16-pin dual in-line plastic package (Case 612).

FIGURE 12.13 **A BCD-to-seven-segment decoder-driver (the 7447) for the LED or incandescent lamp.**

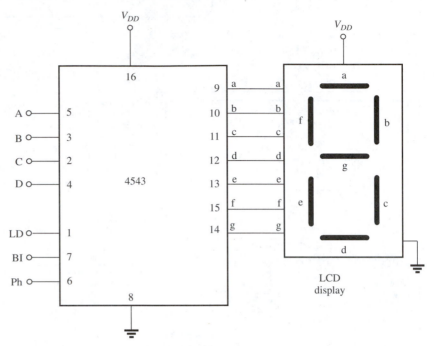

FIGURE 12.14 **BCD-to-seven-segment latch/decoder/driver for LCD display.**

output logic); blanking the display (Bl—a high blanks the display); and latch disable (LD—placing a low on this pin stores the BCD input code into a latch, thereby fixing the output states). The driver can be used with other types of displays, such as LEDs. When used with LCD displays, a square wave is applied to the phase pin and the common backplane of the LCD. Although one digit is shown in the figure, multiple drivers can be used to drive panel displays.

Manufacturers are beginning to package LCD displays along with associated drive circuitry to be used directly as counters or voltmeters. One such counting unit made by Red Lion Controls is called the SUB CUB II. It is easy to use, requires merely a 5 V power supply and has a number of controls including latching, reset and clock inputs. The SUB CUB II can be used at relatively high frequencies.

12-4 EXAMPLE: A "COUNT-TO-16" BINARY COUNTER

A number of important concepts, together with some practical rules for using digital IC circuitry, can be brought together with an illustration of a digital counter circuit using LED read-outs. TTL logic devices are used to put together a 4-bit binary counter. Commonly available ICs are used; the 7473 is a dual JKFF. The schematic, showing the pin connections for this IC, is given in Figure 12.15. The 7473 is a negative edge-triggered FF, which means the device is clocked when the clock signal returns from 1 to 0. The small circle at the C inputs symbolizes this requirement. The requirements for correct use indicate that a logical "1" can be obtained by connecting the J and K inputs

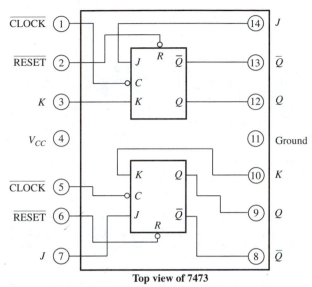

Top view of 7473

FIGURE 12.15   **A dual JKFF (the 7473).**

through a 1 kΩ resistor to V_{CC}. The 7473 does not contain a SET input, but both FFs have $\overline{RESET}$ inputs. Like the clock inputs, this is symbolized by a small circle. When $\overline{RESET} = 0$, Q for that FF is set to 0 and $\overline{Q}$ to 1. A switch must be provided so that the FF may be reset ($\overline{RESET} = 0$) and then $\overline{RESET}$ must be returned to 1 to enable the FF to be clocked.

The "count-to-16 circuit" is given in Figure 12.16A. Notice that the Q output of one stage becomes the clock input for the next stage. Because the J and K inputs are held at 1, each stage is a TFF. The input "toggles" through the FF array, producing the outputs at Q_a, Q_b, Q_c and Q_d as shown in Figure 12.16B. Notice that the changes are always initiated by a signal $1 \rightarrow 0$. The waveforms shown are produced when the FFs have been previously reset. Examination of the output conditions resulting from counts 1 and 17 shows they are the same, as are counts 2 and 18. Thus, the counter cycles through 16 input pulses and begins to count all over again. A fifth FF added to the circuit would have permitted a count up to 32, and so on.

In order to obtain the complete circuit, several other details must be added to the circuit of Figure 12.16. The J and K inputs must be connected to V_{CC} through a 1 kΩ resistor to insure that they remain at a 1. A reset pushbutton switch must be incorporated into the circuit. Finally, since LED read-outs are to be used, provision must be made for their inverted operation. That is, when an output is a 1, the LED is off; when an output is a 0, the LED is on. We can easily solve this problem by using the $\overline{Q}$ outputs to control the LEDs. Current-limiting resistors are used to protect the output transistors. The complete circuit, also showing the LED conditions, is given in Figure 12.17. For simplicity, the V_{CC} and ground connections to each FF have been omitted.

The read-out for this circuit may be improved by the incorporation of a BCD-to-seven-segment decoder driver and a seven-segment LED display.

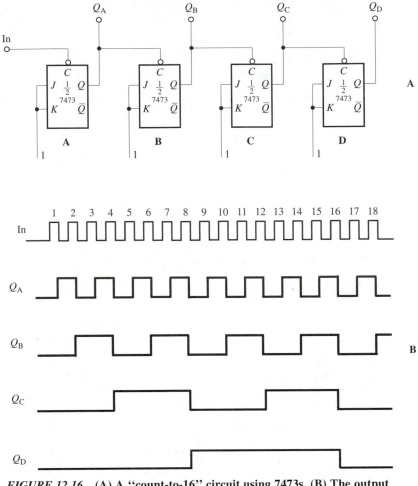

FIGURE 12.16   (A) A "count-to-16" circuit using 7473s. (B) The output waveforms.

This is illustrated in Figure 12.18. Notice that the number of counts displayed is in hexadecimal notation, as shown in the table accompanying the figure.

12-5   COUNTERS AND REGISTERS

Almost all digital signal processing requires the storage of information, however momentary that storage might be. The fundamental storage unit is the flip-flop (FF). Arrays of FFs in many different configurations perform the vast majority of the operations called for by a digital computer. In addition, FF arrays are used for counting. Part of the novice's perceptual problem is that different arrays perform different functions and, as a consequence, have very different names. Remember, they are all FF arrays. *Counters* record the number of events. *Registers* transfer data from one storage element to another. Once we understand the basic operation of counters, we begin to gain insight into registers, so we shall begin with counters.

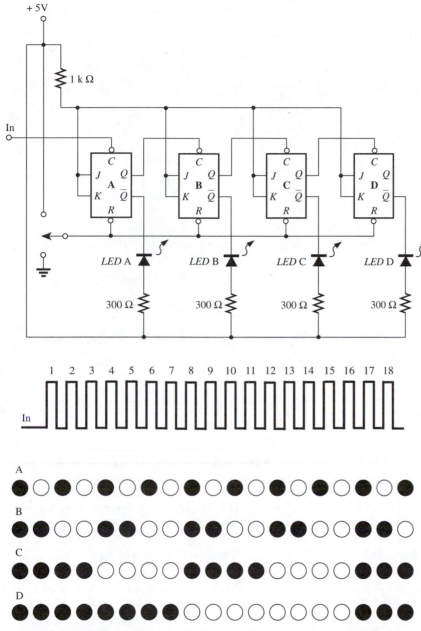

FIGURE 12.17   **The completely implemented circuit of Figure 12.16 showing LED read-outs.**

Counters

Synchronous Counters. There are three basic categories of counters: synchronous, ripple or ring, and shift. The salient characteristic of all synchronous counters is that each clock pulse is fed simultaneously or synchronously to all of the FFs. The synchronous counter may be wired to count either up or down. A synchronous divide-by-6 up counter is shown in Figure 12.19,

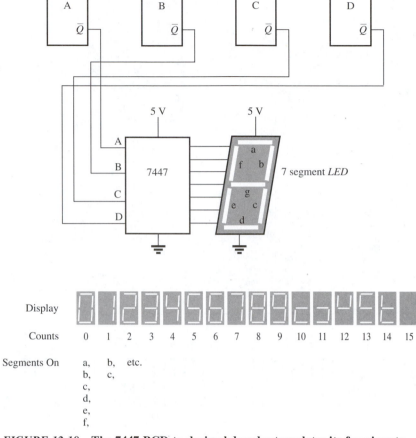

FIGURE 12.18 **The 7447 BCD-to-decimal decoder translates its four inputs, from FFs A to D, into seven outputs that drive the segments of the LED display. Each combination of states of the FFs, corresponding to a number in BCD, causes a unique display. The labeling of the LED segments from *a* to *g* is standard.**

using JKFFs of the 7473 type. The understanding of the circuit operation begins by assuming that all FFs have been reset; that is, $Q = 0$ for each FF. Since $J = K = 1$ for FF *A*, it will behave as a toggle flip-flop (TFF) when clocked. The Q_A output is connected to the K_B and K_C inputs; thus, $K_B = K_C = 0$ initially. The J_B input is controlled by a NOR gate supplied by $\overline{Q_A}$ and Q_C. The J_B input will be 0 unless both inputs are 0. In a corresponding manner, J_C will be 0 unless $\overline{Q_A}$ and $\overline{Q_B}$ are both 0. These value are shown in the truth table associated with the figure. Remember that the initial J and K values result from resetting (state 1). The application of the first negative-going edge of the clock signal (state 2) produces the changes controlled by the J and K values. In order to predict the output changes with clocking, it is necessary to first assess the J and K conditions which result from the previous clocking. The reader should carefully work through this truth table to ensure full understanding. Notice that the seventh state is exactly the same as the first state. This means that the FF array will cycle through six states or "divide by 6."

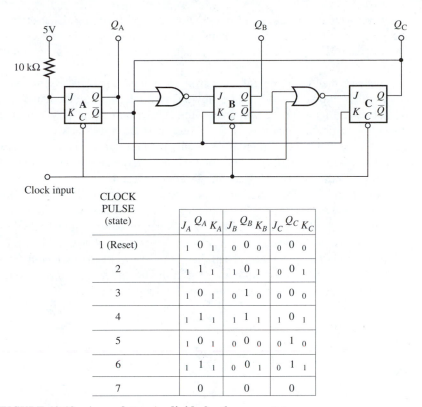

CLOCK PULSE (state)	J_A	Q_A	K_A	J_B	Q_B	K_B	J_C	Q_C	K_C
1 (Reset)	1	0	1	0	0	0	0	0	0
2	1	1	1	1	0	1	0	0	1
3	1	0	1	0	1	0	0	0	0
4	1	1	1	1	1	1	1	0	1
5	1	0	1	0	0	0	0	1	0
6	1	1	1	0	0	1	0	1	1
7		0			0			0	

FIGURE 12.19   A synchronous divide-by-6 up counter.

If it is necessary to know each time six counts have been stored, we must examine the outputs for the conditions $Q_A = Q_C = 1$ and $Q_B = 0$. Alternatively, a simple gating arrangement for the Boolean expression $A \cdot B \cdot C = 1$ can be connected to the outputs. This is shown in Figure 12.20, together with

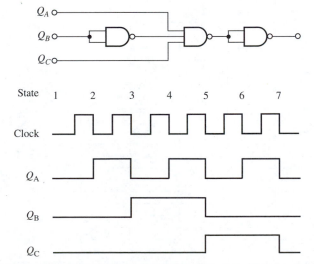

FIGURE 12.20   Gating array to be used in conjunction with the circuit of Figure 11.1 to signal completion of a cycle.

the waveforms for the circuit of the previous figure. Each time the output of the gate array becomes 1, six counts have been processed by the counter. Several exercises at the end of the chapter illustrate other synchronous counters.

As was mentioned before, synchronous counters can be assembled to count either "up" or "down." The previous example was an up counter; that is, it counted up from "empty" to "full." An example of a synchronous divide-by-16 down counter is given in Figure 12.21, together with its truth table. Notice that the operation again begins with reset (state 1); however, the next state places 1's at all Q's. The counting for the remaining states proceeds downward to the last state, where all Q's equal 0 again.

In certain counting situations it may be appropriate to have a synchronous counter which is capable of counting either up or down depending on a control input. The control is established via a gating array. Figure 12.22 illustrates this case for a synchronous divide-by-16 counter. When the control is 1, the circuit is an up counter; when the control is 0, it is a down counter. A change in the control need not occur just at the beginning or end of a counting cycle. It may be changed at any time, so long as the clock is low when

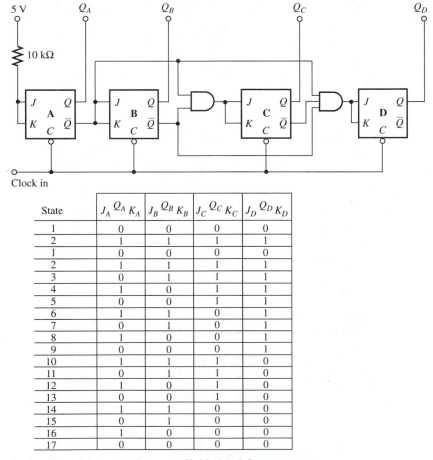

State	J_A Q_A K_A	J_B Q_B K_B	J_C Q_C K_C	J_D Q_D K_D
1	0	0	0	0
2	1	1	1	1
1	0	0	0	0
2	1	1	1	1
3	0	1	1	1
4	1	0	1	1
5	0	0	1	1
6	1	1	0	1
7	0	1	0	1
8	1	0	0	1
9	0	0	0	1
10	1	1	1	0
11	0	1	1	0
12	1	0	1	0
13	0	0	1	0
14	1	1	0	0
15	0	1	0	0
16	1	0	0	0
17	0	0	0	0

FIGURE 12.21   A synchronous divide-by-6 down counter.

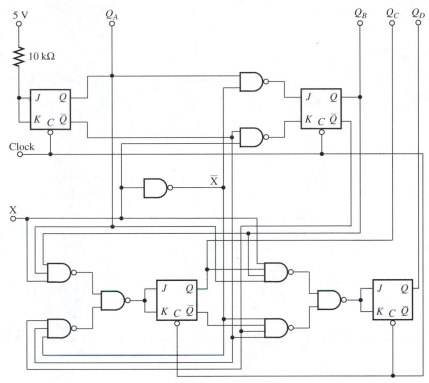

When X=1, the system performs the *UP* count
When X=0, the system performs the *DOWN* count

FIGURE 12.22   A synchronous divide-by-10 up/down counter.

the change is made. The incorporation of the control array reduces the maximum counting frequency.

Ripple Counters. Ripple counters derive their name from the fact that the clock pulse is applied only to the first FF in the array. The output of the first FF then serves as the clock pulse for the second, and so on. In other words, the clock ripples through the FF array rather than feeding every FF simultaneously. This means that a ripple counter is slower than a synchronous counter. On the other hand, the gating needed is reduced considerably. This latter point is well illustrated by Figure 12.23, which shows the circuit for a ripple divide-by-10 up counter. (All of the *K* inputs are continuously supplied with 1's.) Its truth table is included in the figure. A ripple divide-by-10 down counter circuit is given in Figure 12.24. Note the similarities between this circuit and that in Figure 12.23. The step-by-step analysis to determine the behavior of ripple counters is somewhat more complex than that for synchronous counters. As usual, the *J* and *K* inputs to a given FF control the output state when the FF is clocked. However, the nature of the clocking has to be kept in mind. Both of these counters are built with negative edge triggered FFs. Thus, only when a signal intended for clocking changes from 1 to 0 does it actually clock the device.

Shift or Ring Counters. Shift counters move the signal through the FF array each time the system is clocked. Since the outputs of each stage be-

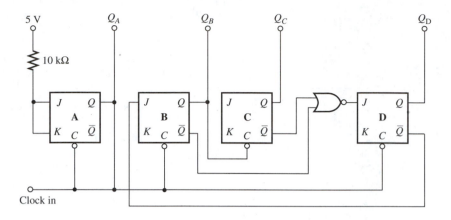

FIGURE 12.23 A divide-by-10 up counter.

State	J_A Q_A K_A	J_B Q_B K_B	J_C Q_C K_C	J_D Q_D K_D
1	0	0	0	0
2	1	0	0	0
3	0	1	0	0
4	1	1	0	0
5	0	0	1	0
6	1	0	1	0
7	0	1	1	0
8	1	1	1	0
9	0	0	0	1
10	1	0	0	1
11	0	0	0	0

come the inputs for the next, this FF array is sometimes referred to as a ring counter. A divide-by-8 shift counter is illustrated in Figure 12.25. The reset is used to load 0's at all Q's. Since $\overline{Q}_D$ is fed into J_A and Q_D into K_A, the next clock pulse sets Q of FF A at 1. When this occurs, the inputs to FF B prepare it for a change in output at the next clock pulse. When FF D is clocked to produce $Q_D = 1$, the J and K inputs of FF A are inverted and $Q_A = 0$. The cycle is completed when the set of 1's has moved through the array.

IC Counters. For the vast majority of counting situations a complete circuit, composed of the necessary FFs, gates and their interconnections, is available on a single IC chip. This level of circuit organization is termed medium scale integration (MSI). A representative example of this sophistication is the 7490 device, a decade counter. Its specification sheets, given in Figure 12.26 on page 281, completely describe its functionality. (The initial point of confusion is likely to be the need to sort out the relevant information, since several devices are included on the same set of specification sheets.) The input-output characteristics for the 7490 (and the low-power version, the 74L90) are given in the tables headed by '90A, 'L90. In order to use this device as a decade counter, the Q_A output is connected to input B. This one external connection creates a decade counter in which the count applied to input A produces a BCD coded output. This configuration thus behaves as a binary-to-BCD encoder. If the counts are applied to input B and Q_D is connected to input A, the configuration becomes a divide-by-10 counter when the output is taken at Q_A. These two configurations are illustrated in Figure 12.27 on page 282.

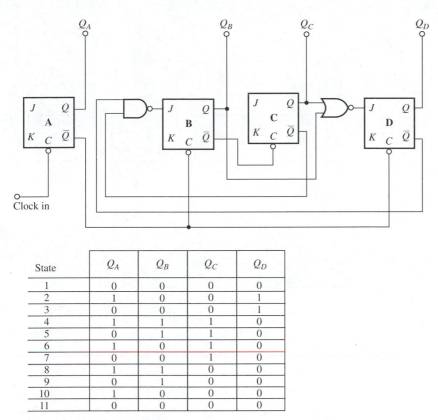

State	Q_A	Q_B	Q_C	Q_D
1	0	0	0	0
2	1	0	0	1
3	0	0	0	1
4	1	1	1	0
5	0	1	1	0
6	1	0	1	0
7	0	0	1	0
8	1	1	0	0
9	0	1	0	0
10	1	0	0	0
11	0	0	0	0

FIGURE 12.24   A ripple divide-by-10 down counter. Note that unconnected inputs are assumed HIGH.

The reset controls, $R_{0(1)}$ and $R_{0(2)}$, and the clock (count) controls, $R_{9(1)}$ and $R_{9(2)}$, are described in a second table. In short, the counter is reset (Q's $= 0$) when $R_{0(1)} = R_{0(2)} = 1$ and either $R_{9(1)}$ or $R_{9(2)} = 0$. Counting (clocking) is enabled when at least one R_9 and one $R_0 = 0$.

Modulo-*N* Counter

The *modulus* of a counter is simply the number of counts or divisions produced by it. In other words, a decade counter has a modulus of 10. The Modulo-*N* Counter is a circuit which allows the operator to select *N*. The 74192 synchronous 4-bit Up/Down Counter is used in this circuit. Its specification sheets are shown as Figure 12.28 on page 283. By itself, the 74192 is a powerful counter, capable of counting up or down and providing indications of *overflow* (the carry output) when counting up or *underflow* (the borrow output) when counting down. Moreover, clear (reset) and preset (to whatever is on the inputs) controls are available. Data are applied in a parallel manner and parallel outputs are available. Using the single 74192 device, a Modulo-*N* circuit for which $1 \leq N \leq 9$ can be achieved. If two 74192 devices are cascaded together, as in Figure 12.29 on page 286, a Modulo-*N* circuit is achieved for a modulus between 1 and 99. The desired value of *N* is set by applying 1's to the appropriate data inputs. For example, a divide-by-53 circuit requires that the data inputs to the first 74192 be $A = 1$, $B = 1$, $C = 0$

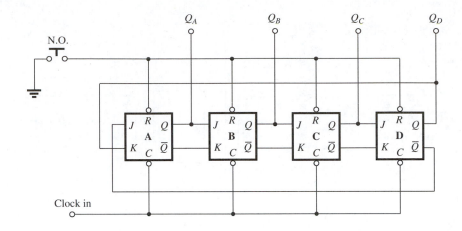

State	Q_A			Q_B			Q_C			Q_D		
	J_A		K_A	J_B		K_B	J_C		K_C	J_D		K_D
1		0			0			0			0	
	1		0	0		1	0		1	0		1
2		1			0			0			0	
	1		0	1		0	0		1	0		1
3		1			1			0			0	
	1		0	1		0	1		0	0		1
4		1			1			1			0	
	1		0	1		0	1		0	1		0
5		1			1			1			1	
	0		1	1		0	1		0	1		0
6		0			1			1			1	
	0		1	0		1	1		0	1		0
7		0			0			1			1	
	0		1	0		1	0		1	1		0
8		0			0			0			1	
	0		1	0		1	0		1	0		1
9		0			0			0			0	

FIGURE 12.25   **A divide-by-8 shift counter.**

and $D = 0$, while the data inputs to the second 74192 are $A = 1$, $B = 0$, $C = 1$ and $D = 0$. Put together in right-to-left sequence, this is 0101 0011, the BCD notation for 53. The output frequency, f_o, is related to the input frequency, f_{in}, as

$$f_o = f_{in}/W$$

where

$$W = (1A_1 + 2B_1 + 4C_1 + 8D_1 + 10A_2 + 20B_2 + 40C_2 + 80D_2) \quad \textbf{(12-1)}$$

The first IC's operation can be thought of as a synchronous divide-by-n down counter, where n is set by the data inputs. When the input counts have decreased the counter to zero, it begins the cycle again and a borrow signal appears. The second IC uses this borrow signal as its input and operates upon it, counting down by 1 each time the first IC generates another borrow signal. The borrow output of the second IC is the output frequency.

'90A, 'L90 . . . DECADE COUNTERS

'92A . . . DIVIDE-BY-TWELVE COUNTER

'93A, 'L93 . . . 4-BIT BINARY COUNTERS

description

Each of these monolithic counters contains four master-slave flip-flops and additional gating to provide a divide-by-two counter and a three-stage binary counter for which the count cycle length is divide-by-five for the '90A and 'L90, divide-by-six for the '92A, and divide-by-eight for the '93A and 'L93.

All of these counters have a gated zero reset and the '90A and 'L90 also have gated set-to-nine inputs for use in BCD nine's complement applications.

To use their maximum count length (decade, divide-by-twelve, or four-bit binary) of these counters, the B input is connected to the Q_A output. The input count pulses are applied to input A and the outputs are as described in the appropriate function table. A symmetrical divide-by-ten count can be obtained from the '90A or 'L90 counters by connecting the Q_D output to the A input and applying the input count to the B input which gives a divide-by-ten square wave at output Q_A.

functional block diagrams

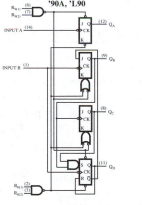

'90A, 'L90

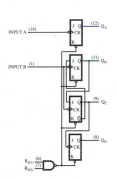

'92A

'93A, 'L93

('93A) ['L93]

⎓⊳ . . . dynamic input activated by transition from a high level to a low level.

The J and K inputs shown without connection are for reference only and are functionally at a high level.

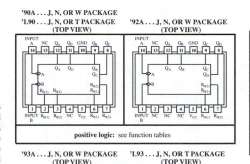

'90A . . . J, N, OR W PACKAGE
'L90 . . . J, N, OR T PACKAGE
(TOP VIEW)

'92A . . . J, N, OR W PACKAGE
(TOP VIEW)

positive logic: see function tables

'93A . . . J, N, OR W PACKAGE
(TOP VIEW)

'L93 . . . J, N, OR T PACKAGE
(TOP VIEW)

positive logic: see function tables

NC–No internal connection

TYPES	TYPICAL POWER DISSIPATION
'90A	145 mW
'L90	20 mW
'92A, '93A	130 mW
'L93	16 mW

'90A, 'L90
BCD COUNT SEQUENCE
(See Note A)

COUNT	OUTPUT			
	Q_D	Q_C	Q_B	Q_A
0	L	L	L	L
1	L	L	L	H
2	L	L	H	L
3	L	L	H	H
4	L	H	L	L
5	L	H	L	H
6	L	H	H	L
7	L	H	H	H
8	H	L	L	L
9	H	L	L	H

'90A, 'L90
BI-QUINARY (5-2)
(See Note B)

COUNT	OUTPUT			
	Q_A	Q_D	Q_C	Q_B
0	L	L	L	L
1	L	L	L	H
2	L	L	H	L
3	L	L	H	H
4	L	H	L	L
5	H	L	L	L
6	H	L	L	H
7	H	L	H	L
8	H	L	H	H
9	H	H	L	L

'90A, 'L90
RESET/COUNT FUNCTION TABLE

RESET INPUTS				OUTPUT			
$R_{0(1)}$	$R_{0(2)}$	$R_{9(1)}$	$R_{9(2)}$	Q_D	Q_C	Q_B	Q_A
H	H	L	X	L	L	L	L
H	H	X	L	L	L	L	L
X	X	H	H	H	L	L	H
X	L	X	L	COUNT			
L	X	L	X	COUNT			
L	X	X	L	COUNT			
X	L	L	X	COUNT			

'92A
COUNT SEQUENCE
(See Note C)

COUNT	OUTPUT			
	Q_D	Q_C	Q_B	Q_A
0	L	L	L	L
1	L	L	L	H
2	L	L	H	L
3	L	L	H	H
4	L	H	L	L
5	L	H	L	H
6	H	L	L	L
7	H	L	L	H
8	H	L	H	L
9	H	L	H	H
10	H	H	L	L
11	H	H	L	H

'93A, 'L93
COUNT SEQUENCE
(See Note C)

COUNT	OUTPUT			
	Q_D	Q_C	Q_B	Q_A
0	L	L	L	L
1	L	L	L	H
2	L	L	H	L
3	L	L	H	H
4	L	H	L	L
5	L	H	L	H
6	L	H	H	L
7	L	H	H	H
8	H	L	L	L
9	H	L	L	H
10	H	L	H	L
11	H	L	H	H
12	H	H	L	L
13	H	H	L	H
14	H	H	H	L
15	H	H	H	H

'92A, '93A, 'L93
RESET/COUNT FUNCTION TABLE

RESET INPUTS		OUTPUT			
$R_{0(1)}$	$R_{0(2)}$	Q_D	Q_C	Q_B	Q_A
H	H	L	L	L	L
L	X	COUNT			
X	L	COUNT			

NOTES: A. Output Q_A is connected to input B for BCD count.
B. Output Q_D is connected to input A for bi-quinary count.
C. Output Q_A is connected to input B.
D. H = high level, L = low level, X = irrelevant

FIGURE 12.26 **Manufacturer's specification sheets for the 7490A, a decade counter.**

This represents a powerful approach to the syntheses of particular counters. Complete counting systems are easily made by adding the appropriate read-out devices and clocking and gating controls. This approach largely reduces the need for exhaustive circuit analysis.

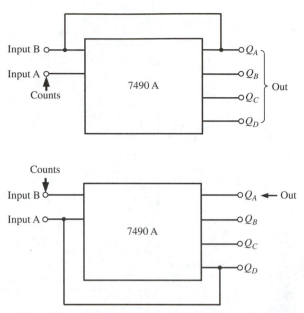

FIGURE 12.27   **Counter configurations for the 7490A.**

Shift Registers

A register is a storage element for information. Thus, an appropriately controlled FF is a register. Generally it is assumed that a register has a capacity equal to the digital word length. Registers are used extensively in digital computers, often going under other descriptive names such as accumulators, program counters, memory addressers and instruction registers. Thus, understanding how a register performs is vital in understanding how computers operate.

Shift registers are FF arrays into which information is moved or shifted as the FFs are clocked. The information may be entered in a sequential or *serial* (S) manner or in a *parallel* (P) manner. Likewise, the information may be shifted out of the register in either a serial or parallel manner. When information is added at the input and taken at the output, the serial shift register is said to shift right, since by convention left is equivalent to the input and right is equivalent to the output.

A 4-bit shift register capable of serial input (SI) or parallel input (PI) and serial output (SO) or parallel output (PO) is given in Figure 12.30 on page 287. The *mode* input selects the method of operation. When a 1 is applied to the mode input, the parallel inputs P_1, P_2, P_3 and P_4 are disabled. Under this condition the register is designated SISO or SIPO; it could be either, since the mode control does not control the outputs. The actual output method depends on whether the next stage in the information-processing circuitry is attached to the serial output terminal or to the parallel output terminals (or, conceivably, to all of them). When a 0 is applied to the mode control, the serial input is disabled and the parallel inputs are enabled. The shift register designation for this case may be either PIPO or PISO. Notice that all of the JKFFs are clocked synchronously. Furthermore, the NAND gate at each *J*

- **Cascading Circuitry Provided Internally**
- **Synchronous Operation**
- **Individual Preset to Each Flip-Flop**
- **Fully Independent Clear Input**

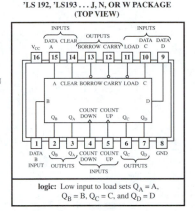

'192, '193 . . . J, N, OR W PACKAGE
'L192, 'L193 . . . J OR N PACKAGE
'LS 192, 'LS193 . . . J, N, OR W PACKAGE
(TOP VIEW)

logic: Low input to load sets Q_A = A,
Q_B = B, Q_C = C, and Q_D = D

TYPES	TYPICAL MAXIMUM COUNT FREQUENCY	TYPICAL POWER DISSIPATION
'192, '193	32 MHz	325 mW
'L192, 'L193	7 MHz	43 mW
'LS192, 'LS193	32 MHz	85 mW

description

These monolithic circuits are synchronous reversible (up/down) counters having a complexity of 55 equivalent gates. The '192, 'L192, and 'LS192 circuits are BCD counters and the '193, 'L193 and 'LS193 are 4-bit binary counters. Synchronous operation is provided by having all flip-flops clocked simultaneously so that the outputs change coincidently with each other when so instructed by the steering logic. This mode of operation eliminates the output counting spikes which are normally associated with asynchronous (ripple-clock) counters.

The outputs of the four master-slave flip-flops are triggered by a low-to-high-level transition of either count (clock) input. The direction of counting is determined by which count inputs is pulsed while the other count input is high.

All four counters are fully programmable; that is, each output may be preset to either level by entering the desired data at the data inputs while the load input is low. The output will change to agree with the data inputs independently of the count pulses. This feature allows the counters to be used as modulo-N dividers by simply modifying the count length with the preset inputs.

A clear input has been provided which forces all outputs to the low level when a high level is applied. The clear function is independent of the count and load inputs. The clear, count, and load inputs are buffered to lower the drive requirements. This reduces the number of clock drivers, etc., required for long words.

These counters were designed to be cascaded without the need for external circuitry. Both borrow and carry outputs are available to cascade both the up- and down-counting functions. The borrow output produces a pulse equal in width to the count-down input when the counter underflows. Similarly, the carry output produces a pulse equal in width to the count-down input when an overflow condition exists. The counters can then be easily cascaded by feeding the borrow and carry outputs to the count-down and count-up inputs respectively of the succeeding counter.

absolute maximum ratings over operating free-air temperature range (unless otherwise noted)

	SN54'	SN54L'	SN54LS'	SN74'	SN74L'	SN74LS'	UNIT
Supply voltage, V_{CC} (see Note 1)	7	8	7	7	8	7	V
Input voltage	5.5	5.5	7	5.5	5.5	7	V
Operating free-air temperature range	−55 to 125			0 to 70			°C
Storage temperature range	−65 to 150			−65 to 150			°C

NOTE 1: Voltage values are with respect to network ground terminal.

FIGURE 12.28   **Manufacturer's specification sheets for the 74192, a synchronous 4-bit up/down counter.**

input converts the JKFF into a DFF. In PIPO operation, one clocking places all 4 bits of data at the parallel outputs Q_1, Q_2, Q_3 and Q_4, whereas in the SISO operation, 4 clock pulses are required to enter the 4 bits and 4 more clock pulses are needed to shift them out.

IC Shift Register. The 74199 is an excellent example of a "universal" 8-bit shift register on a single MSI chip. Its specification sheets are shown in Figure 12.31 on page 288. Controls are provided for clearing or resetting and for shifting right (S_1 = 1 and S_0 = 0) or shifting left (S_1 = 0 and S_0 = 1).

functional block diagrams

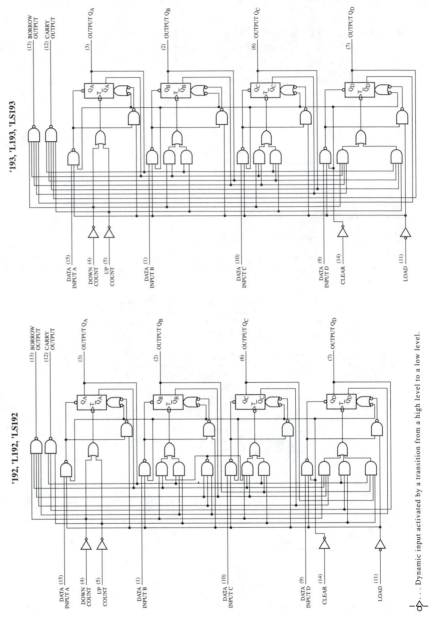

FIGURE 12.28 *(continued)*

Data may be input in either a parallel or a serial manner. The waveforms, included in the figure, illustrate the versatility of this device.

12-6 MULTIPLEXING

A *multiplexer* is a circuit which behaves like a controlled rotary switch. That is, any one of a number of inputs may be selected as the output. In digital

typical clear, load, and count sequences

Illustrated below is the following sequence:

1. Clear outputs to zero.
2. Load (preset) to BCD seven.
3. Count up to eight, nine, carry, zero, one, and two.
4. Count down to one, zero, borrow, nine, eight, and seven.

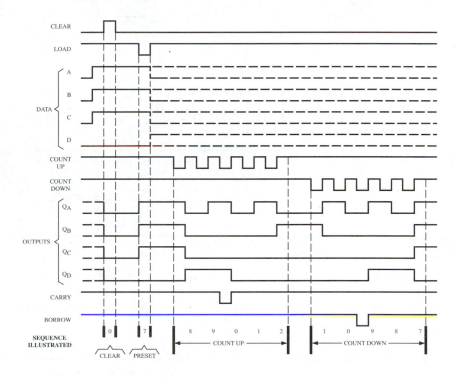

NOTES: A. Clear overrides load, data, and count inputs.
        B. When counting up, count-down input must be high; when counting down, count-up input must be high.

FIGURE 12.28   (continued)

circuitry, a multiplexer is designed to read (or pass on) the digital information on any selected input channel. The selection of the channel to be read is controlled by supplying a specific digital word. Particular arrays of gates are combined to perform this function, as illustrated in Figure 12.32 on page 290. In this circuit, eight input channels, D_0 through D_7, are fed into AND gates. The AND gates also receive the control inputs A, B and C, each of which is inverted either once or twice (recall that, for example, $\overline{\overline{A}} = A$). The gate controls are arranged in such a manner that only a single gate is enabled by any one combination of the control inputs, as shown on the truth table. Because all of the AND gate outputs are NORed together, the *inverse* of the opened channel input appears at the output.

MSI multiplexer circuits are available which can select one of 16 channels with a 4-bit control. The 74151 is such a device. If the control signals move sequentially from 0000 to 1111, the device selects each channel in turn, thus acting to convert the parallel data into a serial form for all 16 channels.

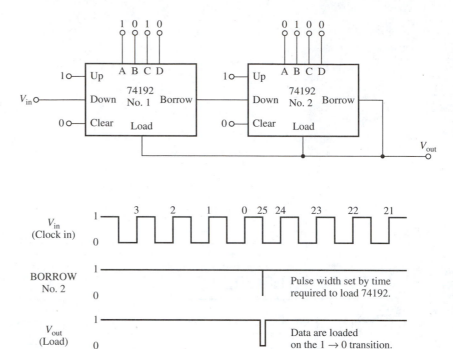

FIGURE 12.29 **A Modulo 25 counter using 74192s.**

Demultiplexers, which accept a single input and cause it to appear at the selected output channel, are also available. The 74156 is an example of such a device. It may be used either as a dual 1-line-to-4-line demultiplexer or as a 1-line-to-8-line demultiplexer. The schematic for the device is given in Figure 12.33 on page 291. The truth table given pertains to the 1-line-to-8-line operation, which is achieved by connecting the Data 1 and Data 2 inputs together to form a Select C input. The two Strobe inputs are connected together, and the input line is supplied to that connection. The particular output at which the signal appears is controlled by the Select settings, as shown in the truth table.

One advantage that multiplexers and demultiplexers offer is a reduction in the number of parallel wires needed to transmit digital information over a long distance. If eight input channels were to be sent directly, eight wires would be needed. If the eight channels are multiplexed and sent sequentially over a single line to a demultiplexer that is supplied with the same digital control word, the same transfer of information is accomplished. The select or control inputs to the two devices must be held at the same 3-bit signal sequence throughout the transmission. Thus, the requirement has been reduced from eight parallel wires to four—the data line plus the three control or select lines.

Digital multiplexers which act as analog switches are also available. One of the most widely used is the CMOS 4066, a quad bilateral switch containing four independent switches. The ON resistance of the switch is rated at 80 Ω and the output current leakage in the OFF state is a mere 50 nA maximum with a typical rating of 0.1 nA. Although the device can pass analog signals, it can be used to control digital ones as well. A pin-out of this device is shown in Figure 12.34 on page 293.

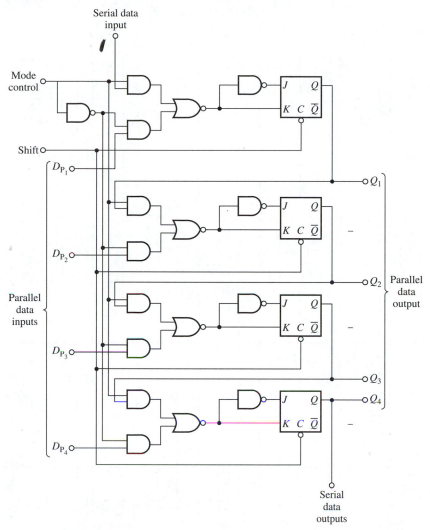

FIGURE 12.30 **Shift register configurations.**

12-7 MONOSTABLE MULTIVIBRATORS: ONE-SHOTS

Often in digital circuits a need arises for a pulse to be created from a transition. Along these lines, a single pulse may need to be stretched, or lengthened in time. Components which accomplish this are called monostable multivibrators, or *one-shots*. Monostable multivibrator refers to the fact that the circuit has a stable state, usually $Q = $ low, and another state which is produced by a low-to-high transition, or vice-versa, on the input. The name one-shot refers to the fact that once an output pulse begins, further inputs are ignored, until the output pulse is complete. We have already seen one example of a monostable multivibrator in an earlier chapter when we discussed the 555 timer chip.

Here we introduce two monostable multivibrators, the 4538 from the CMOS family and the 74123 from the TTL family. Both circuits, along with

SN54199, SN74199

typical clear, shift, load, and inhibit sequences

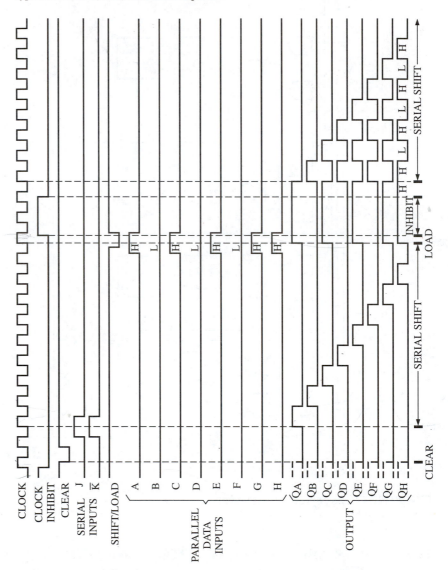

FIGURE 12.31 **Manufacturer's specification sheets for the 74199, a "universal" 8-bit shift register.**

their associated truth tables, are shown in Figure 12.35 on page 294. The width of the output pulse is controlled by an external resistor and capacitor. For the 4538 a value of $R_{ext} = 100 \text{ k}\Omega$ and $C_{ext} = 10.0 \text{ }\mu\text{F}$, the width of the pulse will be 1.0 s. Although the formulae presented in the figure are approximations, the 4538 is generally accurate to 10% regardless of power-supply voltages.

Figure 12.36 on page 295 shows an example of a 4538 circuit, used to detect when a ball has been dropped on a metal plate. The piezoelectric transducer is attached to the plate. When a steel ball is dropped on the plate,

SN54199 and SN74199

These registers feature parallel inputs, parallel outputs, J-K̄ serial inputs, shift/load control input, a direct overriding clear line, and gated clock inputs. The register has three modes of operation:

Parallel (Broadside) Load
Shift (In the direction Q_A toward Q_H)
Inhibit Clock (Do nothing)

Parallel loading is accomplished by applying the eight bits of data and taking the shift/load control input low when the clock input is not inhibited. The data is loaded into the associated flip-flop and appears at the outputs after the positive transition of the clock input. During loading, serial data flow is inhibited.

Shifting is accomplished synchronously when shift/load is high and the clock input is not inhibited. Serial data for this mode is entered at the J-K̄ inputs. See the function table for levels required to enter data into the first flip-flop.

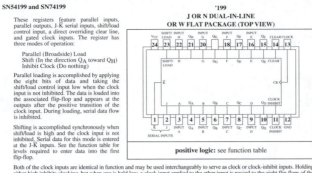

'199
J OR N DUAL-IN-LINE
OR W FLAT PACKAGE (TOP VIEW)

positive logic: see function table

Both of the clock inputs are identical in function and may be used interchangeably to serve as clock or clock-inhibit inputs. Holding either high inhibits clocking, but when one is held low, a clock input applied to the other input is passed to the eight flip-flops of the register. The clock-inhibit input should be changed to the high level only while the clock input is high.

These shift registers contain the equivalent of 79 TTL gates. Average power dissipation per gate is typically 4.55 mW.

'199
FUNCTION TABLE

	INPUTS					OUTPUTS			
CLEAR	SHIFT/ LOAD	CLOCK INHIBIT	CLOCK	SERIAL J K̄	PARALLEL A...H	Q_A	Q_B	Q_C ...	Q_H
L	X	X	X	X X	X	L	L	L	L
H	X	L	L	X X	X	Q_A0	Q_B0	Q_C0	Q_H0
H	L	L	↑	X X	a...h	a	b	c	h
H	H	L	↑	L H	X	Q_A0	Q_A0	Q_B0	Q_G0
H	H	L	↑	L L	X	L	Q_An	Q_Bn	Q_Gn
H	H	L	↑	H H	X	H	Q_An	Q_Bn	Q_Gn
H	H	L	↑	H L	X	Q̄_A0	Q_An	Q_Bn	Q_Gn
H	X	H	↑	X X	X	Q_A0	Q_B0	Q_B0	Q_H0

H = high level (steady state), L = low level (steady state)
X = irrelevant (any state, including transitions)
↑ = transition from low to high level
a . . . h = the level of steady-state input at inputs A thru H, respectively.
Q_A0, Q_B0, Q_C0 . . . Q_H0 = the level of Q_A, Q_B, or Q_C thru Q_H, respectively, before the indicated steady-state input conditions were established.
Q_An, Q_Bn . . . Q_Gn = the level of Q_A or Q_B thru Q_G, respectively, before the most-recent ↑ transition of the clock.

'199

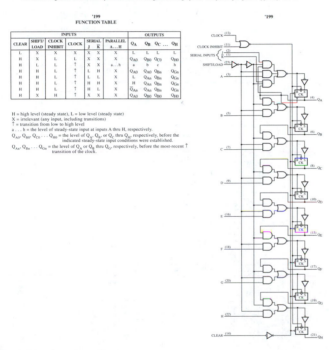

FIGURE 12.31   (continued)

the piezoelectric transducer produces a voltage, since the ball causes the plate and the attached transducer to vibrate, and the vibration of the crystal produces an oscillating voltage. The plate and transducer vibrate for a brief period of time and the ball bounces, producing multiple events as indicated in the figure. The 311 comparator circuit produces a high signal when the input is greater than the level set by the voltage divider on its negative terminal. The first low-to-high transition of the pulse triggers the 4538, which produces a pulse whose period is set by the timing capacitor and resistor attached to it. This signal is fed into the 4050 buffer which drives the 2N3904 npn transistor. The transistor turns on while the signal is high, thus lighting the LED. In this example, the LED stays lit for 10 seconds. This circuit can be used to drive a timer circuit on or off as well, but the time constant, *RC,* should probably be reduced to a pulse length which is just longer than the vibration and bounce.

'152A, 'LS152
FUNCTION TABLE

SELECT INPUTS			OUTPUT W
C	B	A	
L	L	L	$\overline{D0}$
L	L	H	$\overline{D1}$
L	H	L	$\overline{D2}$
L	H	H	$\overline{D3}$
H	L	L	$\overline{D4}$
H	L	H	$\overline{D5}$
H	H	L	$\overline{D6}$
H	H	H	$\overline{D7}$

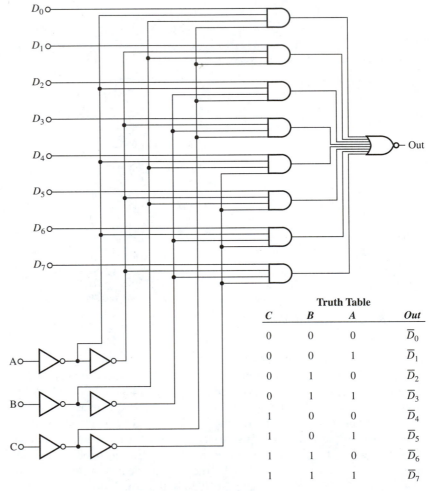

Truth Table

C	B	A	Out
0	0	0	$\overline{D}_0$
0	0	1	$\overline{D}_1$
0	1	0	$\overline{D}_2$
0	1	1	$\overline{D}_3$
1	0	0	$\overline{D}_4$
1	0	1	$\overline{D}_5$
1	1	0	$\overline{D}_6$
1	1	1	$\overline{D}_7$

FIGURE 12.32 **A multiplexer circuit using gates.**

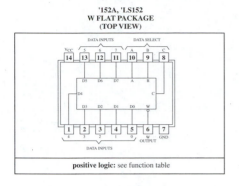

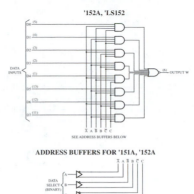

FIGURE 12.32   (continued)

- **Applications:**
  - Dual 2-to-4-Line Decoder
  - Dual 1-to-4-Line Demultiplexer
  - 3-to-8-Line Decoder
  - 1-to-8-Line Demultiplexer
- **Individual Strobes Simplify Cascading for Decoding or Demultiplexing Larger Words**
- **Input Clamping Diodes Simplify System Design**
- **Choice of Outputs:**
  - Totem Pole ('155, 'LS155)
  - Open-Collector ('156)

TYPES	TYPICAL AVERAGE PROPAGATION DELAY 3 GATE LEVELS	TYPICAL POWER DISSIPATION
'155, '156	21 ns	125 mW
'LS155	18 ns	31 mW

description

These monolithic transistor-transistor-logic (TTL) circuits feature dual 1-line-to-4-line demultiplexers with individual strobes and common binary-address inputs in a single 16-pin package. When both sections are enabled by the strobes, the common binary-address inputs sequentially select and route associated input data to the appropriate output of each section. The individual strobes permit activating or inhibiting each of the 4-bit sections as desired. Data applied to input 1C is inverted at its outputs and data applied at 2C is not inverted through its outputs. The inverter following the 1C data input permits use as a 3-to-8-line decoder or 1-to-8-line demultiplexer without external gating. Input clamping diodes are provided on all of these circuits to minimize transmission-line effects and simplify system design.

Series 54 and 54LS are characterized for operation over the full military temperature range of −55°C to 125°C; Series 74 and 74LS are characterized for operation from 0°C to 70°C.

FIGURE 12.33   Manufacturer's specification sheets for the 74156, a demultiplexer.

12-8   MECHANICAL SWITCHES AND LOGIC CIRCUITS

All mechanical switches suffer from a phenomenon called switch bounce. This occurs when the mechanical switch contacts close, and literally bounce, such that the switch contact is opened and closed quite rapidly many times (often up to 100) in a brief period of time (up to 5 ms) before remaining in a desired state. Such occurrences can wreak havoc on digital circuits. For example, a clock circuit which is meant to make one transition from high to low via a mechanical switch can in fact make the transition a multiple number of times. To alleviate switch bounce, a switch-debouncing circuit must be used whenever mechanical switches are connected to a digital circuit where

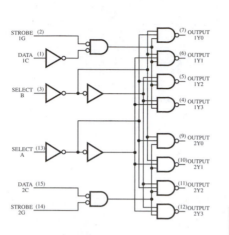

FUNCTION TABLES
2-LINE-TO-4-LINE DECODER
OR 1-LINE-TO-4-LINE DEMULTIPLEXER

INPUTS				OUTPUTS			
SELECT		STROBE	DATA				
B	A	1G	1C	1Y0	1Y1	1Y2	1Y3
X	X	H	X	H	H	H	H
L	L	L	H	L	H	H	H
L	H	L	H	H	L	H	H
H	L	L	H	H	H	L	H
H	H	L	H	H	H	H	L
X	X	X	L	H	H	H	H

INPUTS				OUTPUTS			
SELECT		STROBE	DATA				
B	A	2G	2C	2Y0	2Y1	2Y2	2Y3
X	X	H	X	H	H	H	H
L	L	L	L	L	H	H	H
L	H	L	L	H	L	H	H
H	L	L	L	H	H	L	H
H	H	L	L	H	H	H	L
X	X	X	H	H	H	H	H

FUNCTION TABLE
3-LINE-TO-8-LINE DECODER
OR 1-LINE-TO-8-LINE DEMULTIPLEXER

INPUTS				OUTPUTS							
SELECT			STROBE OR DATA	(0)	(1)	(2)	(3)	(4)	(5)	(6)	(7)
C†	B	A	G‡	2Y0	2Y1	2Y2	2Y3	1Y0	1Y1	1Y2	1Y3
X	X	X	H	H	H	H	H	H	H	H	H
L	L	L	L	L	H	H	H	H	H	H	H
L	L	H	L	H	L	H	H	H	H	H	H
L	H	L	L	H	H	L	H	H	H	H	H
L	H	H	L	H	H	H	L	H	H	H	H
H	L	L	L	H	H	H	H	L	H	H	H
H	L	H	L	H	H	H	H	H	L	H	H
H	H	L	L	H	H	H	H	H	H	L	H
H	H	H	L	H	H	H	H	H	H	H	L

†C = inputs 1C and 2C connected together
‡G = inputs 1G and 2G connected together
H = high level, L = low level, X = irrelevant

FIGURE 12.33   *(continued)*

multiple transitions cannot be tolerated. Several schemes may be used to debounce switches. Here we present several of the most common.

Often, all that is needed is a Schmitt Trigger circuit as shown in Figure 12.37 on page 295. When a Schmitt Trigger gate changes states, the threshold voltage at which the change of states occurs also changes. This behavior acts such that the logic gate is impervious to small voltage fluctuations. In the circuit shown, when the switch is closed, the input to the 4093 CMOS NAND gate becomes high, and thus the output goes low. When the switch is opened, the output goes low. This circuit is fine, if the bounces produce voltage levels which are not beyond the varying voltage trip level of the Schmitt Trigger. Since this is not always the case, an improved version of the circuit is shown in Figure 12.38 on page 295. When the switch is closed, the Schmitt Trigger does not change states until the capacitor is discharged through the 1 kΩ resistor to a level sufficient to trigger the Schmitt Trigger 4093 NAND gate. The output goes low and does not change until the switch is open and the capacitor voltage, charged through the 10 kΩ and 1 kΩ resistor, is great enough to change the state of the NAND gate. Since the 4093 NAND gate is connected as an inverter, a Schmitt Trigger inverter buffer circuit can work in this circuit just as well. The first circuit is more useful with momentarily closed (pushbutton) switches, such as those found in stopwatches, whereas the latter is superior with double-throw switches (toggle).

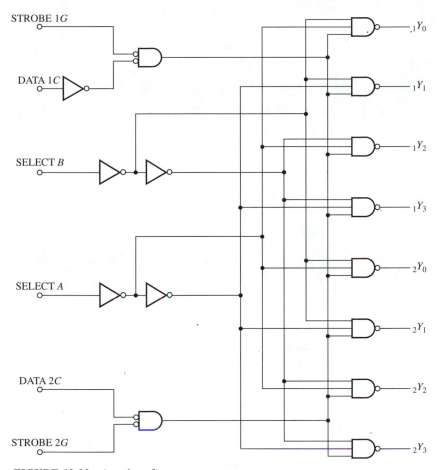

FIGURE 12.33 *(continued)*

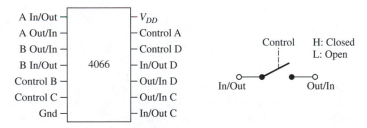

FIGURE 12.34 **Pin-outs for the 4066 quad bilateral switch.**

Both of the above schemes have associated problems. The first is only useful if the switch bounce is not severe, and the latter has a time lag between the switch being open or closed and the output changing states. In some applications, these are not problems, but in others they are. The ultimate simple debounce circuit is shown in Figure 12.39 on page 296 which uses a 4049 CMOS inverting buffer. When the switch is in the 1 position, gate 1 receives

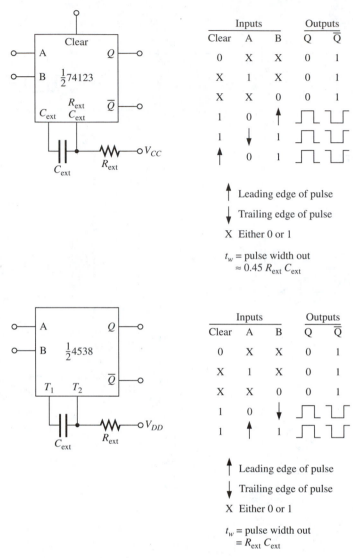

FIGURE 12.35 **74123 and 4538 monostable multibrators.**

a low input and thus outputs a high which feeds gate 2 and drives its output low with low feedback to the input of gate 1, and vice-versa for the switch in the 0 position. When the switch bounces, the output feedback maintains the logic level which is needed to drive the output buffer at its desired state.

12-9 CIRCUIT EXAMPLE: MILLISECOND TIMER/COUNTER

The circuit shown in Figure 12.40 is a millisecond timer and pulse counter. Note the module design of the timer; the Switch Debounce, $\div 2$, Oscilator, Gate and display sections are all stand-alone circuits combined in a way to function as a whole. Gate control input, point A, can be connected to photo-gate or other digital output to allow timing of other events. $\div 2$ output, point

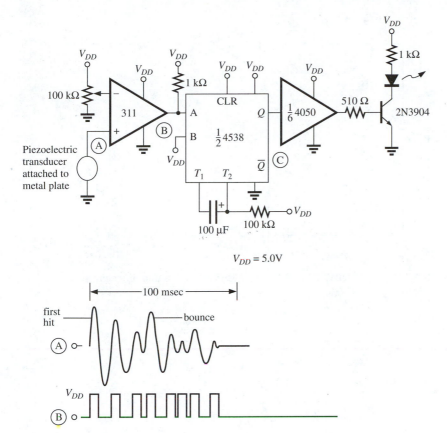

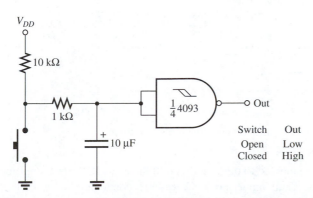

FIGURE 12.36 Circuit using monostable multivibrator to detect ball drop.

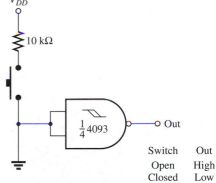

FIGURE 12.37 Schmitt Trigger switch debounce circuit.

FIGURE 12.38 Improved 4093 Schmitt Trigger switch debounce circuit.

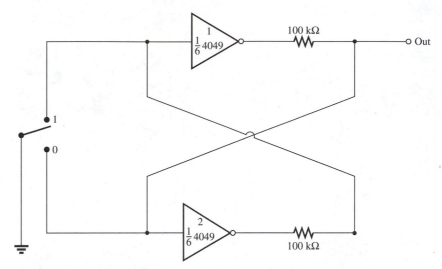

FIGURE 12.39   **4049 CMOS Inverter Buffer switch debounce circuit.**

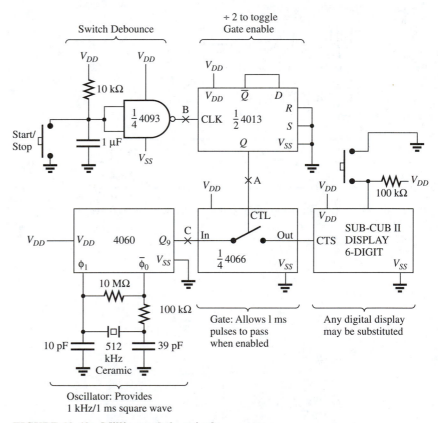

FIGURE 12.40   **Millisecond timer/pulse counter.**

B, can be connected to other digital outputs as well, to measure time between two events. Gate input, point C, can be connected to other digital pulse sources, such as a geiger counter, for the device to be used as a pulse counter.

12-10   QUESTIONS AND PROBLEMS

1.   Develop an RSFF using NOR gates.

2.   Verify that the circuit in Figure 12A is a DFF.

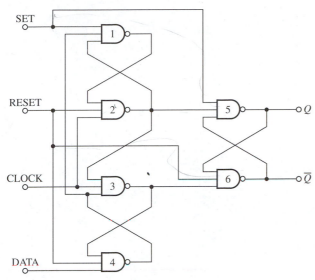

FIGURE 12A   **Circuit for problems 2, 3 and 4. Enabling conditions: SET = RESET = 1; override conditions: SET = RESET = 0 produces Q = 1.**

3.   Which gates in Figure 12A are the master and which are the slave?

4.   What are the conditions needed for SET and RESET in Figure 12A to allow the DATA input to control the output when clocked with a positive-going pulse?

5.   Design a count-by-8 circuit using DFFs.

6.   Cite the advantages and disadvantages of each type of counter.

7.   Design divide-by-4 counters which operate in each of the three modes.

8.   For the synchronous-mode divide-by-4 up counter, add appropriate gating so that it may be made to count down by use of a mode control change.

9.   Determine the truth table and modulus for the synchronous counter of Figure 12B.

10.   Using the 7490A decade counter, develop the appropriate gating so that every tenth count is signaled by an LED.

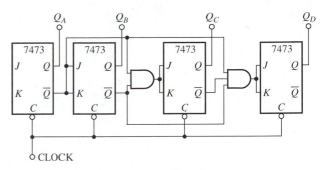

FIGURE 12B   **Circuit for problem 9.**

11.   Show the external circuit connections to achieve a divide-by-5 counter configuration using the 7490A.

12.   Describe the external controls needed to convert a series of 7490A decade counters into a counting system capable of 10^6 counts. This should include appropriate timing circuitry.

13.   How many clock pulses are needed to shift an 8-bit word into the 74199 shift register?

14.   Show the external connections needed to convert the 74156 demultiplexer into 1-line-to-8-line operation.

15.   In block diagram form, indicate how 16 parallel lines may be multiplexed, transmitted serially and then demultiplexed back into 16 lines. How many parallel lines between multiplexer and demultiplexer are needed?

16.   With reference to Question 15, what sort of synchronous signal control is needed?

17.   With reference to Questions 15 and 16, assuming a 1 MHz clock control, what is the maximum data rate for the system?

18.   Explain why $V_{ON} \neq V_{OFF}$ for Schmitt Triggers.

19.   Design a circuit which turns on an LED for two seconds when a loud one-shot noise (e.g., clap of hands) is made.

20.   Design a circuit which clocks 4 analog inputs sequentially to 1 analog output at a rate of 100 Hz.

21.   Design a circuit using flip-flop(s) and other logic which converts 60 Hz 120 V AC line voltage into 1 second pulses (safely).

13

Microprocessor Basics

13-1 INTRODUCTION

For centuries, humans have relentlessly attempted to produce calculating machines. Blaise Pascal, in the seventeenth century, is generally credited with the development of the first calculating machine, although the abacus has been in use for thousands of years. A nineteenth century eccentric, Charles Babbage, drew up plans for a mechanical computing machine. The underlying architecture of that machine is recognized as the same as that used in present-day computers.

Computers as we know them today are the result of two important milestones. The first is the use of stored programs or sets of instructions contained within the computer. The other precondition for the modern computer was the development of large-scale integrated (LSI) circuit manufacturing techniques, and the adaptation of these techniques to what we now call microprocessor and related support circuits.

It is impossible to predict where this continuing evolution will ultimately lead; however, microprocessors have clearly already had an enormous impact upon all facets of our society, from appliance control, to automobiles, to home-entertainment systems and, of course, the personal microcomputer. Hardly a day goes by that an individual in a developed country does not come in contact, often unknowingly, with a number of microprocessor-controlled devices. In the typical modern family household, there are usually at least ten such devices alone. In retrospect, it is interesting to note that in 1969, when the first human walked on the moon, the microprocessor had not yet been introduced commercially. The commercial introduction of the microprocessor was undoubtedly the most significant technological advance of the 1970s.

This chapter is intended to provide only an introduction to microprocessors. Numerous books have been devoted solely to the topic, and the reader is referred to them in the bibliography for further details. The concepts developed in early chapters should be helpful in understanding and appreciating the microprocessor operation. In fact, the developments and concepts of this chapter rest upon those which have preceded it. The emphasis here will be upon the basic functional operation rather than on specific circuits;

however, individual, simple microprocessors are used as foundational units. Nevertheless, the functionality of a data latch is most clearly appreciated when it is recognized as a flip-flop circuit which, in turn, is developed from logic gates.

As was the case in the previous chapters, the reader will continue to be introduced to a new vocabulary. These words have both meaning and functionality associated with them. Anyone who has attempted to read digital-computer textbooks, knows the diversity of this vocabulary. The words should be less abstract in the context of this chapter's organization.

13-2 COMPUTER ARITHMETIC

Bits, Bytes and Nibbles

We have already discussed the binary nature of logic devices in Chapters Eleven and Twelve. Numbering systems were covered in Chapter Eleven; here we complete this discussion by introducing these conventions as they apply to computers. In the microprocessor language, a *bit,* which stands for *bi*nary digi*t,* is a single piece of data, a high or a low. Eight bits constitute a *byte* and four bits are often referred to as a *nibble.* Seemingly contrary to common Systém International practice which defines the prefix of kilo to mean the quantity of 1000, a *kilobyte* (KB or K) is not 1000 bytes, it is 2^{10}, or 1,024 bytes. Therefore, 64K bytes, in computerese, is actually 65,536 bytes, not 64,000. A *megabyte* (MB, M or simply "meg") is 2^{10} times 2^{10} (2^{20}) bytes, or 1,048,576 bytes.

Hexadecimal Representation

Although digital logic consists of signals that are either high or low, represented by 1 or 0, it seems natural to write a byte as a series of eight digits as shown in Figure 13.1. The digit, or bit, on the right is called the least significant bit (LSB) since it has the smallest weight, and the digit on the left is called the most significant bit (MSB) since it has the greatest weight. In this example, the binary, or base-2, representation is 01010011, a number that is rather unwieldy considering it is only an eight-bit word. To convert this number to decimal, base 10, we multiply the value of each digit times the decimal equivalent of that place value. In our example the decimal value is thus computed as

	Binary		*Decimal Value*	*Equivalent*		*Total*
MSB	0	$\times$	128	(2^7)	$=$	0
	1		64	(2^6)		64
	0		32	(2^5)		0
	1		16	(2^4)		16
	0		8	(2^3)		0
	0		4	(2^2)		0
	1		2	(2^1)		2
LSB	1		1	(2^0)		1
					TOTAL $=$	83

Therefore, we can write $01010011_2 = 83_{10}$ where the subscript (sometimes called the radix) indicates the base of the number.

	MSB							LSB
	0	1	0	1	0	0	1	1
BIT NUMBER	7	6	5	4	3	2	1	0
POWERS OF 2	2^7	2^6	2^5	2^4	2^3	2^2	2^1	2^0
DECIMAL VALUE	128	64	32	16	8	4	2	1

FIGURE 13.1   **Representation of a byte, an 8-bit word.**

As we stated before, the binary representation is unwieldy, but the decimal notation is not much better, and in fact often worse, when working with microprocessors. For example, it is not readily apparent from the number 83 what the content of bit 5 is without converting the number back to its original binary equivalent, a somewhat tedious task (although many scientific calculators can do this calculation). For simplicity, hexadecimal notation, base 16, is commonly used in these applications. With hexadecimal notation, or "hex," each digit represents 4 bits and can have the value of 0 through 9 and A through F. Thus, an F represents 1111_2 or 15_{10}. When speaking in hex, it is common practice to place a dollar sign in front of the number to indicate that it is not a decimal or binary value rather than use the subscript "16." The decimal, hexadecimal and binary representations for the numbers 0 to 15 are shown in Table 13.1. To convert the binary number 01010011 to hexadecimal, we break it into two, 4-digit sections, or 0101 0011. Then from the table we see the first section, 0101, is a 5 in hexadecimal and the second section, 0011 is a 3. The value is thus written as 53_{16} or more suitably, $53.

Binary Addition

The rules of binary addition are precisely the same as those used in decimal addition. Figure 13.2 presents the elementary results of adding one bit (the *augend*) to another (the *addend*) to achieve the sum and the carry. The gate

TABLE 13.1   *Decimal, Hexadecimal and Binary Representations of Decimal Numbers 0–15*

Decimal	Hexadecimal	Binary			
0	0	0	0	0	0
1	1	0	0	0	1
2	2	0	0	1	0
3	3	0	0	1	1
4	4	0	1	0	0
5	5	0	1	0	1
6	6	0	1	1	0
7	7	0	1	1	1
8	8	1	0	0	0
9	9	1	0	0	1
10	A	1	0	1	0
11	B	1	0	1	1
12	C	1	1	0	0
13	D	1	1	0	1
14	E	1	1	1	0
15	F	1	1	1	1

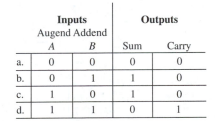

| | **Inputs** | | **Outputs** | |
| | Augend | Addend | | |
	A	*B*	Sum	Carry
a.	0	0	0	0
b.	0	1	1	0
c.	1	0	1	0
d.	1	1	0	1

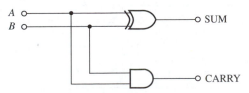

FIGURE 13.2   **A half adder and its truth table.**

circuit used to achieve this is also shown. In the usual mathematical form, the four separate equations of the table are

$$0 + 0 = 00$$
$$0 + 1 = 01$$
$$1 + 0 = 01$$
$$1 + 1 = 10.$$

This circuit is called a half adder. A full-adder circuit is capable of adding two bits and accepting a carry from a previous stage. The gate circuit needed to accomplish this is also given in Figure 13.3. A number of chips are available to perform this function, the 7480 being one example. Further combination of single full adders in parallel produces a circuit capable of adding 4-bit binaries together. Figure 13.4A gives a logical approach to this combination while Figure 13.4B presents the functional block diagram and truth table for the 7483A, a 4-bit full adder. The reader should verify the truth table to ensure full understanding of this operation. An 8-bit full adder can be assembled from two 7483s, where the C_4 output of the four least significant bits is connected to the C_0 input of the four most significant bits.

Binary Subtraction

The gate circuit for a full subtractor is given in Figure 13.5, together with its truth table. The rules of binary subtraction can be deduced from the table. In order to subtract a 1 from a 0, as $0 - 1$, a 1 must be borrowed so that $10 - 1 = 1$. This is the second entry in the table. When the previous

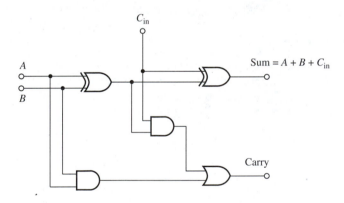

A	B	C_{in}	Sum	Carry
0	0	0	0	0
0	1	0	1	0
1	0	0	1	0
1	1	0	0	1
0	0	1	1	0
0	1	1	0	1
1	0	1	0	1
1	1	1	1	1

FIGURE 13.3 **A full adder and its truth table.**

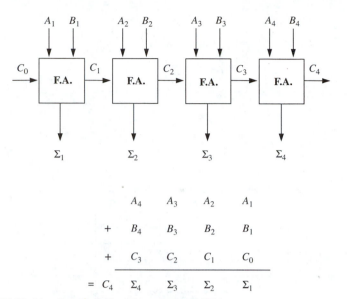

$$
\begin{array}{cccc}
& A_4 & A_3 & A_2 & A_1 \\
+ & B_4 & B_3 & B_2 & B_1 \\
+ & \underline{C_3} & \underline{C_2} & \underline{C_1} & \underline{C_0} \\
= C_4 & \Sigma_4 & \Sigma_3 & \Sigma_2 & \Sigma_1
\end{array}
$$

FIGURE 13.4A   **A 4-bit full adder using a 7483A.**

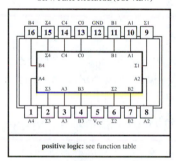

Functional Block Diagrams

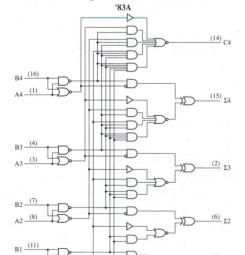

FUNCTION TABLE

	INPUT			OUTPUT					
				WHEN C0 = L			WHEN C0 = H		
					WHEN C2 = L				WHEN C2 = H
A1/A3	B1/B3	A2/A4	B2/B4	Σ1/Σ3	Σ2/Σ4	C2/C4	Σ1/Σ3	Σ2/Σ4	C2/C4
L	L	L	L	L	L	L	H	L	L
H	L	L	L	H	L	L	L	H	L
L	H	L	L	H	L	L	L	H	L
H	H	L	L	L	H	L	H	H	L
L	L	H	L	L	H	L	H	H	L
H	L	H	L	H	H	L	L	L	H
L	H	H	L	H	H	L	L	L	H
H	H	H	L	L	L	H	H	L	H
L	L	L	H	L	H	L	H	H	L
H	L	L	H	H	H	L	L	L	H
L	H	L	H	H	H	L	L	L	H
H	H	L	H	L	L	H	H	L	H
L	L	H	H	L	L	H	H	L	H
H	L	H	H	H	L	H	L	H	H
L	H	H	H	H	L	H	L	H	H
H	H	H	H	L	H	H	H	H	H

H = high level, L = low level

NOTE: Input conditions at A3, A2, B2, and C0 are used to determine outputs Σ1 and Σ2 and the value of the internal carry C2. The values at C2, A3, B3, A4, and B4 are then used to determine outputs Σ3, Σ4, and C4.

FIGURE 13.4B   **Manufacturer's information on the 7483A 4-bit full adder.**

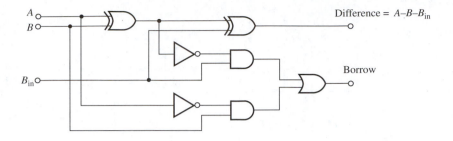

Minuend	Subtrahend	Previous Borrow	Difference	Borrow
A	B	B_{in}		
0	0	0	0	0
0	1	0	1	1
1	0	0	1	0
1	1	0	0	0
0	0	1	1	1
0	1	1	0	1
1	0	1	0	0
1	1	1	1	1

FIGURE 13.5 **A full subtractor and its truth table.**

borrow is included in the consideration of the same simple problem, the solution begins as before. A 1 is borrowed and $10 - 1 = 1$. However, the previous borrow is now subtracted from the result to obtain the difference, $10 - 1 = 1 - 1 = 0$. This is the same as the sixth entry in the table.

Two's-Complement Arithmetic

The two's-complement arithmetic approach offers significant advantages in computer use for subtraction. Most computers designate the sign of a value with the MSB. A 0 as the MSB indicates a positive value, while a 1 signals a negative value. In converting into two's-complement format, the sign of the value is automatically carried. The principle in two's complementation is to add a 1 to the complement of the binary value. For example, -63 in two's-complement 8-bit notation is obtained as follows:

$$
\begin{aligned}
63 &= 0011\ 1111 \\
\overline{63} &= 1100\ 0000 \\
& \underline{+\ 1} \\
-63 &= 1100\ 0001.
\end{aligned}
$$

The method involves writing the 8-bit binary for 63, then complementing it ($\overline{63}$) by changing each 0 to a 1 and vice versa. Finally a 1 is added to obtain the two's complement. Consider $80 - 63$. This is the same as $80 + (-63)$. The 8-bit binary value of 80 is 0101 0000. The problem becomes

$$
\begin{aligned}
80 &= 0101\ 0000 \\
-63 &= +\ \underline{ 1100\ 0001} \\
& 1\ 0001\ 0001.
\end{aligned}
$$

The extra carry bit at the left is discarded. The zero for the MSB indicates a positive value and 0001 0001 is the expected value (17).

If -60 is to be subtracted from -28, the two's-complement method proceeds by writing -28 in two's-complement:

$$
\begin{aligned}
28 &= 0001\ 1100 \\
\overline{28} &= 1110\ 0011 \\
&\quad\ \ + 1 \\
\hline
-28 &= 1110\ 0100.
\end{aligned}
$$

In a like manner, -60 is 1100 0100. Recall the principle of two's-complement subtraction. The subtrahend is two's complemented and added to the minuend. Thus the two's-complement of 1100 0100 is 0011 1100. The "subtraction" is as follows:

$$
\begin{aligned}
-28 &= \quad\ \ 1110\ 0100 \\
\text{two's-comp of } -60 &= +\ 0011\ 1100 \text{ (this is } +60) \\
\hline
& 1\quad 0010\ 0000.
\end{aligned}
$$

The result is $+32$.

If -60 is to be added to -28, the result is obtained:

$$
\begin{aligned}
-28 &= \quad\ \ 1110\ 0100 \\
-60 &= +\ 1100\ 0100 \\
\hline
& 1\quad 1010\ 1000
\end{aligned}
$$

The result is -88. This can be established by obtaining the two's-complement of 88:

$$
\begin{aligned}
88 &= 0101\ 1000 \\
\overline{88} &= 1010\ 0111 \\
&\quad\ \ + 1 \\
\hline
-88 &\quad 1010\ 1000
\end{aligned}
$$

Alternatively, the two's-complement of the initial result, 1010 1000, can be taken directly, to achieve the same result.

A full-adder circuit, such as the 7483A, can be converted to a full adder/subtractor, as shown in Figure 13.6. The Exclusive OR gates act to either pass the B inputs on unchanged or complement them. When the switch is grounded (ADD), 0s are placed at the gates. Thus, a given B input will produce the same value at the output. The 7483A acts to add in the usual manner. When the switch is open (SUBTRACT), 1s are placed at the gates. The complements of the B inputs are applied to the 7483A. The 1 is added to the complemented value since the $C0$ is also held high. The "preconditioning" of the B inputs results in a subtraction of B from A.

BCD Addition

Addition to BCD is somewhat more complicated than addition in pure binary. For example if 0111 and 0100 are added with a 7483A, the result is 1011 which has no BCD meaning. The desired result is really 0001 0001. The full adder can still be used, but its results must have the appropriate correction and carry generation. If the BCD equivalent of 6 (0110 binary) is added to the actual result, it is re-encoded in BCD (1011 + 0110 = 1 0001). Thus a second full adder, hard wired with a BCD 6, makes the appropriate correction and yields the correct sum for the least significant BCD value. In order to properly generate a carry for BCD, a gating circuit is needed. This is shown in Figure 13.7.

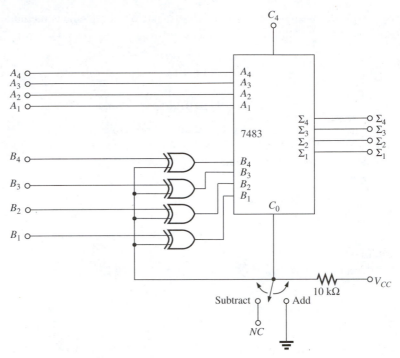

FIGURE 13.6   A 4-bit full adder/subtractor using a 7483A.

An alternative to BCD addition (and subtraction) is to first convert BCD to binary, then use binary techniques, and finally convert the binary result into BCD again. The 74184 and 74185A chips perform these conversions quite readily.

The ALU

The 74181 is a true Arithmetic Logic Unit (ALU). It accepts two 4-bit words, and, by selection, performs all 16 possible logic operations, such as Exclusive OR, compare, AND, NAND, OR, NOR and 16 arithmetic operations, such as add, subtract, double and compare. Its function table and pin layout are shown in Figure 13.8. It is made to perform the logic functions listed when the mode control, M, is high. When $M = 0$, it performs the arithmetic functions listed. The four select line settings determine which one of the 16 functions will be performed.

Two or more of the 74181s can be cascaded together to achieve input word lengths in multiples of 4 bits. For example, if 8-bit words are used, two 74181s are connected in parallel. The C_{n+4} output for the least significant bits is wired to the C_n of the 74181 for the most significant bits. The P and G outputs are used only when a "look-ahead" carry generator is connected. The "look-ahead" carry generator is a gating array which anticipates a carry command before it emerges, thus speeding up the arithmetic operation considerably. One such commercially available logic device is the 74LS182.

Because of the function diversity of the ALU, it is an integral part of a microprocessor. Although the 74181 used in the example here is a separate chip, the ALU is built into most microprocessors.

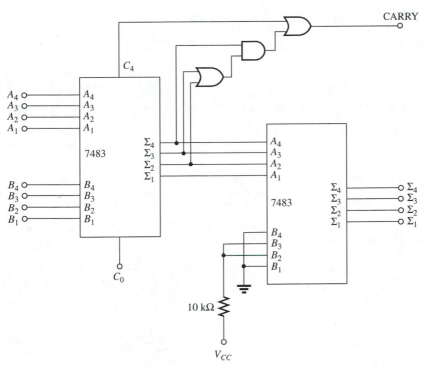

FIGURE 13.7   A 4-bit full adder with BCD conversion.

PIN Designations

DESIGNATION	PIN NOS.	FUNCTION
A3, A2, A1, A0	19, 21, 23, 2	WORD A INPUTS
B3, B2, B1, B0	18, 20, 22, 1	WORD B INPUTS
S3, S2, S1, S0	3, 4, 5, 6	FUNCTION-SELECT INPUTS
C_n	7	INV. CARRY INPUT
M	8	MODE CONTROL INPUT
F3, F2, F1, F0	13, 11, 10, 9	FUNCTION OUTPUTS
A = B	14	COMPARATOR OUTPUT
P	15	CARRY PROPAGATE OUTPUT
C_{n+4}	16	INV. CARRY OUTPUT
G	17	CARRY GENERATE OUTPUT
V_{CC}	24	SUPPLY VOLTAGE
GND	12	GROUND

'181, 'LS181 . . . J, N, OR W PACKAGE
SN54S181 . . . J OR W PACKAGE
SN74S181 . . . J, N, OR W PACKAGE
(TOP VIEW)

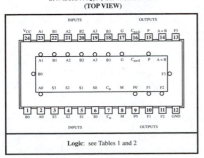

Logic: see Tables 1 and 2

Table 1

SELECTION				M = H LOGIC FUNCTIONS	M = L; ARITHMETIC OPERATIONS	
S3	S2	S1	S0		C_n = H (no carry)	C_n = L (with carry)
L	L	L	L	F = $\overline{A}$	F = A	F = A PLUS 1
L	L	L	H	F = $\overline{A + B}$	F = A + B	F = (A + B) PLUS 1
L	L	H	L	F = $\overline{A}B$	F = A + $\overline{B}$	F = (A + $\overline{B}$) PLUS 1
L	L	H	H	F = 0	F = MINUS 1 (2'S COMPL)	F = ZERO
L	H	L	L	F = $\overline{AB}$	F = A PLUS A$\overline{B}$	F = A PLUS A$\overline{B}$ PLUS 1
L	H	L	H	F = $\overline{B}$	F = (A + B) PLUS A$\overline{B}$	F = (A + B) PLUS A$\overline{B}$ PLUS 1
L	H	H	L	F = A $\oplus$ B	F = A MINUS B MINUS 1	F = A MINUS B
L	H	H	H	F = A$\overline{B}$	F = A$\overline{B}$ MINUS 1	F = A$\overline{B}$
H	L	L	L	F = $\overline{A}$ + B	F = A PLUS AB	F = A PLUS AB PLUS 1
H	L	L	H	F = $\overline{A \oplus B}$	F = A PLUS B	F = A PLUS B PLUS 1
H	L	H	L	F = B	F = (A + $\overline{B}$) PLUS AB	F = (A + $\overline{B}$) PLUS AB PLUS 1
H	L	H	H	F = AB	F = AB MINUS 1	F = AB
H	H	L	L	F = 1	F = A PLUS A*	F = A PLUS A PLUS 1
H	H	L	H	F = A + $\overline{B}$	F = (A + B) PLUS A	F = (A + B) PLUS A PLUS 1
H	H	H	L	F = A + B	F = (A + $\overline{B}$) PLUS A	F = (A + $\overline{B}$) PLUS A PLUS 1
H	H	H	H	F = A	F = A MINUS 1	F = A

ACTIVE HIGH DATA

* Each bit is shifted to the next more significant position.

Description

The '181, 'LS181, and 'S181 are arithmetic logic units (ALU)/function generators which have a complexity of 75 equivalent gates on a monolithic chip. These circuits perform 16 binary arithmetic operations on two 4-bit words as shown in Tables 1 and 2. These operations are selected by the four function-select lines (S0, S1, S2, S3) and include addition, subtraction, decrement, and straight transfer. When performing arithmetic manipulations, the internal carries must be enabled by applying a low-level voltage to the mode control input (M). A full carry look-ahead scheme is made

FIGURE 13.8   Manufacturer's information on the 74S181; a 4-bit arithmetic-logic unit.

TYPES SN54181, SN54LS181, SN54S181, SN74181, SN74LS181, SN74S181 ARITHMETIC LOGIC UNITS/FUNCTION GENERATORS

available in these devices for fast, simultaneous carry generation by means of two cascade-outputs (pins 15 and 17) for the four bits in the package. When used in conjunction with the SN54182, SN54S182, SN74182, or SN74S182, full carry look-ahead circuits, high-speed arithmetic operations can be performed. The typical addition times shown above illustrate the little additional time required for addition of longer words when full carry look-ahead is employed. The method of cascading '182 or 'S182 circuits with these ALU's to provide multi-level full carry look ahead is illustrated under typical applications data for '182 and 'S182.

If high speed is not of importance, a ripple-carry input (C_n) and a ripple-carry output (C_{n+4}) are available. However, the ripple-carry delay has also been minimized so that arithmetic manipulations for small word lengths can be peformed without external circuitry.

The '181, 'LS181 and 'S181 will accommodate active-high or active-low data if the pin designations are interpreted as follows:

PIN NUMBER	2	1	23	22	21	20	19	18	9	10	11	13	7	16	15	17
Active-high data (Table I)	A_0	B_0	A_1	B_1	A_2	B_2	A_3	B_3	F_0	F_1	F_2	F_3	$\overline{C}_n$	$\overline{C}_{n+4}$	X	Y
Active-low data (Table II)	$\overline{A}_0$	$\overline{B}_0$	$\overline{A}_1$	$\overline{B}_1$	$\overline{A}_2$	$\overline{B}_2$	$\overline{A}_3$	$\overline{B}_3$	$\overline{F}_0$	$\overline{F}_1$	$\overline{F}_2$	$\overline{F}_3$	C_n	C_{n+4}	$\overline{P}$	$\overline{G}$

Subtraction is accomplished by 1's complement addition where the 1's complement of the subtrahend is generated internally. The resultant output is A–B–1 which requires an end-around or forced carry to provide A–B.

The '181, 'LS181 or 'S181 can also be utilized as a comparator. The A = B output is internally decoded from the function outputs (F0, F1, F2, F3) so that when two words of equal magnitude are applied at the A and B inputs, it will assume a high level to indicate equality (A = B). The ALU should be in the subtract mode with C_n = H when performing this comparison. The A = B output is open-collector so that it can be wire-AND connected to give a comparison for more than four bits. The carry output (C_{n+4}) can also be used to supply relative magnitude information. Again, the ALU should be placed in the subtract mode by placing the function select inputs S3, S2, S1, S0 at L, H, H, L, respectively.

INPUT C_n	OUTPUT C_{n+4}	ACTIVE-HIGH DATA (FIGURE 1)	ACTIVE-LOW DATA (FIGURE 2)
H	H	A ⩽ B	A ⩾ B
H	L	A > B	A < B
L	H	A < B	A > B
L	L	A ⩾ B	A ⩽ B

These circuits have been designed to not only incorporate all of the designer's requirements for arithmetic operations, but also to provide 16 possible functions of two Boolean variables without the use of external circuitry. These logic functions are selected by use of the four function-select inputs (S0, S1, S2, S3) with the mode-control input (M) at a high level to disable the internal carry. The 16 logic functions are detailed in Tables 1 and 2 and include exclusive-OR, NAND, AND, NOR, and OR functions.

FIGURE 13.8   (continued)

Binary Multiplication

The binary multiplication process is the one we are familiar with in dealing with decimal numbers. Conceptually, multiplying a number, the *multiplicand,* by another, the *multiplier,* is the same as adding the multiplicand *n* times, where *n* is the value of the multiplier. This is one way some computers multiply. However, this process is very slow for large values of the multiplier. The direct approach is outlined below with the example of multiplying 5 and 6.

$$
\begin{aligned}
(5) &= 0000\,0101 \text{ (binary 5)}\\
(6) &= \underline{0000\,0110} \text{ (binary 6)}\\
&\quad\ \ 0000\,0000\ (0 \times 0000\,0101)\\
&\quad\ \ 0000\,101\ \ \ (1 \times 0000\,0101)\\
&\quad\ \ 0001\,01\ \ \ \ (1 \times 0000\,0101)\\
&\quad\ \ \underline{0000\,0}\ \ \ \ \ \ (0 \times 0000\,0101)\\
(30) &= 0001\,1110 \text{ (sum } = \text{ binary 30).}
\end{aligned}
$$

Functionally, what occurs in this process is that each bit of the multiplier is examined. When the bit is a 1, the multiplicand is multiplied by the correct power of 2 and stored in a temporary file. When the process of searching all the bits of the multiplier, from LSB to MSB, is completed, all the file contents are added together to achieve the result. This functionality is contained in the 74284/5 LSI chips. Two 4-bit binaries can be multiplied using these chips to achieve an 8-bit product as shown in Figure 13.9.

13-3   COMPUTER ORGANIZATION

Our focus here is on basic microcomputer use. We will not discuss particular commercial computer applications in detail, such as how the IBM PC/AT or the Macintosh IIci work; rather the discussion involves general microprocessor, not microcomputer, use. This use may be the foundation of a microcomputer, or the basis of an instrument, and often, in application consists of both. The microprocessor and its support components will be referred to hereafter as a computer, but the reader is cautioned not to think of a commercial microcomputer such as those mentioned above, although the basic operation is the same.

General

All digital computers possess the same functional units. There are three essential units in any computer: the controller (the "brains") or *central processing unit* (CPU), the memory unit (RAM and ROM) and the input/output control circuitry. Often, the CPU is referred to as the microprocessor control unit, or main processing unit (MPU). Figure 13.10 shows the typical interconnections of these components. Communication with the outside world is accomplished with the *input/output* (I/O) control circuitry. This can consist of a keyboard input, a display output driver, a hard-drive controller which directs data to be stored and read on some magnetic medium, a printer driver and an instrumentation interface for connecting transducers of the type we have been discussing throughout the text. The information at the input is passed to the CPU, which directs it to the memory unit where it is stored. At this point, the CPU directs further operation. It may process the information and return it to memory, or it may direct the I/O circuitry to store it on a computer disk. Ultimately, the result is transferred to the I/O unit where it becomes available to the user.

- **Fast Multiplication of Two Binary Numbers**
  8-Bit Product in 40 ns Typical

- **Expandable for N-Bit-by-n-Bit Applications:**
  16-Bit Product in 70 ns Typical
  32-Bit Product in 103 ns Typical

- **Fully Compatible with Most DTL and**
  TTL Circuits

- **Diode-Clamped Inputs Simplify System**
  Design

Description

These high-speed TTL circuits are designed to be used in high-performance parallel multiplication applications. When connected as shown in Figure A, these circuits perform the positive-logic multiplication of two 4-bit binary words The eight-bit binary product is generated with typically only 40 nanoseconds delay.

This basic four-by-four multiplier can be utilized as a fundamental building block for implementing larger multipliers. For example, the four-by-four building blocks can be connected as shown in Figure B to generate submultiple partial products. These results can then be summed in a Wallace tree, and, as illustrated, will produce a 16-bit product for the two eight-bit words typically in 70 nanoseconds. SN54H183/SN74H183 carry-save adders and SN54S181/SN74S181 arithmetic logic units with the SN54S182/SN74S182 look-ahead generator are used to achieve this high performance. The scheme is expandable for implementing N × M bit multipliers.

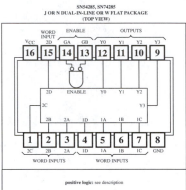

The SN54284 and SN54285 are characterized for operation over the full military temperature range of –55°C to 125°C; the SN74284 and SN74285 are characterized for operation from 0°C to 70°C.

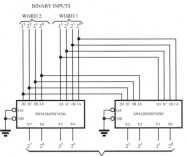

FIGURE 13.9 Manufacturer's information on a 4-bit multiplier, the 74284 and 74285, with correct interconnection.

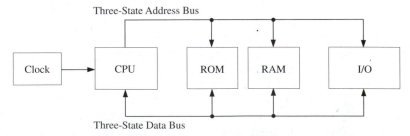

FIGURE 13.10 Block diagram of computer architecture.

The CPU can do nothing without instructions. The basic operating program which instructs the microprocessor is stored in *Read Only Memory* (ROM). It is read from ROM and acted upon by the CPU. ROM information is permanent; it is not lost when the unit is turned off. *Random Access Memory* (RAM), can also store program information, *and* can be written to by the CPU, but its contents are erased after the power is turned off and therefore the basic CPU management program is stored in ROM. The program stored in ROM is often called firmware, since it is a physical entity which does not change. This is in contrast to computer software, which is usually stored on magnetic medium, and is loaded into RAM, where it can be accessed. Hardware is defined as the physical components of the computer, such as the disk drive, keyboard, CPU, and of course the ICs that support the CPU, such as memory, the clock and power supply.

Program instructions and data information in RAM or ROM are contained in "pigeonhole" locations, each of which has a specific address. Additionally, *memory-mapped* I/O devices, such as an input port, have addresses as well. The addresses are used to unequivocally locate and store the information. The data at the particular address being examined by the CPU is transmitted back and forth via the *data bus.* The term "bus" refers to a collection of one or more conductors in parallel that are used for electrical signal transmission. Address and data bus lines are shown in Figure 13.10, where the buses, consisting of multiple conductors, are shown as a single line. When the CPU wants to read or write data to or from a particular address, it sends the binary representation of that address over the *address bus.* The device, be it RAM, ROM or I/O unit, responds by sending or receiving the contents of the address over the data bus. The status of a separate read/write line (not shown) dictates whether data are fetched or stored. The address bus is therefore unidirectional; its signal originates from the CPU, whereas the data bus is bidirectional, transferring data between CPU and another device. For this scheme to work, the data bus must be of the three-state logic type. The logic is enabled by receiving the appropriate address request.

The CPU

The CPU is the heart of the computer or microprocessor-based instrument. The degree of flexibility inherent in a system is largely dependent upon the CPU's characteristics. The CPU is actually composed of a number of subunits which are interconnected to perform the activities in a controlled manner. The basic functions of the CPU are timing (for sequencing), storage of information (data and instructions), and interpretation and manipulation of the information.

A wide variety of microprocessors are available today. Two main differences among these are found in the speed at which the microprocessor runs, the clock speed, and the word length which it can handle. The clock speed is an indication of how fast data are fetched from a storage device and acted upon. Another measure of speed is the number of instructions a microprocessor can process per second. This speed is given in MIPS, or millions of instructions per second. In this text, "speed" refers to the clock speed, but the two are, of course, related. Microprocessors which run at speeds in the kilohertz range are available and still useful in relatively slow applications such as burglar alarms and digital meters, but units with high-speed operation of 25 MHz and more are becoming common, particularly in microcomputer applications. Word length, the bit length a computer can process at any one time, can range from 4 bits to 32 bits and even more. The two main lines of microprocessors which dominate the personal computer industry are the Intel 8086 (IBM) and the Motorola 68000 (Apple Macintosh) families.

When first introduced in the latter part of the 1970s, microprocessor-based devices such as microcomputers and other instrumentation were commonly based upon 8-bit microprocessors. Typical to these is the Motorola 6800 which has an 8-bit data bus and 16-bit address bus and can run at a clock speed of up to 1 MHz. Although this microprocessor is still in design use today, it has been replaced in the microcomputer world by the 68000 family. An example from this family is the 68030 (the CPU for a number of Macintosh II computers) which has a 32-bit data bus and a 32-bit address bus and can operate at a 30-MHz clock rate. Parallel to Motorola's development, Intel introduced the 8086 (the CPU for the IBM/PC) which was followed by the more powerful 80286 and then the 80386 with a 32-bit data bus, a 32-bit address bus and clock speed of 33 MHz. Development of these lines, at the time of printing, has reached the next level, namely the 68040 and the 80486. Both units retain the 32-bit bus structure and run at about the same clock speed, but refinements have included on-board *cache* memory for storing newly fetched data, and built-in *floating-point processors (FPUs),* which significantly increase the speed at which the microprocessor can handle arithmetic manipulations. Both of these features increase the MIP rate of the units.

The 8-bit microprocessor with 8-bit data bus and 16-bit address bus can access 2^{16} = 65,536 (64K) address locations. This limits the amount of information that can be collected and stored, or the length of a program that can be rapidly executed by the microprocessor. In contrast, the 68030 with its 32-bit address bus can access 2^{32} (over 4×10^9) address locations, a virtually limitless quantity for most applications.

RAM

One of the most frequently encountered memory schemes used in memory devices is called *Random Access Memory,* or RAM (pronounced as a single word), in which each word of memory is equally accessible for either reading or writing. The RAM can be either static or dynamic. A *static* RAM, or SRAM (pronounced es-ram), stores information as long as the chip is powered. This is accomplished by holding the information in an array of flip-flops as introduced in the previous chapter. The *dynamic* RAM, or DRAM (pronounced dee-ram), requires that the information stored in memory be refreshed periodically, since it stores information in an array of capacitors. The voltages across the capacitor units representing the binary states decay with time, due to leakage currents. On a periodic basis, on the order of milliseconds, the voltages must be re-established or refreshed. While this need to refresh the DRAM periodically may seem to argue against its use, this is not the case. Standard refresher circuits are widely available on chips; therefore, their availability overcomes this potential weakness. In general, DRAMs are more economical and space- and power-efficient bit for bit than SRAMs, but the latter can operate at higher speeds and are easier to use.

ROM

Read Only Memory, or ROM (pronounced as a single word), is a special kind of memory—one which allows only inspection of its contents, and prohibits writing into its contents. The information in a ROM is often supplied during its manufacture. Naturally, the particular information to be inserted is judged to be important. Once it is inserted into a ROM, it is there to stay; the user can read its contents as often as appropriate without changing the contents. Obviously, ROMs are used for permanently stored programs. The most common use of ROMs is to store the basic instructions which tell the microprocessor how to function on start-up. When the computer or microprocessor-based instrument is turned on, the microprocessor looks at a particular address in ROM which leads it through a start-up, or boot, procedure. Beyond start-up instructions, ROMs are sometimes used to hold programs which run the entire instrument, such as in a microprocessor-controlled meter, and virtually every new car controlled by microprocessor today has its operating instructions stored in ROM.

In addition to instruction sets, ROMs are often used to store data tables which are used in computation. These are called *look-up tables.* An example would be a set of ROM-stored voltages as a function of temperature for a particular thermocouple used in a temperature-measurement instrument. The instrument would read the thermocouple voltage, look it up in ROM and display the corresponding temperature. Since different thermocouple compositions have different look-up tables, the instrument could be used for a variety of thermocouples if the corresponding ROM was inserted.

User-Programmable ROM

ROMs are also available that can be user programmed. These are called *Programmable Read Only Memories,* or PROMs (pronounced prahm). The 74186 is a 512-bit PROM in which every location is initially low. The locations are held low by a metal strip connection. When a current of about 120 mA is applied to a particular location, the metal strip melts, thus opening the circuit and making the output high. By judicious use of a pulse generator with appropriate addressing, the user's program can be permanently placed in the PROM.

To alleviate the problem of incorrect or faulty programming and to increase the versatility of the programmable read only memory chip, the *Erasable* PROM, or EPROM (pronounced ee-prahm) was developed. The device is programmable electronically by application of voltage pulses, and the entire contents can be erased by shining an ultraviolet light on the EPROM's quartz window. Thus, the EPROM can be re-programmed again and again, to a limit of about 100 cycles. Special EPROM programming devices are commercially available. Most of these connect to a microcomputer and download (transfer) a program from the computer to the EPROM inserted in the device using the appropriate voltages. Further in this development is the *electrically erasable* PROM, or EEPROM (pronounced ee-ee-prahm), which can be erased and programmed electronically, often without being removed from the circuit. The EEPROM is the latest wave in PROM technology, but it currently has trade-offs in terms of size and cost.

13-4 AN INSIDE LOOK AT THE MICROPROCESSOR

The microprocessor is nothing more than a collection of logic devices such as those we discussed in the last two chapters. Its function is to fetch, decode and execute program instructions, to transfer data from memory and I/O devices, to respond to external interrupt signals and to provide control and timing. The basic components which a microprocessor needs to execute these functions are the instruction decoder, the arithmetic and logic unit (ALU), a number of registers, control and timing unit, and bus interface for data and address lines. For simplicity, the remainder of the discussion will focus on Motorola's 6800 microprocessor. More advanced families of microprocessors share the basic components of the simple 6800, so an understanding of this device will provide a firm foundation for future study.

The pin-out connections of the 40-pin MC6800 microprocessor chip are shown in Figure 13.11. A brief explanation of the function of each input/output is given below. The overbar signifies the input is active when a low signal is applied.

Pins 1, 8 and 21: These are power-supply lines: ground, + 5 V and ground, respectively.

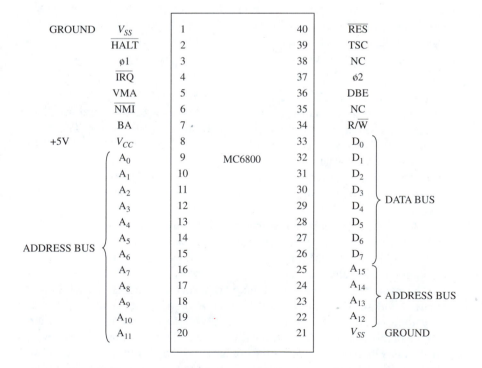

NC — no connection

FIGURE 13.11 **Pin-out of an MC6800 microprocessor.**

Pin 2 ($\overline{HALT}$) This is an input line that allows normal microprocessor operation when high. When a low signal is input to this line from an external source, the microprocessor completes execution of the instruction it is processing, then stops. The BA line (Bus Available) goes high and the VMA (Valid Memory Address) line goes low and the address bus, data bus and R/$\overline{W}$ (Read/Write) lines go into a high-impedance state (they are three-state lines). The $\overline{HALT}$ line is useful in controlling execution of one instruction at a time for such purposes as program debugging.

Pins 3 and 37: These are input clock lines, $\phi 1$ and $\phi 2$, representing digital signals that are 180° out of phase.

Pin 4: The Interrupt Request ($\overline{IRQ}$) line is an input line. When it is high, the microprocessor continues its process, but when it goes low the microprocessor puts the contents of addresses FFF8 and FFF9 on the address bus. The contents of these addresses contain the address to which the microprocessor should go to begin execution of a ROM program designed to run when an interrupt is requested.

Pin 5: The Valid Memory Address (VMA) goes high when a valid address is on the address bus, indicating to other devices that this is so, otherwise it is low.

Pin 6: The Nonmaskable Interrupt ($\overline{NMI}$) is similar in nature to the $\overline{IRQ}$ line, but the contents of addresses FFFC and FFFD are placed on the address bus telling the microprocessor where to go to execute the $\overline{NMI}$ sequence program.

Pin 7: The Bus Available (BA) line is normally in the low state, signalling to other devices that the address and data buses are under microprocessor control. When the line is high, the buses are in a high-impedance state and available to the other devices.

Pins 9–20 and 22–25: These are Address (A) output lines used to address memory devices and other circuits connected to the microprocessor. They are three-state lines as previously mentioned, capable of driving one TTL load.

Pins 26–33: These are bi-directional Data (D) lines which allow the transfer of data between the microprocessor, memory and other devices. As mentioned earlier they are three-state lines and are capable of driving one TTL load.

Pin 34: This is the Read/Write (R/$\overline{W}$) line which indicates if the data on the Data bus are to be read or written. When high, the signal indicates that the microprocessor will read data, and when low that the microprocessor intends to write data. It is a three-state line.

Pin 36: The Data Bus Enable (DBE) line is an input line that enables data to be output on the data bus when it is high. It is a three-state line.

Pin 39: The Three-State Control (TSC) line is an input line which when made high causes the address lines and the R/$\overline{W}$ line to go into a high-impedance mode. It does not affect the data bus lines.

Pin 40: The Reset ($\overline{RES}$) line is held low for at least eight clock cycles after the power is applied and reaches 4.75 V. When the line is made high, after this time, the contents of addresses FFFE and FFFF are placed on the address bus. This information tells the microprocessor where in memory the restart sequence is stored, and likewise begins the sequence. To reset the microprocessor during operation, this line is brought low.

To further understand the workings of the microprocessor and how it handles data and instructions, consider the block diagram in Figure 13.12 which also shows the pin-outs just discussed. The six internal registers (recall from Chapter Twelve that a register is a set of flip-flops used to store data) are used to process data and aid in the execution of a program set. Their functions are summarized below.

Accumulator A, Accumulator B: 8-bit registers used as temporary storage of data.

Index Register: 16-bit register used to hold and modify address numbers. The information in the register can be compared, loaded, stored, incremented and decremented.

Program Counter: 16-bit register used to store the address of the next instruction to be obtained from memory. When the address is placed on the address bus, the contents of the register is incremented.

Stack Pointer: 16-bit register which tells the microprocessor where to store the contents of the other registers should it have other functions to execute, such as during an interrupt. It contains one address which indicates the

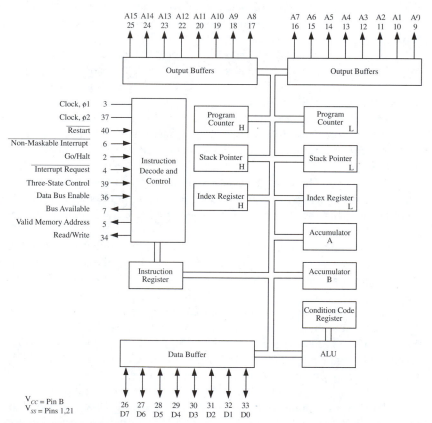

FIGURE 13.12   **Block diagram of MC6800 microprocessor with pin-out connections. Adapted from *The Complete Motorola Microcomputer Data Library*, Motorola 1978.**

starting point to sequentially store the necessary information, since 7 bytes are needed.

Condition Code Register: An 8-bit register in which each of the bits has a different significant meaning. It does not contain address information, rather it contains information that is necessary in executing instruction sets. Two of the bits are not used by the microprocessor.

The ALU, or *Arithmetic and Logic Unit,* performs a number of operations, arithmetic or logic as the name implies, on data. In some sense, it is the heart of the microprocessor, but it is dependent upon the other components in the chip. The Instruction Decode and Control and the Instruction Register handle all control signals in and out of the microprocessor and tell the ALU what function to perform and on what data, dependent upon the instruction being executed.

Keep in mind that the 6800 is being presented here merely as an example of a microprocessor. The more developed 68000 family and others share similar traits, but there are some significant differences. For example, the word length of the 68020 is much longer than that of the 6800. This means the registers in the 68000 are not 8 and 16 bits in length as described above; they are larger, and the number of registers is different. The 68020 has eight 32-bit Data Registers, seven 32-bit Address Registers, a 32-bit Stack Pointer, 32-bit Program Counter and an 8-bit Status Register. There are basic similarities, and that is why we are presenting a simple example.

13-5   INSTRUCTION SETS

All microprocessors, regardless of size, have a set of instructions by which they operate. These are composed of a series of logical 1s and 0s which cause the CPU to take the appropriate action. These instructions are said to be written in machine language since they are immediately recognized by the CPU. Unfortunately, machine language is largely unrecognizable by most humans. Clearly, it is tedious, confusing and prone to human error. Many of the problems associated with machine language can be overcome by the use of instruction sets which contain mnemonics which assist the programmer in understanding the instruction. These instructions are referred to as Assembly language. Assembly language consists of a group of letters and numbers which suggest the operation, such as LDA which means "Load the Accumulator" or SUB M which means "subtract contents of register M." In other words, Assembly language is a human-microprocessor dictionary. Once written, the Assembly language program is translated to machine language, or object code, using a translator, or assembler, program.

Each instruction has an object code, which is usually represented in hex for humans and, of course, binary for computers, and an Assembly language mnemonic for humans. However, most Assembly language instructions, such as LDA, are executed in a variety of ways depending on how the accumulator is to be loaded and with what. For example the code LDA A #$30 means load accumulator A with the hex number 30, whereas LDA A $30 means load accumulator A with the contents of memory location hex 30. The first instruction is an example of immediate addressing and the latter, direct

TABLE 13.2 Examples of Assembly Language and Machine Code for the 6800 Microprocessor

Instruction	Assembly Language	Object Code (hex)	Example
Load Accumulator	LDA	86 (immediate)	LDA A #$23
	Load accumulator A with the hex number 23		
Jump to Subroutine	JSR	BD (extended)	JSR $F3DE
	Jump to the subroutine located at address $F3DE		
Subtract Accumulators	SBA	10	SBA
	Subtract contents of accumulator B from that of accumulator A and place the result in accumulator A		
Subtract	SUB	E0 (indexed)	SUB B $A2,X
	Subtract the content of the address in the index register plus A2 from accumulator B and place the result in accumulator B		
Increment	INC	7C (extended)	INC $2D21
	Add 1 to the content of address 2D21		

addressing. The object code for LDA is different for both. In hex notation the immediate addressing mode LDA is represented by the number 86, and the latter by 96. Table 13.2 lists some instructions for the Motorola 6800 microprocessor.

Higher level languages, such as BASIC, FORTRAN or C are even more convenient, allowing people to use the computer without knowing much about how it works. All this is done at a cost. The cost is found in the amount of memory which must be used for these language conversions and the speed at which they will run. In addition, hardware interfacing often requires direct access and precise control of the microprocessor and specific address locations which can sometimes only be done with Assembly language.

Each microprocessor manufacturer supplies a particular instruction set adapted to that system. As of yet, no uniformity of instruction sets has appeared among different microprocessor units; however, there are similarities. Knowing Assembly language for one microprocessor makes it much easier to learn the dialect of others.

Programming Example

Many textbooks have been written on Assembly language programming for virtually every microprocessor commercially available. In fact, for specific computers, such as the IBM and Macintosh lines, books are readily available in local bookstores. Several of these are listed in the bibliography at the end of the text. Here we present a simple Assembly language program written for the Motorola 6809E microprocessor, which is the heart of the inexpensive Radio Shack Color Computer. Although the Radio Shack Color Computer may be viewed as a "toy" computer, this example uses the 6809E as a programming example since that microprocessor is an enhanced version of the Motorola 6800.

Signal Averager

In many experimental measurements, it is necessary to recover a time-dependent, repetitive signal in the presence of electronic noise (unwanted signals). By sampling the signal multiple times at evenly spaced time intervals, we can decrease the effect of noise and better obtain the signal. This is done by averaging the readings at each time signal. If noise is random, its effect should average to zero (if ample measurements are taken), while the signal of interest should remain constant. A multichannel signal averager is a device which accomplishes this. Noise will be discussed in greater detail in Chapter Fifteen; here we present the signal averager as a typical, simple use of a microprocessor instrument.

The program has two parts: an Assembly language subroutine which controls data input and timing and a BASIC program which drives the subroutine and controls input parameters and data display. Assembly language is essential in this application since timing is critical, something which higher level languages are not proficient at. Combining Assembly language programs with higher level language programs is very common in computer-based instrumentation since the latter are usually easier to use. A block diagram illustrating the program's action is shown in Figure 13.13. The

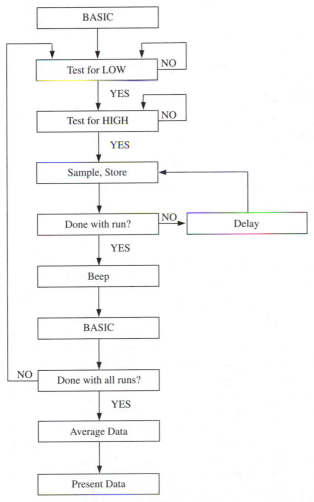

FIGURE 13.13   **Block diagram of multichannel signal averager programming example.**

```
100  REM  MULTICHANNEL SIGNAL AVERAGER BASIC DRIVER
110  DIM X(255)
120  FOR K=1 TO 255 : X(K)=0: NEXT K
125  DEFUSRO=15000
130  CLS
135  INPUT"NUMBER OF SIGNAL REPETITIONS";NR
140  INPUT "NUMBER OF POINTS IN RUN";PT
150  POKE 14402, PT
160  INPUT "DELAY TIME BETWEEN POINTS";DL
170  POKE 14403, DL
173  CLS
175  FOR KL=1 TO NR
176  PRINT KL
180  XD=USRO(Y)
190  FOR I=0 TO PT-1:X(I)=X(I)
     +PEEK(14500+I):NEXTI
205  NEXT KL
210  FOR P=0 TO PT-1:X(P)=X(P)/NR:NEXT P
220  FOR L=0 TO PT-1: PRINT X(L); NEXT L
230  END
```

FIGURE 13.14   **BASIC driver program for multichannel signal averager programming example.**

BASIC program shown in Figure 13.14 defines (to itself) where in memory the Assembly language subroutine resides, then asks the user how many times to take readings from the signal (NR), how many data points should be taken in each run (PT) and what the time delay is between data points (DL). The variables PT and DL are stored in RAM memory locations 14402 and 14403 (decimal). The remainder of the program is a loop which calls the Assembly language subroutine and stores and processes data.

The Assembly language routine shown in Figure 13.15 requires more explanation, since the mnemonics used are not as easy to understand as those of higher level languages. Lines 100 to 130 initialize a variety of parameters, namely ORG 15000, which defines where in RAM memory this program will reside, LDB 14402, which reads the contents of memory location 14402 containing the value of PT placed there by the BASIC program into Accumulator register B, and LDY #14500, which loads the value 14500 into Index Register Y. This register is the start address where the data will go sequentially. Lines 131 to 136 examine memory location $FF00, the location of an input

```
INITIALIZE PARAMETERS
010   *        SIMPLE MULTICHANNEL AVERAGER
100            ORG   15000
110            LDB   14402    #TIMES TO TAKE READING
120            STB   14401    STORE 14402 INTO 14401
130            LDY   #14500   BEGINNING ADD.DATA ARRAY

TRIGGER DETECT
131  SCOOP LDA   $FF00     TEST FOR LO, IF LO GO IF NOT STAY
132            ANDA  #1
133            BNE   SCOOP
134  SCOPE LDA   $FF00     TEST FOR HI, IF HI GO IF LO STAY
135            ANDA  #1
136            BEQ   SCOPE

SAMPLE AND STORE
140  TOP   LBSR  $A9DE     CALL A/D CONVERTER SUBROUTINE
150            LDB   $015A     GET BINARY VOLTAGE
160            STB   ,Y+       PUT VOLTAGE IN THIS ADDRESS
170            LDB   14401     GET CURRENT TIMES THROUGH
180            DECB            DECREMENT TIMES THROUGH COUNT
190            TSTB            DONE?
200            BEQ   DONE      BRANCH IF DONE
210            STB   14401     STORE TIMES THROUGH AT 14401

DELAY LOOPS TO WAIT IN BETWEEN READINGS
220            LDA   14403     GET DELAY COUNTER VALUE
230  DELA  LDB   14404     MAIN DELAY LOOP
240  DEL   NOP             NESTED DELAY LOOP
320            DECB
330            TSTB            DONE FOR NESTED DELAY?
340            BNE   DEL       IF NOT DONE, GO UP TO DEL
350            DECA
360            TSTA            DONE FOR DELAY?
370            BNE   DELA      IF NOT DONE, GO UP TO DELA

END DATA TAKING
380            BRA   TOP       BACK UP TO TOP
385  DONE  NOP

MAKE A BEEP SOUND FROM SPEAKER
390  WAIT  LDA   $FF23     BEEP ROUTINE TO 600
400            ANDA  #$F3
410            STA   $FF23
420            LDA   $FF22
430            ORA   #$02
440            STA   $FF22
450            LDA   $FF23
```

FIGURE 13.15   **Assembly language subroutine of multichannel signal averager programming example.**

```
460             ORA     #$04
470             STA     $FF23
471             LDA     #$44
472             STA     <1
473    SHO      LDA     <1
474             CMPA    #0
475             BEQ     DONE1
476             DECA
477             STA     <1
480    REP      LDB     $FF22
490             ORB     #$02
500             STB     $FF22
510             LDA     #$88
520    WS1      DECA
530             BNE     WS1
540             LDB     $FF22
550             ANDB    #$FD
560             STB     $FF22
570             LDA     #$88
580    WS2      DECA
590             BNE     WS2
600             BRA     SHO


RETURN TO BASIC PROGRAM
610    DONE1    RTS               RETURN TO BASIC
620             END
```

FIGURE 13.15   (continued)

device that provides a trigger signal to start a data run. When the low signal is received, the program (lines 131, 132, and 133) branches to line 134 which in turn waits for a high signal at $FF00.

When received, control passes out of the loop to line 140 that tells the microprocessor to go to another internal Assembly language subroutine (LBSR) located in ROM, at memory location $A9DE. The ROM subroutine drives circuitry which reads the voltage signal on the input and returns a binary value into RAM memory location $015A. Lines 150 through 210 take this information and store it in an appropriate memory location. When finished with storing the data, the program checks to see if it has completed the run (BEQ DONE) or continues to lines 220 through 370, a delay loop which causes the program to wait a certain program time, depending on the value of DL, before taking another reading. Lines 380 and 385 guide the program into either taking another data point (back to line 140), or finishing. If finished, the program goes through lines 390 to 600 which is merely code to drive the computer speaker to make a beeping sound (in BASIC a beep is made by a simple one-line command!). Line 610 causes the program to end and return to the BASIC program which called it.

Each Assembly language command takes an exact, known number of clock, or machine, cycles to complete. Since the length of a clock cycle is known, accurate timing can be accomplished. The sample program utilizes this information in lines 230 through 370 listed below. Each command is shown below with the corresponding number of machine cycles it takes to complete.

Line #		Instruction		# of machine cycles to execute
220		LDA	14403	2
230	DELA	LDB	14404	2
240	DEL	NOP		2
320		DECB		2
330		TSTB		2
340		BNE	DEL	3
350		DECA		2
360		TSTA		2
370		BNE	DELA	3

The minimum delay time is once through these commands without looping through the main loop (lines 230 through 370) or the nested loop (lines 240 through 340). One time through requires 20 machine cycles. If the microprocessor is running with a clock yielding $1\mu s$ machine cycles, then data could be taken every 20 μs. Since Addresses 14403 and 14404 contain the number of times the program will loop through the main loop and nested loop respectively, the equation for the number of machine cycles as a function of these parameters is given by

$$\text{machine cycles} = 2 + (14403)*9 + (14403)*(14404)*9.$$

Since these are 8-bit addresses, the maximum number they can hold is 255, giving a maximum delay of 5,852,252 machine cycles, or just under 6 seconds using the machine cycle of $1\mu s$.

Memory Mapping

Every microprocessor device, particularly the personal computer, has address locations allocated to certain functions. Basically these are RAM, ROM and I/O devices. A list of address locations and their allocation is called a memory map. Memory maps can be quite general, merely outlining which portions of memory are for RAM, which are for ROM and which are for general input and output devices. Alternatively, they can be very specific, listing the content of each address in ROM, such as the starting location of restart and interrupt procedures. It is often useful to understand the memory map of a particular computer, particularly if you are designing your own input/output devices or programming in Assembly language. General memory maps for the Macintosh II, Apple IIGS, and Radio Shack Color Computer are shown in Figure 13.16. Notice that although these machines vary tremendously in power, their maps share common attributes.

13-6   THE INTEL 8080—AN EXAMPLE

Several major semiconductor manufacturers have introduced microprocessor systems. One of the earliest introduced was the Intel 8080 microprocessor. It

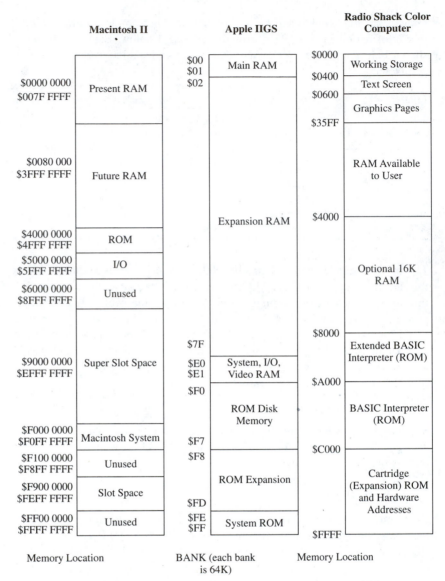

FIGURE 13.16 General memory maps of Macintosh II, Apple IIGS and Radio Shack Color computers.

is now quite dated, but in order to acquaint the student with 8-bit microprocessors available commercially, the Intel 8080 system is discussed as a simple, basic example.

The 8080 itself is the CPU portion of a microcomputer. Its functional block diagram is shown in Figure 13.17. The reader should notice that the subunits are not unfamiliar. The timing and control and the stacks are included for completeness. The timing cycle permits the fetching of one, two, or three 8-bit bytes of instruction. The instruction cycle for the 8080 can be as long as five machine cycles. In order to achieve this type of versatility, a two-phase clock (ϕ_1, ϕ_2) is used. These two inputs are provided by means of a 20 Mhz crystal oscillator and the appropriate control circuitry, as shown in

Figure 13.18. Note that the input machine cycle is reduced to 2 Mhz (ϕ_{1A}). The instruction-timing cycle of the 8080 is given in Figure 13.19. Using the Intel nomenclature, the timing cycle can be interpreted by means of the flow diagram of Figure 13.20, which illustrates the three basic functions: READY, HOLD and INTERRUPT. In essence, the timing control provides the means for full versatility of the CPU operation, so that normal, interrupted or direct memory access operation may be used. A status latch, such as the 8212, must be used in conjunction with the 8080 to achieve these modes. The connections between the two chips are illustrated in Figure 13.21, but this connection is unique to the 8080.

In order to carry the 8080 microprocessor to the microcomputer state, RAMs and ROMs must be added. In addition, I/O interfaces will certainly be needed. Such a system is illustrated in Figure 13.22. The LSI chips shown in the figure are fully compatible with the 8080 CPU. This development represents a fully augmented microcomputer system, although in this case, a rather limited one by today's standards.

As with all computer systems, the Intel system provides an instruction set. The 8080 instruction set covers data transfer, arithmetic operations, logic and program branching, as well as a machine control set. The interested reader is referred to Intel literature for a full discussion of the instruction sets.

The state of the art has developed to a point where most semiconductor manufacturers that produce microprocessors have become, in all but name, computer manufacturers. For example, the components added to the 8080 system just described are manufactured by Intel for use with the 8080.

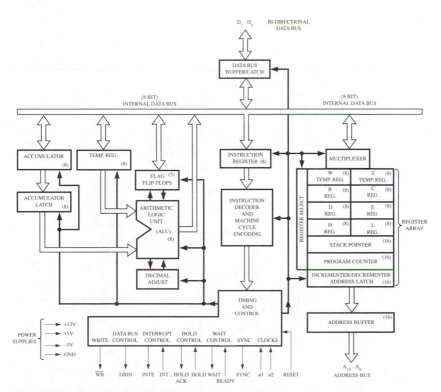

FIGURE 13.17   **The Intel 8080 organization. (Courtesy of Intel Corp.)**

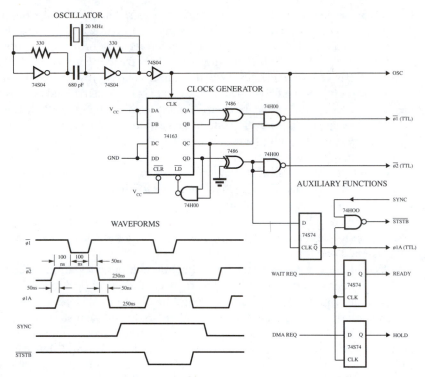

FIGURE 13.18 **The clock control circuitry needed for the Intel 8080.
(Courtesy of Intel Corp.)**

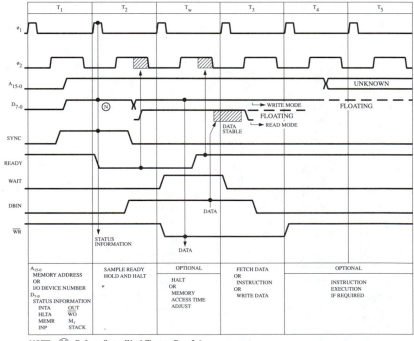

NOTE: Ⓝ Refer to Status Word Chart on Page 2-6.

FIGURE 13.19 **The timing cycle for the 8080. (Courtesy of Intel Corp.)**

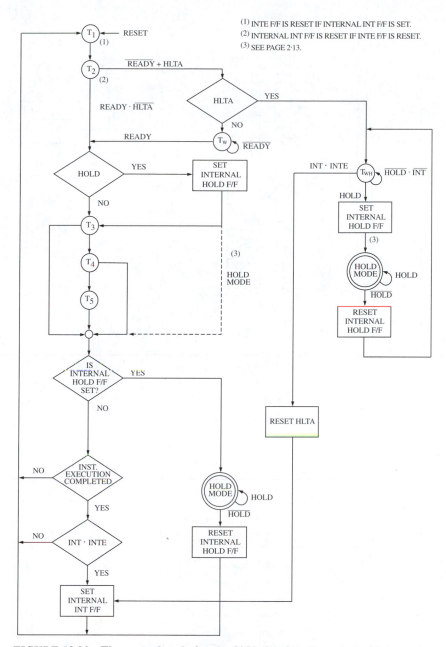

(1) INTE F/F IS RESET IF INTERNAL INT F/F IS SET.
(2) INTERNAL INT F/F IS RESET IF INTE F/F IS RESET.
(3) SEE PAGE 2·13.

FIGURE 13.20   **The control cycle for the 8080 showing the means of interrupt action. (Courtesy of Intel Corp.)**

Similarly, Motorola produces a large array of chip components to be used with its 6800 family such as the 6821 Peripheral Interface Adapter (discussed in the next chapter), the 6840 Programmable Timer Module, 6843 Floppy Disk Controller, 6845 CRT Controller, 6850 Asynchronous Communications Interface Adapter (discussed in the next chapter), and the 6854 Advanced Data Link Controller. The newer 68000 family of microprocessors includes not only the 68020, 68030 and 68040 microprocessors, but these 6800 family

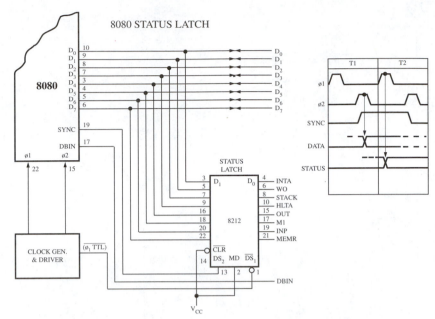

FIGURE 13.21 **The 8212 Status Latch; used in conjunction with the 8080. (Courtesy of Intel Corp.)**

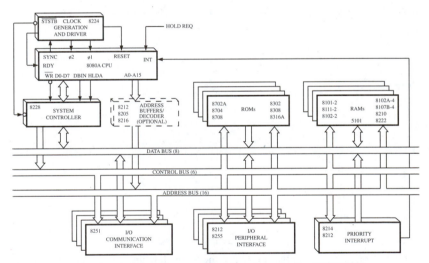

FIGURE 13.22 **A microcomputer system based upon the 8080 CPU. (Courtesy of Intel Corp.)**

peripheral chips and some newer ones such as the 68121 Intelligent Peripheral Controller as well. In a loose sense, from the hardware aspect, the availability of these components makes microprocessor-based instrument and personal computer design a matter of choosing which components to incorporate in the device, although other manufacturer chips performing similar functions can be adapted as well.

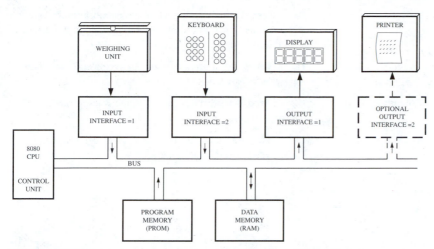

FIGURE 13.23   **A microprocessor system. (Courtesy of Intel Corp.)**

13-7   A MICROPROCESSOR APPLICATION

While this chapter began with the evolution of the microprocessor, it has turned to a discussion of the microcomputer. This accurately represents the current industry trends and accomplishments. At the same time, it is decidedly unfair to relegate the microprocessor to merely a part of the computer, albeit an important one. This chapter began with the recognition of the impact of microprocessors in our everyday lives. It is, therefore, most appropriate to end with an example of a microprocessor application which most of us have encountered unknowingly, the automatic scale found in food stores. The system is illustrated in Figure 13.23. There are two input devices: the scale which signals the weight of the material, and the keyboard which signals the cost per unit weight. The only necessary output device is a decimal display. The CPU accepts these inputs and refers to the PROM, which acts as a multiplication look-up table. The CPU then routes that result to the output device. The system can be made more sophisticated by also storing each output in RAM so that at the end of a time period, the amount sold can be known. There is, in fact, no end to the "fine-tuning" one can perform on this basic system. It is a good example of microprocessor use.

13-8   QUESTIONS AND PROBLEMS

1.   Explain why three-state logic is necessary in microprocessor design.

2.   Perform the following arithmetic operations in binary or in 2's-complement.

  a. $32_{10} - 18_{10}$
  b. $11_{10} - 15_{10}$
  c. $77_8 - 66_8$
  d. $101010_2 + 010101_2$
  e. $55_8 + 26_8$
  f. $1111101_2 - 1010100_2$
  g. $BC_{16} + 9A_{16}$
  h. $19_{16} - 20_{16}$

3.   Distinguish between full and half adders.

4.   Give the binary results for the following binary numbers:

  a. 1111×1100
  b. 1001×1001
  c. 0101×1010
  d. 0011×0010

5. Explain the reason(s) for the "look ahead" in multiplication.

6. Using the 74181 with $A = 0101$ and $B = 1010$, develop an output table for all 16 control settings for $M = 0$ and 1 (with and without carry).

7. Explain how two 74181s may be wired together to handle 8-bit words.

8. Justify the inclusion on the 74181 chip of a separate output for $A = B$.

9. The Motorola 68040 microprocessor has a 32-bit address bus. How many memory locations can it access?

10. Describe the events that take place in a microcomputer when two numbers stored in RAM are added and the result is stored in RAM.

11. Discuss what are meant by the terms software, hardware and firmware.

12. Discuss the difference between a microprocessor and a microcomputer.

13. What are the advantages (if any) of using dynamic RAM over static RAM? What are the disadvantages?

Digital and Analog I/O

14-1 INTRODUCTION

Until now we have only alluded to connecting a computer to the real world for use as an instrument. The world of digital electronics and microprocessors, as powerful as it is, would be virtually inaccessible to us if we could not get information in and out of it as we wish. Ours is still an analog world. We must pass through the analog-digital input/output (I/O) portal.

The I/O section provides the only means the computer has for interaction with the outside world and vice versa. The devices which provide input and output are often designated as peripherals since they are added to the basic computer. For most people, the printer, keyboard and monitor, or cathode ray tube (CRT), are the I/O devices which are commonly associated with computers. In fact, there is a vast array of devices which are properly classified as peripherals, including disk drives, A/D and D/A converters, relays, switches, lamps, motors and many others. An input peripheral device presents information to the computer while the output device provides information from the computer; many devices serve both functions. For example, the keyboard of a personal computer is generally regarded as an input device, but if it contains an LED to indicate the status of the "caps lock" button, then it is also an output device, in some sense.

In this chapter, we will explore the realm of computer input/output as it relates to instrumentation and process control. We begin with a look at parallel and serial data transmission and then move to electronic converters which take analog signals and convert them to digital signals and vice versa.

14-2 PARALLEL DATA TRANSMISSION

The microprocessors we examined in the last chapter generally communicate over parallel data and address buses. Although parallel communication requires more connections and lines, this type of communication is much faster than serial communication, where a signal is transmitted bit by bit over one line (and the associated ground line).

But most peripheral circuits cannot be connected directly to the data and address buses. Some type of adapter is needed to interface the microprocessor, or computer, to the circuitry in question. A chip which accomplishes this is called a peripheral I/O IC (PIO), or peripheral interface adapter (PIA). They contain the necessary circuitry and registers which allow interfacing between the microprocessor and peripheral devices. Some, such as Motorola's 68121 Intelligent Peripheral Controller (IPC), go beyond the simple by including ROM, RAM and a separate microprocessor in the chip.

To understand the function of a typical PIA, we introduce Motorola's 6821 Peripheral Interface Adapter. Although this PIA was originally designed to interface with Motorola's 6800 microprocessor family, it has found use in many other applications as well. The 6821 has an 8-bit data bus, two bi-directional 8-bit buses for peripheral interfacing, two programmable control registers, two programmable data direction registers, four interrupt lines, and handshake control logic. An expanded block diagram with pin-outs is shown in Figure 14.1, and pin functions are as follows.

V_{CC} and V_{SS} (pins 20 and 1): 5 V power input and ground input respectively.

$\overline{IRQA}$, $\overline{IRQB}$ (pins 38 and 37): PIA output interrupt lines.

CA1 and CB1 (pins 40 and 18): Input lines that set the interrupt flags of the control registers and are programmed by the control registers.

CA2 and CB2 (pins 39 and 19): Peripheral control lines each can be programmed through the corresponding control register (A or B) to be an interrupt input or peripheral control output.

$\overline{Reset}$ (pin 34): When low, the reset acts to zero all register bits, normally held high.

Enable (pin 25): Timing signal usually connected to ϕ_2 of MPU clock.

CS0, CS1 and $\overline{CS2}$ (pins 22, 24, 23): For data transfer to occur, the chip must be selected. This is the same as the enable input on a tristate logic device. To select the PIA, CS0 and CS1 are made high and $\overline{CS2}$ is made low. This can be done directly from the address bus or via logic chips connected to the address bus of the MPU.

RS0 and RS1: Register Select lines used to control which register is being written to or read from. These can be connected directly to the address bus or via some external logic circuitry.

$R/\overline{W}$ (pin 21): Read/$\overline{Write}$, when the signal is low, data is transferred to the PIA; when the signal is high, data is transferred from the PIA onto the data bus. This occurs only when the chip is selected via CS0, CS1 and $\overline{CS2}$.

D0–D7 (pins 26–33): Bi-directional data lines.

PA0–PA7 (pins 2–9): Peripheral data lines of A. Each line can be programmed individually to act as either an input or an output line. A high in the data direction register bit that corresponds to the peripheral data bit makes the line an output, and a low causes the line to become an input. Can drive two TTL loads and CMOS.

PB0–PB7 (pins 10–17): Peripheral data lines of B. Identical in function to PA0–PA7 with the exception that they can drive two TTL loads only, and may be used to directly drive the base of a transistor to source up to 1mA at 1.5 V.

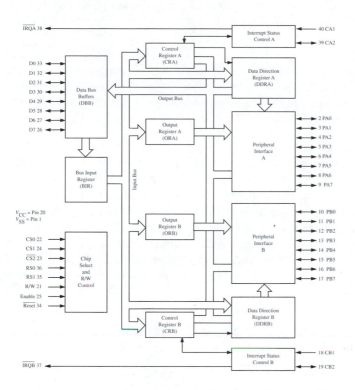

FIGURE 14.1   Expanded block diagram of Motorola's 6821 Peripheral Interface Adapter.

At first glance the function of the PIA, with its six registers, two sides (A and B) and various control lines, may seem difficult to understand. The basic operation of the PIA is actually rather easily understood if we keep in mind that the main function of the chip is to provide two sets of eight I/O lines. With the exception of some drive capability, the operation of the A side is identical to the operation of the B side. The two data direction registers, DDRA and DDRB, have eight bits each which control whether their corresponding peripheral data lines will be inputs or outputs. For example, setting all eight bits of DDRA to high makes all eight lines of the corresponding peripheral data lines, PA0–PA7, outputs. Setting bits 0, 1, 2 and 3 of DDRA high and bits 4, 5, 6 and 7 low will cause PA0–PA3 to be outputs and PA4–PA7 to be inputs.

For data to be read from or written to the peripheral data lines, the appropriate peripheral register (PRA or PRB) must be selected. When data is to be transferred to the peripheral data line, PRA or PRB is selected and data is entered into the PIA through the data lines from the MPU, and is transferred into the appropriate peripheral register. It is then moved by the PIA so it appears on the peripheral data line. When data are read from the peripheral data line, PRA or PRB is selected and data from the peripheral data line are moved into the peripheral register where they are placed on the data bus and read by the MPU. In either case, the MPU sends the PIA the appropriate read/write signal.

RS1	RS0	CRA2	CRB2	Register Selected
0	0	1	x	Peripheral Register A
0	0	0	x	Data Direction Register A
0	1	x	x	Control Register A
1	0	x	1	Peripheral Register B
1	0	x	0	Data Direction Register B
1	1	x	x	Control Register B

FIGURE 14.2 **Addressing truth table for 6821 PIA chip showing bit 2 of control register A and B.**

The function of the control registers, CRA and CRB, is to program the function of the interrupt request lines ($\overline{\text{IRQA}}$ and $\overline{\text{IRQB}}$), peripheral control lines (CA2 and CB2) and interrupt input lines (CA1 and CB1), and to allow for access from the data bus into the data direction registers (DDRA and DDRB) or the peripheral registers (PRA and PRB). Bit 2 of control register A and of control register B is the control bit for the latter. When it is high, access is made to the peripheral register and when it is low, access is to the data direction register. The internal addressing truth table is shown in Figure 14.2.

14-3 PERIPHERAL INTERFACE ADAPTER PROGRAMMING EXAMPLE

A programming example will help to clarify the functions of the various registers. To keep the numbers simple, we will program the A peripheral data lines to be outputs and the B lines to be inputs, thus placing a 0 (0000 0000) in DDRB will cause all bits to be low and placing a 255_{10} (1111 1111) in DDRA will make all its bits high. There is no need to make all the lines on any side inputs or outputs. For example, placing the number 85_{10} (0101 0101) into DDRA will make every other peripheral data line an input (PA7, PA5, PA3 and PA1) while the others will be outputs. For simplicity's sake, we will stay with A lines as outputs and B as inputs.

First we access the control register A by making RS1 low and RS0 high, as shown in the truth table of Figure 14.2. Next, we place the number 0 (0000 0000) onto the data bus from the MPU to allow access to DDRA since this will set bit 2 of CRA low corresponding to DDRA. We then select DDRA by making RS0 and RS1 low, and then place 255_{10} (1111 1111) into DDRA via the data bus from the MPU to make the peripheral data lines outputs. To finish we must select the peripheral register A by first programming the control register again by making RS1 low and RS0 high and loading a 4 (bit 2, 0000 0100) onto the data bus and into CRA. Last, we make RS1 and RS0 low again, but since bit 2 of CRA is high, the peripheral register will be selected and the peripheral data lines of A are now outputs. Data that appear on the data bus will now appear on the peripheral data lines.

To program port B, we access the control register B by making RS1 and RS0 high. Next, we place the number 0 onto the data bus from the MPU to allow access to DDRB (recall this makes bit 2 of CRB low). We then select DDRB by making RS0 high and RS1 low, and then place 0 (0000 0000) into DDRB via the data bus from the MPU to make the peripheral data lines inputs. To finish we must select the peripheral register B by first programming the control register again by making RS1 and RS0 high and loading a 4 (bit 2, 0000 0100) onto the data bus and into CRB. Last, we make RS1 high and RS0 low again, but since bit 2 of CRB is high, the peripheral register will be selected and the peripheral data lines of A are now inputs. Data that appear on the peripheral data lines will now be transferred to the data bus.

In BASIC, the command POKE X, Y places the value of Y into memory location X while the command $Z = PEEK$ (X) places the value of memory location X into the variable Z. Thus, a BASIC program which accomplishes the programming example above would read as follows (assuming that R0 is connected to address bus bit 0, R1 is connected to address bus bit 1 and the remaining address bits are used to select the PIA when they are all low):

```
125   POKE 1, 0      RS0 high, RS1 low, bit 2 of CRA low:
126                   select DDRA

127   POKE 0, 255    RS0 and RS1 high, move 255 into DDRA

      POKE 1, 2      RS0 high, RS1 low, bit 2 of CRA high:
128                   select PRA

129   Z=PEEK (0)     Read PRA and place into Z

      POKE 3, 0      RS0 high, RS1 high, bit 2 of CRB low:
130                   select DDRB

131   POKE 2, 0      RS0 low and RS1 high, move 0 into DDRB

      POKE 3, 2      RS0 high, RS1 high, bit 2 of CRB high:
132                   select PRB

      POKE 2, Y      Place value Y into PRB to appear on
                     peripheral data lines
```

The PIA is an enormously powerful chip in the realm of programmable input and output. It is relatively simple to use and can form the foundation of a general-purpose interface board. Such a board, the Dual PIA Interface that uses two PIAs to provide 32 lines of programmable I/O, is shown in Figure 14.3. The Dual PIA Interface was originally designed for the Radio Shack Color Computer, a simple 8-bit machine based on Motorola's 6809E microprocessor, but the interface can be easily adapted to any other computer.

To better understand the programming of the PIA, an examination of the connections to the address bus is in order. Note that address lines A8–A15 are connected to a 7430 8-input NAND gate, while A4–A7 are connected to a 7485 4-bit magnitude comparator. The 7485 compares its inputs B0–B3 with A0–A3 and returns a high signal when they are equal. Thus the interface can be moved in memory by setting switches 1–4 appropriately. If switches 1–4 are open, a high is returned and sent to the connected 1/3 7410 NAND gate. When A4–A7 are high, and when A8–A15 are high, a high is placed into the input of the middle 1/3 7410 3-input NAND gate. The third input of the 7410 is address line A3. Thus, when A3 through A15 are high, $\overline{CS2}$ for both PIAs is low. Address bus line A2 allows selection of the individual PIAs. When it is low, PIA A (on the left) is selected and when it is

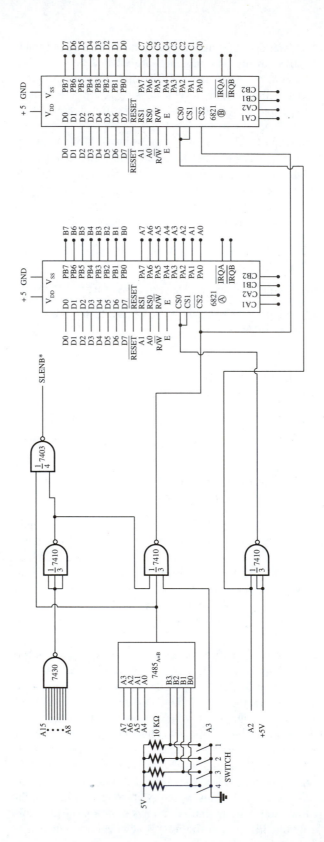

FIGURE 14.3 Dual PIA interface board with 32 lines of programmable I/O (8-bit microprocessor with 16-bit address bus connection shown).

high, PIA B is selected. A1 and A0 are directly connected to the Register Select inputs allowing register selection.

The interface is said to be memory mapped, since access to it is gained the same as in a memory location of RAM or ROM. Thus simple PEEKs and POKEs in BASIC can program and perform read/write functions with the device. Memory-mapped devices are simple to use and common in real-world applications.

14-4   SERIAL DATA TRANSMISSION

Parallel communication is by far the fastest means of transmitting digital data, but numerous applications cannot make use of parallel communication, due mostly to the large number of lines necessary for its operation. Serial transmission, bit by bit rather than all bits at one time, is used when parallel transmission is not possible (Figure 14.4). One application is that of communication between computers over large distances. Here a telephone line is used to send data between a host and remote computer. Clearly the telephone line does not contain the necessary number of connections for parallel communication; rather it uses two lines—signal and ground. The data are trans-

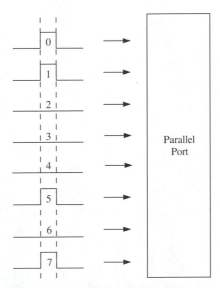

FIGURE 14.4   **Serial data transmission and parallel data transmission. The serial port receives data one bit at a time while the parallel port receives all bits at once.**

mitted one bit at a time. Conversion of parallel data, used internally by the computer, to serial, and vice versa is usually accomplished by a Universal Asynchronous Receiver/Transmitter, or UART, chip. When this is done, the digital signal is modulated into an audio signal, sent over the telephone line, and then demodulated back into the digital signal by a *modem* (modulator-demodulator).

The most common standard used for serial data transmission was set by the Electronics Industry Association and is called RS232C. It is an asynchronous format, meaning that timing signals of the transmitter and receiver are not correlated. Whereas parallel communication can occur in the MHz range, RS232C communication operates in the kHz range, much slower. There are other major differences in RS232C. A true, or marking, signal is in the voltage range of -1.5 V to -36 V and a false (spacing) signal is in the 1.5 V to 36 V range, although most RS232C devices operate in the ± 5 V to ± 14 V range. This is opposite in sign of what one might expect (here a logical low is a high voltage and a logical high is a low voltage) and the range of voltages is not common to the internal operation of computers. Since the signal is usually transmitted over a large distance, iR losses and noise dictate the wide voltage ranges in which true and false signals must be accepted. Drivers which convert TTL signals to RS232C signals are readily available (National Semiconductor's DS1488 Quad Line Driver and DS1489 Quad Line Receiver). The RS232C pin connections for 25-pin and 9-pin D-type connectors are shown in Table 14.1. Pin numbers are shown in Figure 14.5.

TABLE 14.1 25-Pin DIN and 9-Pin DIN RS232C Connections

Pin Number 25-pin DIN	Pin Number 9-pin DIN	Name	Name Function
1	NA	FG	Frame Ground
2	3	TD	Transmitted Data
3	2	RD	Received Data
4	7	RTS	Request to Send
5	8	CTS	Clear to Send
6	6	DSR	Data Set Ready
7	5	SG	Signal Ground
8	1	DCD	Data Carrier Detect
9	NA		Positive Voltage
10	NA		Negative Voltage
11	NA		Not Defined
12	NA		Secondary Received Line Detector
13	NA		Secondary Clear to Send
14	NA		Secondary Transmitted Data
15	NA		DCE Transmitted Signal Element Timing
16	NA		Secondary Received Data
17	NA		Receiver Signal Element Timing
18	NA		Not Defined
19	NA		Secondary Request to Send
20	4	DTR	Data Terminal Ready Detector
21	NA		Signal Quality Detector
22	9	RI	Ring Indicator
23	NA		Data Signal Rate Selector
24	NA		DTE Transmitter Signal Element Timing
25	NA		Not Defined

NA = not available

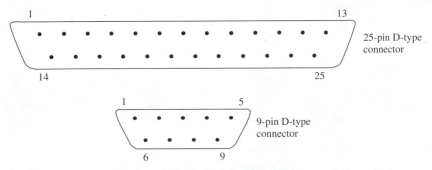

FIGURE 14.5   Numbering of 25-pin and 9-pin D-type connectors.

Most UARTs are 40-pin chips, such as National Semiconductor's 8250 that is used in IBM personal computers. USARTs, Universal Synchronous/ Asynchronous Receiver/Transmitters serve the same function as UARTs, with the exception that they can be operated in synchronous mode as well. Typical units are Zilog's 8530 used in Macintosh computers, and Intel's 8251. Most applications of UARTs do not use all 40 connections available on the chip. An example block connection diagram of National Semiconductor's 8250 as an RS232C driver is shown in Figure 14.6. Additionally, power connections and clock connections are provided. Generally, the transmit/receive rate can be set by an external oscillator, such as a crystal, which may be divided by a user-controlled divisor. For example, the 8250 has a maximum input frequency of 3.1 MHz that is divided internally to a desired 50–38,400 baud (bits per second), including the common baud rates of 1200, 2400, 9600 and 19,200. For further information on the design of UART circuits, the reader is referred to the references listed at the end of the text.

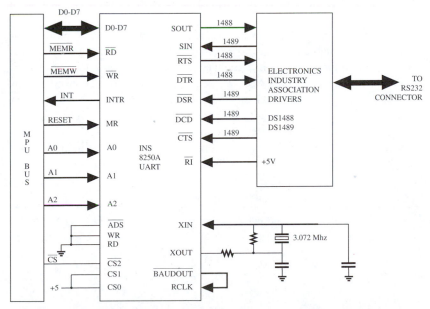

FIGURE 14.6   UART connection for National Semiconductor INS 8250A as RS232 driver. Adapted from National Semiconductor's *Data Communication, Local Area Networks, UARTs Handbook*.

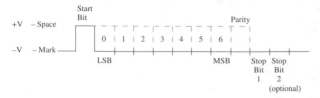

FIGURE 14.7A   **Asynchronous transmission of 7-bit data.**

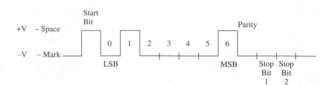

FIGURE 14.7B   **Asynchronous transmission in ASCII format using two stop bits and even parity (− V is high).**

An RS232C device, or more generally an asynchronous device, begins to receive a character when it sees a transition, or start bit, followed by seven (or eight) data bits, a parity bit and then one or two (depending on the device) stop bits, as shown in Figure 14.7A. Note that when not active, the signal is in the negative voltage range; a transition to positive voltage signals the device to begin receiving the set of bits, or character. The least significant bit is sent just after the start bit. Since noise may cause a false reading of one of the bits, a scheme using a parity bit is incorporated to help detect such an erroneous reading. When the number of high bits representing a character is even, the representation is termed even parity, and vice versa for odd parity. If the receiving device is set to look for even parity, then it expects every character, with the parity bit added, to have even parity. The parity bit is set high or low by the sender to make the transmission even. Thus a character with even parity of its seven bits is accompanied by a low-parity bit and a character with odd parity is accompanied by a high-parity bit. The scheme catches the displacement of one bit of data (the most common error) in the transmission, but not two. The American Standard Code for Information Interchange (ASCII) is the language most often spoken between these devices over RS232C. All ASCII characters are 7 bits in length, as shown in Table 14.2. Asynchronous transmission of an equals sign (=) in ASCII using even parity is shown in Figure 14.7B. More often than not, parity is not used in data transmission.

RS232C communication protocol is almost always used for communications between computer and keyboards, plotters, digitizers and printers. Data Terminal Equipment, or DTEs, and Data Communications Equipment, or DCEs are the devices which RS232C was first designed to connect. In this scheme a DTE is a terminal which is connected to a computer via a DCE, or modem. With the advent of microcomputers, the definition of DCE and DTE has become somewhat confused, and interconnection of RS232C devices does not always work as simply as one might expect from an industry standard interface protocol. A problem which occurs with RS232C devices is that

TABLE 14.2   *ASCII Codes in Decimal and Hexadecimal*

Hex	*Decimal*	*Character*	*Hex*	*Decimal*	*Character*
00	00	CTRL-@	40	64	@
01	01	CTRL-A	41	65	A
02	02	CTRL-B	42	66	B
03	03	CTRL-C	43	67	C
04	04	CRTL-D	44	68	D
05	05	CTRL-E	45	69	E
06	06	CTRL-F	46	70	F
07	07	CTRL-G	47	71	G

TABLE 14.2   (CONTINUED)

Hex	Decimal	Character	Hex	Decimal	Character
08	08	CTRL-H	48	72	H
09	09	CTRL-I	49	73	I
0A	10	CTRL-J	4A	74	J
0B	11	CTRL-K	4B	75	K
0C	12	CTRL-L	4C	76	L
0D	13	CTRL-M	4D	77	M
0E	14	CTRL-N	4E	78	N
0F	15	CTRL-O	4F	79	O
10	16	CTRL-P	50	80	P
11	17	CTRL-Q	51	81	Q
12	18	CTRL-R	52	82	R
13	19	CTRL-S	53	83	S
14	20	CTRL-T	54	84	T
15	21	CTRL-U	55	85	U
16	22	CTRL-V	56	86	V
17	23	CTRL-W	57	87	W
18	24	CTRL-X	58	88	X
19	25	CTRL-Y	59	89	Y
1A	26	CTRL-Z	5A	90	Z
1B	27	CTRL-[	5B	91	[
1C	28	CTRL-\	5C	92	\
1D	29	CTRL-]	5D	93	]
1E	30	CTRL-^	5E	94	^
1F	31	CTRL-_	5F	95	_
20	32	SP	60	96	'
21	33	!	61	97	a
22	34	"	62	98	b
23	35	#	63	99	c
24	36	$	64	100	d
25	37	%	65	101	e
26	38	&	66	102	f
27	39	'	67	103	g
28	40	(	68	104	h
29	41	)	69	105	i
2A	42	*	6A	106	j
2B	43	+	6B	107	k
2C	44	,	6C	108	l
2D	45	−	6D	109	m
2E	46	.	6E	110	n
2F	47	/	6F	111	o
30	48	0	70	112	p
31	49	1	71	113	q
32	50	2	72	114	r
33	51	3	73	115	s
34	52	4	74	116	t
35	53	5	75	117	u
36	54	6	76	118	v
37	55	7	77	119	w
38	56	8	78	120	x
39	57	9	79	121	y
3A	58	:	7A	122	z
3B	59	;	7B	123	{
3C	60	<	7C	124	|
3D	61	=	7D	125	}
3E	62	>	7E	126	~
3F	63	?	7F	127	DEL

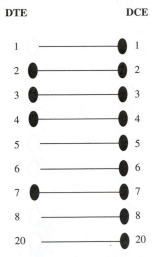

DTE   DCE

FIGURE 14.8A **RS232C connections between DTE and DCE devices with handshaking.**

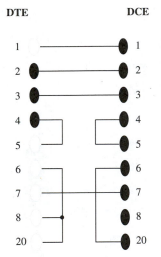

DTE   DCE

FIGURE 14.8B **RS232C connections between DTE and DEC devices with self-handshaking.**

some require handshaking signals and others do not. Examples of a handshaking signal are the RTS (request to send) and CTS (clear to send) signals. The DTE sends an RTS signal to the DCE, which responds by sending a CTS signal back to the DTE. If both devices act accordingly, the communication link will operate as expected; however, if one device is designed to operate without handshaking, while the other is not, the communication link will not work. Again, this is not a problem one would normally expect from an industry standard!

In general, when both devices operate with handshaking, the connection scheme shown in Figure 14.8A will work. If the devices expect no handshaking, then sender and receiver must signal themselves to send or receive. Connections that accomplish such self-handshaking are shown in Figure 14.8B. If two computers are to communicate with each other over short distances, sometimes a null-modem connection is useful. This link between two computers requires no modem, just a cable, hence the name null-modem. Connections for this setup are shown in Figure 14.8C. When all else fails, device function can be diagnosed with a tool called an RS232 breakout box. The breakout box has LEDs which indicate who is sending what signals and allows for interconnections between various pins via switches.

14-5   OTHER STANDARD COMMUNICATION PROTOCOLS

IEEE-488

Over 15 years ago, Hewlett-Packard, one of the industry giants in instrumentation, saw a need for a communication standard specific to laboratory instruments. They designed and developed the Hewlett-Packard Interface Bus (HPIB), also known as the General Purpose Instrumentation Bus (GPIB), that was later adopted as an industry standard by the Institute of Electrical and Electronics Engineers and designated IEEE-488 ("I, triple-E, 488"). The IEEE-488 bus can connect up to 14 instruments (and computer), with maximum total cable length of less than 20 meters and individual cable length of no greater than 4 meters, making it ideal for most laboratories. Data communication rate can be as high as 1 MB per second, but is usually slower. Virtually all major manufacturers of research-grade instruments offer IEEE-488 devices, and software and interfaces for most personal computers are readily available.

Generally speaking, IEEE-488 devices fit one of four classifications: listen only, talk only, talk-listen, and talk-listen-control. The bus contains 16 bidirectional lines; eight are data lines and the other eight are control lines. Eight additional lines used for logical ground returns and shielding complete the connections to the 24-pin IEEE-488 connector. Parameters such as waveform, frequency and amplitude of a function generator, voltage and current ranges of a voltmeter, and voltage of a power supply can be simply controlled by a computer on the IEEE-488 instrumentation bus.

MIDI

In the early 1980s the music industry recognized the need for an industry standard to connect microprocessor-based instruments and computers together. The result was the design and implementation of the Musical Instrument Digital Interface, or *MIDI*. The MIDI operates serially, similar in some respects to RS232C but with one start bit, eight data bits and two stop bits, at 31.25 kilobaud. It also uses only two lines for input devices and three lines for output devices. In order to prevent ground loops (stray signals caused by current flowing through a ground wire), there is no ground common to the connected devices; in fact, no physical electrical connection per se exists between the devices. The controlling device sends a signal through a UART to the 5-pin DIN (MIDI out) connector. As shown in Figure 14.9, pins 1 and 3 are not used, pin 4 is a 5 V power connection, pin 5 the signal out and pin 2 is connected from ground to the metal shield which surrounds the wire of the cable. On the input side, pins 4 and 5 are used to drive the LED of an *optoisolator*. An optoisolator is nothing more than an input LED next to an output phototransistor, but it has the important feature of involving no direct electrical connection. Light is the communication medium by which the two ends communicate. One note of caution when using MIDI cables: Do not substitute cheap 5-pin DIN cables for MIDI 5-pin DIN cables unless you disconnect the ground shield from the metal chassis connection at the input. Otherwise ground loops will more than likely change the sound information (considerably) that you wish to transmit.

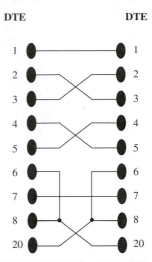

FIGURE 14.8C RS232C null-modem connections between DTE and DTE devices good for connections less than 100 feet.

14-6   DIGITAL-TO-ANALOG CONVERTERS: DACs

Until now, we have discussed circuits that contain both digital and analog signals in a peripheral manner. For example, oscillators such as the 555 and comparators such as the LM311 combine digital and analog technologies. But we have not explored the direct conversion of a digital signal to an accurate analog signal and vice-versa. These conversions are at the heart of microprocessor-based instrumentation. Although you may not wish to design conversion circuits since they are commercially available, you should understand the basic properties of conversion circuits in order to make intelligent decisions on their purchase and use.

Weighted Current Source DAC

The *digital-to-analog converter,* or DAC, does just what its name implies: it takes a digital signal and converts it to an analog signal. Consider the 4-bit DAC circuit shown in Figure 14.10 that uses 5 resistors in an operational amplifier adder circuit. Switch 1 represents the least significant bit and switch 4 the most significant bit. When all four switches are open (the digital equivalent of 0000), no voltage is applied to the non-inverting input of the amplifier and V_{out} should be equal to zero. Closing switch 1 applies -10 V to the

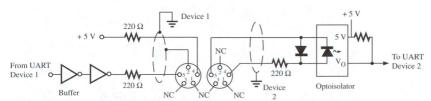

FIGURE 14.9 Typical MIDI connection from Device 1, the sender, to Device 2, the receiver.

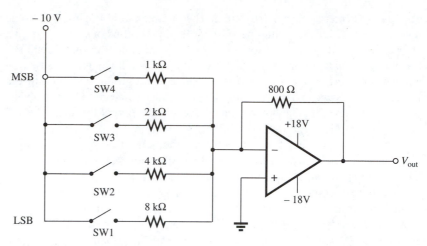

FIGURE 14.10 **Digital-to-analog converter using resister network and op-amp.**

input of the 8 kΩ resistor, yielding an output voltage of $[-V_{in} (800 \ \Omega/ 8000 \ \Omega)] = 1$ V. Likewise, closing switches 2, 3 and 4 individually gives an output voltage of 2, 4 and 8 V respectively. Since the circuit is an adder, closing a combination of switches causes an output of the addition of the voltages associated with each switch. Table 14.3 gives the sixteen possible combinations of the four switches and respective output voltages.

Another way of looking at this example is that by closing switch 1, 1.25 mA of current flows from the non-inverting input of the operational amplifier through the 8 kΩ resistor. Since the junction at the op-amp input is a virtual ground, the current source must be the output of the amplifier. Looking at the circuit in this manner suggests that the resistors and the -10 V supply act as a current source with weighting determined by the value of the resistor. Switching the direction of current, replacing the mechanical switches and

TABLE 14.3 *Output Voltage as a Function of Switch Status in DAC of Figure 14.10*

MSB SW4	SW3	SW2	LSB SW1	Output Voltage (Volts)
0	0	0	0	0
0	0	0	1	1
0	0	1	0	2
0	0	1	1	3
0	1	0	0	4
0	1	0	1	5
0	1	1	0	6
0	1	1	1	7
1	0	0	0	8
1	0	0	1	9
1	0	1	0	10
1	0	1	1	11
1	1	0	0	12
1	1	0	1	13
1	1	1	0	14
1	1	1	1	15

resistors with logic-controlled switches and current sources results in the op-amp circuit shown in Figure 14.11 which is a current-to-voltage converter circuit. This configuration is one of the more popular digital-to-analog circuits used today and is called a Weighted Current Source Digital-to-Analog Converter. It should be noted at this time that the circuits shown here are 4-bit DACs used for simplicity of example. The same circuitry can be expanded for DACs with more than 4 bits. Figure 14.12 shows a schematic

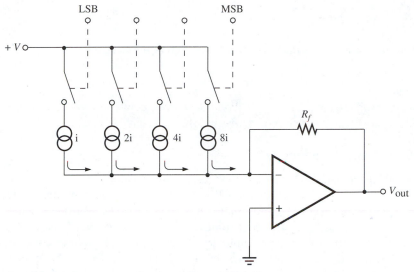

FIGURE 14.11   **DAC with current-to-voltage operational amplifier circuit. (Here we assume that the current sources are off when the corresponding switch is open.)**

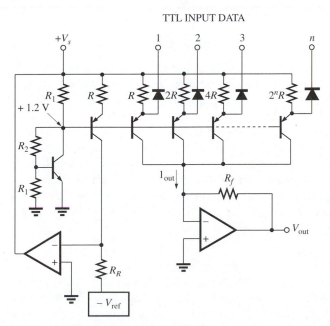

FIGURE 14.12   **Weighted current source DAC.**

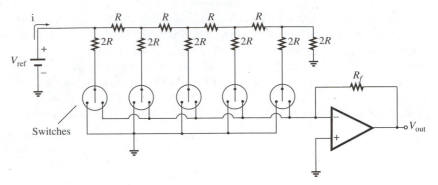

FIGURE 14.13   5-bit R-2R ladder method digital-to-analog converter.

diagram of a general weighted digital-to-analog converter with transistors, diodes and resistors that provide the weighted current sources.

R-2R Ladder Method DAC

Another popular DAC circuit, known as the R-2R ladder method, is shown in Figure 14.13. Here the status of the switches is controlled by digital signals, and the converter features 5-bit conversion, but additional stages can be added to offer more resolution. The gist is the same as the weighted current source DAC; the operational amplifier acts as a current-to-voltage converter and the resistor network provides weighting of current applied to the op-amp. R-2R converters are easier to manufacture since the resistor network requires only two values of resistors regardless of the number of bits of resolution, as opposed to the previous weighted converter which requires many more different values of resistance and other components. The R-2R method allows excellent results for minimal costs.

Multiplying and De-glitched DACs

The R-2R ladder method of digital-to-analog conversion is specifically used in DACs which span both the positive and negative voltage ranges. The term *multiplying DAC* comes from the fact that the output is the product of a reference voltage and the logical input driving the switching circuitry. The input voltage, or reference voltage, can be either positive or negative.

Most DACs suffer from *glitching*. Glitches are dips in the output voltage caused by the real time delay it takes the switches to close and current to flow into the op-amp output. *De-glitched DACs* are available with circuitry that minimizes glitching. Popular applications of these relatively expensive de-glitched converters are video circuits that drive monitors and oscilloscopes.

Summary

In general, commercial digital-to-analog converters consist of the drive circuitry needed for an op-amp current-to-voltage converter. Many also include the feedback resistor, which must be fairly accurate for predictable results.

A voltage reference is sometimes included as well, but more than likely an external accurate reference must be provided by the designer. Voltage reference sources available from most DAC manufacturers have specifications that meet the demanding requirements of DAC circuits. For example, Analog Devices, a leader in instrumentation circuits, offers the AD581 High Precision 10 V IC Reference that provides 10.000 V $\pm$ 5 mV at a low cost.

Commercial digital-to-analog converters are readily available at low cost. The circuits shown above are for example only; we do not imply that you design your own with discrete components. Three important parameters when selecting commercial DACs are resolution, settling time and relative accuracy. Resolution is the number of bits the DAC converts. An 8-bit DAC has a resolution of 8 bits, which implies that each bit, or the smallest quantization of voltage, has the value of $V_{ref}/2^8$ or $V_{ref}/256$. The relative accuracy of the device is at best equal to $\pm 1/2$ the voltage value for the least significant bits because of the nature of quantization. Some DACs, particularly the inexpensive ones, have numerically greater, and thus poorer, relative accuracy. Settling time is the time it takes the converter to produce the output determined by the logical input in the range dictated by the relative accuracy (commonly $\pm$ 1/2 LSB). In general, DACs are fast, in the range of μs or ns. Overall, the speed of operation is usually determined by the op-amp connected to the DAC. Typical circuits are shown in Figures 14.14A and B, respectively, using Analog Devices' 12-bit multiplying DAC, the AD 7545, in unipolar and bipolar operation. A table of common DACs with their specifications is given in Table 14.4.

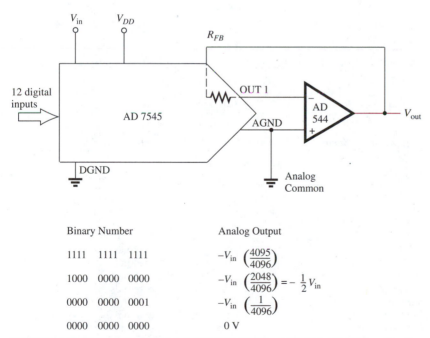

Binary Number	Analog Output
1111 1111 1111	$-V_{in}\left(\dfrac{4095}{4096}\right)$
1000 0000 0000	$-V_{in}\left(\dfrac{2048}{4096}\right)=-\dfrac{1}{2}V_{in}$
0000 0000 0001	$-V_{in}\left(\dfrac{1}{4096}\right)$
0000 0000 0000	0 V

FIGURE 14.14A   **Analog Devices' AD7545 12-bit multiplying DAC in unipolar operation.**

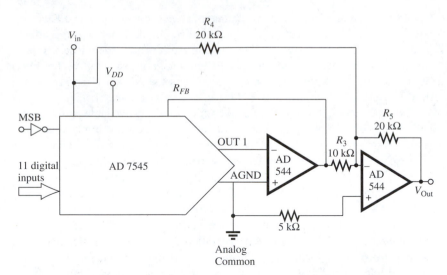

R$_3$, R$_4$ and R$_5$ should be wire-wound or metal foil and match to 0.01%.

Binary Number	Analog Output
0111 1111 1111	$+V_{in}\left(\dfrac{2047}{2048}\right)$
0000 0000 0001	$+V_{in}\left(\dfrac{1}{2048}\right)$
0000 0000 0000	0 V
1111 1111 1111	$-V_{in}\left(\dfrac{1}{2048}\right)$
1000 0000 0000	$-V_{in}\left(\dfrac{2048}{2048}\right)=-V_{in}$

(2's complement)

FIGURE 14.14B   Analog Devices' AD7545 12-bit multiplying DAC in bipolar operation.

TABLE 14.4   Some Common Digital-to-Analog Converters

Name	# Bits	Settling Time (μs)	Voltage Reference	Output
AD DAC80-V	12	3	6.3 V, internal	V
AD1139	18	40	6 V, internal	V
AD7545	12	2	external	I
AD563	12	1.5	2.5 V, internal	I
ICL7134	14	3	external	I
DAC0800	8	0.1	external	I
DAC0830	8	1	external	I
DAC1000	10	0.5	external	I
Multiple DACs				
AD7225	8	5	external	V
AD7547	12	1.5	external	I

AD: Analog Devices; ICL: Intersil; DAC: National Semiconductor

14-7   ANALOG-TO-DIGITAL CONVERTERS: ADCs

Converting analog voltages to digital signals is more complicated than converting digital voltages to analog signals. Integrated circuits that accomplish this are called *analog-to-digital converters* (ADC). Basically, there are three techniques commonly used to perform this function: parallel (or flash), successive approximation and dual-slope integration. All share the three common parameters of conversion time, absolute accuracy and resolution. Resolution is defined just as it was in the case of the DAC: the number of bits to which the analog signal is converted. Absolute accuracy is usually defined similarly to the DAC relative accuracy, usually in bits. The absolute accuracy, in bits, times the voltage value represented by the least significant bit gives the absolute accuracy in volts. Conversion time is the time it takes the converter to take an analog signal and convert it to the binary representation. Overall, ADCs are slower than DACs.

Parallel Conversion

The fastest, and most technically difficult, means of converting an analog signal to a digital one uses a series of comparator circuits as shown in Figure 14.15. It is also known as flash or simultaneous conversion. The circuit shown in the figure is a 3-bit parallel ADC that uses seven comparators. Higher resolution converters mandate the use of $2^n - 1$ comparators where n is the resolution in number of bits. Hence, a 12-bit parallel converter requires 4,095 comparators!

The comparator trip points are set so that each successive device is spaced a voltage equal to that represented by one bit. In our example circuit, if we assume $V_{ref} = 7.0$ V, then the trip point for comparator A is 0.5 V, B is 1.5 V, C is 2.5 V and so forth. For $V_{in} = 4.9$ V, comparators A through E would be tripped, while F and H would remain unchanged (from $V_{in} = 0$ V). The encoding circuitry would thus return a logical 101 representing a voltage in the range of 4.50 to 5.49 V as shown in Table 14.5. Obviously the resolution of a 3-bit ADC leaves much to be desired, but commercial circuits are readily available with much greater resolution. Ten-bit converters are fairly common these days, with conversion frequencies in the 100s of MHz range.

Successive Approximation

One of the most popular types of conversion is known as successive approximation. A successive-approximation ADC uses a DAC to compare the voltage in question to an internally generated voltage, as shown in the block diagram of Figure 14.16. The scheme is fairly simple. First, a voltage of 1/2 the full scale value is generated by the DAC by making the most significant bit of the digital inputs (and ADC digital outputs) a logical 1. If the comparator input voltage is greater than the reference (zero and assuming we are speaking of positive voltages) the output is not tripped and the bit in question is left high. If, on the other hand, the comparator is tripped, the bit is returned to a logical 0. The procedure is repeated each clock period until all the bits have been tested in this manner. Figure 14.17 shows the comparator

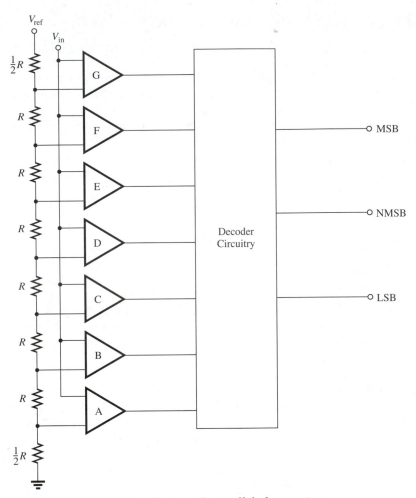

FIGURE 14.15 **Parallel or flash, analog-to-digital converter.**

TABLE 14.5 *Voltage Input, Comparator and Encoded Binary Output of Parallel Analog-to-Digital Converter Shown in Figure 14.15*

V_{in} Range (Volts)	Comparator							Encoded Binary		
	A	B	C	D	E	F	G	MSB		MSB
0–0.49	0	0	0	0	0	0	0	0	0	0
0.5–1.49	1	0	0	0	0	0	0	0	0	1
1.5–2.49	1	1	0	0	0	0	0	0	1	0
2.5–3.49	1	1	1	0	0	0	0	0	1	1
3.5–4.49	1	1	1	1	0	0	0	1	0	0
4.5–5.49	1	1	1	1	1	0	0	1	0	1
5.5–6.49	1	1	1	1	1	1	0	1	1	0
>6.5	1	1	1	1	1	1	1	1	1	1

input voltage as a function of clock periods. Successive-approximation converters are relatively fast, in the range of μseconds. Typical conversion times are $1\mu s$ for a 10-bit ADC and $2\mu s$ for a 12-bit unit.

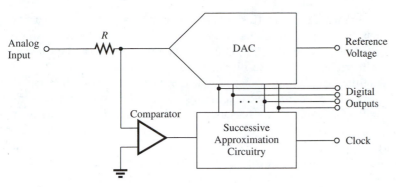

FIGURE 14.16   Block diagram of successive-approximation ADC.

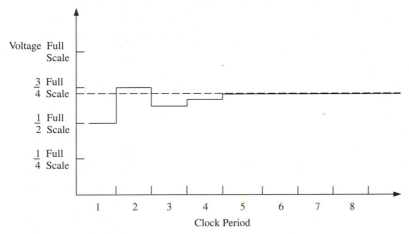

FIGURE 14.17   Typical DAC output of successive-approximation ADC converter as a function of clock cycle.

A cruder method of using the comparator and DAC circuit is realized in the counter, or encoder, ADC. Here, the DAC drive circuitry adds one LSB at a time until the comparator is tripped. When the trip occurs, the conversion ends and the result is available on the digital output lines. Counter ADCs suffer from slow speed and the random nature associated with the conversion time. This technique is not widely used.

Integrating Type: Dual-Slope Conversion

Successive-approximation and flash converters suffer from their inability to deal well with noisy signals. A slight perturbation in the input voltage while the IC is converting will cause an erroneous reading (although the speed with which they convert does help to minimize this problem). The dual-slope ADC minimizes this problem in another manner, but a rapid time of conversion is given up in the process.

The dual-slope ADC works in a relatively simple manner. An external capacitor is charged from zero volts by a current proportional to the input voltage for a given time. Then the input voltage is disconnected and the capacitor is discharged back to zero volts by a constant current source whose value is known. Figure 14.18 shows the integrator capacitor voltage as it goes through the charge and discharge cycle. Since the discharge time is proportional to the value of the constant current and the voltage on the capacitor, a digital representation of the discharge time (in clock cycles) is the digital value of the input voltage. Since accuracy does not depend on the stability of the clock or the integrating capacitor (as long as the value remains constant during the integration), high accuracy can be obtained. Additionally, noise which has a repetitive nature can be rejected if the time for the capacitor to charge is a multiple of the period of the noise. For example, an integration time of 16.667 ms will vastly improve the noise rejection of 60-cycle line voltage noise since it will average to zero over this period.

Typical conversion times are in the range of ms or s. These converters are often used in digital voltmeters where accuracy is important, but speed is generally not an issue. They are usually the least expensive of the converter varieties.

Table 14.6 lists some common ADCs and their specifications.

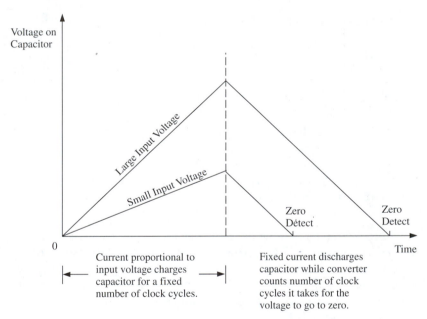

FIGURE 14.18   Charge and discharge cycle for integrating ADC.

TABLE 14.6   Some Common Analog-to-Digital Converters

Name	Bits	Speed	Input Range	Remarks
ADC0802	8	110 μs	5 V	differential input
ADC0809	8	100 μs	5 V	8-input multiplexed
ICL7115	14	50 μs	$\pm$ 5 V	successive approximation
ICL7136	3.5 digits	333 ms	$\pm$ 200 mV	for voltmeter display
AD7576	8	10 μs	2.46 V	low cost
AD ADC80	12	30 μs	$\pm$ 10 V	industry standard

ADC: National Semiconductor; ICL: Intersil; AD: Analog Devices

14-8   SUPPORT CIRCUITRY

Generally, there are two additional circuits found in analog-to-digital conversion circuits: multiplexers and sample-and-holds. Often one will use an amplifier before the ADC, but we have covered their use in an earlier chapter.

The multiplexer allows the user to select an input from a number of possible sources, usually under microprocessor control. Since the ADC is generally the most expensive component in an instrumentation circuit, it makes economical sense to share its use when possible. A typical multiplexer, Analog Devices' AD 7501, is shown in Figure 14.19. Switch selection is accomplished through three TTL inputs.

Caution should be exercised when selecting a multiplexer for instrumentation purposes. Errors can be generated from a number of sources. Ratings which are important in multiplexer selection are transfer error, settling time and crosstalk. Transfer error is the ratio of output voltage to input voltage and is usually expressed in percent of input voltage. One major source of transfer error is the resistance of the switch. Generally, CMOS circuitry is used in multiplexer chips, which yields an ON resistance of 50 Ω to 2 kΩ depending on the IC. To minimize transfer error, high input impedance circuitry, such as an operational amplifier, should follow the multiplexer. Settling time is the time it takes the switch to turn on. Obviously the ADC following the multiplexer should not begin conversion until the switch has settled. Crosstalk is the ratio of output to input voltage with all channels connected and the switches off. Ideally, the output voltage under this condition should not be related to the input voltages, but in the real world such is not the case. Crosstalk is usually expressed in dB as the input-to-output attenuation ratio.

A typical sample-and-hold circuit is shown in Figure 14.20. Amplifier 1 is a high-impedance circuit with a gain of 1, that ensures that the driving circuitry, possibly a multiplexer, is not loaded. When the switch is closed, the amplifier charges the capacitor to a voltage equal to the input voltage. Amplifier 2 must be of high impedance so that when the switch is open, the capacitor is not drained through the amplifier. The output voltage of amplifier 2 thus remains identical to the original input voltage at the instant the switch

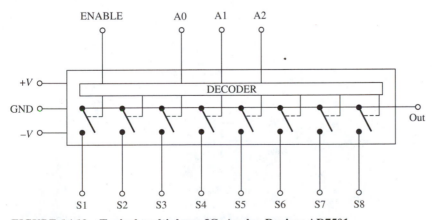

FIGURE 14.19   Typical multiplexer IC, Analog Devices AD7501.

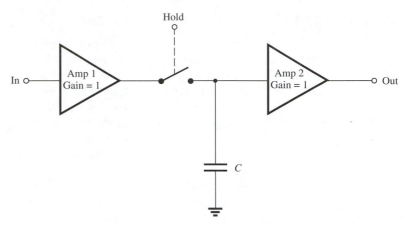

FIGURE 14.20 Typical sample-and-hold circuit using high-impedance amplifiers.

was opened, regardless of any further change the input voltage may undergo. The circuit holds the voltage value at the time of sampling; therefore the name sample-and-hold. Often this circuit is referred to as a sample-and-hold or track/hold circuit.

Design of sample-and-hold circuits appears to be relatively easy. One merely needs a logic-controlled switch such as the CMOS 4066, a capacitor and two operational amplifiers. But, switches and capacitors have leakage currents and operational amplifiers do not always have ideal characteristics; recall offset voltages and the like. Inexpensive, easy-to-use, commercial sample-and-hold ICs are available at minimal cost from virtually every ADC manufacturer. Two such circuits using National Semiconductor's LF198 and Analog Devices' AD1154 are shown in Figure 14.21. Note that the LF198 requires an external capacitor whereas the AD1154 has an internal unit. Accuracy of the two units is quite different—0.02% for the LF198 and 0.00076% (16-bit accurate) for the AD1154, in spite of their apparent similarities. The AD1154 features an offset adjustment to account for total system offsets.

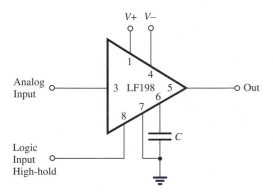

FIGURE 14.21A National Semiconductor's LF198 sample-and-hold IC.

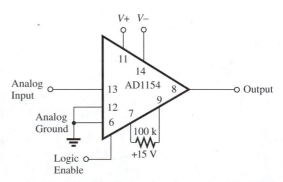

FIGURE 14.21B Analog Devices' AD1154 16-bit accurate sample-and-hold IC.

Important parameters associated with sample-and-hold circuits are the same as those of amplifier circuits previously discussed (input offset voltage, input and output impedance, etc.) with the addition of acquisition time, the time it takes to charge the capacitor, and droop rate, the discharge voltage as a function of time when in hold mode. Superior ratings of commercial sample-and-hold ICs dictate the rare use of discrete component design for these circuits.

Some circuit manufacturers offer integrated circuits that incorporate multiplexers and an ADC on one chip. An example of this is the popular ADC0808/9 offered by National Semiconductor. This IC features an 8-bit successive-approximation ADC with 8 multiplexed analog inputs. Figure 14.22 shows the pin-out connections of the IC. The ADC0816/17 offered by National is a similar chip with 16 multiplexed inputs. Conversion time is relatively slow at 100 μs and the analog voltage range is 0 to 5 V. Normal operation calls for the address latch bits to be set first, then a positive TTL

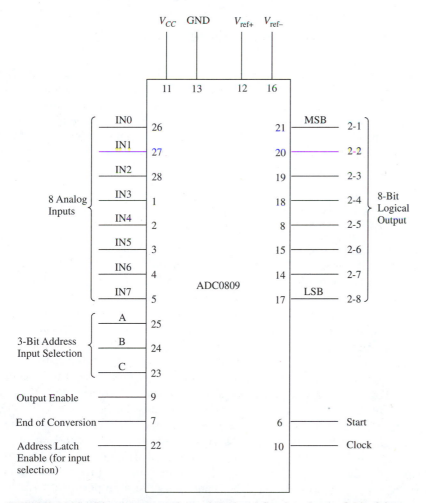

FIGURE 14.22 **Pin-out connections of National Semiconductor 8-bit, 8-input multiplexed successive-approximation ADC.**

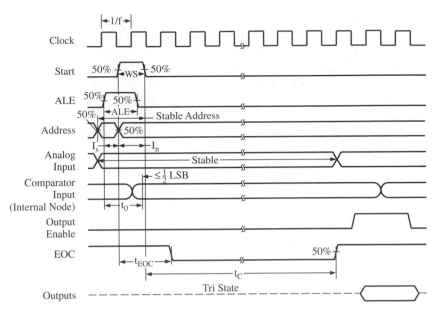

FIGURE 14.23 **Timing diagram of ADC0809 8-bit, 8-input ADC.**

signal sent to the address latch enable input locking in the analog selection, followed by a positive TTL start pulse to the start input. When the conversion is complete, the end of conversion output will go high (it is low while converting). This is followed by a positive TTL signal to the output enable, which allows the reading of the 8-bit logical outputs representing the analog input. Clock frequency may range from 10 kHz to 1280 kHz. This timing scheme is shown in Figure 14.23.

14-9   VOLTAGE-TO-FREQUENCY AND FREQUENCY-TO-VOLTAGE CONVERTERS: V/F AND F/V

The transmission of analog voltages over long distances is wrought with many pitfalls. Noise and line resistance are among the major contributors to the degradation of the analog signal. Recent development of *voltage-to-frequency converter* ICs, V/Fs, that take an analog voltage in and put a stream of pulses out, with frequency linearly proportional to the input voltage, have minimized the problems associated with long-distance transmission of data. Additionally, circuits that require a frequency output, possibly in the audio range, as a function of the input voltage, from say a transducer, are sometimes desired. And vice-versa, a frequency pulse may need to be converted to an analog voltage via a *frequency-to-voltage converter,* or F/V.

The standard IC used in such an application is the LM331, a V/F. The circuit requires a number of discrete resistors and capacitors and is not the most simple to use. Analog Devices' AD654 is another voltage-to-frequency converter that is easier to use. It is an 8-pin IC that can output frequencies up to 500 kHz. Standard connections for use are shown in Figure 14.24. The AD654 V/F converter converts the input voltage to a proportional current. This current drives an oscillator which charges and discharges a timing capacitor at a rate dependent upon the magnitude of the input current, or input

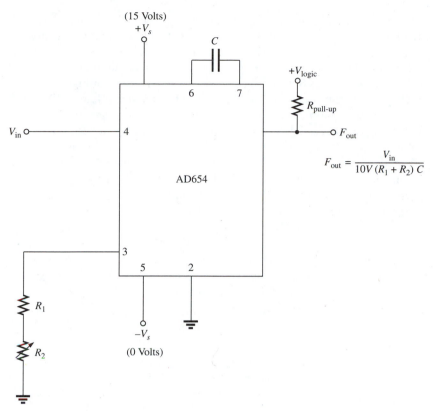

$$F_{out} = \frac{V_{in}}{10V\,(R_1 + R_2)\,C}$$

FIGURE 14.24   Standard connections for Analog Devices' AD654 V/F converter.

voltage. The output frequency is governed by the rate at which the capacitor charges and discharges. Virtually all V/F converters are, in this respect, current-to-frequency converters. Since the V/F and F/V converters are fairly new, we can expect the rapid introduction of new ICs to accomplish these functions.

14-10   COMMERCIAL DATA ACQUISITION SYSTEMS/ CONCLUSION

Chances are that most of us will not design and construct data-acquisition systems consisting of PIAs, ADCs, DACs, multiplexers, instrumentation amplifiers and sample-and-hold circuits for critical applications. Software design, from ground zero, to accompany such computer peripherals is also probably not high on our list either. But the understanding of the operation of such circuits and software is vitally important to anyone in the laboratory. Only the feeblest experimentalist does not understand the instrument he or she is using, particularly its limitations.

Numerous commercial data-acquisition systems that simply plug into a personal computer are available from manufacturers along with easy-to-use driving software. A leader in this field is National Instruments. An example of one of its many data acquisition boards is the NB-MIO-16 for the Apple

Macintosh II computer that features a 12-bit ADC, 16 analog ground-referenced inputs (or 8 analog differential inputs), programmable gain of 1, 10, 100, 500 (or 1, 2, 4, 8) two multiplying 12-bit DACs, 8 programmable digital I/O lines, and more. All this is included on one plug-in board!

Control of the NB-MIO-16 can come from a number of high-level software packages including C, Pascal, BASIC, Hypercard and LabVIEW, which is a National Instruments program that allows easy use and control of the board and other IEEE-488-, RS232- and RS422-attached devices. This program is icon driven, which means the program is developed through the manipulation of pictures representing the various components of the experiment. Similar packages are available for the IBM line as well. The National Instruments catalog should be a standard reference of all experimentalists. The company's address is listed in Appendix C.

14-11 QUESTIONS AND PROBLEMS

1. Explain the use and need for parity in asynchronous serial data transmission.

2. What is meant by even or odd parity? Give examples.

3. Describe the programming steps needed to program the peripheral data lines (ports A and B) of a 6821 PIA to be alternatively inputs and outputs; that is, PA0 an input, PA1 an output, etc.

4. Write a BASIC program which accomplishes the task set in problem 3.

5. Write a BASIC program for the 6821 described in problems 3 and 4 which makes all outputs low when the signal from any of the inputs is high.

6. Design a complete data logger system using the 6821 along with a microcomputer, that records light and temperature readings over a 24-hour period. Use a block diagram.

7. Write a BASIC computer program that would accompany the data logger system of problem 6.

8. Describe the function of the IRQA and IRQB lines of the 6821.

9. Describe why RS232C uses the logic ranges it does, instead of the common scheme 0 V/5 V for low/high.

10. Prepare a comparison table of the different types of DACs presented in this chapter which takes into account complexity, resolution, accuracy and speed.

11. Using logic gates, design a 4-bit digital-to-analog converter that can produce voltages in the range of 0 to 5 V.

12. Using logic gates, design a 4-bit analog-to-digital converter that can input voltages in the range of 0 to 5 V. Use LEDs to indicate output status.

13. How fast does a DAC have to be (what is the maximum conversion time) to convert 16-bit signals from a Compact Disc to audible sound?

14. Prepare a comparison table of the different types of ADCs presented in this chapter which takes into account complexity, resolution, accuracy and speed.

15. Using circuits given in this chapter and the previous chapters, design a system that measures temperature from 0 to 128° K and displays the temperature value in binary via 7 LEDs. Each binary digit should therefore correspond to an interval of 1° K.

16. Design a ramp signal generator circuit using an 8-bit DAC. Choose a fixed frequency of 1 kHz.

17. A 1.0000 V signal is converted to an 8-bit binary signal through an ADC. The signal is then fed back into an 8-bit DAC. Assuming that both the DAC and ADC range from 0 to 5.0 V, what would the output voltage of the DAC be?

Noise

15-1 INTRODUCTION

Our increased sophistication and understanding of the physical world owes much to the advances in measurement science made possible by electronics. The phenomena being pursued are increasingly obscure. Quantification of these relatively small-effect phenomena depends upon sensitive energy-matter interactions. The "better mousetrap" in measurement science may cease to exist as we approach the fundamental limitation of measurement which nature has placed upon the system. It is important to understand these natural limitations: one is called the *uncertainty principle,* and the other, *noise.* In addition to the fundamental sources of noise, man contributes to the electronic noise pollution. Anyone who has tuned a radio receiver and heard static or turned on a television set and seen "snow" has encountered noise. By turning on fluorescent lights or by starting a car we add to electrical noise. In other words, noise is here to stay. The designer and the user of electronic equipment should be cognizant of both kinds of noise, fundamental and environmental. Moreover, he or she should take steps to avoid the latter while not attempting to better the former.

Numerous occasions are encountered when a perfectly good piece of equipment does not operate at its full capacity because noise is forcibly introduced into it—often out of ignorance. Sound wiring practices can significantly reduce the effect of environmental noise.

This chapter examines the various sources of noise and how they can be minimized. Further, several techniques currently available for noise reduction are presented. This is called signal-to-noise (*S/N*) enhancement. The signal-to-noise ratio is defined as

$$\text{SNR} = 10 \log_{10}(v_{\text{signal}}^2/v_{\text{noise}}^2)\text{dB} \qquad \textbf{(15-1)}$$

where v_{signal} is the rms voltage of the signal and v_{noise} is the rms voltage of the noise. For example if a 1.0 V (rms) signal has 0.1 V (rms) of noise, then the signal-to-noise ratio is computed as

$$\text{SNR} = 10 \log_{10}(1^2/0.1^2)\text{dB}$$

$$= 20 \text{ dB}.$$

If the same 1.0 V signal has a smaller 0.01 V of noise then the signal-to-noise ratio is given by

$$SNR = 10 \log_{10}(1^2/0.01^2)dB$$

$$= 40 \text{ dB}.$$

Thus a greater numerical value for the SNR implies less noise in the signal.

15-2   ORIGINS OF NOISE

Noise arises both from within the system (endogenous) and from sources outside of the system (exogenous). For reasons of clarity, the word *interference* is preferred to environmental noise. Interference properly denotes that something is hindering a measurement. Furthermore, the source of interference can often be identified and eliminated. This is not true of fundamental (or endogenous) noise, which is random and unavoidable in nature.

Two factors of nature force a lower limit on us regarding noise. Often we disregard them until they are forced on us in a dramatic manner. All matter that is not at absolute zero temperature has thermal fluctuations associated with it. Second, a number of phenomena with which we work are quantized. That is, such phenomena as charge, energy and light do not change smoothly but change in steps. The net result is that there is always (unless at absolute zero temperature) agitation associated with these processes which cannot be eliminated. This agitation is the fundamental source of noise.

Three types of noise are generally associated with transistors, other solid state devices and the components involved in circuitry: thermal noise, shot noise and flicker noise.

Thermal Noise

Thermal noise arises from the random movement of electrons and other free carriers. Other names have been given to this source of noise: Johnson noise, after the discoverer of the phenomenon, and Nyquist noise, after the man who showed that this noise was a consequence of the Second Law of Thermodynamics. If the measurement of thermal noise is made over a sufficiently long time to virtually eliminate random fluctuations, the noise will, of course, be zero. At any instant of time, however, a net movement exists in a given direction. This net current produces a voltage which is observed and designated thermal noise, v_N. The thermal noise voltage (*rms*) is related to the resistance of the material, the temperature and the bandwidth:

$$v_N^2 = 4kTRB \tag{15-2}$$

where k is Boltzmann's constant (1.38×10^{-23} joule/°K). The temperature, T, is in °K, and R is in ohms. The bandwidth, B (in Hertz), is the frequency band of the measuring instrument ($f_{max} - f_{min}$). The meaning of the above equation is clear. Thermal noise increases with the temperature, the bandwidth and the resistance. Since there is no frequency term in the expression, thermal noise is frequency independent; i.e., it appears at *all* frequencies of electronic interest. As a result, it is often referred to as white noise. The magnitude of thermal noise is not insignificant. For example, a 100 kΩ resistor at room temperature has an *rms* noise value of 4 μV when measured by an instrument with a 10 kHz bandwidth. This magnitude in itself is hardly large. However, when that resistor forms the input to an amplifier whose gain is 10^5, the noise at the output of the amplifier is nearly half a volt!

The available noise power, P_N, can be obtained from the above equation by dividing both sides by R:

$$P_N = kTB \qquad (15\text{-}3)$$

Note that the 4 of equation 15-2 is lost through the application of the maximum power transfer theorem. The noise power depends only upon the temperature and the bandwidth of the measuring instrument. This quantity is useful in defining a commonly used figure of merit for electronic devices, the noise figure, F; the definition is

$$F = \frac{\left(\dfrac{\text{Signal Power}}{\text{Noise Power}}\right)_{\text{input}}}{\left(\dfrac{\text{Signal Power}}{\text{Noise Power}}\right)_{\text{output}}} \qquad (15\text{-}4)$$

This expression can be rewritten, making use of the definition of power gain, as

$$F = \frac{P_{N\text{ output}}}{P_{N\text{ input}}} \cdot \frac{1}{A_p} \qquad (15\text{-}5)$$

It is clear that $F = 1$ for an ideal noise-free amplifier, whereas $F > 1$ for a real amplifier.

The total noise power of an amplifier, regardless of the sources of noise, can be expressed in terms of the contributions from each of the n stages of the amplifier.

$$P_{N(\text{total})} = P_{N_1}(A_{P_1} \ldots A_{P_n}) + P_{N_2}(A_{P_2} \ldots A_{P_n}) + \ldots + P_{N_n} A_{P_n} \quad (15\text{-}6)$$

This expression can be simplified by using the definition of F. This yields

$$F_{\text{total}} = F_1 + \frac{F_2 - 1}{A_{P_1}} + \frac{F_3 - 1}{A_{P_1} A_{P_2}} + \cdots \qquad (15\text{-}7)$$

This result has immediate significance for any designer or user of electronic equipment. Recall that the power gain is usually quite large. This means that equation 15-6 may be approximated as

$$F_{\text{total}} \approx F_1 \qquad (15\text{-}8)$$

Thus, in most practical circuits, the major contribution to the system noise figure comes from that of the first stage. This result has a clear warning for the user when considering a signal pickup and subsequent amplification. Since the total noise figure is essentially that of the first stage, the first stage of amplification (often the pre-amplifier) should be as free from noise as possible. Since connections from the signal source to the amplifier assembly often introduce a significant amount of noise, the pre-amplifier should be placed as close to the signal source as possible. The output of the pre-amplifier can then be carried to the main amplifier.

Thermal noise, in transistors, comes from the resistance of the device, principally from the base area as the charge carrier is depleted.

Shot Noise

Shot noise results from the statistical fluctuations of charge carriers across a junction. The expression, derived by Schottky for the vacuum diode, is applicable to all junctions or interfaces:

$$i_N^2 = 2qIB \qquad \qquad \textbf{(15-9)}$$

where i_N is the *rms* noise current in amperes, q is the charge of an electron (1.59×10^{-19} coulombs), I is the DC current in amperes, and B has its earlier definition. Since shot noise is associated with random movement of charge carriers, this means that vacuum tubes (including phototubes and photomultipliers), diodes, transistors and the like *all* evidence the effect. In transistors, the currents through the emitter-base and collector-base diodes are the sources of shot noise. If necessary, the noise power for shot noise can be expressed in terms of the square of the shot current times the resistance:

$$P_N = i_N^2 R = 2qIRB \qquad \qquad \textbf{(15-10)}$$

where R is the equivalent resistance of the junction.

As with thermal noise, shot noise is independent of frequency. This means that it appears at all frequencies. The difference between thermal noise and shot noise is that the latter is related to the DC current through the junction. Immediately, this suggests that reduction in shot noise is accomplished by limiting the current through the device.

Flicker Noise

The third source of fundamental noise, flicker noise, increases with decreasing frequency. Because of this, it is often called l/f noise. The phenomenon giving rise to flicker noise is not well understood. Partly because of this, the third name euphemistically given this noise is *excess noise*. Several observations characterize the noise, including the fact that l/f noise is significantly greater in solid-state devices than in vacuum devices, and its effect is seldom important above 1 kHz. One important conclusion may be drawn from l/f noise behavior: when sensitive measurements are to be made, avoid DC.

These three fundamental sources of noise are seen to be operative in a graph of F versus frequency for a transistor, as in Figure 15.1. The frequency region below 1 kHz is due, almost completely, to flicker noise. The magnitude of F in the intermediate region is clearly affected by thermal noise, while the steep increase in F at high frequencies is a combination of both shot and thermal noise effects.

As often as not, fundamental noise sources are not at the root of measurement problems. Rather, the most important noise is of environmental origin. The most common source of interference in the United States is 60 Hz power distribution. Most people are cognizant of this, but ignore the higher harmonics at 120 Hz, 180 Hz, and even 240 Hz. Another troublesome source of interference is the AM radio frequency band. Because a conductor acts as an antenna for this band, this is often a problem. Also, noise from the brushes in nearby electric motors will invariably be noticed on sensitive

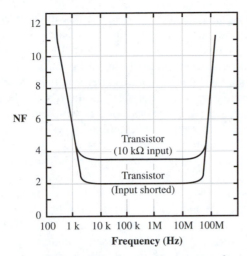

FIGURE 15.1   The noise figure (*F*) of a transistor *versus* frequency.

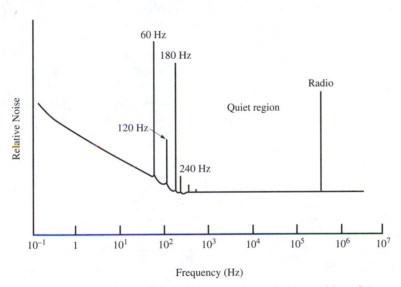

FIGURE 15.2   The noise spectrum, showing regions of quiet and interface.

equipment. Even X-ray equipment can contribute greatly to environmental noise. Figure 15.2 summarizes these various sources of noise and, more importantly, shows the frequency regions that are relatively free of noise.

15-3 REDUCTION OF EXOGENOUS NOISE

From the preceding discussion, it is evident that care should be taken to keep unwanted signals from entering the measurement system. It is just as important to avoid inadvertent addition of a stray noise. Consider the circuit in Figure 15.3, containing a voltage source and a load, *Z*, which is coupled to the source by a stray capacitance, *C,* an unwanted component. This is, of course, undesirable. The magnitude of the stray voltage developed across *Z* depends upon the relative values of *C* and *Z* and on the frequency of the

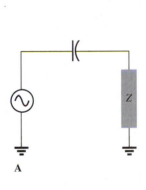

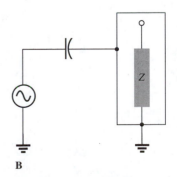

FIGURE 15.3   Capacitive coupling of a source to a load. **(A)** Unshielded; **(B)** shielding to eliminate the coupling.

source. As the magnitude of Z increases, the possibility of a stray noise affecting further stages in the circuitry also increases. If the load is placed inside a conducting box, as shown in Figure 15.3B, the problem can be reduced; the shield is grounded, and the stray capacitance is coupled to ground. This model for shielding, often called a Faraday shield, is not unreasonable. Radiative energy is all around us in the form of 60 Hz power. The capacitively coupled container exists, too; as you read this, you are in the middle of a capacitor. One plate is the ceiling (or walls) containing the power lines, and the other plate is the floor. A typical value of capacitance (to ground) for a person standing on an insulator is over 500 pF. To witness pickup of 60 Hz noise, often referred to as "60-cycle noise," a simple experiment is composed of holding the input lead of an oscilloscope with your hand. The oscilloscope will display significant 60-cycle noise picked up by your body, a very good antenna, along with other stray noise.

Shielding a sensitive piece of equipment, such as an amplifier, "remedies" the problem if, and only if, the shielding is carried out sensibly. Consider the amplifier in a container illustrated in Figure 15.4. Stray capacitances exist between the input and the shield, between the output and the shield, and between the common and the shield. The latter is the most serious, since the common to shield capacitance forms the critical arm of a capacitive feedback system. The capacitive feedback is harmful, but it can be eliminated by connecting the common to the shield (which eliminates the stray capacitance completely).

Several pieces of equipment are often interconnected. The problem then becomes how to ground the system. One method is shown in Figure 15.5, in

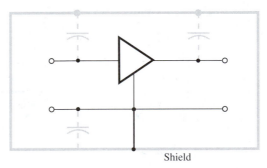

Shield

FIGURE 15.4 **Stray capacitances between shield and amplifier. The solid lead indicates a connection between the shield and the common connection of the amplifier. It is used to eliminate capacitive feedback.**

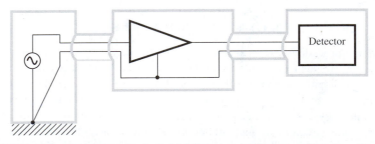

Detector

FIGURE 15.5 **Correct method of shielding several items together to a single ground point.**

which a signal source, an amplifier and a detector are connected together. Notice that each is shielded separately but that the shields are commonly connected. This is the correct approach. Inferior approaches under the same conditions are shown in Figures 15.6A and B. In Figure 15.6A, the common lead of the amplifier is connected to the shield. Since the shield is not grounded, an electrostatic coupling exists between the shield and ground. This can be a source by which spurious signals are injected. A more complex problem is shown in Figure 15.6B. In this situation, the several shields are not connected together. Rather, each is taken to a different grounding point. Any differences (there surely will be differences) are injected into the system, electrostatically if not directly. These are called ground loops. They result wherever there is more than one common point in the circuit. Often the user of the equipment does not recognize that the chassis connection forms a ground through the power line. In order to avoid this possibility, it is expedient to establish a common ground bus to all chassis and to make a single connection from this bus to the power line ground. If at all possible, the signal source should be fed into a first stage differential amplifier.

The block diagram of a differential amplifier is given in Figure 15.7. The gain of amplifier 1 is A while that of amplifier 2 is $-A$. Generally, the gain of the third amplifier is $A/2$. This means that any noise signal common to the inputs of amplifiers 1 and 2 will be cancelled out by the time the signal reaches amplifier 3. This situation is shown in Figure 15.8, where the noise is larger in amplitude and lower in frequency than the signal. The ability of a differential amplifier to reject a common signal is designated as the com-

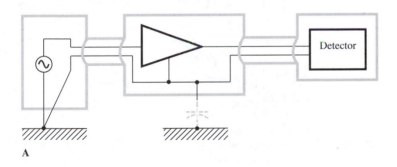

A

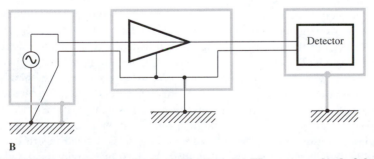

B

FIGURE 15.6   Improper methods of shielding. (A) The common lead of the amplifier is connected to the shield, producing capacitive coupling to ground at that point. (B) Each item is grounded individually, producing ground loops.

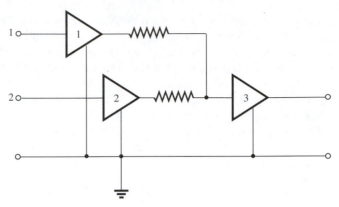

FIGURE 15.7 Differential amplifier configuration.

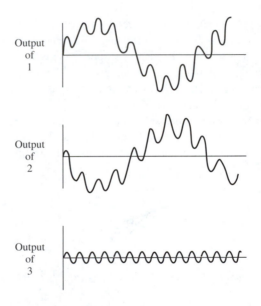

FIGURE 15.8 Waveforms of outputs of first stages of differential amplifier (1 and 2) and resulting output waveform (3).

mon mode rejection ratio and was defined in Chapter 9. As a figure of merit, a differential amplifier should have a $CMRR \geq 100$. Differential amplifiers are commonly encountered in oscilloscopes. The quoted $CMRR$ values may be used as one guide to the quality of a particular model.

Coaxial cable, as shown in Figure 15.9, is frequently used to shield connection cable. The concentric braided-wire shielding completely surrounds the central wire. Often overlooked is the need to ground all the shields to a common point. In critical situations, if this is not done, the electrostatic coupling is only increased. Alternatively, the various shields may be grounded to different reference points, creating the same situation described earlier in relation to Figure 15.6B. Coaxial cable is sometimes needed in critical locations within a device to shield a given lead from certain local radiation sources, like transformers. The magnetic fields produced by a transformer

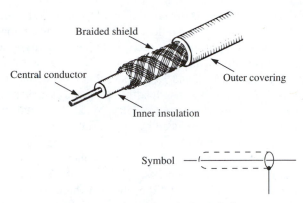

Braided shield

Central conductor

Outer covering

Inner insulation

Symbol

FIGURE 15.9   Coaxial cable. Lower: electrical symbol.

generally have a short range. Any conductor, unless it is shielded, is susceptible to induction of an alternating signal by the alternating magnetic field.
One way to avoid this is to use transformers shielded with a material of high
magnetic permeability. Any wires carrying more than approximately 10 mA
of alternating current should be braided together. This tends to cancel out the
magnetic fields. Better still, critical wires and components should be placed
well away from the wires carrying the alternating current. A partial solution
is to place the wires carrying alternating current perpendicular to the sensitive or critical wires.

Mechanical vibrations may be an overlooked source of environmental
noise or interference. This is more a problem when vacuum tubes are used.
The mechanical vibrations are picked up by the electrodes within the enve-
lope. The electrodes vibrate, and this vibration modulates the signal being
processed. More rigid electrode structures within the tube and shock mounting of the tubes have done much to reduce this problem, called *microphonics.*
Any signal source should be examined for effects of vibration upon it. Transistors, because of their construction, generally are not susceptible to
microphonics.

A number of suggestions have been made in this section as to how the
user of electronic equipment might minimize the effects of noise. For a person building his own equipment, certain practices have been mentioned
which, if followed, will produce a low-noise system. In the latter instance,
ignorance of these practices will often produce a noise generator and little
else. The situation where a source of noise is inescapable is yet to be faced.
The remainder of the chapter will deal with those problems.

15-4 PULLING A SIGNAL FROM NOISE

Assuming that the equipment on hand is well built and consistent with relatively noise-free operation, are there ways in which a noisy signal can be
treated? The answer is a qualified yes.

Filtering

The obvious way around the noise problem at this stage is filtering. The
chapter on operational amplifiers presented the low-pass, high-pass and notch

filters. These are good filters if the noise is at a frequency different from that of the signal. This requirement is inflexible. The effects of thermal noise and shot noise can only be reduced with proper filtering. The basic idea behind the use of a filter is to drastically reduce the bandwidth, *B*, of the system. Re-examination of equations 15-2 and 15-9 shows that the amount of noise from those sources is reduced as the bandwidth is reduced. Furthermore, the effect of particular sources of environmental noise, if their frequencies differ from the signal frequency, is reduced by appropriate filtering.

For example, if the signal you are looking at has a bandwidth of 100 Hz, a low-pass filter could be used to eliminate noise above this frequency. Likewise, if you are interested in a signal of 100 Hz, but have a large noise component below that frequency and little above, a high-pass filter could be employed. If noise is present both above and below the frequency of interest, a notch filter could be used. Keep in mind, however, that the filters we have previously discussed are not perfect in the sense that they pass a certain frequency without attenuation and perfectly filter other frequencies. If the signal you are looking at has a frequency of 59 Hz, but there is considerable 60 Hz noise (even after proper shielding), then a 60 Hz notch filter would probably be of little value since it would significantly attenuate the signal of interest as well.

Signal Averaging

A technique which is particularly useful when signals are literally swamped with noise is called *signal averaging.* This approach is valuable when the bandwidths of the signal and the noise are the same, a feature which precludes the approaches mentioned above. The only requirement for signal averagers is that the signal be repetitive. Signal averagers work both with analog and digital circuits. The latter is preferred when extremely long times (low frequencies) are needed or where the signal is to be processed further after recovery from the noise. It is somewhat easier to understand the operation on an analog basis, so that mode will be developed here. Consider the circuit of Figure 15.10. At the beginning of the cycle, the first switch is closed and the remaining ones are open. If the duration of the cycle is *T,* the

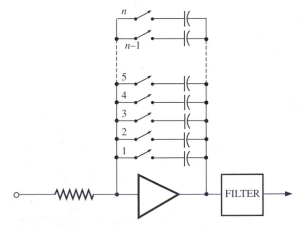

FIGURE 15.10   **Schematic diagram of a signal averager.**

first switch remains closed for T/n seconds; then the second switch is closed as the first is opened. The second switch remains closed for the same amount of time, after which it opens and the third closes, and so on through the complete cycle. The signal (plus the noise) has been sliced into n segments, and the segments are stored individually on the capacitors. This process is repeated as many times as deemed necessary. The signal adds arithmetically each time, while the noise adds as its *rms* value. For example, after 100 repetitions, the signal is 100 times stronger, while the noise is $\sqrt{100} = 10$ times stronger. This is a 10-fold increase in *S/N*. The number of slices into which the signal is divided can be quite large; 1024 is typical. Greater improvement in *S/N* is obtained by increasing the number of repetitions. A 10 kHz signal repeats 10^4 times in a second. This represents an increase in *S/N* of 100 for one second of signal averaging. The problem associated with long times (either a large number of repetitions or a very low signal frequency) is that the time-delay may shift somewhat and delete part or all of the previously stored signal. Signal averagers are marketed as accessories for a number of sophisticated instruments under a variety of names, such as the computer of average transients (CAT). The striking results obtained with this approach are illustrated in Figure 15.11. On the whole, as increased storage time is added, the cost increases for the reason mentioned above.

Ingenious ways have been used to avoid the requirement for a repetitive signal. Many non-repetitive signals can be made repetitive by an external stimulus. An interesting adaptation of signal averaging involves the counting of micrometeorites striking a satellite. Transmission of the data was nearly impossible because of noise. It was suggested that the information be recorded on tape in the satellite and repetitively broadcast. Thus, a single series

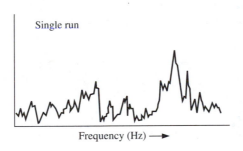

Single run

Frequency (Hz) ⟶

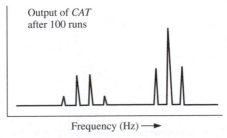

Output of *CAT*
after 100 runs

Frequency (Hz) ⟶

FIGURE 15.11   **Illustration of signal improvement using averaging (CAT).**

of recorded events was made repetitive and sent bathed in noise. A signal averager connected to the receiver on earth plucked the information out.

Lock-In Amplifiers

One of the most useful techniques for extracting a signal from noise, even of the same bandwidth, is the use of a *lock-in amplifier* (LIA). This technique is also known as *phase sensitive detection* or *phase discrimination*. It carries the requirement that the signal be repetitive, as in signal averaging. Most DC measurements can be modified by chopping to meet this stipulation, so it does not impose a severe limitation on the applications of the LIA. The heart of the technique is the phase-sensitive detector, as shown in Figure 15.12. The output, V_0, is proportional to the amplitude of V_S, the input signal, if V_S is the same frequency as V_R, the reference signal. In addition, the DC output is proportional to the cosine of the phase angle between the input and reference signals. This is illustrated in Figure 15.13 for phase angles of 0°, 45°, 90° and 180°. The output is actually composed of the sum of the signal and reference frequencies, as well as the difference between those frequencies. The term "beat" is applied to the sum frequency, while the term "hererodyne" is used to describe the frequency difference. In this application the difference frequency is zero because V_S and V_R are the same frequency. The sum frequency ($2f_S = 2f_R$) is removed by the low-pass RC filter.

In addition to the phase-sensitive detector, most LIAs provide for internal amplification of both the input and output signals. Quite often, an internal reference oscillator is provided. A block diagram of one such LIA is given in Figure 15.14. The signal path through the LIA is also given. The clearest idea of the use of the LIA is gained by examining the signal at various stages within the system. When the reference signal is internally provided, it may be used to drive a power amplifier which, in turn, drives the signal modulator. In cases where a light signal is being detected with an optical transducer,

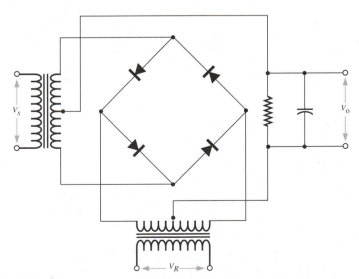

FIGURE 15.12   Phase-sensitive detector using a diode detection system with a filtered output.

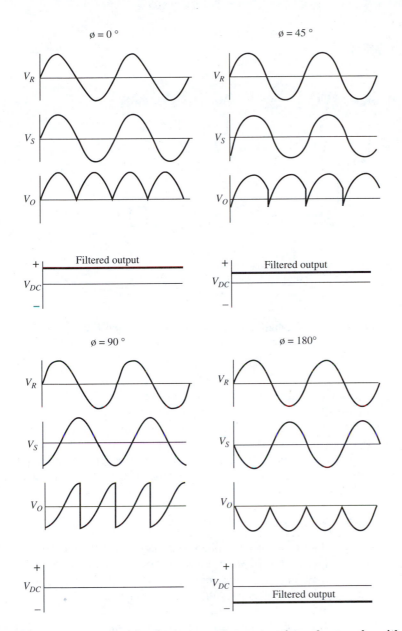

FIGURE 15.13 **Signal and reference waveforms at various phase angles with resulting DC output.**

the modulation may take the form of optical chopping. In this approach, the light beam is interrupted periodically to convert a slowly varying light signal into a square wave signal of known frequency. Slight drift in the reference oscillator is automatically compensated, since it is both the internal reference and the source of the modulation frequency.

Use of the LIA embodies two of the principles developed in this chapter. First, the measurement frequency is moved, via modulation, from DC (where l/f noise is extreme) to a frequency region which is relatively free of

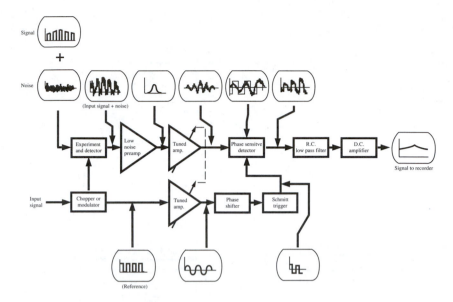

FIGURE 15.14   Block diagram of a lock-in amplifier and the signals associated with the several portions. (Courtesy of Princeton Applied Research, Inc.)

noise. The upper frequency limit is generally set by the transducer response and not by the LIA. Second, the principle of narrow bandwidths for reduced noise is used, resulting in the narrow bandwidth tuned amplifiers noted in Figure 15.14. The high gain characteristic (as much as 10^{10}) of the LIA is also useful. With reasonable care, the LIA is capable of giving a clear output, essentially noise free, from an original signal that was as much as 50 to 60 dB below the noise level!

Correlation Techniques

Correlation techniques can be applied to situations where the *S/N* ratio is such that the signal is lost in noise. These techniques rest on a mathematical foundation. Essentially, when two signals, $V_1(t)$ and $V_2(t)$, are "mixed" or correlated, the cross-correlated result, R_{12}, is obtained according to

$$R_{12} = \lim_{T \to \infty} \frac{1}{2T} \int_{-T}^{+T} V_1(t)V_2(t - \tau)dt \qquad \textbf{(15-11)}$$

where τ is a delay intentionally introduced into the system. Generally $V_2(t)$ is taken as the reference signal. The auto-correlation, R_{11}, approach makes use of the same mathematical foundation but is applied to the case in which $V_1(t) = V_2(t)$ (in other words, the signal becomes its own reference) or

$$R_{11} = \lim_{T \to \infty} \frac{1}{2T} \int_{-T}^{+T} V_1(t)v_1(t - \tau)dt \qquad \textbf{(15-12)}$$

A block diagram of an autocorrelator is given in Figure 15.15. The result, R_{11}, as a function of τ is displayed. The number of correlations performed depends on how deeply the signal is buried in the noise. The output identifies

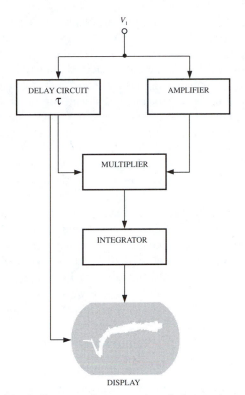

FIGURE 15.15   Block diagram of an auto-correlation system.

the period of the signal but not necessarily the waveform. In essence, pure noise produced no correlation with itself except when $\tau = 0$; therefore there is no output.

The cross-correlation technique requires the use of a second or modulating signal, for much the same reason that was evident in an LIA. Again a delay is introduced and the display is the result, R_{12}, as a function of the delay. This latter technique imposes no restriction upon the wave shape of the reference. The cross-correlated result is capable of ascertaining the *waveform* of the signal. Again, the number of repetitions can be controlled to extract the signal from the noise.

Computer-Assisted Noise Reduction

Averaging With the onslaught of computer-based instrumentation comes the power of manipulating data from transducers numerically. The simplest form of noise reduction using a computer-based instrument and repetitious signal is to take many readings and have the computer average their value. Here noise, which is assumed to be random, will be reduced proportionately to the number of readings taken for the average. In Chapter 13 we described a simple signal averager of this type. Commercial software for data acquisition nearly always contains subroutines to accomplish this calculation.

Sampling Frequency When using an analog-to-digital converter and computer to record data from a transducer, the question always arises as to how many samplings of the signal must be made in order to reconstruct the signal.

The answer to this question is given by the Nyquist frequency, which is equal to twice the value of the highest frequency component in the signal of interest. If the sampling is at this rate, or higher, the signal of interest can be faithfully reconstructed from the data collected. This does not imply that a square wave with a frequency of 100 Hz must be sampled at a frequency of 200 Hz in order to be reconstructed. You must keep in mind that the square wave is composed of frequency components of higher order given by Fourier analysis described in Chapter Four. In fact, for a square wave it would be wise to have the sampling frequency be at least 11 times the fundamental frequency of 100 Hz in order to reconstruct the square wave.

If a sampling rate is chosen to be lower than the Nyquist frequency, an error called aliasing may occur. An example of aliasing is shown in Figure 15.16. Here the time between data readings is 75% of the original signal period. In order to eliminate aliasing, a sampling time of 50% of the signal period or less must be made. Generally speaking, the more data taken, the better the reconstruction of the original signal. It is common to sample at a frequency of 10 times that of the highest signal frequency. Thus, if possible, one might sample the above 100 Hz square wave at a frequency of 11 kHz. This rule of thumb holds true for one-shot events. If the signal is repetitious, or can be made repetitious, then signal averaging can reduce the frequency requirements of sampling rates as well as improve the SNR.

Smoothing Once a good sampling is made, smoothing calculations can be made. Smoothing involves averaging a given number of points over some given time. Since noise is assumed to be random, this localized averaging helps to eliminate noise spikes—essentially it "smoothes" the data and thus the name. An example will help explain the procedure. If we have 10 data points at equal time intervals from a signal and choose to average the data in groups of three, then we add the first three, average and create a new data point at the average time, or for our example, the time at which data point 2 was collected. Next, we average data points 2, 3 and 4 and add this number to the new data set at the average time of that set, or in our case, the time at which data point 3 was taken. The analysis continues until all data are aver-

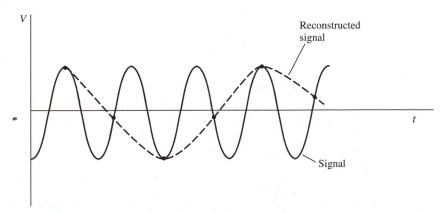

FIGURE 15.16 Aliasing caused by inadequate sampling rate.

aged and the new data set is complete. Caution must be exercised when choosing the number of data points to average. While increasing the number of data points in the average improves the signal-to-noise ratio, it can also act to smooth out the original signal of interest. Table 15.1 shows a smoothing calculation with 3-data-point weighting.

From Figure 15.17 the original signal can be seen along with the reconstructed smoothed signal. Note that the original signal has a fundamental period of about 3.5 ms. If the signal we wished to analyze from our transducer was this signal, then signal smoothing has essentially eliminated any information about that signal. If, on the other hand, the signal we expected from the transducer was a DC voltage of about 0.4 V, then signal smoothing has helped eliminate considerable noise, but we could have accomplished the same noise reduction by simply averaging the 10 original data points. Consider another example in which the data in Table 15.2 are collected and smoothed using a 3-point averaging. The original signal is shown in Figure 15.18 while the smoothed signal is shown in Figure 15.19. Notice that the

TABLE 15.1 Data from Signal Shown in Figure B and 3-Point Signal Smoothing

Time (ms)	V (Volts)	Averaged Signal (Volts)
0	0.40	
1	0.64	0.43
2	0.25	0.44
3	0.41	0.44
4	0.65	0.45
5	0.29	0.37
6	0.18	0.35
7	0.59	0.48
8	0.66	0.47
9	0.16	0.39
10	0.35	

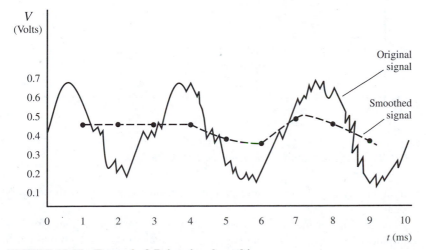

FIGURE 15.17 Example 3-Point signal soothing.

TABLE 15.2   Data Shown in Figure 15.8 with 3-Point Signal Smoothing

Time (ms)	Voltage (V)	3-Point Smooth (V)
0	−2	
10	5	15
20	41	24
30	27	44
40	65	47
50	48	66
60	84	66
70	66	76
80	77	86
90	115	98
100	102	95
110	69	77
120	59	67
130	73	56
140	36	55
150	55	36
160	17	27
170	8	16
180	23	

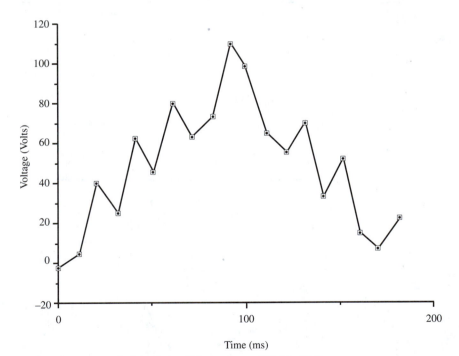

FIGURE 15.18   Original signal from data of Table 15.2.

smoothed signal is void of any of the noise peaks present in the original signal.

A good experimentalist knows what type of signal to expect from the equipment before applying any type of numerical smoothing. This is not to say that one should always know what to expect from an experiment. Some of the most interesting modern discoveries have been made when analyzing

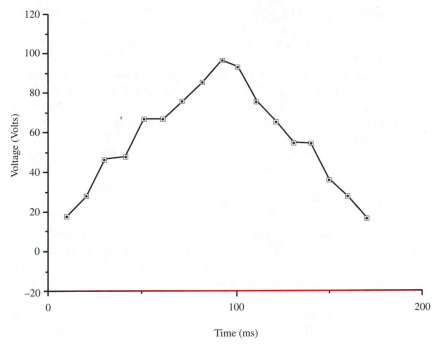

FIGURE 15.19   Smoothed signal from data of Table 15.2.

what at first appeared to be noise signals (examples $-3°K$ background radiation and pulsars).

Fourier Transform

Another very useful computer-based technique used in reducing noise is to take the data, obtain its Fourier components (Fourier spectrum), apply digital software-based frequency filters, and reconstruct the signal. Electronic filters previously discussed suffer from a number of drawbacks. For example, a notch filter which passes a single, exact frequency is nearly impossible to construct, as is a low-pass filter with similar ideal characteristics. Digital filters consisting of computer algorithms do not suffer from these limitations. Virtually all commercial, research-grade data-acquisition software packages have built-in digital filtering capacity or provide for their addition. For example of digital filtering, consider an experiment which produces a 100 Hz signal with inherent noise of higher frequencies. If the data are collected and analyzed into their Fourier components, you will be able to tell at what frequencies the noise is occurring and simply eliminate these frequencies from the Fourier spectrum. The signal can then be reconstructed without noise. This is a very powerful technique, since the digital filters are equivalent to perfect electronic filters.

15-5   SOME PRAGMATICS

The spec sheets of the more thoroughly characterized devices contain information about the noise figure as well as other data of general interest. It is appropriate to consider this information and how it can assist the designer or user.

This chapter has dealt with the sources or origins of noise and how it can be reduced. For the sake of completeness, we must address a few more noise "traps" into which most users of electronic equipment fall on occasion. When the need for as noise-free a system as possible is anticipated, wire-wound resistors should be used instead of the carbon-film or composition types. This is because $1/f$ noise is inherently present only in semiconductors, and is entirely absent in pure metals. This factor generally outweighs the higher cost of wire-wound resistors for precise applications. High-quality capacitors should be used for the same reason.

The need for sound wiring practices has already been mentioned, but bears repeating. The poorly soldered connection is an insidious source of noise which is often difficult to trace. Therefore, make sure no "cold solder" joints exist. At the same time, however, it is equally important to avoid overheating semiconductor components during the soldering process; the standard practice of attaching a heat sink (such as a solder-filled alligator clip) to the lead between the soldering point and the component is usually a sufficient precaution.

As the power used in a device increases, so does the need to keep it cool. Blowers or fans are often incorporated in a unit for that purpose. Be sure, in following this approach, that the fan itself does not introduce noise. By maintaining the unit at or near room temperature, thermal noise generation is reduced. At times, cooling a transducer to low temperatures improves the signal-to-noise ratio. This approach has been widely used to improve the *S/N* ratio in optical transducers, and sensitive radio telescopes routinely employ liquid nitrogen cooling of the focal pickup for this reason. For troubleshooting of noise problems, chilling of suspected components often shows which one is the source of noise; this approach has the advantage that the components need not be unsoldered. Generally, the chilling of a noisy component reduces or eliminates the noise.

Whenever possible, operate the instrumental system in such a manner that the noise level does not interfere with the measurement. As simplistic as this may sound, it is a useful way to get around noise problems. For example, the HTL (high threshold logic) line of digital ICs is expressly designed to have a minimum of 5 V difference between the 0 and 1 states. This permits operation in a noise environment in which a smaller difference in levels would lead to erroneous information because of noise.

15-6   QUESTIONS AND PROBLEMS

1.   A diode is sometimes used as a white noise generator. Assuming that 100 V is applied to a diode in series with a 10 Ω resistor, how much shot noise is generated when measured with an instrument whose bandwidth is 100 kHz?

2.   If the shot noise signal in the previous problem is taken across the resistor and amplified ($A_p = 10^5$), what will be the shot noise at the amplifier output?

3.   Under the conditions of problems 1 and 2, compute the thermal noise at the output of the amplifier.

4.   The $1/f$ noise has essentially a 3 dB/octave response at low frequencies. Using Figure 15.1, determine the frequency at which $1/f$ noise is twice that of the other fundamental noises.

5. What is the difference between white noise and pink noise?

6. What advantage does coaxial cable have over individual leads in terms of noise?

7. What amount of thermal noise do you expect from a 10 MΩ resistor measured at room temperature with an instrument that has a 1 kHz bandwidth (e.g., a multimeter)?

8. Can you determine which would be greater, the thermal noise measured in problem 4 or the $1/f$ noise that would be present in the measurement as well? Explain your answer.

9. Describe four practical ways to minimize 60-cycle noise.

10. It is a good practice to place $0.1\,\mu\text{F}$ capacitors across the power supply inputs of digital ICs if the chips are located relatively far from the power supply capacitor. Can you think of a reason or reasons why this is true?

11. Describe what a ground loop is and why it should be avoided.

12. When an operational amplifier follower is laid out on a printed circuit board, it is common to encircle the non-inverting input lead with the output/inverting-input trace connection. Why?

The Method of Determinants

The basis for solving simultaneous linear equations by determinants is Cramer's rule, which is stated as follows:

If the determinant of the coefficients of the equations is different from zero, then the value of each unknown is given by the quotient of two determinants. The determinant in the denominator is that of the coefficients of the unknowns. The determinant in the numerator is the same as that in the denominator *except* that the coefficients of the unknown whose value is being sought are replaced by the corresponding constant terms.

While this rule may sound rather complicated in words, its use is quite simple in practice, as an example will show. Consider the set of simultaneous equations:

$$6i_1 - 3i_2 - 2i_3 = 50 \qquad \text{(A-1)}$$

$$-3i_1 + 17i_2 - 6i_3 = 0 \qquad \text{(A-2)}$$

$$-2i_1 - 6i_2 + 13i_3 = 0 \qquad \text{(A-3)}$$

The determinant of the coefficients, D, which appears in the denominator of each quotient, is constructed by placing each coefficient in the place it appears in the set of equations:

$$D = \begin{vmatrix} 6 & -3 & -2 \\ -3 & 17 & -6 \\ -2 & -6 & 13 \end{vmatrix} \qquad \text{(A-4)}$$

The three unknowns may then be expressed as

$$i_1 = \frac{\begin{vmatrix} 50 & -3 & -2 \\ 0 & 17 & -6 \\ 0 & -6 & 13 \end{vmatrix}}{D} \qquad \text{(A-5)}$$

$$i_2 = \frac{\begin{vmatrix} 6 & 50 & -2 \\ -3 & 0 & -6 \\ -2 & 0 & 13 \end{vmatrix}}{D} \tag{A-6}$$

$$i_3 = \frac{\begin{vmatrix} 6 & -3 & 50 \\ -3 & 17 & 0 \\ -2 & -6 & 0 \end{vmatrix}}{D} \tag{A-7}$$

Before evaluating these determinants, it is desirable to list a number of the properties of determinants, which are often helpful in reducing the amount of labor involved.

(a) A second-order (two-by-two square) determinant may be evaluated by cross-multiplication, defined by example as

$$\begin{vmatrix} a & b \\ c & d \end{vmatrix} = ad - bc \tag{A-8}$$

(b) A third-order (three-by-three square) determinant may be evaluated by taking triple products with the proper signs. This is defined by example as

$$\begin{vmatrix} a & b & c \\ d & e & f \\ g & h & i \end{vmatrix} = aei + bfg + cdh - ceg - bdi - afh \tag{A-9}$$

A convenient way to remember this process is to imagine the first two columns of the determinant replicated to the right of the original square; each of the three diagonals going to the *right* and down are *added,* while each of the three diagonals going to the *left* and down are *subtracted,* as in the diagram:

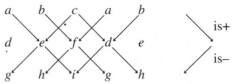

(c) The *minor* of any given element of a determinant is the determinant of next lower order obtained by deleting the row and column in which the element stands. Thus, for the determinant

$$\begin{vmatrix} a & b & c \\ d & e & f \\ g & h & i \end{vmatrix}$$

the minor of a is $\begin{vmatrix} e & f \\ h & i \end{vmatrix}$, the minor of e is $\begin{vmatrix} a & c \\ g & i \end{vmatrix}$, and the minor

of f is $\begin{vmatrix} a & b \\ g & h \end{vmatrix}$.

(d) The *cofactor* of any element in the mth row and n^{th} column of a determinant is $(-1)^{m+n}$ times the minor of that element. Thus, for the determinant just given, the cofactor of a is

$$(-1)^2 \begin{vmatrix} e & f \\ h & i \end{vmatrix} = \begin{vmatrix} e & f \\ h & i \end{vmatrix} \tag{A-10}$$

the cofactor of e is

$$(-1)^4 \begin{vmatrix} a & c \\ g & i \end{vmatrix} = \begin{vmatrix} a & c \\ g & i \end{vmatrix} \tag{A-11}$$

and the cofactor of f is

$$(-1)^5 \begin{vmatrix} a & b \\ g & h \end{vmatrix} = - \begin{vmatrix} a & b \\ g & h \end{vmatrix} \tag{A-12}$$

(e) The value of a determinant is the sum of the results obtained by multiplying the elements of any one row or column by their cofactors. For example,

$$\begin{vmatrix} a & b & c \\ d & e & f \\ g & h & i \end{vmatrix} = a \begin{vmatrix} e & f \\ h & i \end{vmatrix} - d \begin{vmatrix} b & c \\ h & i \end{vmatrix} + g \begin{vmatrix} b & c \\ e & f \end{vmatrix} \tag{A-13}$$

(f) The sign of a determinant is changed by interchanging two rows or two columns. Thus,

$$\begin{vmatrix} a & b & c \\ d & e & f \\ g & h & i \end{vmatrix} = - \begin{vmatrix} d & e & f \\ a & b & c \\ g & h & i \end{vmatrix}$$

and

$$\begin{vmatrix} a & b & c \\ d & e & f \\ g & h & i \end{vmatrix} = - \begin{vmatrix} a & c & b \\ d & f & e \\ g & i & h \end{vmatrix}$$

(g) The value of the determinant is unchanged if each column is interchanged with the corresponding row. That is,

$$\begin{vmatrix} a & b & c \\ d & e & f \\ g & h & i \end{vmatrix} = \begin{vmatrix} a & d & g \\ b & e & h \\ c & f & i \end{vmatrix}$$

(h) If all of the elements of a row or column are multiplied by the same number x, then the value of the determinant is multiplied by x.

(i) If the product of any quantity x and the elements of one row (or column) is added to the elements of another row (or column), the value of the determinant is not changed:

$$\begin{vmatrix} a & b & c \\ d & e & f \\ g & h & i \end{vmatrix} = \begin{vmatrix} a & (b + ax) & c \\ d & (e + dx) & f \\ g & (h + gx) & i \end{vmatrix}$$

Let us now evaluate the determinants in the first example. The first step is to evaluate D_x, using equation (A-9):

$$D = (6)(17)(13) + (-3)(-6)(-2) + (-2)(-3)(-6)$$

$$- (-2)(17)(-2) - (6)(-6)(-6) - (-3)(-3)(13)$$

$$= 1326 - 36 - 36 - 68 - 216 - 117$$

$$= 1326 - 473 = 853$$

The first unknown, given by equation (A-5), is

$$i_1 = \frac{11,050 - 1800}{853} = 10.84$$

The second unknown is

$$i_2 = \frac{1950 + 600}{853} = 2.99$$

and the third is

$$i_3 = \frac{900 + 1700}{853} = 3.04.$$

A-1 THE j OPERATOR

A convenient method of designating vectors incorporates the use of the j operator. It is also called i (for imaginary numbers) but j will be used herein to avoid confusion with the designation of instantaneous current. The definition of j is

$$j \equiv \sqrt{-1}. \tag{A-14}$$

The following definitions also apply:

$$j \cdot \text{real number} = \text{imaginary number}$$

$$\text{real number} + \text{imaginary number} = \text{complex number}$$

The j operator is useful in defining direction, since multiplication of a number by j rotates the vector 90° ($\pi/2$). Multiplying the number by j a second time rotates it another 90°, or a total of 180° (π). A real, negative number is obtained as a result of this second multiplication by j. The coordinate system corresponding to these operations is given in Figure A.1. Several examples of this designation are given in Figures A.2 and A.3.

The mathematical operations with j are straightforward. The necessary elements are reviewed below.

Powers of j:

$$j^2 = (\sqrt{-1})^2 = -1 \qquad \textbf{(A-15)}$$

$$j^3 = j \cdot -1 = -j \qquad \textbf{(A-16)}$$

$$j^4 = j^2 \cdot j^2 = +1 \qquad \textbf{(A-17)}$$

Multiplication by j:

$$-j \cdot j = -j^2 = +1 \qquad \textbf{(A-18)}$$

$$j^2 \cdot j^3 = +1 \cdot j = +j \qquad \textbf{(A-19)}$$

Complex number arithmetic is an important part of impedance problem solving. The arithmetic operations are also reviewed below.

Addition:

$$(5 + j3) + (6 - j8) = (5 + 6) + j(3 - 8) = 11 - j5 \qquad \textbf{(A-20)}$$

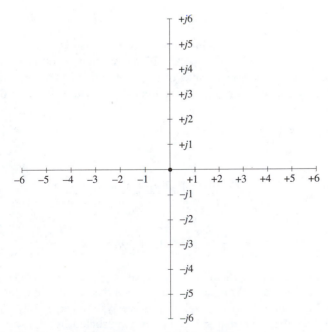

FIGURE A.1   **Cartesian coordinates using *j* notation.**

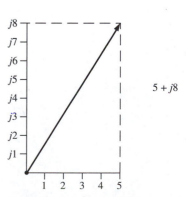

FIGURE A.2   **Vector representation of 5 + *j*8.**

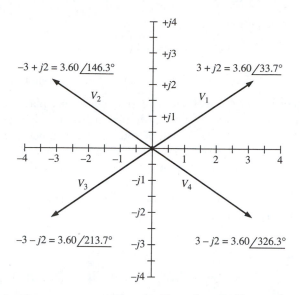

FIGURE A.3   Vector representations for $V_1 = 3 + j2$, $V_2 = -3 + j2$, $V_3 = -3 - j2$, and $V_4 = 3 - j2$. The polar coordinate notation is also given.

Subtraction:

$$(5 + j3) - (6 - j8) = (5 - 6) + j(3 - (-8)) = -1 + j11 \quad \text{(A-21)}$$

Multiplication:

$$(5 + j3)(6 - j8) = (5)(6) + j[(3)(6) + (5)(-8)] - j^2(3)(8)$$

$$= 30 + j(18 - 40) - (-1)(24)$$

$$= 54 - j22 \quad \text{(A-22)}$$

Division: The quotient is multiplied by the *complex conjugate* of the divisor divided by itself (of course, anything divided by itself is unity, so this operation does not change the value of the quotient). The complex conjugate is obtained by changing the sign of the imaginary term, so that for $6 - j8$ it is $6 + j8$. Thus,

$$\frac{5 + j3}{6 - j8} = \frac{5 + j3}{6 - j8} \cdot \frac{6 + j8}{6 + j8}$$

$$= \frac{(5)(6) + j[(3)(6) + (5)(8)] + j^2(3)(8)}{(6)(6) + j[(6)(8) - (6)(8)] - j^2(8)(8)}$$

$$= \frac{6 + j58}{100} = 0.06 + j0.58 \quad \text{(A-23)}$$

The applicability of this kind of analysis to AC circuit calculations was described in Chapter 3. There it was mentioned that resistance is represented by a pure real number, that capacitance is always represented by a *negative imaginary* number, and that inductance is represented by a *positive imagi-*

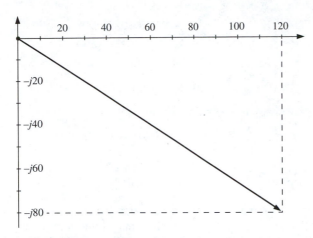

FIGURE A.4   Vector representation of impedance of a series combination of a 120 Ω resistor and a 2 μF capacitor.

nary number. Reactances, of course, can be mixed, and are represented by complex numbers. For example, consider a resistor (120 Ω) and a capacitor (2 μF) in series. The impedance of this circuit at 1000 Hz is calculated directly. The capacitive reactance, X_c, is

$$X_c = \frac{-1}{2\pi f C} = \frac{-1}{(2\pi)(1 \times 10^3)(2 \times 10^{-6})} = -79.3 \; \Omega \qquad \textbf{(A-24)}$$

This reactance is, of course, at 90° to the resistance, so that the total impedance, Z, is $120 - j79.3$ ohms. This result is graphed in Figure A.4. In order to evaluate the magnitude and phase angle of the impedance vector, the procedures developed in Chapter 3 are followed:

$$|Z| = \sqrt{R^2 + X_c^2} = [(120)^2 + (79.3)^2]^{1/2} = 144 \; \Omega \qquad \textbf{(A-25)}$$

and

$$\phi = -\tan^{-1}\left(\frac{|X_c|}{R}\right) = -\tan^{-1}\left(\frac{79.3}{120}\right)$$

$$= -\tan^{-1}(0.66) = -33.4° \qquad \textbf{(A-26)}$$

These results may be stated in polar notation as $144 \underline{/-33.4°}$.

As another example, consider a parallel combination of an inductor (0.2 H) and a capacitor (0.2 μF) at a frequency of 500 Hz. The reactances are

$$X_L = (2\pi)(5 \times 10^2)(2 \times 10^{-1}) = 630 \; \Omega \qquad \textbf{(A-27)}$$

$$X_c = \frac{-1}{(2\pi)(5 \times 10^2)(2 \times 10^{-7})} = -1590 \; \Omega \qquad \textbf{(A-28)}$$

In *j* notation, because they are 180° out of phase with each other, they become $Z_L = j630$ and $Z_c = -j1590$. The total impedance, Z, of the parallel combination is thus

$$Z = \frac{Z_L Z_C}{Z_L + Z_C} = \frac{(j630)(j1590)}{j630 - j1590}$$

$$= \frac{10^6}{-j960} = \frac{-10^6}{j960} \qquad \textbf{(A-29)}$$

Since $1/j = -j$, we have as a final result

$$Z = -(-j)(1040) = j1040 \qquad \textbf{(A-30)}$$

A-2 PHASORS

A phasor is a vector which rotates in a counterclockwise manner at a constant velocity. Thus, a given point has both a magnitude and a phase angle. A phasor voltage, V, is

$$V = V \angle \theta. \qquad \textbf{(A-31)}$$

where V is the magnitude of the vector in rms units and θ is the phase angle. In a like manner, a phasor current, I, is written as

$$I = I \angle \theta. \qquad \textbf{(A-32)}$$

Phasor representation is useful for describing capacitors and inductors. The impedance of a capacitor, Z_c, by definition, is

$$Z_c = \frac{V}{I} = \frac{V \angle \theta}{I \angle \theta + 90°} = \frac{V}{I} \angle -90°. \qquad \textbf{(A-33)}$$

Because Z_c is a property of the physical capacitor and not dependent on the notation, this must be equivalent to the j notation form, or

$$Z_c = -j|X_c|. \qquad \textbf{(A-34)}$$

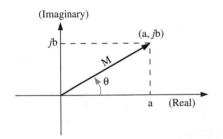

(Imaginary)

jb

(a, jb)

M

θ

a (Real)

FIGURE A.5   **Cartesian-polar coordinate conversion.**

Conversions from cartesian to polar coordinates are a simple matter; in fact, many calculators will perform the conversion. With reference to Figure A.5,

$$M = M \angle \theta \qquad \textbf{(A-35)}$$

and

$$M = a + jb. \qquad \textbf{(A-36)}$$

Thus

$$M = \sqrt{a^2 + b^2} \qquad \textbf{(A-37)}$$

and

$$\theta = \tan^{-1}\left(\frac{b}{a}\right). \qquad \textbf{(A-38)}$$

Alternatively,

$$a = M \cos \theta \qquad \textbf{(A-39)}$$

and

$$b = M \sin \theta. \tag{A-40}$$

Therefore,

$$\boldsymbol{M} = M(\cos \theta + j \sin \theta). \tag{A-41}$$

It turns out that the cartesian coordinate system (j notation) is more useful when two complex numbers are to be added or subtracted. The polar coordinate system (phasor notation) is more useful when complex numbers are to be multiplied or divided. Consider two complex numbers in phasor notation,

$$\boldsymbol{A} = A \underline{/\theta_1} \tag{A-42}$$

and

$$\boldsymbol{B} = B \underline{/\theta_2}. \tag{A-43}$$

The product, $\boldsymbol{A} \cdot \boldsymbol{B}$, is

$$\boldsymbol{A} \cdot \boldsymbol{B} = A \cdot B \underline{/\theta_1 + \theta_2}. \tag{A-44}$$

If $\boldsymbol{A}$ is to be divided by $\boldsymbol{B}$, the result is

$$\frac{\boldsymbol{A}}{\boldsymbol{B}} = \frac{A}{B} \underline{/\theta_1 - \theta_2}. \tag{A-45}$$

Since it is relatively easy to convert from one form to another, you can use whichever form is more convenient.

A-3 BODE DIAGRAMS

Bode diagrams or plots are simply graphic representations of the response of a given circuit over a range of frequencies. Since the response change and the frequency change may be large, a logarithmic plot is used. This approach has the added advantage that the response may be plotted directly in decibels (dB). The definition of the decibel, given in the text, is repeated here:

$$dB = 20 \log (V_2/V_1) \tag{A-46}$$

Since the dB is already logarithmically related to voltage, a linear dB plot is equivalent to a logarithmic voltage plot. Still another advantage of this kind of representation is that a gradually changing response can be adequately represented by a straight line segment that approaches the actual curve asymptotically. This approach finds extensive use in describing circuit response.

A-4 FOURIER ANALYSIS OF WAVEFORMS

Any periodic and well-behaved waveform can be represented as a Fourier series of sine and cosine waves of varying amplitudes and harmonically related frequencies. The Fourier series for a wave, $v(t)$, is

$$v(t) = \frac{A_0}{2} + \sum_{n=1}^{\infty} A_n \cos n\,\omega t + \sum_{m=1}^{\infty} B_m \sin m\,\omega t. \tag{A-47}$$

where $A_0, A_1, \ldots A_n$, and $B_1, \ldots B_m$ are coefficients which, when evaluated, determine the amplitudes of the contributing terms. Each term is a sine or cosine waveform of the fundamental ($n = m = 1$) or higher harmonics ($n, m > 1$).

The coefficients are determined by the relationships

$$A_n = \frac{2}{T} \int_{-T/2}^{T/2} v(t) \cos n\, \omega t\, dt \qquad \text{(A-48)}$$

and

$$B_m = \frac{2}{T} \int_{-T/2}^{T/2} v(t) \sin m\, \omega t\, dt. \qquad \text{(A-49)}$$

The first term in equation (A-47), $A_0/2$, is the DC, or average, value of the waveform. If $v(t)$ is known, then each coefficient may be evaluated directly. Alternatively, the character of the waveform may be used to simplify the coefficient evaluation. If the function is *even*, that is, $v(t) = v(-t)$, all of the B coefficients equal zero and the Fourier series contains only cosine terms. Figure A.6 illustrates an even function waveform. Alternatively, if the function is *odd*, that is, $v(-t) = -v(t)$, all of the A coefficients are zero and the Fourier series contains only sine terms. An example of an odd function waveform is given in Figure A.7.

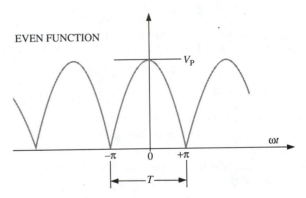

FIGURE A.6   Even function waveform in Fourier analysis.

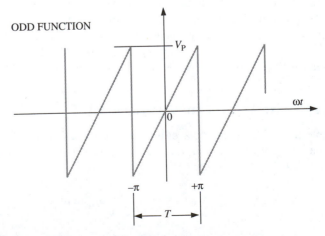

FIGURE A.7   Odd function waveform in Fourier analysis.

Transient RLC Circuit Operational Amplifiers

B-1 INTRODUCTION

In Chapter Two, Capacitors and Inductors, we explored a number of *RL* and *RC* circuits and the transient response they undergo. Here we combine a resistor, capacitor and inductor to form an *RLC* circuit and examine, using differential equations, the transient response it undergoes.

In Chapter Nine, Operational Amplifiers, we used a number of rules to derive the output voltages of an inverting and a non-inverting amplifier. Here we derive, in a more formal manner, the behavior of these circuits using the amplification of the operational amplifier. Both techniques of analysis, of course, yield the same result.

B-2 TRANSIENT RLC CIRCUIT RESPONSE

Consider the *RLC* circuit in Figure (B.1). When the switch is closed, a voltage of V_b is applied to the circuit. The battery/switch combination can be replaced with a function generator set to produce a pulse with voltage V_b. By applying Kirchoff's Law, we can write the voltage as a function of time ($t = 0$ when the switch is closed) as

$$V(t) = L(di/dt) + q/C + Ri. \qquad \textbf{(B-1)}$$

Differentiating and re-arranging Equation (B-1) yields

$$0 = (d^2i/dt^2) + (R/L)(di/dt) + (1/LC)i. \qquad \textbf{(B-2)}$$

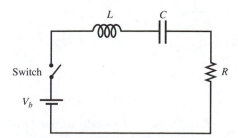

FIGURE B.1 Transient *RLC* circuit.

The solution of this differential equation is given by

$$i(t) = i_o e^{-ct} \sin(\omega t + \theta) \tag{B-3}$$

where

$$c = R/2L \tag{B-4}$$

$$\omega^2 = (1/LC) - (R^2/4L^2) \tag{B-5}$$

$$\omega = 2\pi f \tag{B-6}$$

$$\theta = \text{constant} \tag{B-7}$$

$$i_o = V_b/R \tag{B-8}$$

When $(R^2/4L^2)$ is greater than $(1/LC)$ the circuit is overdamped and there is no oscillation. When the two products are equal, the circuit is critically damped. If $(R^2/4L^2)$ is less than $(1/LC)$, the circuit is underdamped and Equation (B-3) represents an exponentially damped oscillating signal. Current as a function of time for overdamped, critically damped and underdamped circuits is shown in Figure (B.2). In circuits which have $(R^2/4L^2) \ll (1/LC)$ ω is simply $(1/LC)^{1/2}$.

Since the voltage across the resistor is simply $V_R = iR$, the voltage as a function of time curves is identical in form. Equation (B-3) becomes

$$V_R(t) = V_b e^{-ct} \sin(\omega t + \theta). \tag{B-9}$$

Sometimes this effect is brought about unintentionally, particularly when working with high frequencies. Stray inductance and capacitance can easily cause this phenomena which is often referred to as "ringing." To reduce the ringing of a circuit, one must locate and eliminate, or at least reduce, the stray capacitances and inductances which often arise from long signal cables and even from signals on PC board traces interacting with each other.

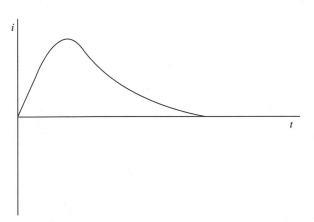

FIGURE B.2a   Overdamped *RLC* response, current as a function of time ($\theta = 0$).

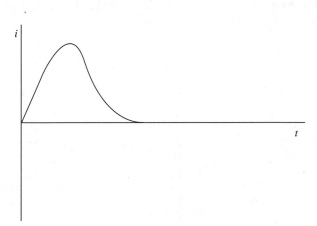

FIGURE B.2b   Critically damped *RLC* response, current as a function of time ($\theta = 0$).

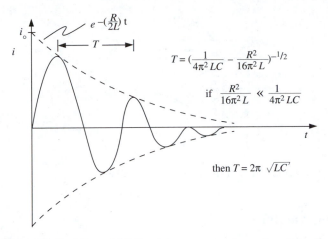

FIGURE B.2c   Underdamped *RLC* response, current as a function of time ($\theta = 0$).

B-3 OPERATIONAL AMPLIFIER

Follower

When operating within specified low frequencies (generally less than 1 Mhz) an operational amplifier has a large gain A. This gain is, at low frequencies, commonly found to be on the order of 10^5. The output of the amplifier is the product of the gain and the difference between the non-inverting and inverting inputs, or

$$V_{out} = A(V_{non} - V_{inv}). \tag{B-10}$$

Consider the voltage follower circuit of Figure (B.3). Since $V_{in} = V_+$ and $V_{out} = V_-$, we can write Equation (B-10) as

$$V_{out} = A(V_{in} - V_{out}) \tag{B-11}$$

or

$$V_{out}(1 + A) = A\,V_{in}. \tag{B-12}$$

Since $A \gg 1$, Equation (B-12) simply becomes

$$V_{out} = V_{in} \tag{B-13}$$

which is what was expected from Chapter Nine. Notice that as the gain, A, decreases with higher frequencies, Equation (B-13) no longer holds, but Equation (B-12) is valid.

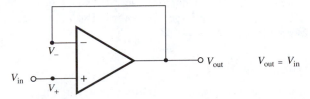

FIGURE B.3   Voltage follower operational amplifier circuit.

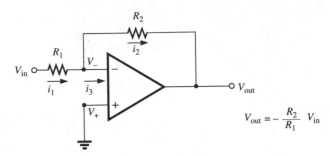

FIGURE B.4   **Inverting operational amplifier circuit.**

Inverting Amplifier

Consider the inverting amplifier shown in Figure (B.4). From Kirchoff's Current Equation we can write

$$i_1 = i_2 + i_3 \qquad \text{(B-14)}$$

and since no current flows into the ideal operational amplifier input, $i_2 = 0$, and we can write

$$i_1 = i_2. \qquad \text{(B-15)}$$

Writing this equation in terms of voltages and resistances we find

$$(V_{in} - V_-)/R_1 = (V_- - V_{out})/R_2. \qquad \text{(B-16)}$$

Recall from Equation (B-10) that $V_{out} = A(V_+ - V_-)$, which in this case, since $V_{+-} = 0$,

$$V_{out} = -A\,V_- \qquad \text{(B-17)}$$

or

$$V_- = V_{out}/A \qquad \text{(B-18)}$$

but since A is very large

$$V_- = 0 \ . \qquad \text{(B-19)}$$

and Equation (B-16) becomes

$$V_{in}/R_1 = -V_{out}/R_2 \qquad \text{(B-20)}$$

or

$$V_{out}/V_{in} = -R_2/R_1 \qquad \text{(B-21)}$$

which is the same as that from Chapter Nine. Note that when frequencies are high, and A is no longer very large, Equations (B-16) and (B-18) must be combined to predict V_{out}.

Non-inverting Amplifier

Finally, consider the non-inverting amplifier shown in Figure (B.5). Since $i_3 = 0$ in a similar manner to the inverting amplifier, we can write

$$(V_{out} - V_-)/R_2 = (V_- - 0)/R_1 \qquad \text{(B-22)}$$

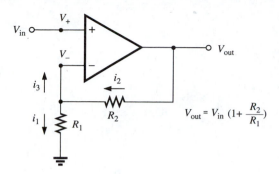

FIGURE B.5 **Non-inverting operational amplifier circuit.**

and from Equation (B-10)

$$V_{out} = A(V_{in} - V_-) \qquad \textbf{(B-23)}$$

or

$$V_- = V_{in} \qquad (\text{since } (V_{out}/A) = 0) \qquad \textbf{(B-24)}$$

we can write

$$(V_{out} - V_{in})/R_2 = V_{in}/R_1 \qquad \textbf{(B-25)}$$

and finally

$$V_{out} = V_{in}(1 + R_2/R_1) \qquad \textbf{(B-26)}$$

which is what we expected from Chapter Nine. Again, at high frequencies and thus low gain, A, Equation (B-24) cannot be used to simplify Equation (B-22) and Equation (B-26) is no longer valid.

C

Bibliography
Suggested Reading
Parts Sources

TEXTBOOKS

Horowitz, P. and W. Hill. *The Art of Electronics*. New York: Cambridge University Press, 1989. A difficult book for the beginner, but a classic "must" reference textbook for any laboratory.

Brophy, J. J. *Basic Electronics for Scientists*. New York: McGraw Hill, Inc., 1990.

Barnaal, D. *Digital and Microprocessor Electronics for Scientific Application*. Prospect Heights, Ill.: Wayland Press, Inc., 1982.

Malmstadt, H. V., C. G. Enke and S. R. Crouch. *Electronics and Instrumentation for Scientists*. Menlo Calif.: Benjamin Cummings, 1981.

Gates, S. C. and J. Becker. *Laboratory Automation Using the IBM PC*. Englewood Cliffs, NJ: Prentice Hall, 1989.

Soclof, S. *Analog Integrated Circuits*. Englewood Cliffs, NJ: Prentice Hall, 1985.

Faissler, W. L. *An Introduction to Modern Electronics*. New York: John Wiley and Sons, Inc., 1991.

HANDBOOKS

Fredrickson, R. M. *Intuitive IC Op Amps From Basics to Useful Applications*. National Semiconductor Corporation, Santa Clara, CA, 1984. A "must" for solid operational amplifier design.

Zuch, E. L. *Data Acquisition and Conversion Handbook*. Datel, Mansfield, MA, 1979.

Berlin, H. M. *The 555 Timer Applications Sourcebook with Experiments*. Indianapolis, Ind.: Howard W. Sams & Co., Inc., 1976.

Tolkheim, R. L. *Microprocessor Fundamentals*. New York: McGraw Hill Book Company, 1983.

Bishop, R. *Basic Microprocessors and the 6800*. Rochelle, NJ: Hayden Book Company, 1979.

DATA BOOKS AND COMPONENT MANUFACTURERS

Data Conversion Products Databook 1989/90
Linear Products Databook 1990/91
 Analog Devices
 Two Technology Way
 PO Box 280
 Norwood, MA 02062

Hot Ideas in CMOS, 1983
 Intersil, Inc.
 10710 N. Tantan Ave.
 Cupertino, CA 95014

Motorola CMOS Data
Motorola Optoelectronics Device Data
Motorola Small Signal Transistor Data
Motorola Power MOSFET Transistor Data
 Motorola, Literature Distribution Center
 PO Box 20912
 Phoenix, AZ 85036

Special Purpose Linear Devices Databook, 1989
LS/S/TTL Logic Databook, 1989
CMOS Logic Databook, Rev. 1
Linear Applications Handbook, 1986
General Purpose Linear Devices Databook, 1989
Data Acquisition Linear Devices Databook, 1989
Discrete Semiconductor Products Databook, 1989
 National Semiconductor
 2900 Semiconductor Drive
 Santa Clara, CA 95051

 Texas Instruments, Literature Response Center
 PO Box 401560
 Dallas, TX 75240

MAGAZINES

Omega Engineering, Inc. PO Box 2284, Stamford, CT 06906.
Popular Electronics, Gemsback Publications, Inc. 500-B, Bi-County Blvd., Farmingdale, NY 11735.
Radio Electronics, Gemsback Publications, Inc. 500-B, Bi-County Blvd., Farmingdale, NY 11735.
BYTE, One Phoenix Mill Lane, Peterborough, NH 03458.
EDN, Cahners Publishing Co., 275 Washington St., Newton, MA 02158.
Electronic Design, Hayden Publishing Co., Inc., 50 Essex St., Rochelle Park, NJ 07662.

Electronic Products, Hearst Business Communications, Inc., 645 Stewart Ave., Garden City, NY 11530.

VENDORS

It is well worth the while of buying an issue of *Radio Electronics* magazine (listed above) from the local convenience store and writing to the advertised commercial vendors and manufacturers to obtain their catalogs. Chances are if they are advertising in this hobbyist magazine that they will deal directly with you on small quantity orders. Prices for mail-order components are generally much cheaper than those of local vendors. Additionally, a number of surplus vendors advertise in this publication from which you can obtain unusual components at relatively good prices (such as laser tubes). Surplus vendors go in and out of business rapidly, so none are listed below. Equipment and component manufacturers will usually tell you where the nearest local vendor is in your area when asked.

Allied Radio
401 E. 8th Street
Fort Worth, TX 76102

Newark Electronics
4801 N. Ravenswood Ave.
Chicago, IL 60640-4496

Mouser Electronics
2401 Highway 287 North
Mansfield, TX 76063-4827

Jamesco Electronics
1355 Shoreway Rd.
Belmont, CA 94002

Digi-Key Corp.
701 Brooks Ave. South
PO Box 677
Thief River Falls, MN 56701

MISCELLANEOUS

The local Radio Shack store is probably the most convenient source of components and literature the novice first encounters. Even the expert will purchase components there and take note of new publications. A number of excellent, very inexpensive paperback books are available for any electronics user, beginner and expert alike. The handbooks listed below, written by Forrest M. Mimms, III are simply a must for everyone interested in electronics. Projects in these can spark the interest of the grade-school student to the university professor. Radio Shack also publishes the *Semiconductor Refer-*

ence Guide covering the carefully selected components they sell. The guide is available for just a few dollars. Books by Forrest M. Mimms, III include:

Getting Started in Electronics (Author's note—Buy It!)
Engineer's Notebook, A Handbook of Integrated Circuit Applications
   (Versions I and II)
Engineer's Mini-Notebook: Formulas, Tables and Basic Circuits
Engineer's Mini-Notebook: Digital Logic Circuits
Engineer's Mini-Notebook: 555 Timer Circuits
Engineer's Mini-Notebook: Basic Semiconductor Circuits

D

Data Sheets

Sources

Material on pages 3–14, 22–24, 49–79, 86–92, and 105–164 reprinted with permission of National Semiconductor Corporation.

Material on pages 15–17, 18–21, and 35–39 copyright of Motorola, Inc. Used by permission.

Material on pages 40–48, 80–85, 93–104, and 165–170 reprinted with permission of Analog Devices, Inc.

National Semiconductor

Diode Data

Computer Diodes (Glass Package)

Device No.	Package No.	V_{RRM} V Min	I_R nA Max	@	V_R V	V_F V Min	Max	@	I_F mA	C pF Max	t_{rr} ns Max	Test Cond.	Proc. No.
1N625	DO-35	30	1000		20		1.5		4		1000	(Note 1)	D4
1N914	DO-35	100	25 5000		20 75		1.0		10		4	(Note 2)	D4
1N914A	DO-35	100	25 5000		20 75		1.0		20		4	(Note 2)	D4
1N914B	DO-35	100	25 5000		20 75		0.72 1.0		5 100		4	(Note 2)	D4
1N916	DO-35	100	25 5000		20 75		1.0		10		4	(Note 2)	D4
1N916A	DO-35	100	25 5000		20 75		1.0		20		4	(Note 2)	D4
1N916B	DO-35	100	25 5000		20 75		0.73 1.0		5 30		4	(Note 2)	D4
1N3064	DO-35	75	100		50		0.575 0.650 0.710 1.0		0.250 1.0 2.0 10.0	2	4	(Note 3)	D4
1N3600	DO-35	75	100		50	0.54 0.66 0.76 0.82 0.87	0.62 0.74 0.86 0.92 1.0		1.0 10.0 50.0 100.0 200.0	2.5	4	(Note 4)	D4
1N4009	DO-35	35	100		25		1.0		30	4	2	(Note 2)	D4
1N4146	DO-35	See Data for 1N914A/914B											
1N4147	DO-35	See Data for 1N914A/914B											
1N4148	DO-35	See Data for 1N914											
1N4149	DO-35	See Data for 1N916											
1N4150	DO-35	See Data for 1N3600											
1N4151	DO-35	75	50		50		1.0		50	4	2	(Note 2)	D4
1N4152	DO-35	40	50		30	0.49 0.53 0.59 0.62 0.70 0.74	0.55 0.59 0.67 0.70 0.81 0.88		0.1 0.25 1.0 2.0 10.0 20.0	4	2	(Note 2)	D4
1N4153	DO-35	75	50		50	See 1N4152				4	2	(Note 2)	D4
1N4154	DO-35	35	100		25		1.0		30	4	2	(Note 2)	D4

Diode Process Characteristics

Curve Set Number D4

Typical Electrical Characteristic Curves 25°C Ambient Temperature unless otherwise noted

Forward Voltage vs Forward Current

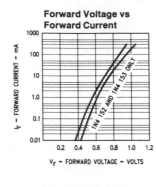

Forward Current vs Temperature Coefficient

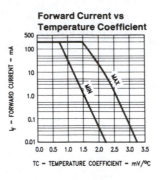

Capacitance vs Reverse Voltage

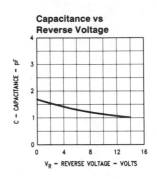

Reverse Current vs Reverse Voltage

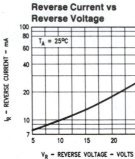

Reverse Current vs Ambient Temperature

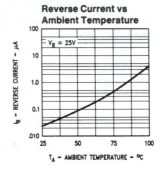

Dynamic Impedance vs Forward Current

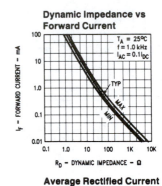

Reverse Recovery Time vs Forward Current (F = I_R)

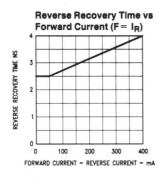

Power Derating Curve

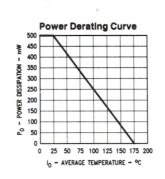

Average Rectified Current and Forward Current vs Ambient Temperature

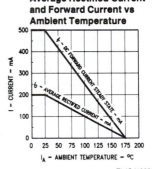

TL/G/10033-4

Diode Data

Zener Diodes (Glass Package)

Device No.	Package No.	V_Z V Nom	Tol. $\pm V_Z$ %	Z_Z Ω Max	@ I_Z mA	I_R μA Max	@ V_R V	T.C. %/°C Typ (Max)	P_D mW $T_A = 25°C$	Proc. No.
1N746A	DO-35	3.3	5.0	28.0	20	10	1.0	−0.070	500	D13
1N747A	DO-35	3.6	5.0	24.0	20	10	1.0	−0.065	500	D13
1N748A	DO-35	3.9	5.0	23.0	20	10	1.0	−0.060	500	D13
1N749A	DO-35	4.3	5.0	22.0	20	2	1.0	−0.055	500	D13
1N750A	DO-35	4.7	5.0	19.0	20	2	1.0	−0.043	500	D13
1N751A	DO-35	5.1	5.0	17.0	20	1	1.0	−0.030	500	D13
1N752A	DO-35	5.6	5.0	11	20	1.0	1.0	+0.028	500	D13
1N753A	DO-35	6.2	5.0	7.0	20	0.1	1.0	+0.045	500	D13
1N754A	DO-35	6.8	5.0	5.0	20	0.1	1.0	+0.050	500	D13
1N755A	DO-35	7.5	5.0	6.0	20	0.1	1.0	+0.058	500	D13
1N756A	DO-35	8.2	5.0	8.0	20	0.1	1.0	+0.062	500	D13
1N757A	DO-35	9.1	5.0	10	20	0.1	1.0	+0.068	500	D13
1N758A	DO-35	10	5.0	17	20	0.1	1.0	+0.075	500	D13
1N759A	DO-35	12	5.0	30	20	0.1	1.0	+0.077	500	D13
1N957B	DO-35	6.8	5.0	4.5	18.5	150	5.2	+0.050	500	D13
1N958B	DO-35	7.5	5.0	5.5	16.5	75	5.7	+0.058	500	D13
1N959B	DO-35	8.2	5.0	6.5	15	50	6.2	+0.062	500	D13
1N960B	DO-35	9.1	5.0	7.5	14	25	6.9	+0.068	500	D13
1N961B	DO-35	10	5.0	8.5	12.5	10	7.6	+0.072	500	D13
1N962B	DO-35	11	5.0	9.5	11.5	5.0	8.4	+0.073	500	D13
1N963B	DO-35	12	5.0	11.5	10.5	5.0	9.1	+0.076	500	D13
1N964B	DO-35	13	5.0	13	9.5	5.0	9.9	+0.079	500	D13
1N965B	DO-35	15	5.0	16	8.5	5.0	11.4	+0.082	500	D13
1N966B	DO-35	16	5.0	17	7.8	5.0	12.2	+0.083	500	D13
1N967B	DO-35	18	5.0	21	7.0	5.0	13.7	+0.085	500	D13
1N968B	DO-35	20	5.0	25	6.2	5.0	15.2	+0.086	500	D13
1N969B	DO-35	22	5.0	29	5.6	5.0	16.7	+0.087	500	D13
1N970B	DO-35	24	5.0	33	5.2	5.0	18.2	+0.088	500	D13
1N971B	DO-35	27	5.0	41	4.6	5.0	20.6	+0.090	500	D13
1N972B	DO-35	30	5.0	49	4.2	5.0	22.8	+0.091	500	D13
1N973B	DO-35	33	5.0	58	3.8	5.0	25.1	+0.092	500	D13

Curve Set Number D13

Typical Electrical Characteristic Curves 25°C Ambient Temperature unless otherwise noted

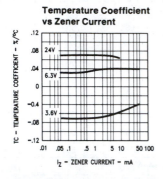

Temperature Coefficient vs Zener Current

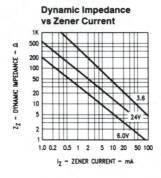

Dynamic Impedance vs Zener Current

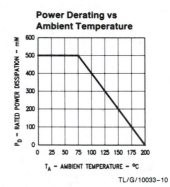

Power Derating vs Ambient Temperature

TL/G/10033–10

Test Circuit

NOISE DENSITY MEASUREMENT CIRCUIT

1N4099–1N4121
1N4620–1N4627

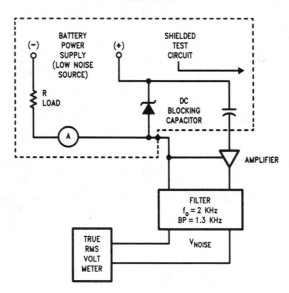

TL/G/10033–11

NPN Transistors

General Purpose Amplifiers and Switches (Continued)

Type No.	Case Style	V_{CBO} (V) Min	V_{CEO} (V) Min	V_{EBO} (V) Min	I_{CBO} (nA) Min	@ V_{CB} (V)	h_{FE} Min	Max	@ I_C (mA)	& V_{CE} (V)	$V_{CE(SAT)}$ (V) Max	$V_{BE(SAT)}$ (V) Min	Max	@ I_C (mA)	C_{ob} (pF) Max	f_T (MHz) Min Max	@ I_C (mA)	t_{off} (ns) Max	NF (dB) Max	Test Conditions	Process No.
2N3903	TO-92 (92)	60	40	6			15		100	1	0.2	0.6	0.85	10	4	250	10	225	6	(Notes 6 & 7)	23
							30	150	50	1	0.3		0.95	50							
							50		10	1											
							35		1	1											
							20		100 μA	1											
2N3904	TO-92 (92)	60	40	6		30	30		100	1	0.2	0.65	0.85	10	4	300	10	250	5	(Notes 6 & 7)	23
							60	300	50	1	0.3		0.95	50							
							100		10	1											
							70		1	1											
							40		100 μA	1											
2N3946	TO-18	60	40	6			20		50	1	0.2	0.6	0.9	10	4	250	10	375	5	(Notes 6 & 7)	23
							50	150	10	1	0.3		1.0	50							
							45		1	1											
							30		100 μA	1											
2N3947	TO-18	60	40	6			40		50	1	0.2	0.6	0.9	10	4	300	10	450	5	(Notes 6 & 7)	23
							100	300	10	1	0.3		1.0	50							
							90		1	1											
							60		100 μA	1											
2N4123	TO-92 (92)	40	30	5	50	20	25		50	1	0.3		0.95	50	4	250	10		6	(Note 7)	23
							50	150	2	1											
2N4124	TO-92 (92)	30	25	5	50	20	60		50	1	0.3		0.95	50	4	300	10		5	(Note 7)	23
							120	360	2	1											
MPQ3904	TO-116 (39)	60	40	6	50	40	30		0.1	1	0.2		0.85	10	4	250	10				23
							50		1	1											
							75		10	1											
MPQ6700	TO-116 (39)	40	40	5	50	30	30		0.1	1	0.25		0.1	10	4.5	200	10				23/66
							50		1	1											
							70		10	1											
MPS2711	TO-92 (92)	18	18	5	500	18	30	90	2	4.5					4						23
MPS2712	TO-92 (92)	18	18	5	500	18	75	225	2	4.5					4						23

National Semiconductor

Process 23
NPN Small Signal

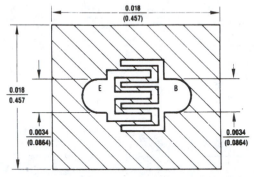

0.018
(0.457)

0.018
0.457

0.0034
(0.0864)

0.0034
(0.0864)

E   B

TL/G/10034–47

DESCRIPTION

Process 23 is an overlay, double-diffused, gold doped, silicon epitaxial device. Complement to Process 66.

APPLICATION

This device is designed as a general purpose amplifier and switch. The useful dynamic range extends to 100 mA as a switch and to 100 MHz as an amplifier.

PRINCIPAL DEVICE TYPES

TO-92 EBC: 2N3904, 2N4124

TO-236: MMBT3904, MMBT4124

TO-116: MPQ3904

16-SOIC: MMPQ3904

ELECTRICAL CHARACTERISTICS (T_A = 25°C)

Symbol	Conditions	Min	Typ	Max	Units
t_{ON}	I_C = 10 mA, I_{B1} = 1 mA *(Figure 1)*		30	70	ns
t_{OFF}	I_C = 10 mA, I_{B2} = 1 mA *(Figure 2)*		150	250	ns
C_{ob}	V_{CB} = 5V, f = 1 MHz		2.7	4.0	pF
C_{ib}	V_{EB} = 0.5V, f = 1 MHz			8.0	pF
NF	V_{CE} = 5V, I_C = 100 μA, R_S = 1 kΩ, P_{BW} = 15.7 kHz		2.0		dB
h_{fe}	I_C = 10 mA, V_{CE} = 20V, f = 100 MHz	2.5	4.5		
h_{FE}	I_C = 100 μA, V_{CE} = 5V I_C = 1 mA, V_{CE} = 5V I_C = 10 mA, V_{CE} = 5V I_C = 50 mA, V_{CE} = 5V I_C = 100 mA, V_{CE} = 5V	40 90 60 40 20	150	360	
$V_{CE(SAT)}$	I_C = 10 mA, I_B = 1 mA			0.15	V
$V_{BE(SAT)}$	I_C = 10 mA, I_B = 1 mA			0.80	V
$V_{CE(SAT)}$	I_C = 50 mA, I_B = 5 mA			0.25	V
$V_{BE(SAT)}$	I_C = 50 mA, I_B = 5 mA			0.85	V
BV_{CBO}	I_C = 10 μA	60			V
BV_{CEO}	I_C = 1 mA	30			V
BV_{EBO}	I_E = 10 μA	6.0			V
I_{CBO}	V_{CB} = 30V			100	nA
I_{EBO}	V_{EB} = 4V			100	nA

Process 23

Symbol	Conditions	Min	Typ	Max	Units
$P_{D(max)}$					
TO-92	$T_A = 25°C$	600			mW
TO-116	$T_A = 25°C$				
	(Total)	900			mW
	(Each Transistor)	500			mW
TO-236	$T_C = 25°C$	350			mW
$T_{j(max)}$	All Plastic Parts	150			°C

DC Current Gain vs Collector Current
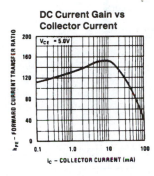

Base-Emitter ON Voltage vs Collector Current
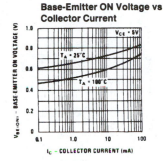

Maximum Power Dissipation vs Ambient Temperature
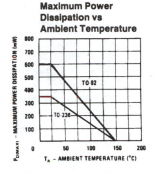

Contours of Constant Gain Bandwidth Product (f_T)
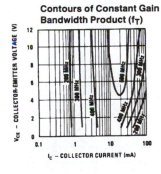

Collector Saturation Voltage vs Collector Current
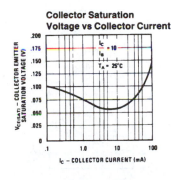

Base Saturation Voltage vs Collector Current
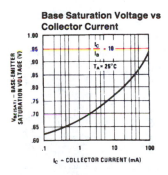

Capacitance vs Reverse Bias Voltage
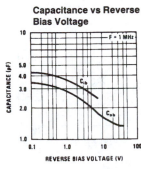

Collector Cutoff Current vs Ambient Temperature
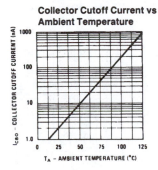

Current Gain and Phase Angle vs Frequency
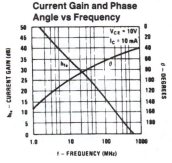

TL/G/10034–48

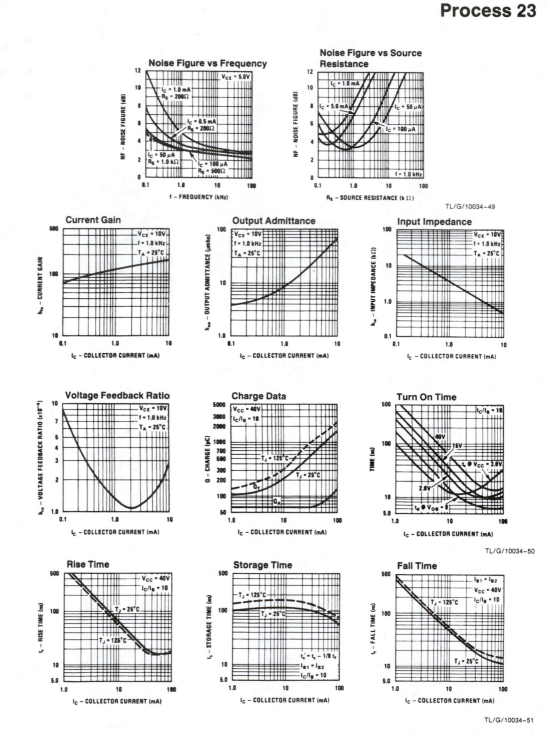

Process 23

PNP Transistors

General Purpose Amplifiers and Switches (Continued)

Type No.	Case Style	V_{CBO} (V) Min	V_{CEO} (V) Min	V_{EBO} (V) Min	I_{CES}^* I_{CBO} (nA) Max	@ V_{CB} (V)	h_{FE} Min	Max	@ I_C (mA)	& V_{CE} (V)	$V_{CE(SAT)}$ (V) Max	$V_{BE(SAT)}$ (V) Min	Max	@ I_C (mA)	C_{OB} (pF) Max	f_T (MHz) Min Max	@ I_C (mA)	t_{OFF} (ns) Max	NF (dB) Max	Test Conditions	Process No.
TN2905A	TO-237 (91)	60	60	5	10	50	50 100 100 100 75	300	500 150 10 1 0.1	10 10 10 10 10	0.4 1.6		1.3 2.6	150 500	8	200	50	100		(Note 2)	63
2N3905	TO-92 (92)	40	40	5			15 30 50 40 30	150	100 50 10 1 0.1	1 1 1 1 1	0.25 0.4	0.65	0.85 0.95	10 50	4.5	200	10	260	5	(Notes 5, 8)	66
2N3906	TO-92 (92)	40	40	5			30 80 100 80 60	300	100 50 10 1 0.1	1 1 1 1 1	0.25 0.4	0.65	0.85 0.95	10 50	4.5	250	10	300	4	(Notes 5, 8)	66
2N4121	Same as PN4121																				66
2N4122	Same as PN4122																				66
2N4125	TO-92 (92)	30	30	4	50	20	25 50	150	50 2	1 1	0.4		0.95	50	4.5	200	10		5	(Note 8)	66
2N4126	TO-92 (92)	25	25	4	50	20	60 120	360	50 2	1 1	0.4		0.95	50	4.5	250	10		4	(Note 8)	66
2N4916	Same as PN4916																				66
2N4917	Same as PN4917																				66
2N5138	Same as PN5138																				66
2N5139	Same as PN5139																				66
MPQ3906	TO-116	60	40	6	50	30	40 60 75		0.1 1 10	1 1 1	0.25		0.85	10	4.5						66

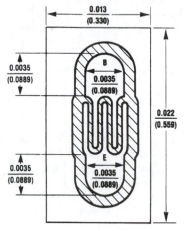

TL/G/10038-19

Process 66
PNP Small Signal

DESCRIPTION
Process 66 is an overlay, double-diffused, silicon epitaxial device. Complement to Process 23.

APPLICATION
This device was designed for general purpose amplifier and switching applications at collector currents of 10 μA to 100 mA.

PRINCIPAL DEVICE TYPES
TO-92 EBC: 2N3906, 4126
TO-236: MMBT3906
TO-116: MPQ3906
16-SOIC: MMPQ3906

ELECTRICAL CHARACTERISTICS (T_A = 25°C)

Symbol	Conditions	Min	Typ	Max	Units
t_{OFF}	I_C = 10 mA, I_{B2} = 1 mA		150	300	ns
t_{ON}	I_C = 10 mA, I_{B1} = 1 mA		30	70	ns
C_{ob}	V_{CB} = 5V		3.0	4.5	pF
C_{ib}	V_{EB} = 0.5V			15	pF
h_{fe}	f = 100 MHz, V_{CE} = 20V, I_C = 10 mA	2.5	4.5		
NF (wideband)	I_C = 100 μA, V_{CE} = 5V, R_S = 1 kΩ		2.0		dB
h_{FE}	I_C = 0.1 mA, V_{CE} = 1V	40			
	I_C = 1 mA, V_{CE} = 1V	50			
	I_C = 10 mA, V_{CE} = 1V	50	150	350	
	I_C = 50 mA, V_{CE} = 1V	40			
	I_C = 100 mA, V_{CE} = 1V	20			
$V_{CE(SAT)}$	I_C = 10 mA, I_B = 1 mA			0.25	V
	I_C = 50 mA, I_B = 5 mA			0.40	V
$V_{BE(SAT)}$	I_C = 10 mA, I_B = 1 mA			0.85	V
	I_C = 50 mA, I_B = 5 mA			0.95	V
BV_{CEO}	I_C = 1 mA	35			V
BV_{CBO}	I_C = 10 μA	45			V
BV_{EBO}	I_C = 10 μA	5.0			V
I_{CBO}	V_{CB} = 25V			100	nA
I_{EBO}	V_{EB} = 4V			100	nA

Process 66

Process 66

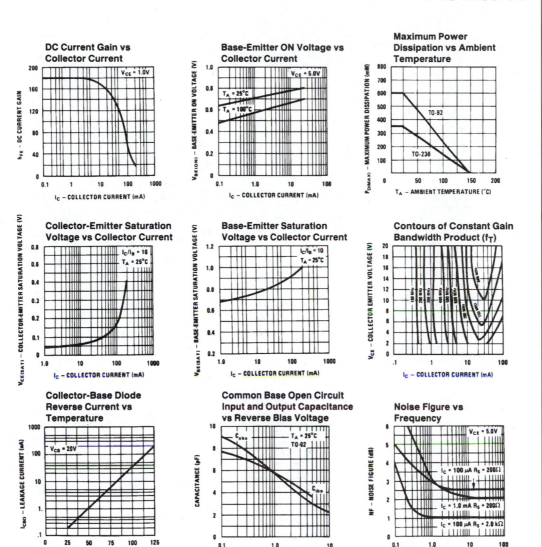

TL/G/10038–21

Process 66

Noise Figure vs Source Resistance

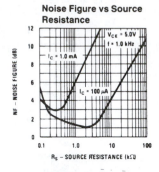

Input Impedance

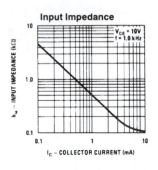

Output Admittance

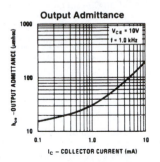

Current Gain

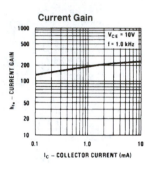

Voltage Feedback Ratio

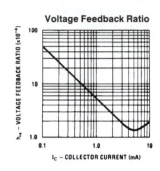

Turn On and Turn Off Times vs Collector Current

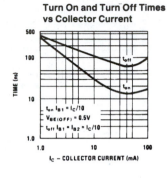

Switching Times vs Collector Current

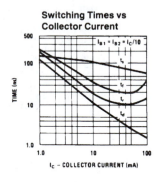

TL/G/10038–20

NPN   PNP
2N3055 MJ2955

COMPLEMENTARY SILICON POWER TRANSISTORS

. . . designed for general-purpose switching and amplifier applications.

- DC Current Gain — h_{FE} = 20-70 @ I_C = 4 Adc

- Collector-Emitter Saturation Voltage —
  $V_{CE(sat)}$ = 1.1 Vdc (Max) @ I_C = 4 Adc

- Excellent Safe Operating Area

15 AMPERE
POWER TRANSISTORS
COMPLEMENTARY SILICON

60 VOLTS
115 WATTS

MAXIMUM RATINGS

Rating	Symbol	Value	Unit
Collector-Emitter Voltage	V_{CEO}	60	Vdc
Collector-Emitter Voltage	V_{CER}	70	Vdc
Collector-Base Voltage	V_{CB}	100	Vdc
Emitter-Base Voltage	V_{EB}	7	Vdc
Collector Current — Continuous	I_C	15	Adc
Base Current	I_B	7	Adc
Total Power Dissipation @ T_C = 25°C Derate above 25°C	P_D	115 0.657	Watts W/°C
Operating and Storage Junction Temperature Range	T_J, T_{stg}	-65 to +200	°C

THERMAL CHARACTERISTICS

Characteristic	Symbol	Max	Unit
Thermal Resistance, Junction to Case	$R_{\theta JC}$	1.52	°C/W

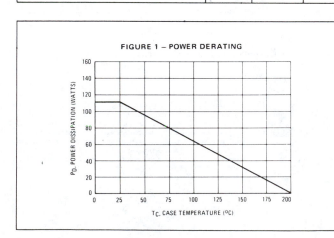

FIGURE 1 — POWER DERATING

P_D, POWER DISSIPATION (WATTS)

T_C, CASE TEMPERATURE (°C)

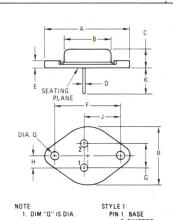

NOTE:
1. DIM "Q" IS DIA.

STYLE 1:
   PIN 1. BASE
        2. EMITTER
   CASE: COLLECTOR

DIM	MILLIMETERS		INCHES	
	MIN	MAX	MIN	MAX
A	–	39.37	–	1.550
B	–	21.08	–	0.830
C	6.35	7.62	0.250	0.300
D	0.99	1.09	0.039	0.043
E	–	3.43	–	0.135
F	29.90	30.40	1.177	1.197
G	10.67	11.18	0.420	0.440
H	5.33	5.59	0.210	0.220
J	16.64	17.15	0.655	0.675
K	11.18	12.19	0.440	0.480
Q	3.84	4.09	0.151	0.161
R	–	26.67	–	1.050

Collector connected to case.
CASE 11-01
(TO-3)

2N3055 NPN/MJ2955 PNP

ELECTRICAL CHARACTERISTICS ($T_C = 25^{\circ}C$ unless otherwise noted)

Characteristic	Symbol	Min	Max	Unit
***OFF CHARACTERISTICS**				
Collector-Emitter Sustaining Voltage (1) (I_C = 200 mAdc, I_B = 0)	$V_{CEO(sus)}$	60	–	Vdc
Collector-Emitter Sustaining Voltage (1) (I_C = 200 mAdc, R_{BE} = 100 Ohms)	$V_{CER(sus)}$	70	–	Vdc
Collector Cutoff Current (V_{CE} = 30 Vdc, I_B = 0)	I_{CEO}	–	0.7	mAdc
Collector Cutoff Current (V_{CE} = 100 Vdc, $V_{BE(off)}$ = 1.5 Vdc) (V_{CE} = 100 Vdc, $V_{BE(off)}$ = 1.5 Vdc, T_C = 150°C)	I_{CEX}	– –	1.0 5.0	mAdc
Emitter Cutoff Current (V_{BE} = 7.0 Vdc, I_C = 0)	I_{EBO}	–	5.0	mAdc
***ON CHARACTERISTICS (1)**				
DC Current Gain (I_C = 4.0 Adc, V_{CE} = 4.0 Vdc) (I_C = 10 Adc, V_{CE} = 4.0 Vdc)	h_{FE}	20 5.0	70 –	–
Collector-Emitter Saturation Voltage (I_C = 4.0 Adc, I_B = 400 mAdc) (I_C = 10 Adc, I_B = 3.3 Adc)	$V_{CE(sat)}$	– 	1.1 3.0	Vdc
Base-Emitter On Voltage (I_C = 4.0 Adc, V_{CE} = 4.0 Vdc)	$V_{BE(on)}$	–	1.5	Vdc
SECOND BREAKDOWN				
Second Breakdown Collector Current with Base Forward Biased (V_{CE} = 40 Vdc, t = 1.0 s; Nonrepetitive)	$I_{s/b}$	2.87	–	Adc
DYNAMIC CHARACTERISTICS				
Current Gain – Bandwidth Product (I_C = 0.5 Adc, V_{CE} = 10 Vdc, f = 1.0 MHz)	f_T	2.5	–	MHz
*Small-Signal Current Gain (I_C = 1.0 Adc, V_{CE} = 4.0 Vdc, f = 1.0 kHz)	h_{fe}	15	120	–
*Small-Signal Current Gain Cutoff Frequency (V_{CE} = 4.0 Vdc, I_C = 1.0 Adc, f = 1.0 kHz)	f_{hfe}	10	–	kHz

* Indicates Within JEDEC Registration. (2N3055)
(1) Pulse Test: Pulse Width ≤ 300 μs, Duty Cycle ≤ 2.0%.

FIGURE 2 – ACTIVE REGION SAFE OPERATING AREA

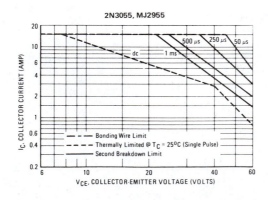

2N3055, MJ2955

There are two limitations on the power handling ability of a transistor: average junction temperature and second breakdown. Safe operating area curves indicate I_C-V_{CE} limits of the transistor that must be observed for reliable operation; i.e., the transistor must not be subjected to greater dissipation than the curves indicate.

The data of Figure 2 is based on $T_C = 25^{\circ}C$; $T_{J(pk)}$ is variable depending on power level. Second breakdown pulse limits are valid for duty cycles to 10% but must be derated for temperature according to Figure 1.

2N3055 NPN/MJ2955 PNP

NPN	**PNP**
2N3055	MJ2955

FIGURE 3 – DC CURRENT GAIN

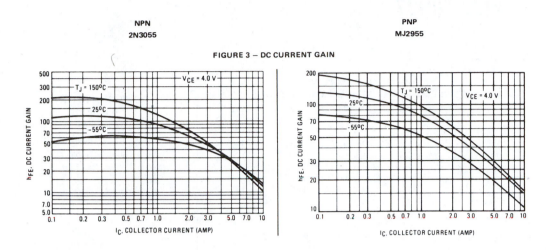

FIGURE 4 – COLLECTOR SATURATION REGION

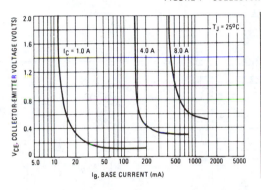

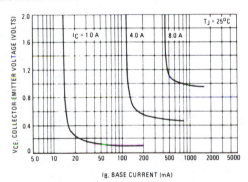

FIGURE 5 – "ON" VOLTAGES

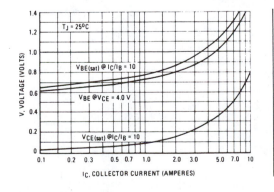

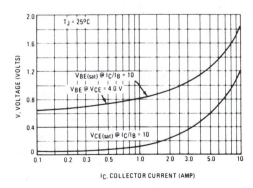

NPN   PNP
TIP120 TIP125
TIP121 TIP126
TIP122 TIP127

 **MOTOROLA**

PLASTIC MEDIUM-POWER
COMPLEMENTARY SILICON TRANSISTORS

. . . designed for general-purpose amplifier and low-speed switching applications.

- High DC Current Gain —
  h_{FE} = 2500 (Typ) @ I_C = 4.0 Adc

- Collector-Emitter Sustaining Voltage — @ 100 mAdc
  $V_{CEO(sus)}$ = 60 Vdc (Min) — TIP120, TIP125
        = 80 Vdc (Min) — TIP121, TIP126
        = 100 Vdc (Min) — TIP122, TIP127

- Low Collector-Emitter Saturation Voltage —
  $V_{CE(sat)}$ = 2.0 Vdc (Max) @ I_C = 3.0 Adc
        = 4.0 Vdc (Max) @ I_C = 5.0 Adc

- Monolithic Construction with Built-In Base-Emitter Shunt Resistors

- TO-220AB Compact Package

- TO-66 Leadform Also Available

DARLINGTON
8 AMPERE
COMPLEMENTARY SILICON
POWER TRANSISTORS

60-80-100 VOLTS
65 WATTS

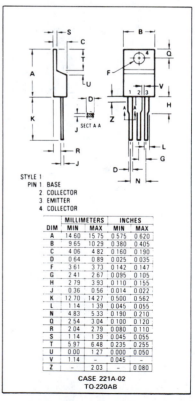

---

*MAXIMUM RATINGS

Rating	Symbol	TIP120, TIP125	TIP121, TIP126	TIP122, TIP127	Unit
Collector-Emitter Voltage	V_{CEO}	60	80	100	Vdc
Collector-Base Voltage	V_{CB}	60	80	100	Vdc
Emitter-Base Voltage	V_{EB}	← 5.0 →			Vdc
Collector Current — Continuous	I_C	← 5.0 →			Adc
Peak		← 8.0 →			
Base Current	I_B	← 120 →			mAdc
Total Power Dissipation @ T_C = 25°C	P_D	← 65 →			Watts
Derate above 25°C		← 0.52 →			W/°C
Total Power Dissipation @ T_A = 25°C	P_D	← 2.0 →			Watts
Derate above 25°C		← 0.016 →			W/°C
Unclamped Inductive Load Energy (1)	E	← 50 →			mJ
Operating and Storage Junction, Temperature Range	T_J, T_{stg}	← −65 to +150 →			°C

THERMAL CHARACTERISTICS

Characteristics	Symbol	Max	Unit
Thermal Resistance, Junction to Case	$R_{\theta JC}$	1.92	°C/W
Thermal Resistance, Junction to Ambient	$R_{\theta JA}$	62.5	°C/W

(1) I_C = 1 A, L = 100 mH, P.R.F. = 10 Hz, V_{CC} = 20 V, R_{BE} = 100 Ω.

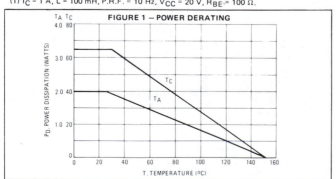

FIGURE 1 — POWER DERATING

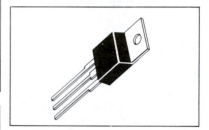

STYLE 1
PIN 1 BASE
    2 COLLECTOR
    3 EMITTER
    4 COLLECTOR

DIM	MILLIMETERS		INCHES	
	MIN	MAX	MIN	MAX
A	14.60	15.75	0.575	0.620
B	9.65	10.29	0.380	0.405
C	4.06	4.82	0.160	0.190
D	0.64	0.89	0.025	0.035
F	3.61	3.73	0.142	0.147
G	2.41	2.67	0.095	0.105
H	2.79	3.93	0.110	0.155
J	0.36	0.56	0.014	0.022
K	12.70	14.27	0.500	0.562
L	1.14	1.39	0.045	0.055
N	4.83	5.33	0.190	0.210
Q	2.54	3.04	0.100	0.120
R	2.04	2.79	0.080	0.110
S	1.14	1.39	0.045	0.055
T	5.97	6.48	0.235	0.255
U	0.00	1.27	0.000	0.050
V	1.14	–	0.045	–
Z	–	2.03	–	0.080

CASE 221A-02
TO-220AB

TIP120, TIP121, TIP122, NPN, TIP125, TIP126, TIP127, PNP

ELECTRICAL CHARACTERISTICS   (T_C = 25°C unless otherwise noted)

Characteristic		Symbol	Min	Max	Unit
OFF CHARACTERISTICS					
Collector-Emitter Sustaining Voltage (1)		$V_{CEO(sus)}$			Vdc
(I_C = 100 mAdc, I_B = 0) TIP120, TIP125			60	—	
TIP121, TIP126			80	—	
TIP122, TIP127			100	—	
Collector Cutoff Current		I_{CEO}			mAdc
(V_{CE} = 30 Vdc, I_B = 0) TIP120, TIP125			—	0.5	
(V_{CE} = 40 Vdc, I_B = 0) TIP121, TIP126			—	0.5	
(V_{CE} = 50 Vdc, I_B = 0) TIP122, TIP127			—	0.5	
Collector Cutoff Current		I_{CBO}			mAdc
(V_{CB} = 60 Vdc, I_E = 0) TIP120, TIP125			—	0.2	
(V_{CB} = 80 Vdc, I_E = 0) TIP121, TIP126			—	0.2	
(V_{CB} = 100 Vdc, I_E = 0) TIP122, TIP127			—	0.2	
Emitter Cutoff Current		I_{EBO}	—	2.0	mAdc
(V_{BE} = 5.0 Vdc, I_C = 0)					
ON CHARACTERISTICS (1)					
DC Current Gain		h_{FE}		—	—
(I_C = 0.5 Adc, V_{CE} = 3.0 Vdc)			1000	—	
(I_C = 3.0 Adc, V_{CE} = 3.0 Vdc)			1000	—	
Collector-Emitter Saturation Voltage		$V_{CE(sat)}$			Vdc
(I_C = 3.0 Adc, I_B = 12 mAdc)			—	2.0	
(I_C = 5.0 Adc, I_B = 20 mAdc)			—	4.0	
Base-Emitter On Voltage		$V_{BE(on)}$			Vdc
(I_C = 3.0 Adc, V_{CE} = 3.0 Vdc)			—	2.5	
DYNAMIC CHARACTERISTICS					
Small-Signal Current Gain		$|h_{fe}|$			—
(I_C = 3.0 Adc, V_{CE} = 4.0 Vdc, f = 1.0 MHz)			4.0	—	
Output Capacitance		C_{ob}			pF
(V_{CB} = 10 Vdc, I_E = 0, f = 0.1 MHz) TIP125, TIP126, TIP127			—	300	
TIP120, TIP121, TIP122			—	200	

(1) Pulse Test: Pulse Width ≤ 300 μs, Duty Cycle ≤ 2%.

FIGURE 2 – SWITCHING TIMES TEST CIRCUIT

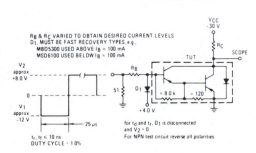

FIGURE 3 – SWITCHING TIMES

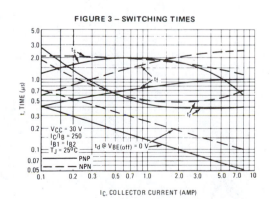

I_C, COLLECTOR CURRENT (AMP)

TIP120, TIP121, TIP122, NPN, TIP125, TIP126, TIP127, PNP

FIGURE 4 — THERMAL RESPONSE

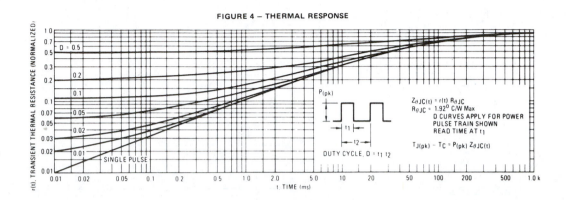

FIGURE 5 — ACTIVE-REGION SAFE OPERATING AREA

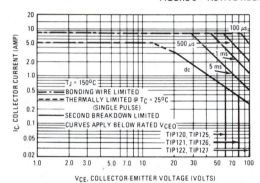

There are two limitations on the power handling ability of a transistor: average junction temperature and second breakdown. Safe operating area curves indicate $I_C - V_{CE}$ limits of the transistor that must be observed for reliable operation; i.e., the transistor must not be subjected to greater dissipation than the curves indicate.

The data of Figure 5 is based on $T_{J(pk)} = 150^\circ C$; T_C is variable depending on conditions. Second breakdown pulse limits are valid for duty cycles to 10% provided $T_{J(pk)} < 150^\circ C$. $T_{J(pk)}$ may be calculated from the data in Figure 4. At high case temperatures, thermal limitations will reduce the power that can be handled to values less than the limitations imposed by second breakdown

FIGURE 6 — SMALL-SIGNAL CURRENT GAIN

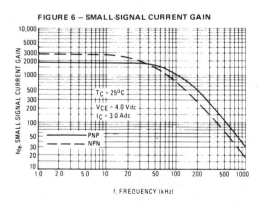

FIGURE 7 — CAPACITANCE

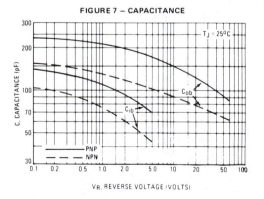

TIP140, TIP141, TIP142 NPN, TIP145, TIP146, TIP147 PNP

ELECTRICAL CHARACTERISTICS (T_C = 25°C unless otherwise noted)

Characteristic	Symbol	Min	Typ	Max	Unit
OFF CHARACTERISTICS					
Collector-Emitter Sustaining Voltage (1)	$V_{CEO(sus)}$				Vdc
(I_C = 30 mA, I_B = 0) TIP140, TIP145		60	—	—	
TIP141, TIP146		80	—	—	
TIP142, TIP147		100	—	—	
Collector Cutoff Current	I_{CEO}				mA
(V_{CE} = 30 Vdc, I_B = 0) TIP140, TIP145		—	—	2.0	
(V_{CE} = 40 Vdc, I_B = 0) TIP141, TIP146		—	—	2.0	
(V_{CE} = 50 Vdc, I_B = 0) TIP142, TIP147		—	—	2.0	
Collector Cutoff Current	I_{CBO}				mA
(V_{CB} = 60 V, I_E = 0) TIP140, TIP145		—	—	1.0	
(V_{CB} = 80 V, I_E = 0) TIP141, TIP146		—	—	1.0	
(V_{CB} = 100 V, I_E = 0) TIP142, TIP147		—	—	1.0	
Emitter Cutoff Current V_{BE} = 5.0 V	I_{EBO}	—	—	2.0	mA
ON CHARACTERISTICS (1)					
DC Current Gain	h_{FE}				—
(I_C = 5.0 A, V_{CE} = 4.0 V)		1000	—	—	
(I_C = 10 A, V_{CE} = 4.0 V)		500	—	—	
Collector-Emitter Saturation Voltage	$V_{CE(sat)}$				Vdc
(I_C = 5.0 A, I_B = 10 mA)		—	—	2.0	
(I_C = 10 A, I_B = 40 mA)		—	—	3.0	
Base-Emitter Saturation Voltage	$V_{BE(sat)}$				Vdc
(I_C = 10 A, I_B = 40 mA)		—	—	3.5	
Base-Emitter On Voltage	$V_{BE(on)}$				Vdc
(I_C = 10 A, V_{CE} = 4.0 Vdc)		—	—	3.0	
SWITCHING CHARACTERISTICS					
Resistive Load (See Figure 1)					
Delay Time (V_{CC} = 30 V, I_C = 5.0 A,	t_d	—	0.15	—	μs
Rise Time I_B = 20 mA, Duty Cycle ⩽ 2.0%,	t_r	—	0.55	—	μs
Storage Time I_{B1} = I_{B2}, R_C & R_B Varied, T_J = 25°C)	t_s	—	2.5	—	μs
Fall Time	t_f	—	2.5	—	μs

(1) Pulse Test: Pulse Width = 300 μs, Duty Cycle ⩽ 2.0%.

FIGURE 1 — SWITCHING TIMES TEST CIRCUIT

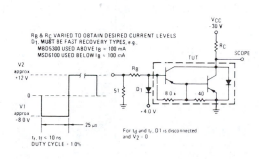

R_B & R_C VARIED TO OBTAIN DESIRED CURRENT LEVELS
D1 MUST BE FAST RECOVERY TYPES, e.g.,
MBD5300 USED ABOVE I_B = 100 mA
MSD6100 USED BELOW I_B = 100 mA

For t_d and t_r, D1 is disconnected and V2 = 0

For NPN test circuit reverse diode and voltage polarities.

FIGURE 2 — SWITCHING TIMES

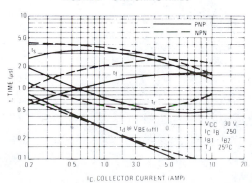

National Semiconductor

RF, VHF, UHF Amplifiers

N-Channel JFETs

Type No.	Case Style	BVGSS (V) Min	@ IG (μA)	IGSS (nA) Max	@ VDG (V)	VP (V) Min	Max	@ VDS (V)	ID (nA)	IDSS (mA) Min	Max	@ VDS (V)	Re|Yfs| (mmho) Min	@ Freq (MHz)	Re(Yos) (μmho) Max	@ f (MHz)	Ciss (pF) Max	@ VDS (V)	VGS (V)	Crss (pF) Max	@ VDS (V)	VGS (V)	NF (dB) @ RG=1k Max	Freq (MHz)	Process No.	Pkg. No.
2N3819	TO-92	25	1	2	15		8	15	2	2	20	15	1.6	100			8	15	0	4	15	0			50	94
2N3823	TO-72	30	1	0.5	20		8	15	0.5	4	20	15	3.2	200	200	200	6	15	0	2	15	0	2.5	100	50	25
2N4223	TO-72	30	10	0.25	20	0.1	8	15	0.25	3	18	15	2.7	200	200	200	6	15	0	2	15	0			50	25
2N4224	TO-72	30	10	0.5	20	0.1	8	15	0.5	2	20	15	1.7	200	200	200	6	15	0	2	15	0	5	200	50	25
2N4416	TO-72	30	1	0.1	20		6	15	1	5	15	15	4	400	100	400	4	15	0	0.8	15	0	4	400	50	25
2N4416A	TO-72	35	1	0.1	20	2.5	6	15	1	5	15	15	4	400	100	400	4	15	0	0.8	15	0	4	400	50	25
2N5078	TO-72	30	1	0.25	20	0.5	8	15		4	25	15	4	200	150	200	6	15	0	2	15	0	3	200	50	25
2N5245	TO-92	30	1	1	20	1	6	15	10	5	15	15	4	400	100	400	4.5	15	0	1	15	0	4	400	90	97
2N5246	TO-92	30	1	1	20	0.5	4	15	10	1.5	7	15	2.5	400	100	400	4.5	15	0	1	15	0			90	97
2N5247	TO-92	30	1	1	20	1.5	8	15	10	8	24	15	4	400	150	400	4.5	15	0	1	15	0			90	97
2N5248	TO-92	30	1	5	20	1	8	15	10	4	20	15	3	200	200	200	6	15	0	2	15	0	3.5	450	50	94
2N5397	TO-72	25	1	0.1	15	1	6	10	1	10	30	10	5.5	450	200	450	5	10	10 mA	1.2	10	10 mA	3.2	450	90	29
2N5398	TO-72	25	1	0.1	15	0.3	6	10	1	5	40	10	5.0	450	400	450	5.5	10	0	1.3	10	0	3	100	90	29
2N5484	TO-92	25	1	1	20	0.3	3	15	10	1	5	15	2.5	100	75	100	5	15	0	1	15	0			50	92
2N5485	TO-92	25	1	1	20	1	4	15	10	4	10	15	3	400	100	400	5	15	0	1	15	0	4	400	50	92
2N5486	TO-92	25	1	1	20	2	6	15	10	8	20	15	3.5	400	100	400	5	15	0	1	15	0	4	400	50	92
2N5468	TO-92	25	10	2	15	0.2	4	14	10	1	5	15	1	100	50	100	7	15	0	3	15	0	2.5	100	50	92
2N5469	TO-92	25	10	2	15	1	6	15	10	4	10	15	1.6	100	100	100	7	15	0	3	15	0	2.5	100	50	92
2N5470	TO-92	25	10	2	15	2	8	15	10	8	20	15	2.5	100	150	100	7	15	0	3	15	0	2.5	100	50	92
2N5949	TO-92	30	1	1	15	3	7	15	100	12	18	15	3.0	100	75	100	6	15	0	2	15	0	5	100	50	97
2N5950	TO-92	30	1	1	15	2.5	6	15	100	10	15	15	3.0	100	75	100	6	15	0	2	15	0	5	100	50	97
2N5951	TO-92	30	1	1	15	2	5	15	100	7	13	15	3.0	100	75	100	6	15	0	2	15	0	5	100	50	97
2N5952	TO-92	30	1	1	15	1.3	3.5	15	100	4	8	15	1.0	100	75	100	6	15	0	2	15	0	5	100	50	97
2N5953	TO-92	30	1	1	15	0.8	3	15	100	2.5	5	15	1.0	100	50	100	6	15	0	2	15	0	5	100	50	97
J300	TO-92	25	1	0.5	15	1	6	10	1	6	30	10	4.5	0.001	200	0.001	5.5	10	5 mA	1.7	10	5 mA			90	92
J304	TO-92	30	1	0.1	20	2	6	15	1	5	15	15	t4.2	400	t80	100	7.5	0	0	2.5	0	0			50	92
J305	TO-92	30	1	0.1	20	0.5	3	15	1	1	8	15	t3.0	400	t80	100	7.5	0	0	2.5	0	0			50	92
J308	TO-92	25	1	1	15	1	6.5	10	1	12	60	10	8	0.001	200	0.001	7.5	0	−10	2.5	0	−10			92	92
J309	TO-92	25	1	1	15	1	4.0	10	1	12	30	10	10	0.001	200	0.001	7.5	0	−10	2.5	0	−10			92	92
J310	TO-92	25	1	1	15	2	6.5	10	1	24	60	10	8	0.001	200	0.001	7.5	0	−10	2.5	0	−10			92	92

t = typical value

Process 50

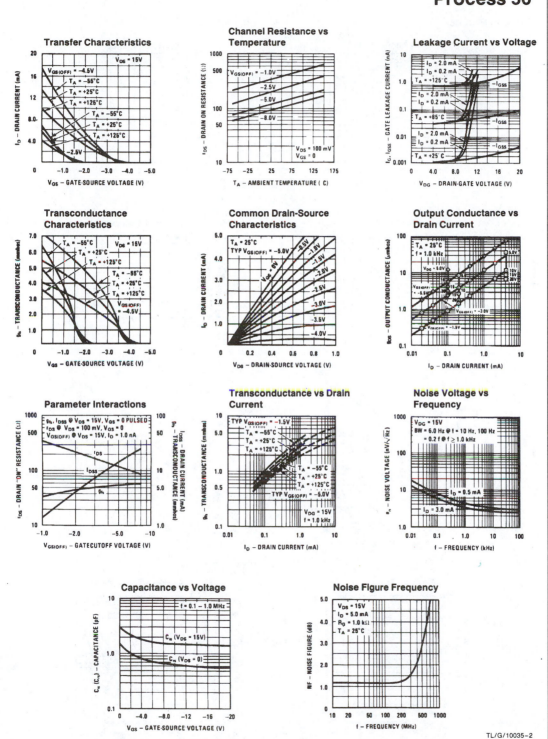

TL/G/10035–2

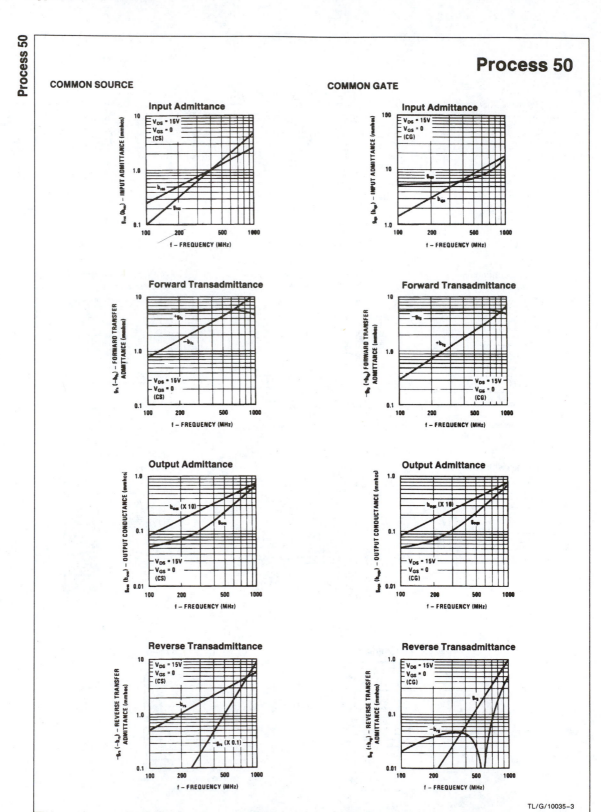

IRF510
IRF511
IRF512
IRF513

 **MOTOROLA**

Advance Information

Part Number	V$_{DS}$	r$_{DS(on)}$	I$_D$
IRF510	100 V	0.6 Ω	4.0 A
IRF511	60 V	0.6 Ω	4.0 A
IRF512	100 V	0.8 Ω	3.5 A
IRF513	60 V	0.8 Ω	3.5 A

N-CHANNEL ENHANCEMENT MODE SILICON GATE TMOS POWER FIELD EFFECT TRANSISTOR

These TMOS Power FETs are designed for low voltage, high speed power switching applications such as switching regulators, converters, solenoid and relay drivers.

- Silicon Gate for Fast Switching Speeds
- Rugged — SOA is Power Dissipation Limited
- Source-to-Drain Diode Characterized for Use With Inductive Loads

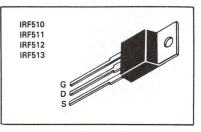

IRF510
IRF511
IRF512
IRF513

 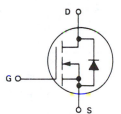

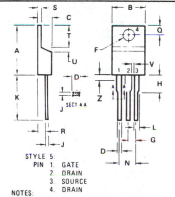

STYLE 5:
PIN 1. GATE
2. DRAIN
3. SOURCE
4. DRAIN

NOTES:
1. DIMENSION H APPLIES TO ALL LEADS.
2. DIMENSION L APPLIES TO LEADS 1 AND 3 ONLY.
3. DIMENSION Z DEFINES A ZONE WHERE ALL BODY AND LEAD IRREGULARITIES ARE ALLOWED.
4. DIMENSIONING AND TOLERANCING PER ANSI Y14.5 1973.
5. CONTROLLING DIMENSION: INCH.

DIM	MILLIMETERS MIN	MAX	INCHES MIN	MAX
A	15.11	15.75	0.595	0.620
B	9.65	10.29	0.380	0.405
C	4.06	4.82	0.160	0.190
D	0.64	0.89	0.025	0.035
F	3.61	3.73	0.142	0.147
G	2.41	2.67	0.095	0.105
H	2.79	3.30	0.110	0.130
J	0.36	0.56	0.014	0.022
K	12.70	14.27	0.500	0.562
L	1.14	1.39	0.045	0.055
N	4.83	5.33	0.190	0.210
Q	2.54	3.04	0.100	0.120
R	2.04	2.79	0.080	0.110
S	1.14	1.39	0.045	0.055
T	5.97	6.48	0.235	0.255
U	0.76	1.27	0.030	0.050
V	1.14		0.045	
Z	—	2.03	—	0.080

CASE 221A-02
TO-220AB

MAXIMUM RATINGS

Rating	Symbol	IRF 510	511	512	513	Unit
Drain-Source Voltage	V$_{DSS}$	100	60	100	60	Vdc
Drain-Gate Voltage (R$_{GS}$ = 1.0 MΩ)	V$_{DGR}$	100	60	100	60	Vdc
Gate-Source Voltage	V$_{GS}$	± 20				Vdc
Continuous Drain Current T$_C$ = 25°C	I$_D$	4.0	4.0	3.5	3.5	Adc
Continuous Drain Current T$_C$ = 100°C	I$_D$	2.5	2.5	2.0	2.0	Adc
Drain Current Pulsed	I$_{DM}$	16	16	14	14	Adc
Gate Current — Pulsed	I$_{GM}$	1.5				Adc
Total Power Dissipation @ T$_C$ = 25°C Derate above 25°C	P$_D$	20 0.16				Watts W/°C
Operating and Storage Temperature Range	T$_J$,T$_{stg}$	− 55 to 150				°C

THERMAL CHARACTERISTICS

Thermal Resistance Junction to Case	R$_{θJC}$	6.4	°C/W
Maximum Lead Temp. for Soldering Purposes, 1/8" from case for 5 seconds	T$_L$	300	°C

IRF510-513

ELECTRICAL CHARACTERISTICS (T_C = 25°C unless otherwise noted)

Characteristic		Symbol	Min	Typ	Max	Unit
OFF CHARACTERISTICS						
Drain-Source Breakdown Voltage (V_{GS} = 0, I_D = 250 μA)	IRF510,512	$V_{(BR)DSS}$	100	—	—	Vdc
	IRF511,513		60	—	—	
Zero Gate Voltage Drain Current (V_{GS} = 0 V, V_{DS} = Rated V_{DSS})		I_{DSS}	—	—	0.25	mAdc
(V_{GS} = 0 V, V_{DS} = 0.8 Rated V_{DSS}, T_C = 125°C)			—	—	1.0	
Forward Gate-Body Leakage Current (V_{GS} = 20 V, V_{DS} = 0)		I_{GSSF}	—	—	100	nAdc
Reverse Gate-Body Leakage Current (V_{GS} = −20 V, V_{DS} = 0)		I_{GSSR}	—	—	−100	nAdc
ON CHARACTERISTICS*						
Gate Threshold Voltage (V_{DS} = V_{GS}, I_D = 250 μA)		$V_{GS(th)}$	2.0	—	4.0	Vdc
On-State Drain Current (V_{DS} = 25 V, V_{GS} = 10 V)	IRF510,511	$I_{D(on)}$	4.0	—	—	Adc
	IRF512,513		3.5	—	—	
Static Drain-Source On-Resistance (V_{GS} = 10 V, I_D = 2.0 A)	IRF510,511	$r_{DS(on)}$	—	—	0.6	Ohms
	IRF512,513		—	—	0.8	
Forward Transconductance (V_{DS} = 15 V, I_D = 2.0 A)		g_{fs}	1.0	—	—	mhos
DYNAMIC CHARACTERISTICS						
Input Capacitance	(V_{DS} = 25 V, V_{GS} = 0, f = 1.0 MHz)	C_{iss}	—	—	150	pF
Output Capacitance		C_{oss}	—	—	100	
Reverse Transfer Capacitance		C_{rss}	—	—	25	
SWITCHING CHARACTERISTICS* (T_J = 100°C)						
Turn-On Delay Time	$V_{DD} \approx$ 0.5 V_{DSS}, I_D = 2.0 A Z_o = 50 Ω	$t_{d(on)}$	—	—	20	ns
Rise Time		t_r	—	—	25	
Turn-Off Delay Time		$t_{d(off)}$	—	—	25	
Fall Time		t_f	—	—	20	

SOURCE DRAIN DIODE CHARACTERISTICS*

Characteristic		Symbol	Typ	Unit
Forward On-Voltage	(I_S = Rated I_D, V_{GS} = 0)	V_{SD}	2.0	Vdc
Reverse Recovery Time		t_{rr}	230	ns

*Pulse Test: Pulse Width ≤ 300 μs, Duty Cycle ≤ 2.0 %.

RESISTIVE SWITCHING

FIGURE 1 — SWITCHING TEST CIRCUIT   **FIGURE 2 — SWITCHING WAVEFORMS**

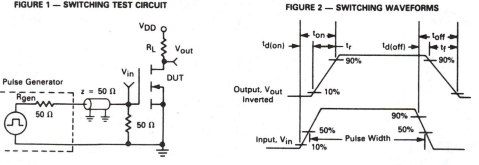

 **MOTOROLA**

MLED930

INFRARED-EMITTING DIODE
900 nm
PN GALLIUM ARSENIDE
250 MILLIWATTS

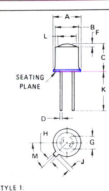

CONVEX LENS

INFRARED-EMITTING DIODE

. . . designed for applications requiring high power output, low drive power and very fast response time. This device is used in industrial processing and control, light modulators, shaft or position encoders, punched card readers, optical switching, and logic circuits. It is spectrally matched for use with silicon detectors.

- High-Power Output — 650 μW (Typ) @ I_F = 100 mA
- Infrared-Emission — 900 nm (Typ)
- Low Drive Current — 10 mA for 70 μW (Typ)
- Popular TO-18 Type Package for Easy Handling and Mounting
- Hermetic Metal Package for Stability and Reliability

MAXIMUM RATINGS

Rating	Symbol	Value	Unit
Reverse Voltage	V_R	6 0	Volts
Forward Current-Continuous	I_F	150	mA
Total Device Dissipation @ T_A = 25°C Derate above 25°C	$P_D(1)$	250 2.5	mW mW/°C
Operating and Storage Junction Temperature Range	T_J, T_{stg}	-65 to +125	°C

THERMAL CHARACTERISTICS

Characteristics	Symbol	Max	Unit
Thermal Resistance, Junction to Ambient	θ_{JA}	400	°C/W

(1) Printed Circuit Board Mounting

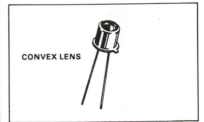

STYLE 1:

PIN 1. ANODE
PIN 2. CATHODE

NOTES:
1. PIN 2 INTERNALLY CONNECTED TO CASE
2. LEADS WITHIN 0.13 mm (0.005) RADIUS OF TRUE POSITION AT SEATING PLANE AT MAXIMUM MATERIAL CONDITION.

DIM	MILLIMETERS		INCHES	
	MIN	MAX	MIN	MAX
A	5.31	5.84	0.209	0.230
B	4.52	4.95	0.178	0.195
C	5.08	6.35	0.200	0.250
D	0.41	0.48	0.016	0.019
F	0.51	1.02	0.020	0.040
G	2.54 BSC		0.100 BSC	
H	0.99	1.17	0.039	0.046
J	0.84	1.22	0.033	0.048
K	12.70	–	0.500	–
L	3.35	4.01	0.132	0.158
M	45° BSC		45° BSC	

CASE 209-01

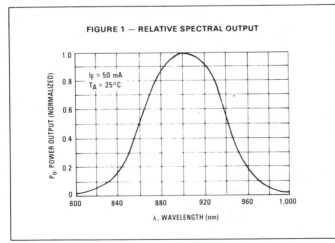

FIGURE 1 — RELATIVE SPECTRAL OUTPUT

I_F = 50 mA
T_A = 25°C

MLED930

ELECTRICAL CHARACTERISTICS (T_A = 25°C unless otherwise noted)

Characteristic	Fig. No.	Symbol	Min	Typ	Max	Unit
Reverse Leakage Current (V_R = 3.0 V)	—	I_R	—	2.0	—	nA
Reverse Breakdown Voltage (I_R = 100 μA)	—	$V_{(BR)R}$	6.0	20	—	Volts
Forward Voltage (I_F = 50 mA)	2	V_F	—	1.25	1.5	Volts
Total Capacitance (V_R = 0 V, f = 1.0 MHz)	—	C_T	—	150	—	pF

OPTICAL CHARACTERISTICS (T_A = 25°C unless otherwise noted)

Characteristic	Fig. No.	Symbol	Min	Typ	Max	Unit
Total Power Output (Note 1) (I_F = 100 mA)	3, 4	P_O	200	650	—	μW
Radiant Intensity (Note 2) (I_F = 100 mA)	—	I_O	—	1.5	—	mW/steradian
Peak Emission Wavelength	1	λP	—	900	—	nm
Spectral Line Half Width	1	$\Delta \lambda$	—	40	—	nm

NOTE:
1. Power Output, P_O, is the total power radiated by the device into a solid angle of 2π steradians. It is measured by directing all radiation leaving the device, within this solid angle, onto a calibrated silicon solar cell.
2. Irradiance from a Light Emitting Diode (LED) can be calculated by:

$H = \dfrac{I_O}{d^2}$ where H is irradiance in mW/cm²; I_O is radiant intensity in mW/steradian; d^2 is distance from LED to the detector in cm.

FIGURE 2 — FORWARD CHARACTERISTICS

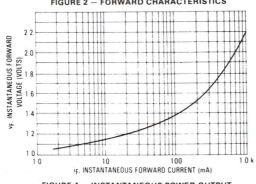

FIGURE 3 — POWER OUTPUT versus JUNCTION TEMPERATURE

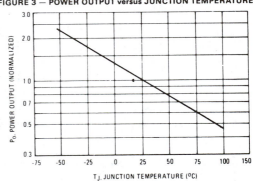

FIGURE 4 — INSTANTANEOUS POWER OUTPUT versus FORWARD CURRENT

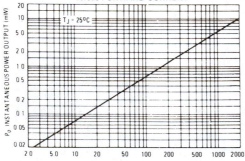

FIGURE 5 — SPATIAL RADIATION PATTERN

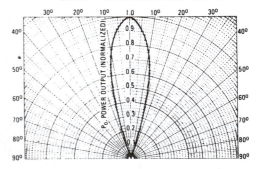

 **MOTOROLA**

4N25, 4N25A
4N26
4N27
4N28

OPTO COUPLER/ISOLATOR

TRANSISTOR OUTPUT

NPN PHOTOTRANSISTORS AND
PN INFRARED EMITTING DIODES

. . . gallium arsenide LED optically coupled to silicon phototransistors designed for applications requiring electrical isolation, high-current transfer ratios, small package size and low cost; such as interfacing and coupling systems, phase and feedback controls, solid-state relays and general-purpose switching circuits.

- High Isolation Voltage —
  $V_{ISO} = 7500$ V (Min)

- High Collector Output Current
  @ $I_F = 10$ mA —
  $I_C = 5.0$ mA (Typ) — 4N25,A,4N26
  2.0 mA (Typ) — 4N27,4N28

- Economical, Compact, Dual-In-Line Package

- Excellent Frequency Response —
  300 kHz (Typ)

- Fast Switching Times @ $I_C = 10$ mA
  $t_{on} = 0.87$ μs (Typ) — 4N25,A,4N26
  2.1 μs (Typ) — 4N27,4N28
  $t_{off} = 11$ μs (Typ) — 4N25,A,4N26
  5.0 μs (Typ) — 4N27,4N28

- 4N25A is UL Recognized
  File Number E54915

*MAXIMUM RATINGS ($T_A = 25^{o}$C unless otherwise noted).

Rating	Symbol	Value	Unit
INFRARED-EMITTING DIODE MAXIMUM RATINGS			
Reverse Voltage	V_R	3.0	Volts
Forward Current — Continuous	I_F	80	mA
Forward Current — Peak Pulse Width = 300 μs, 2.0% Duty Cycle	I_F	3.0	Amp
Total Power Dissipation @ $T_A = 25^{o}$C Negligible Power in Transistor Derate above 25°C	P_D	150 2.0	mW mW/^{o}C
PHOTOTRANSISTOR MAXIMUM RATINGS			
Collector-Emitter Voltage	V_{CEO}	30	Volts
Emitter-Collector Voltage	V_{ECO}	7.0	Volts
Collector-Base Voltage	V_{CBO}	70	Volts
Total Device Dissipation @ $T_A = 25^{o}$C Negligible Power in Diode Derate above 25°C	P_D	150 2.0	mW mW/^{o}C
TOTAL DEVICE RATINGS			
Total Device Dissipation @ $T_A = 25^{o}$C	P_D	250	mW
Equal Power Dissipation in Each Element Derate above 25°C		3.3	mW/^{o}C
Junction Temperature Range	T_J	−55 to +100	^{o}C
Storage Temperature Range	T_{stg}	−55 to +150	^{o}C
Soldering Temperature (10 s)		260	^{o}C

*Indicates JEDEC Registered Data.

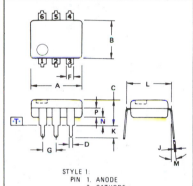

STYLE 1:
PIN 1. ANODE
2. CATHODE
3. NC
4. EMITTER
5. COLLECTOR
6. BASE

NOTES:
1. DIMENSIONS A AND B ARE DATUMS.
2. -T- IS SEATING PLANE.
3. POSITIONAL TOLERANCES FOR LEADS:
   ⊕ ⌀ 0.13 (0.005) Ⓜ T AⓂ BⓂ
4. DIMENSION L TO CENTER OF LEADS WHEN FORMED PARALLEL.
5. DIMENSIONING AND TOLERANCING PER ANSI Y14.5, 1973.

DIM	MILLIMETERS MIN	MILLIMETERS MAX	INCHES MIN	INCHES MAX
A	8.13	8.89	0.320	0.350
B	6.10	6.60	0.240	0.260
C	2.92	5.08	0.115	0.200
D	0.41	0.51	0.016	0.020
F	1.02	1.78	0.040	0.070
G	2.54 BSC		0.100 BSC	
J	0.20	0.30	0.008	0.012
K	2.54	3.81	0.100	0.150
L	7.62 BSC		0.300 BSC	
M	0°	15°	0°	15°
N	0.38	2.54	0.015	0.100
P	1.27	2.03	0.050	0.080

CASE 730A-01

FIGURE 1 — MAXIMUM POWER DISSIPATION

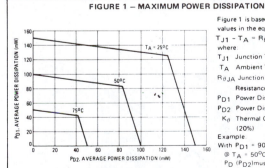

Figure 1 is based upon using limit values in the equation:

$$T_{J1} - T_A = R_{\theta JA}(P_{D1} + K_{\theta} P_{D2})$$

where:

T_{J1} Junction Temperature (100°C)

T_A Ambient Temperature

$R_{\theta JA}$ Junction to Ambient Thermal Resistance (500°C/W)

P_{D1} Power Dissipation in One Chip

P_{D2} Power Dissipation in Other Chip

K_{θ} Thermal Coupling Coefficient (20%)

Example:
With $P_{D1} = 90$ mW in the LED @ $T_A = 50^{o}$C, the transistor P_D (P_{D2}) must be less than 50 mW.

4N25, 4N25A, 4N26, 4N27, 4N28

LED CHARACTERISTICS ($T_A = 25°C$ unless otherwise noted)

Characteristic	Symbol	Min	Typ	Max	Unit
*Reverse Leakage Current ($V_R = 3.0$ V, $R_L = 1.0$ M ohms)	I_R	–	0.005	100	μA
*Forward Voltage ($I_F = 10$ mA)	V_F	–	1.2	1.5	Volts
Capacitance ($V_R = 0$ V, $f = 1.0$ MHz)	C	–	40	–	pF

PHOTOTRANSISTOR CHARACTERISTICS ($T_A = 25°C$ and $I_F = 0$ unless otherwise noted)

Characteristic		Symbol	Min	Typ	Max	Unit
*Collector-Emitter Dark Current ($V_{CE} = 10$ V, Base Open)	4N25, A, 4N26, 4N27	I_{CEO}	–	3.5	50	nA
	4N28		–	–	100	
*Collector-Base Dark Current ($V_{CB} = 10$ V, Emitter Open)		I_{CBO}	–	–	20	nA
*Collector-Base Breakdown Voltage ($I_C = 100 \mu A$, $I_E = 0$)		$V_{(BR)CBO}$	70	–	–	Volts
*Collector-Emitter Breakdown Voltage ($I_C = 1.0$ mA, $I_B = 0$)		$V_{(BR)CEO}$	30	–	–	Volts
*Emitter-Collector Breakdown Voltage ($I_E = 100 \mu A$, $I_B = 0$)		$V_{(BR)ECO}$	7.0	8.0	–	Volts
DC Current Gain ($V_{CE} = 5.0$ V, $I_C = 500 \mu A$)		h_{FE}	–	325	–	–

COUPLED CHARACTERISTICS ($T_A = 25°C$ unless otherwise noted)

Characteristic		Symbol	Min	Typ	Max	Unit
*Collector Output Current (1) ($V_{CE} = 10$ V, $I_F = 10$ mA, $I_B = 0$)	4N25, A, 4N26	I_C	2.0	5.0	–	mA
	4N27, 4N28		1.0	2.0	–	
Isolation Surge Voltage (2, 5) (60 Hz Peak ac, 5 Seconds)		V_{ISO}	7500	–	–	Volts
(60 Hz Peak)	*4N25, A		2500	–	–	
	*4N26, 4N27		1500	–	–	
	*4N28		500	–	–	
(60 Hz RMS for 1 Second) (3)	*4N25A		1775	–	–	
Isolation Resistance (2) (V = 500 V)		–	–	10^{11}	–	Ohms
*Collector-Emitter Saturation ($I_C = 2.0$ mA, $I_F = 50$ mA)		$V_{CE(sat)}$	–	0.2	0.5	Volts
Isolation Capacitance (2) (V = 0, $f = 1.0$ MHz)		–	–	0.5	–	pF
Bandwidth (4) ($I_C = 2.0$ mA, $R_L = 100$ ohms, Figure 11 (2))		–	–	300	–	kHz

SWITCHING CHARACTERISTICS

Characteristic			Symbol	Min	Typ	Max	Unit
Delay Time	($I_C = 10$ mA, $V_{CC} = 10$ V	4N25, A, 4N26	t_d	–	0.07	–	μs
		2N27, 4N28		–	0.10	–	
Rise Time	Figures 6 and 8)	4N25, A, 4N26	t_r	–	0.8	–	μs
		4N27, 4N28		–	2.0	–	
Storage Time	($I_C = 10$ mA, $V_{CC} = 10$ V	4N25, A, 4N26	t_s	–	4.0	–	μs
		4N27, 4N28		–	2.0	–	
Fall Time	Figures 7 and 8)	4N25, A, 4N26	t_f	–	8.0	–	μs
		4N27, 4N28		–	8.0	–	

*Indicates JEDEC Registered Data
(1) Pulse Test: Pulse Width = 300 μs, Duty Cycle $\leqslant$ 2.0%.
(2) For this test LED pins 1 and 2 are common and phototransistor pins 4, 5, and 6 are common.
(3) RMS Volts, 60 Hz. For this test, pins 1, 2, and 3 are common and pins 4, 5, and 6 are common.
(4) I_F adjusted to yield $I_C = 2.0$ mA and $i_c = 2.0$ mA p-p at 10 kHz.
(5) Isolation Surge Voltage, V_{ISO}, is an internal device dielectric breakdown rating.

DC CURRENT TRANSFER CHARACTERISTICS

FIGURE 2 – 4N25,A,4N26

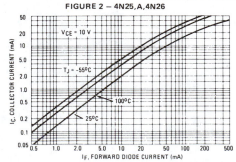

FIGURE 3 – 4N27,4N28

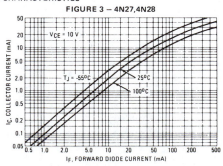

4N25, 4N25A, 4N26, 4N27, 4N28

TYPICAL ELECTRICAL CHARACTERISTICS

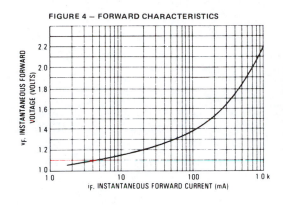

FIGURE 4 — FORWARD CHARACTERISTICS

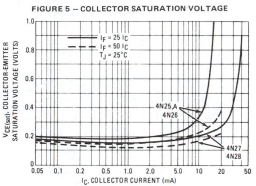

FIGURE 5 — COLLECTOR SATURATION VOLTAGE

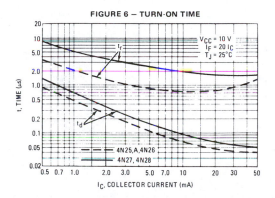

FIGURE 6 — TURN-ON TIME

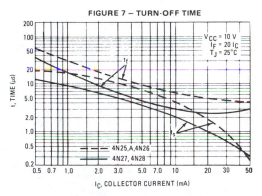

FIGURE 7 — TURN-OFF TIME

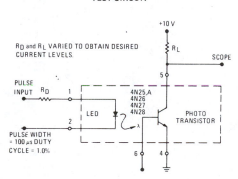

FIGURE 8 — SATURATED SWITCHING TIME TEST CIRCUIT

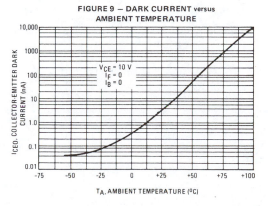

FIGURE 9 — DARK CURRENT versus AMBIENT TEMPERATURE

4N25, 4N25A, 4N26, 4N27, 4N28

FIGURE 11 – FREQUENCY RESPONSE TEST CIRCUIT

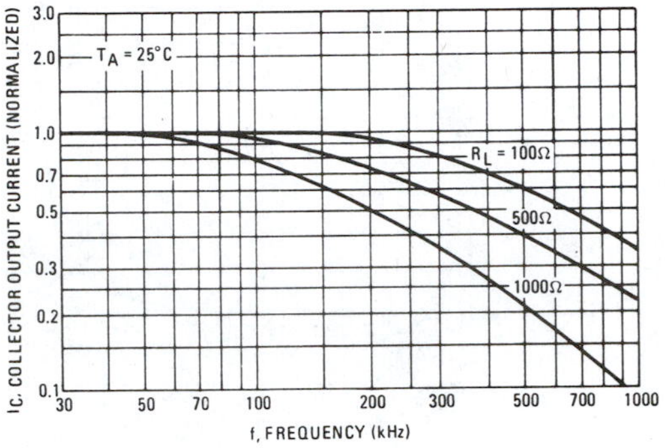

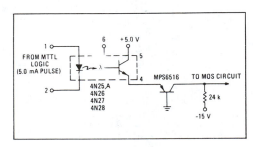

FIGURE 10 – FREQUENCY RESPONSE

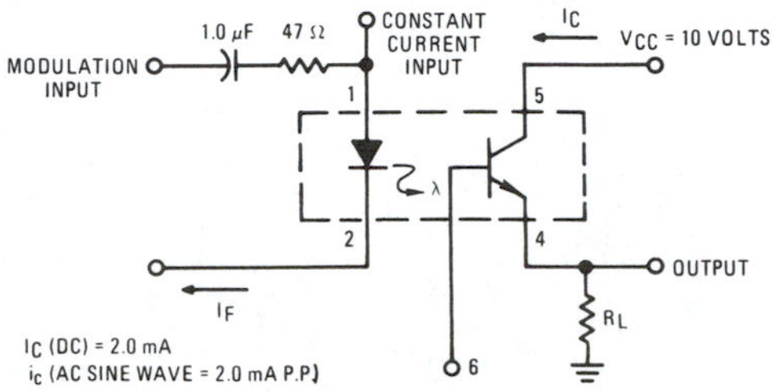

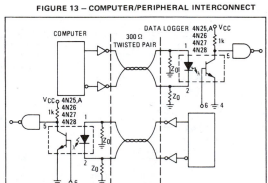

I_C (DC) = 2.0 mA
i_c (AC SINE WAVE = 2.0 mA P.P.)

TYPICAL APPLICATIONS

FIGURE 12 – ISOLATED MTTL
TO MOS (P-CHANNEL) LEVEL TRANSLATOR

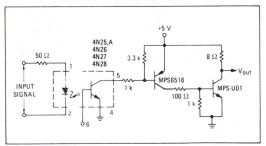

FIGURE 13 – COMPUTER/PERIPHERAL INTERCONNECT

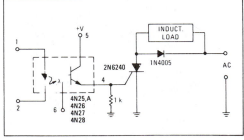

FIGURE 14 – POWER AMPLIFIER

FIGURE 15 – INTERFACE BETWEEN LOGIC AND LOAD

MOTOROLA

MRD300
MRD310

**50 VOLT
PHOTOTRANSISTOR
NPN SILICON**

250 MILLIWATTS

NPN SILICON HIGH SENSITIVITY PHOTOTRANSISTORS

. . . designed for application in industrial inspection, processing and control, counters, sorters, switching and logic circuits or any design requiring radiation sensitivity, and stable characteristics.

- Popular TO-18 Type Package for Easy Handling and Mounting
- Sensitive Throughout Visible and Near Infrared Spectral Range for Wider Application
- Minimum Light Current 4 mA at $H = 5$ mW/cm$_2$ (MRD300)
- External Base for Added Control
- Annular Passivated Structure for Stability and Reliability

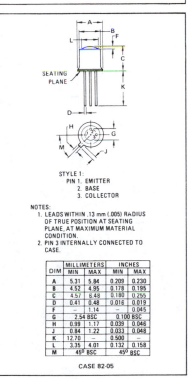

MAXIMUM RATINGS ($T_A = 25^oC$ unless otherwise noted)

Rating (Note 1)	Symbol	Value	Unit
Collector-Emitter Voltage	V_{CEO}	50	Volts
Emitter-Collector Voltage	V_{ECO}	7.0	Volts
Collector-Base Voltage	V_{CBO}	80	Volts
Total Device Dissipation @ $T_A = 25^oC$ Derate above 25^oC	P_D	250 1.43	mW mW/^oC
Operating Junction and Storage Temperature Range	T_J, T_{stg}	–65 to +200	^oC

STYLE 1:
PIN 1. EMITTER
2. BASE
3. COLLECTOR

NOTES:
1. LEADS WITHIN .13 mm (.005) RADIUS OF TRUE POSITION AT SEATING PLANE, AT MAXIMUM MATERIAL CONDITION.
2. PIN 3 INTERNALLY CONNECTED TO CASE.

DIM	MILLIMETERS		INCHES	
	MIN	MAX	MIN	MAX
A	5.31	5.84	0.209	0.230
B	4.52	4.95	0.178	0.195
C	4.57	6.48	0.180	0.255
D	0.41	0.48	0.016	0.019
F	–	1.14	–	0.045
G	2.54 BSC		0.100 BSC	
H	0.99	1.17	0.039	0.046
J	0.84	1.22	0.033	0.048
K	12.70	–	0.500	–
L	3.35	4.01	0.132	0.158
M	45° BSC		45° BSC	

CASE 82-05

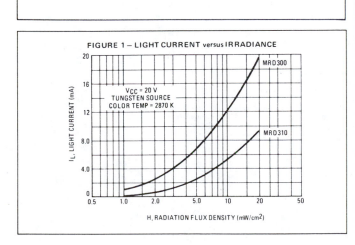

FIGURE 1 – LIGHT CURRENT versus IRRADIANCE

MRD300, MRD310

STATIC ELECTRICAL CHARACTERISTICS ($T_A = 25^{\circ}C$ unless otherwise noted)

Characteristic	Symbol	Min	Typ	Max	Unit
Collector Dark Current ($V_{CC} = 20$ V, $H \approx 0$) $T_A = 25^{\circ}C$ $T_A = 100^{\circ}C$	I_{CEO}	– –	5.0 4.0	25 –	na μA
Collector-Base Breakdown Voltage ($I_C = 100$ μA)	$V_{(BR)CBO}$	80	120	–	Volts
Collector-Emitter Breakdown Voltage ($I_C = 100$ μA)	$V_{(BR)CEO}$	50	85	–	Volts
Emitter-Collector Breakdown Voltage ($I_E = 100$ μA)	$V_{(BR)ECO}$	7.0	8.5	–	Volts

OPTICAL CHARACTERISTICS ($T_A = 25^{\circ}C$ unless otherwise noted)

Characteristic	Device Type	Symbol	Min	Typ	Max	Unit
Light Current ($V_{CC} = 20$ V, $R_L = 100$ ohms) Note 1	MRD300 MRD310	I_L	4.0 1.0	8.0 3.5	– –	mA
Light Current ($V_{CC} = 20$ V, $R_L = 100$ ohms) Note 2	MRD300 MRD310	I_L	– –	2.5 0.8	– –	mA
Photo Current Rise Time (Note 3) ($R_L = 100$ ohms $I_L = 1.0$ mA peak)		t_r	–	2.0	2.5	μs
Photo Current Fall Time (Note 3) ($R_L = 100$ ohms $I_L = 1.0$ mA peak)		t_f	–	2.5	4.0	μs

NOTES:

1. Radiation flux density (H) equal to 5.0 mW/cm^2 emitted from a tungsten source at a color temperature of 2870 K.
2. Radiation flux density (H) equal to 0.5 mW/cm^2 (pulsed) from a GaAs (gallium-arsenide) source at $\lambda \approx 0.9$ μm.
3. For unsaturated response time measurements, radiation is provided by pulsed GaAs (gallium-arsenide) light-emitting diode ($\lambda \approx 0.9$ μm) with a pulse width equal to or greater than 10 microseconds (see Figure 6) $I_L = 1.0$ mA peak.

MRD300, MRD310

TYPICAL ELECTRICAL CHARACTERISTICS

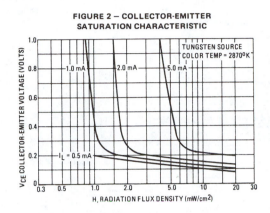

FIGURE 2 — COLLECTOR-EMITTER
SATURATION CHARACTERISTIC

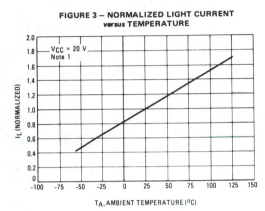

FIGURE 3 — NORMALIZED LIGHT CURRENT
versus TEMPERATURE

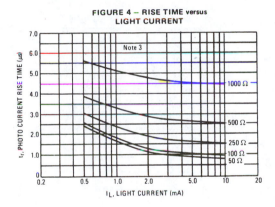

FIGURE 4 — RISE TIME versus
LIGHT CURRENT

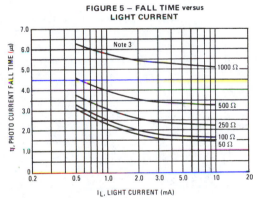

FIGURE 5 — FALL TIME versus
LIGHT CURRENT

FIGURE 6 — PULSE RESPONSE TEST CIRCUIT AND WAVEFORM

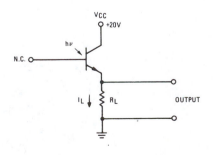

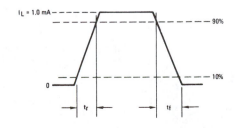

MRD300, MRD310

FIGURE 7 — DARK CURRENT versus TEMPERATURE

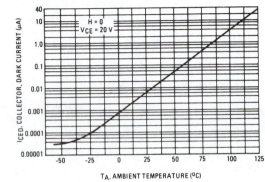

FIGURE 8 — CONSTANT ENERGY SPECTRAL RESPONSE

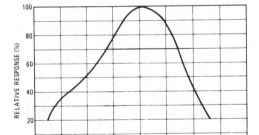

FIGURE 9 — ANGULAR RESPONSE

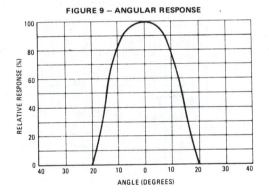

 **MOTOROLA**

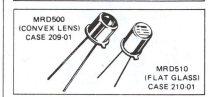

MRD500
MRD510

PIN SILICON PHOTO DIODES

... designed for application in laser detection, light demodulation, detection of visible and near infrared light-emitting diodes, shaft or position encoders, switching and logic circuits, or any design requiring radiation sensitivity, ultra high-speed, and stable characteristics.

- Ultra Fast Response — (<1.0 ns Typ)

- High Sensitivity — MRD500 (1.2 μA/mW/cm^2 Min)
  MRD510 (0.3 μA/mW/cm^2 Min)

- Available With Convex Lens (MRD500) or Flat Glass (MRD510) for Design Flexibility

- Popular TO-18 Type Package for Easy Handling and Mounting

- Sensitive Throughout Visible and Near Infrared Spectral Range for Wide Application

- Annular Passivated Structure for Stability and Reliability

PHOTO DIODES
PIN SILICON
100 VOLTS
100 MILLIWATTS

MRD500
(CONVEX LENS)
CASE 209-01

MRD510
(FLAT GLASS)
CASE 210-01

MAXIMUM RATINGS (T$_A$ = 25°C unless otherwise noted)

Rating	Symbol	Value	Unit
Reverse Voltage	V$_R$	100	Volts
Total Power Dissipation @ T$_A$ = 25°C Derate above 25°C	P$_D$	100 0.57	mW mW/^{o}C
Operating and Storage Junction Temperature Range	T$_J$,T$_{stg}$	-65 to +200	^{o}C

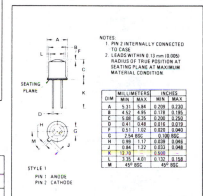

NOTES:
1. PIN 2 INTERNALLY CONNECTED TO CASE
2. LEADS WITHIN 0.13 mm (0.005) RADIUS OF TRUE POSITION AT SEATING PLANE AT MAXIMUM MATERIAL CONDITION.

DIM	MILLIMETERS		INCHES	
	MIN	MAX	MIN	MAX
A	5.31	5.84	0.209	0.230
B	4.52	4.95	0.178	0.195
C	5.08	6.35	0.200	0.250
D	0.41	0.48	0.016	0.019
F	0.51	1.02	0.020	0.040
G	2.54 BSC		0.100 BSC	
H	0.99	1.17	0.039	0.046
J	0.84	1.22	0.033	0.048
K	12.70	–	0.500	–
L	3.35	4.01	0.132	0.158
M	45° BSC		45° BSC	

STYLE 1
PIN 1 ANODE
PIN 2 CATHODE

CASE 209-01

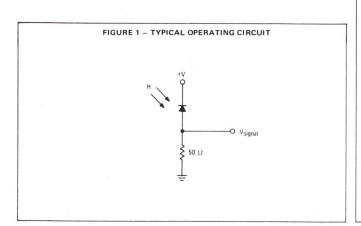

FIGURE 1 – TYPICAL OPERATING CIRCUIT

+V

H

V$_{signal}$

50 Ω

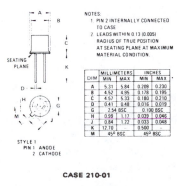

NOTES:
1. PIN 2 INTERNALLY CONNECTED TO CASE
2. LEADS WITHIN 0.13 (0.005) RADIUS OF TRUE POSITION AT SEATING PLANE AT MAXIMUM MATERIAL CONDITION.

DIM	MILLIMETERS		INCHES	
	MIN	MAX	MIN	MAX
A	5.31	5.84	0.209	0.230
B	4.52	4.95	0.178	0.195
C	4.57	5.33	0.180	0.210
D	0.41	0.48	0.016	0.019
G	2.54 BSC		0.100 BSC	
H	0.99	1.17	0.039	0.046
J	0.84	1.22	0.033	0.048
K	12.70		0.500	–
M	45° BSC		45° BSC	

STYLE 1
PIN 1 ANODE
    2 CATHODE

CASE 210-01

MRD500, MRD510

STATIC ELECTRICAL CHARACTERISTICS (T_A = 25°C unless otherwise noted)

Characteristic	Fig. No.	Symbol	Min	Typ	Max	Unit
Dark Current (V_R = 20 V, R_L = 1.0 megohm; Note 2) T_A = 25°C T_A = 100°C	4 and 5	I_D		– –	– 14	2.0 –	nA
Reverse Breakdown Voltage (I_R = 10 μA)	–	$V_{(BR)R}$	100	300	–	Volts
Forward Voltage (I_F = 50 mA)	–	V_F	–	0.82	1.1	Volts
Series Resistance (I_F = 50 mA)	–	R_s	–	1.2	10	ohms
Total Capacitance (V_R = 20 V; f = 1.0 MHz)	6	C_T	–	2.5	4	pF

OPTICAL CHARACTERISTICS (T_A = 25°C)

Characteristic		Fig. No.	Symbol	Min	Typ	Max	Unit
Radiation Sensitivity (V_R = 20 V, Note 1)	MRD500 MRD510	2 and 3	S_R	1.2 0.3	3.0 0.42	– –	μA/mW/cm^2
Sensitivity at 0.8 μm (V_R = 20 V, Note 3)	MRD500 MRD510	– –	$S_{(\lambda = 0.8\,\mu m)}$	–	6.6 1.5	– –	μA/mW/cm^2
Response Time (V_R = 20 V, R_L = 50 ohms)		– –	$t_{(resp)}$		1.0		ns
Wavelength of Peak Spectral Response		7	λ_s	–	0.8	–	μm

NOTES:

1. Radiation Flux Density (H) equal to 5.0 mW/cm^2 emitted from a tungsten source at a color temperature of 2870°K.

2. Measured under dark conditions. (H ≈ 0).

3. Radiation Flux Density (H) equal to 0.5 mW/cm^2 at 0.8 μm.

MRD500, MRD510

TYPICAL ELECTRICAL CHARACTERISTICS

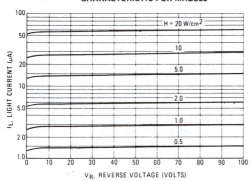

FIGURE 2 – IRRADIATED VOLTAGE – CURRENT
CHARACTERISTIC FOR MRD500

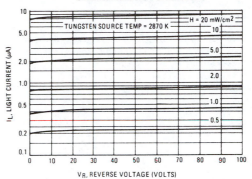

FIGURE 3 – IRRADIATED VOLTAGE – CURRENT
CHARACTERISTIC FOR MRD 510

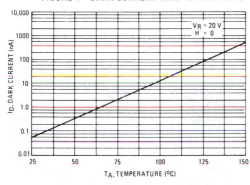

FIGURE 4 – DARK CURRENT versus TEMPERATURE

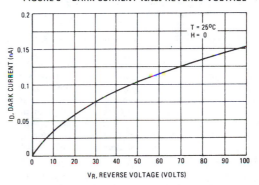

FIGURE 5 – DARK CURRENT versus REVERSE VOLTAGE

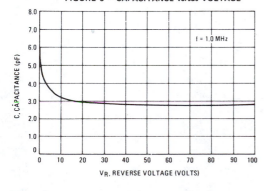

FIGURE 6 – CAPACITANCE versus VOLTAGE

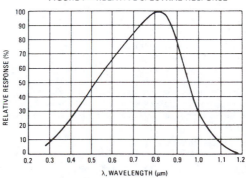

FIGURE 7 – RELATIVE SPECTRAL RESPONSE

ANALOG DEVICES

Two-Terminal IC Temperature Transducer
AD590*

FEATURES
Linear Current Output: 1μA/K
Wide Range: –55°C to +150°C
Probe Compatible Ceramic Sensor Package
Two-Terminal Device: Voltage In/Current Out
Laser Trimmed to ±0.5°C Calibration Accuracy (AD590M)
Excellent Linearity: ±0.3°C Over Full Range (AD590M)
Wide Power Supply Range: +4V to +30V
Sensor Isolation from Case
Low Cost

AD590 PIN DESIGNATIONS

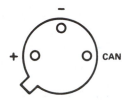

BOTTOM VIEW

PRODUCT DESCRIPTION

The AD590 is a two-terminal integrated circuit temperature transducer which produces an output current proportional to absolute temperature. For supply voltages between +4V and +30V the device acts as a high impedance, constant current regulator passing 1μA/K. Laser trimming of the chip's thin film resistors is used to calibrate the device to 298.2μA output at 298.2K (+25°C).

The AD590 should be used in any temperature sensing application below +150°C in which conventional electrical temperature sensors are currently employed. The inherent low cost of a monolithic integrated circuit combined with the elimination of support circuitry makes the AD590 an attractive alternative for many temperature measurement situations. Linearization circuitry, precision voltage amplifiers, resistance measuring circuitry and cold junction compensation are not needed in applying the AD590.

In addition to temperature measurement, applications include temperature compensation or correction of discrete components, biasing proportional to absolute temperature, flow rate measurement, level detection of fluids and anemometry. The AD590 is available in chip form making it suitable for hybrid circuits and fast temperature measurements in protected environments.

The AD590 is particularly useful in remote sensing applications. The device is insensitive to voltage drops over long lines due to its high impedance current output. Any well-insulated twisted pair is sufficient for operation hundreds of feet from the receiving circuitry. The output characteristics also make the AD590 easy to multiplex: the current can be switched by a CMOS multiplexer or the supply voltage can be switched by a logic gate output.

*Covered by Patent No. 4,123,698.

PRODUCT HIGHLIGHTS

1. The AD590 is a calibrated two terminal temperature sensor requiring only a dc voltage supply (+4V to +30V). Costly transmitters, filters, lead wire compensation and linearization circuits are all unnecessary in applying the device.

2. State-of-the-art laser trimming at the wafer level in conjunction with extensive final testing insures that AD590 units are easily interchangeable.

3. Superior interference rejection results from the output being a current rather than a voltage. In addition, power requirements are low (1.5mW's @ 5V @ +25°C). These features make the AD590 easy to apply as a remote sensor.

4. The high output impedance ($>$10MΩ) provides excellent rejection of supply voltage drift and ripple. For instance, changing the power supply from 5V to 10V results in only a 1μA maximum current change, or 1°C equivalent error.

5. The AD590 is electrically durable: it will withstand a forward voltage up to 44V and a reverse voltage of 20V. Hence, supply irregularities or pin reversal will not damage the device.

SPECIFICATIONS (@ +25℃ and V_S = 5V unless otherwise noted)

Model	AD590J Min	AD590J Typ	AD590J Max	AD590K Min	AD590K Typ	AD590K Max	Units
ABSOLUTE MAXIMUM RATINGS							
Forward Voltage (E + to E −)			+ 44			+ 44	Volts
Reverse Voltage (E + to E −)			− 20			− 20	Volts
Breakdown Voltage (Case to E + or E −)			± 200			± 200	Volts
Rated Performance Temperature Range[1]	− 55		+ 150	− 55		+ 150	℃
Storage Temperature Range[1]	− 65		+ 155	− 65		+ 155	℃
Lead Temperature (Soldering, 10 sec)			+ 300			+ 300	℃
POWER SUPPLY							
Operating Voltage Range	+ 4		+ 30	+ 4		+ 30	Volts
OUTPUT							
Nominal Current Output @ + 25℃ (298.2K)		298.2			298.2		μA
Nominal Temperature Coefficient		1			1		μA/K
Calibration Error @ + 25℃			± 5.0			± 2.5	℃
Absolute Error (over rated performance temperature range)							
Without External Calibration Adjustment			± 10			± 5.5	℃
With + 25℃ Calibration Error Set to Zero			± 3.0			± 2.0	℃
Nonlinearity			± 1.5			± 0.8	℃
Repeatability[2]			± 0.1			± 0.1	℃
Long Term Drift[3]			± 0.1			± 0.1	℃
Current Noise		40			40		pA/√Hz
Power Supply Rejection							
+ 4V ≤ V_S ≤ + 5V		0.5			0.5		μA/V
+ 5V ≤ V_S ≤ + 15V		0.2			0.2		μA/V
+ 15V ≤ V_S ≤ + 30V		0.1			0.1		μA/V
Case Isolation to Either Lead		10^{10}			10^{10}		Ω
Effective Shunt Capacitance		100			100		pF
Electrical Turn-On Time		20			20		μs
Reverse Bias Leakage Current[4]							
(Reverse Voltage = 10V)		10			10		pA
PACKAGE OPTIONS[5]							
TO-52 (H-03A)		AD590JH			AD590KH		
Flat Pack (F–2A)		AD590JF			AD590KF		

NOTES

[1]The AD590 has been used at − 100℃ and + 200℃ for short periods of measurement with no physical damage to the device. However, the absolute errors specified apply to only the rated performance temperature range.

[2]Maximum deviation between + 25℃ readings after temperature cycling between − 55℃ and + 150℃; guaranteed not tested.

[3]Conditions: constant + 5V, constant + 125℃; guaranteed, not tested.

[4]Leakage current doubles every 10℃.

[5]See Section 20 for package outline information.

Specifications subject to change without notice.

Specifications shown in boldface are tested on all production units at final electrical test. Results from those tests are used to calculate outgoing quality levels. All min and max specifications are guaranteed, although only those shown in **boldface** are tested on all production units.

Model	AD590L			AD590M			Units
	Min	Typ	Max	Min	Typ	Max	
ABSOLUTE MAXIMUM RATINGS							
Forward Voltage (E + to E −)			+ 44			+ 44	Volts
Reverse Voltage (E + to E −)			− 20			− 20	Volts
Breakdown Voltage (Case to E + or E −)			± 200			± 200	Volts
Rated Performance Temperature Range[1]	− 55		+ 150	− 55		+ 150	°C
Storage Temperature Range[1]	− 65		+ 155	− 65		+ 155	°C
Lead Temperature (Soldering, 10 sec)			+ 300			+ 300	°C
POWER SUPPLY							
Operating Voltage Range	+ 4		+ 30	+ 4		+ 30	Volts
OUTPUT							
Nominal Current Output @ + 25°C (298.2K)		298.2			298.2		μA
Nominal Temperature Coefficient		1			1		μA/K
Calibration Error @ + 25°C			± 1.0			± 0.5	°C
Absolute Error (over rated performance temperature range)							
Without External Calibration Adjustment			± 3.0			± 1.7	°C
With + 25°C Calibration Error Set to Zero			± 1.6			± 1.0	°C
Nonlinearity			± 0.4			± 0.3	°C
Repeatability[2]			± 0.1			± 0.1	°C
Long Term Drift[3]			± 0.1			± 0.1	°C
Current Noise		40			40		pA√Hz
Power Supply Rejection							
+ 4V ≤ V_S ≤ + 5V		0.5			0.5		μA/V
+ 5V ≤ V_S ≤ + 15V		0.2			0.2		μA/V
+ 15V ≤ V_S ≤ + 30V		0.1			0.1		μA/V
Case Isolation to Either Lead		10^{10}			10^{10}		Ω
Effective Shunt Capacitance		100			100		pF
Electrical Turn-On Time		20			20		μs
Reverse Bias Leakage Current[4] (Reverse Voltage = 10V)		10			10		pA
PACKAGE OPTION[5]							
TO-52 (H-03A)		AD590LH			AD590MH		
Flat Pack (F-2A)		AD590LF			AD590MF		

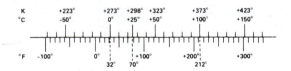

TEMPERATURE SCALE CONVERSION EQUATIONS

$$°C = \frac{5}{9}·(°F - 32) \qquad K = °C + 273.15$$

$$°F = \frac{9}{5}°C + 32 \qquad °R = °F + 459.7$$

The 590H has 60μ inches of gold plating on its Kovar leads and Kovar header. A resistance welder is used to seal the nickel cap to the header. The AD590 chip is eutectically mounted to the header and ultrasonically bonded to with 1 MIL aluminum wire. Kovar composition: 53% iron nominal; 29% ±1% nickel; 17% ±1% cobalt; 0.65% manganese max; 0.20% silicon max; 0.10% aluminum max; 0.10% magnesium max; 0.10% zirconium max; 0.10% titanium max; 0.06% carbon max.

The 590F is a ceramic package with gold plating on its Kovar leads, Kovar lid, and chip cavity. Solder of 80/20 Au/Sn composition is used for the 1.5 mil thick solder ring under the lid. The chip cavity has a nickel underlay between the metalization and the gold plating. The AD590 chip is eutectically mounted in the chip cavity at $410°C$ and ultrasonically bonded to with 1 mil aluminum wire. Note that the chip is in direct contact with the ceramic base, not the metal lid. When using the AD590 in die form, the chip substrate must be kept electrically isolated, (floating), for correct circuit operation.

METALIZATION DIAGRAM

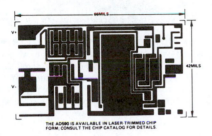

THE AD590 IS AVAILABLE IN LASER TRIMMED CHIP FORM: CONSULT THE CHIP CATALOG FOR DETAILS.

CIRCUIT DESCRIPTION[1]
The AD590 uses a fundamental property of the silicon transistors from which it is made to realize its temperature proportional characteristic: if two identical transistors are operated at a constant ratio of collector current densities, r, then the difference in their base-emitter voltages will be $(kT/q)(\ln r)$. Since both k, Boltzman's constant and q, the charge of an electron, are constant, the resulting voltage is directly proportional to absolute temperature (PTAT).

In the AD590, this PTAT voltage is converted to a PTAT current by low temperature coefficient thin film resistors. The total current of the device is then forced to be a multiple of this PTAT current. Referring to Figure 1, the schematic diagram of the AD590, Q8 and Q11 are the transistors that produce the PTAT voltage. R5 and R6 convert the voltage to current. Q10, whose collector current tracks the collector currents in Q9 and Q11, supplies all the bias and substrate leakage current for the rest of the circuit, forcing the total current to be PTAT. R5 and R6 are laser trimmed on the wafer to calibrate the device at $+25°C$.

Figure 2 shows the typical V–I characteristic of the circuit at $+25°C$ and the temperature extremes.

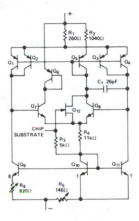

Figure 1. Schematic Diagram

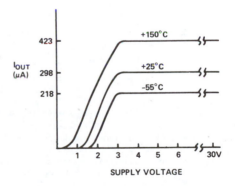

Figure 2. V–I Plot

[1] For a more detailed circuit description see M.P. Timko, "A Two-Terminal IC Temperature Transducer," IEEE J. Solid State Circuits, Vol. SC-11, p. 784-788, Dec. 1976.

Understanding the Specifications – AD590

EXPLANATION OF TEMPERATURE SENSOR SPECIFICATIONS

The way in which the AD590 is specified makes it easy to apply in a wide variety of different applications. It is important to understand the meaning of the various specifications and the effects of supply voltage and thermal environment on accuracy.

The AD590 is basically a PTAT (proportional to absolute temperature)[1] current regulator. That is, the output current is equal to a scale factor times the temperature of the sensor in degrees Kelvin. This scale factor is trimmed to $1\mu A/K$ at the factory, by adjusting the indicated temperature (i.e. the output current) to agree with the actual temperature. This is done with 5V across the device at a temperature within a few degrees of $25°C$ (298.2K). The device is then packaged and tested for accuracy over temperature.

CALIBRATION ERROR

At final factory test the difference between the indicated temperature and the actual temperature is called the calibration error. Since this is a scale factor error, its contribution to the total error of the device is PTAT. For example, the effect of the $1°C$ specified maximum error of the AD590L varies from $0.73°C$ at $-55°C$ to $1.42°C$ at $150°C$. Figure 3 shows how an exaggerated calibration error would vary from the ideal over temperature.

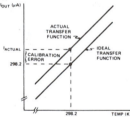

Figure 3. Calibration Error vs. Temperature

The calibration error is a primary contributor to maximum total error in all AD590 grades. However, since it is a scale factor error, it is particularly easy to trim. Figure 4 shows the most elementary way of accomplishing this. To trim this circuit the temperature of the AD590 is measured by a reference temperature sensor and R is trimmed so that $V_T = 1mV/K$ at that temperature. Note that when this error is trimmed out at one temperature, its effect is zero over the entire temperature range. In most applications there is a current to voltage conversion resistor (or, as with a current input ADC, a reference) that can be trimmed for scale factor adjustment.

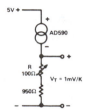

Figure 4. One Temperature Trim

[1] $T(°C) = T(K) - 273.2$; Zero on the Kelvin scale is "absolute zero"; there is no lower temperature.

ERROR VERSUS TEMPERATURE: WITH CALIBRATION ERROR TRIMMED OUT

Each AD590 is also tested for error over the temperature range with the calibration error trimmed out. This specification could also be called the "variance from PTAT" since it is the maximum difference between the actual current over temperature and a PTAT multiplication of the actual current at $25°C$. This error consists of a slope error and some curvature, mostly at the temperature extremes. Figure 5 shows a typical AD590K temperature curve before and after calibration error trimming.

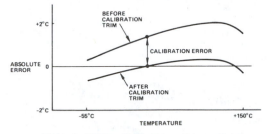

Figure 5. Effect of Scale Factor Trim on Accuracy

ERROR VERSUS TEMPERATURE: NO USER TRIMS

Using the AD590 by simply measuring the current, the total error is the "variance from PTAT" described above plus the effect of the calibration error over temperature. For example the AD590L maximum total error varies from $2.33°C$ at $-55°C$ to $3.02°C$ at $150°C$. For simplicity, only the larger figure is shown on the specification page.

NONLINEARITY

Nonlinearity as it applies to the AD590 is the maximum deviation of current over temperature from a best-fit straight line. The nonlinearity of the AD590 over the $-55°C$ to $+150°C$ range is superior to all conventional electrical temperature sensors such as thermocouples, RTD's and thermistors. Figure 6 shows the nonlinearity of the typical AD590K from Figure 5.

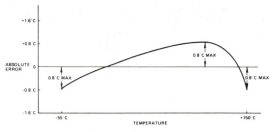

Figure 6. Nonlinearity

Figure 7A shows a circuit in which the nonlinearity is the major contributor to error over temperature. The circuit is trimmed by adjusting R_1 for a 0V output with the AD590 at $0°C$. R_2 is then adjusted for 10V out with the sensor at $100°C$. Other pairs of temperatures may be used with this procedure as long as they are measured accurately by a reference sensor. Note that for +15V output ($150°C$) the V+ of the op amp must be greater than 17V. Also note that V– should be at least –4V: if V– is ground there is no voltage applied across the device.

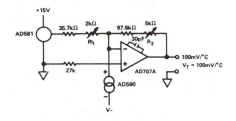

Figure 7A. Two Temperature Trim

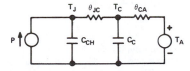

Figure 7B. Typical Two-Trim Accuracy

VOLTAGE AND THERMAL ENVIRONMENT EFFECTS

The power supply rejection specifications show the maximum expected change in output current versus input voltage changes. The insensitivity of the output to input voltage allows the use of unregulated supplies. It also means that hundreds of ohms of resistance (such as a CMOS multiplexer) can be tolerated in series with the device.

It is important to note that using a supply voltage other than 5V does not change the PTAT nature of the AD590. In other words, this change is equivalent to a calibration error and can be removed by the scale factor trim (see previous page).

The AD590 specifications are guaranteed for use in a low thermal resistance environment with 5V across the sensor. Large changes in the thermal resistance of the sensor's environment will change the amount of self-heating and result in changes in the output which are predictable but not necessarily desirable.

The thermal environment in which the AD590 is used determines two important characteristics: the effect of self heating and the response of the sensor with time.

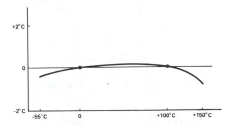

Figure 8. Thermal Circuit Model

Figure 8 is a model of the AD590 which demonstrates these characteristics. As an example, for the TO-52 package, θ_{JC} is the thermal resistance between the chip and the case, about

26°C/watt. θ_{CA} is the thermal resistance between the case and its surroundings and is determined by the characteristics of the thermal connection. Power source P represents the power dissipated on the chip. The rise of the junction temperature, T_J, above the ambient temperature T_A is:

$$T_J - T_A = P (\theta_{JC} + \theta_{CA}).\qquad \text{Eq. 1}$$

Table I gives the sum of θ_{JC} and θ_{CA} for several common thermal media for both the "H" and "F" packages. The heatsink used was a common clip-on. Using Equation 1, the temperature rise of an AD590 "H" package in a stirred bath at $+25^{\circ}$C, when driven with a 5V supply, will be 0.06°C. However, for the same conditions in still air the temperature rise is 0.72°C. For a given supply voltage, the temperature rise varies with the current and is PTAT. Therefore, if an application circuit is trimmed with the sensor in the same thermal environment in which it will be used, the scale factor trim compensates for this effect over the entire temperature range.

MEDIUM	$\theta_{JC} + \theta_{CA}(^{\circ}$C/watt)		τ (sec)(Note 3)	
	H	F	H	F
Aluminum Block	30	10	0.6	0.1
Stirred Oil[1]	42	60	1.4	0.6
Moving Air[2]				
With Heat Sink	45	–	5.0	–
Without Heat Sink	115	190	13.5	10.0
Still Air				
With Heat Sink	191	–	108	–
Without Heat Sink	480	650	60	30

[1] Note: τ is dependent upon velocity of oil; average of several velocities listed above.

[2] Air velocity $\cong$ 9ft/sec.

[3] The time constant is defined as the time required to reach 63.2% of an instantaneous temperature change.

Table I. Thermal Resistances

The time response of the AD590 to a step change in temperature is determined by the thermal resistances and the thermal capacities of the chip, C_{CH}, and the case, C_C. C_{CH} is about 0.04 watt-sec/$^{\circ}$C for the AD590. C_C varies with the measured medium since it includes anything that is in direct thermal contact with the case. In most cases, the single time constant exponential curve of Figure 9 is sufficient to describe the time response, T (t). Table I shows the effective time constant, τ, for several media.

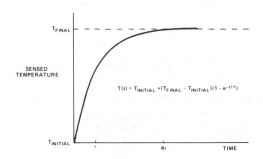

Figure 9. Time Response Curve

Applying the AD590

GENERAL APPLICATIONS

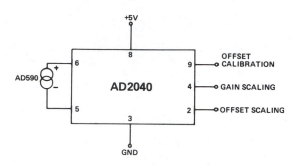

Figure 10. Variable Scale Display

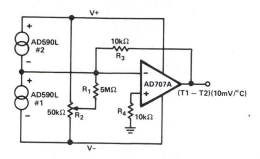

Figure 12. Differential Measurements

Figure 10 demonstrates the use of a low-cost Digital Panel Meter for the display of temperature on either the Kelvin, Celsius or Fahrenheit scales. For Kelvin temperature Pins 9, 4 and 2 are grounded; and for Fahrenheit temperature Pins 4 and 2 are left open.

The above configuration yields a 3 digit display with $1°C$ or $1°F$ resolution, in addition to an absolute accuracy of $\pm2.0°C$ over the $-55°C$ to $+125°C$ temperature range if a one-temperature calibration is performed on an AD590K, L, or M.

a desired temperature difference. For example, the inherent offset between the two devices can be trimmed in. If V+ and V– are radically different, then the difference in internal dissipation will cause a differential internal temperature rise. This effect can be used to measure the ambient thermal resistance seen by the sensors in applications such as fluid level detectors or anemometry.

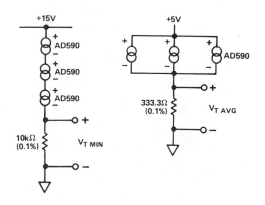

Figure 11. Series & Parallel Connection

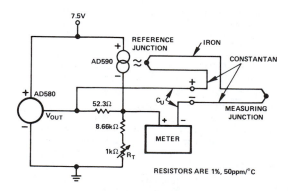

Figure 13. Cold Junction Compensation Circuit for Type J Thermocouple

Connecting several AD590 units in series as shown in Figure 11 allows the minimum of all the sensed temperatures to be indicated. In contrast, using the sensors in parallel yields the average of the sensed temperatures.

The circuit of Figure 12 demonstrates one method by which differential temperature measurements can be made. R_1 and R_2 can be used to trim the output of the op amp to indicate

Figure 13 is an example of a cold junction compensation circuit for a Type J Thermocouple using the AD590 to monitor the reference junction temperature. This circuit replaces an ice-bath as the thermocouple reference for ambient temperatures between $+15°C$ and $+35°C$. The circuit is calibrated by adjusting R_T for a proper meter reading with the measuring junction at a known reference temperature and the circuit near $+25°C$. Using components with the T.C.'s as specified in Figure 13, compensation accuracy will be within $\pm0.5°C$ for circuit temperatures between $+15°C$ and $+35°C$. Other thermocouple types can be accommodated with different resistor values. Note that the T.C.'s of the voltage reference and the resistors are the primary contributors to error.

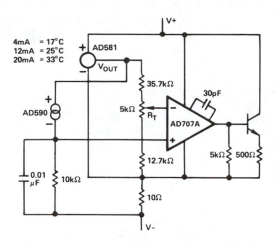

Figure 14. 4 to 20mA Current Transmitter

Figure 14 is an example of a current transmitter designed to be used with 40V, 1kΩ systems; it uses its full current range of 4mA to 20mA for a narrow span of measured temperatures. In this example the 1μA/K output of the AD590 is amplified to 1mA/°C and offset so that 4mA is equivalent to 17°C and 20mA is equivalent to 33°C. R_T is trimmed for proper reading at an intermediate reference temperature. With a suitable choice of resistors, any temperature range within the operating limits of the AD590 may be chosen.

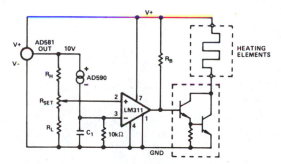

Figure 15. Simple Temperature Control Circuit

Figure 15 is an example of a variable temperature control circuit (thermostat) using the AD590. R_H and R_L are selected to set the high and low limits for R_{SET}. R_{SET} could be a simple pot, a calibrated multi-turn pot or a switched resistive divider. Powering the AD590 from the 10V reference isolates the AD590 from supply variations while maintaining a reasonable voltage (~7V) across it. Capacitor C_1 is often needed to filter extraneous noise from remote sensors. R_B is determined by the β of the power transistor and the current requirements of the load.

Figure 16 shows how the AD590 can be configured with an 8-bit DAC to produce a digitally controlled set point. This

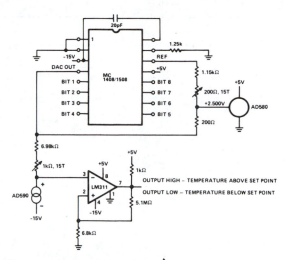

Figure 16. DAC Set Point

particular circuit operates from 0 (all inputs high) to +51°C (all inputs low) in 0.2°C steps. The comparator is shown with 1°C hysteresis which is usually necessary to guard-band for extraneous noise; omitting the 5.1MΩ resistor results in no hysteresis.

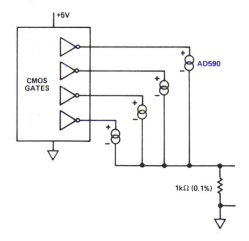

Figure 17. AD590 Driven from CMOS Logic

The voltage compliance and the reverse blocking characteristic of the AD590 allows it to be powered directly from +5V CMOS logic. This permits easy multiplexing, switching or pulsing for minimum internal heat dissipation. In Figure 17 any AD590 connected to a logic high will pass a signal current through the current measuring circuitry while those connected to a logic zero will pass insignificant current. The outputs used to drive the AD590's may be employed for other purposes, but the additional capacitance due to the AD590 should be taken into account.

AD590

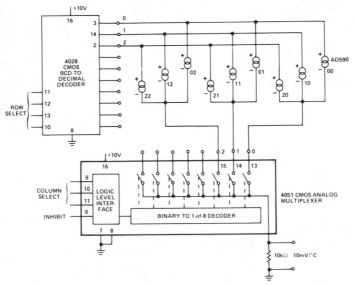

Figure 18. Matrix Multiplexer

CMOS Analog Multiplexers can also be used to switch AD590 current. Due to the AD590's current mode, the resistance of such switches is unimportant as long as 4V is maintained across the transducer. Figure 18 shows a circuit which combines the principal demonstrated in Figure 17 with an 8 channel CMOS Multiplexer. The resulting circuit can select one of eighty sensors over only 18 wires with a 7 bit binary word. The inhibit input on the multiplexer turns all sensors off for minimum dissipation while idling.

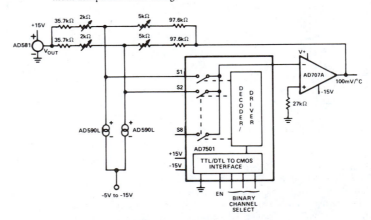

Figure 19. 8-Channel Multiplexer

Figure 19 demonstrates a method of multiplexing the AD590 in the two-trim mode (Figure 7). Additional AD590's and their associated resistors can be added to multiplex up to 8 channels of ±0.5°C absolute accuracy over the temperature range of –55°C to +125°C. The high temperature restriction of +125°C is due to the output range of the op amps; output to +150°C can be achieved by using a +20V supply for the op amp.

National Semiconductor

LM78XX Series Voltage Regulators

General Description

The LM78XX series of three terminal regulators is available with several fixed output voltages making them useful in a wide range of applications. One of these is local on card regulation, eliminating the distribution problems associated with single point regulation. The voltages available allow these regulators to be used in logic systems, instrumentation, HiFi, and other solid state electronic equipment. Although designed primarily as fixed voltage regulators these devices can be used with external components to obtain adjustable voltages and currents.

The LM78XX series is available in an aluminum TO-3 package which will allow over 1.0A load current if adequate heat sinking is provided. Current limiting is included to limit the peak output current to a safe value. Safe area protection for the output transistor is provided to limit internal power dissipation. If internal power dissipation becomes too high for the heat sinking provided, the thermal shutdown circuit takes over preventing the IC from overheating.

Considerable effort was expanded to make the LM78XX series of regulators easy to use and mininize the number of external components. It is not necessary to bypass the output, although this does improve transient response. Input bypassing is needed only if the regulator is located far from the filter capacitor of the power supply.

For output voltage other than 5V, 12V and 15V the LM117 series provides an output voltage range from 1.2V to 57V.

Features

- Output current in excess of 1A
- Internal thermal overload protection
- No external components required
- Output transistor safe area protection
- Internal short circuit current limit
- Available in the aluminum TO-3 package

Voltage Range

LM7805C	5V
LM7812C	12V
LM7815C	15V

Schematic and Connection Diagrams

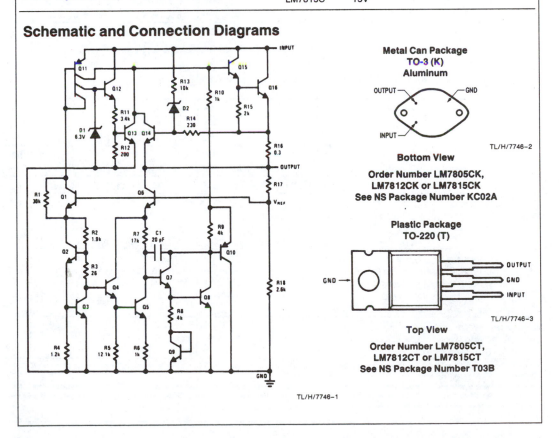

TL/H/7746-2

Metal Can Package
TO-3 (K)
Aluminum

OUTPUT — GND

INPUT

Bottom View

Order Number LM7805CK,
LM7812CK or LM7815CK
See NS Package Number KC02A

Plastic Package
TO-220 (T)

GND →

OUTPUT
GND
INPUT

TL/H/7746-3

Top View

Order Number LM7805CT,
LM7812CT or LM7815CT
See NS Package Number T03B

TL/H/7746-1

Absolute Maximum Ratings

If Military/Aerospace specified devices are required, please contact the National Semiconductor Sales Office/Distributors for availability and specifications.

Input Voltage (V_O = 5V, 12V and 15V)	35V	Maximum Junction Temperature	
Internal Power Dissipation (Note 1)	Internally Limited	(K Package)	150°C
Operating Temperature Range (T_A)	0°C to +70°C	(T Package)	150°C
		Storage Temperature Range	−65°C to +150°C
		Lead Temperature (Soldering, 10 sec.)	
		TO-3 Package K	300°C
		TO-220 Package T	230°C

Electrical Characteristics LM78XXC (Note 2) 0°C ≤ Tj ≤ 125°C unless otherwise noted.

Symbol	Parameter		Conditions	5V			12V			15V			Units
			Input Voltage (unless otherwise noted)	10V			19V			23V			
				Min	Typ	Max	Min	Typ	Max	Min	Typ	Max	
V_O	Output Voltage		Tj = 25°C, 5 mA ≤ I_O ≤ 1A	4.8	5	5.2	11.5	12	12.5	14.4	15	15.6	V
			P_D ≤ 15W, 5 mA ≤ I_O ≤ 1A	4.75		5.25	11.4		12.6	14.25		15.75	V
			V_{MIN} ≤ V_{IN} ≤ V_{MAX}	(7.5 ≤ V_{IN} ≤ 20)			(14.5 ≤ V_{IN} ≤ 27)			(17.5 ≤ V_{IN} ≤ 30)			V
ΔV_O	Line Regulation	I_O = 500 mA	Tj = 25°C		3	50		4	120		4	150	mV
			ΔV_{IN}	(7 ≤ V_{IN} ≤ 25)			(14.5 ≤ V_{IN} ≤ 30)			(17.5 ≤ V_{IN} ≤ 30)			V
			0°C ≤ Tj ≤ +125°C			50			120			150	mV
			ΔV_{IN}	(8 ≤ V_{IN} ≤ 20)			(15 ≤ V_{IN} ≤ 27)			(18.5 ≤ V_{IN} ≤ 30)			V
		I_O ≤ 1A	Tj = 25°C			50			120			150	mV
			ΔV_{IN}	(7.5 ≤ V_{IN} ≤ 20)			(14.6 ≤ V_{IN} ≤ 27)			(17.7 ≤ V_{IN} ≤ 30)			V
			0°C ≤ Tj ≤ +125°C			25			60			75	mV
			ΔV_{IN}	(8 ≤ V_{IN} ≤ 12)			(16 ≤ V_{IN} ≤ 22)			(20 ≤ V_{IN} ≤ 26)			V
ΔV_O	Load Regulation	Tj = 25°C	5 mA ≤ I_O ≤ 1.5A		10	50		12	120		12	150	mV
			250 mA ≤ I_O ≤ 750 mA			25			60			75	mV
		5 mA ≤ I_O ≤ 1A, 0°C ≤ Tj ≤ +125°C				50			120			150	mV
I_Q	Quiescent Current	I_O ≤ 1A	Tj = 25°C			8			8			8	mA
			0°C ≤ Tj ≤ +125°C			8.5			8.5			8.5	mA
ΔI_Q	Quiescent Current Change	5 mA ≤ I_O ≤ 1A				0.5			0.5			0.5	mA
		Tj = 25°C, I_O ≤ 1A				1.0			1.0			1.0	mA
		V_{MIN} ≤ V_{IN} ≤ V_{MAX}		(7.5 ≤ V_{IN} ≤ 20)			(14.8 ≤ V_{IN} ≤ 27)			(17.9 ≤ V_{IN} ≤ 30)			V
		I_O ≤ 500 mA, 0°C ≤ Tj ≤ +125°C				1.0			1.0			1.0	mA
		V_{MIN} ≤ V_{IN} ≤ V_{MAX}		(7 ≤ V_{IN} ≤ 25)			(14.5 ≤ V_{IN} ≤ 30)			(17.5 ≤ V_{IN} ≤ 30)			V
V_N	Output Noise Voltage	T_A = 25°C, 10 Hz ≤ f ≤ 100 kHz			40			75			90		μV
$\dfrac{\Delta V_{IN}}{\Delta V_{OUT}}$	Ripple Rejection	f = 120 Hz { I_O ≤ 1A, Tj = 25°C or I_O ≤ 500 mA 0°C ≤ Tj ≤ +125°C		62 62	80		55 55	72		54 54	70		dB dB
		V_{MIN} ≤ V_{IN} ≤ V_{MAX}		(8 ≤ V_{IN} ≤ 18)			(15 ≤ V_{IN} ≤ 25)			(18.5 ≤ V_{IN} ≤ 28.5)			V
R_O	Dropout Voltage	Tj = 25°C, I_{OUT} = 1A			2.0			2.0			2.0		V
	Output Resistance	f = 1 kHz			8			18			19		mΩ
	Short-Circuit Current	Tj = 25°C			2.1			1.5			1.2		A
	Peak Output Current	Tj = 25°C			2.4			2.4			2.4		A
	Average TC of V_{OUT}	0°C ≤ Tj ≤ +125°C, I_O = 5 mA			0.6			1.5			1.8		mV/°C
V_{IN}	Input Voltage Required to Maintain Line Regulation	Tj = 25°C, I_O ≤ 1A			7.5			14.6			17.7		V

Note 1: Thermal resistance of the TO-3 package (K, KC) is typically 4°C/W junction to case and 35°C/W case to ambient. Thermal resistance of the TO-220 package (T) is typically 4°C/W junction to case and 50°C/W case to ambient.

Note 2: All characteristics are measured with capacitor across the input of 0.22 μF, and a capacitor across the output of 0.1 μF. All characteristics except noise voltage and ripple rejection ratio are measured using pulse techniques (t_w ≤ 10 ms, duty cycle ≤ 5%). Output voltage changes due to changes in internal temperature must be taken into account separately.

LM78XX

Typical Performance Characteristics

Maximum Average Power Dissipation

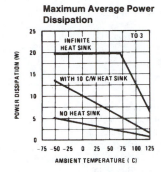

Maximum Average Power Dissipation

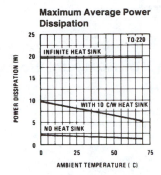

Peak Output Current

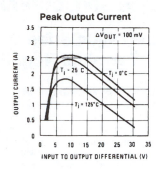

Output Voltage (Normalized to 1V at Tj = 25°C)

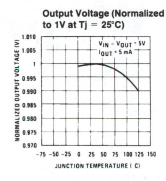

Ripple Rejection

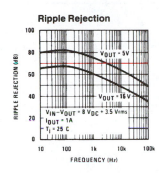

Ripple Rejection

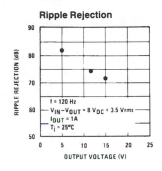

Output Impedance

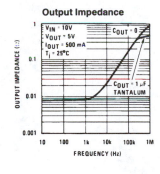

Dropout Voltage

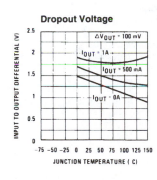

Dropout Characteristics

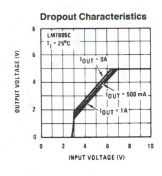

Quiescent Current

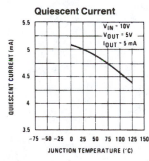

Quiescent Current

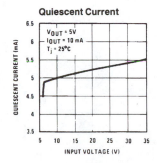

TL/H/7746–4

National Semiconductor

LM7900 Series
3-Terminal Negative Voltage Regulators

General Description

The LM7900 series of monolithic 3-terminal negative regulators are intended as complements to the popular LM7800 series of positive voltage regulators, and they are available in voltage options from −5.0V to −15V. The LM7900 series employ internal current-limiting, thermal shutdown, and safe-area compensation, making them virtually indestructible.

Features

- Output current in excess of 1.0A
- Internal thermal overload protection
- Internal short circuit current-limiting
- Output transistor safe-area compensation
- Available in JEDEC TO-220 and TO-3 packages
- Output voltage of −8V (See Note)

Connection Diagrams

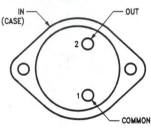

Top View

TL/H/10050−1

Order Number LM7908K and LM7908CK
See NS Package Number K02A

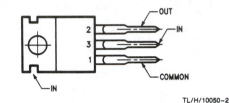

Lead 3 connected to case.

TL/H/10050−2

Top View

Order Number LM7908CT
See NS Package Number T03B

Note: See the LM79xx datasheet in the General Purpose Linear Databook for specifications on products in this series with −5V, −12V, and −15V outputs. Refer to the LM120/LM320 datasheet for −5V, −12V, and −15V regulators specified over extended temperature ranges.

Absolute Maximum Ratings

If Military/Aerospace specified devices are required, please contact the National Semiconductor Sales Office/Distributors for availability and specifications.

Storage Temperature Range
    TO-3 Metal Can $-65°C$ to $+175°C$
    TO-220 Package $-65°C$ to $+150°C$

Operating Junction Temperature Range
    Extended (LM7900) $-55°C$ to $+150°C$
    Commercial (LM7900C) $0°C$ to $+150°C$

Lead Temperature
    TO-3 Metal (soldering, 60 sec.)   300°C
    TO-220 Package (soldering, 10 sec.)   265°C

Power Dissipation   Internally Limited

Input Voltage
    $-5V$ to $-15V$ $-35V$

ESD Susceptibility   2000V

Note 1: The convention for Negative Regulators is the Algebraic value, thus $-15V$ is less then $-10V$.

LM7908
Electrical Characteristics

$-55°C \leq T_A \leq 125°C$, $V_I = -14V$, $I_O = 500$ mA, $C_I = 2.0$ μF, $C_O = 1.0$ μF, unless otherwise specified

Symbol	Parameter		Conditions (Note 1)		Min	Typ	Max	Units
V_O	Output Voltage		$T_J = 25°C$		-7.7	-8.0	-8.3	V
$V_{R\ LINE}$	Line Regulation		$T_J = 25°C$	$-10.5V \leq V_I \leq -25V$		6.0	80	mV
				$-11V \leq V_I \leq -17V$		2.0	40	
$V_{R\ LOAD}$	Load Regulation		$T_J = 25°C$	5.0 mA $\leq I_O \leq 1.5$A		12	100	mV
				250 mA $\leq I_O \leq 750$ mA		4.0	40	
V_O	Output Voltage		$-11.5V \leq V_I \leq -23V$, 5.0 mA $\leq I_O \leq 1.0$A, $P \leq 15$W		-7.6		-8.4	V
I_Q	Quiescent Current		$T_J = 25°C$			3.5	7.0	mA
ΔI_Q	Quiescent Current Change	With Line	$-11.5V \leq V_I \leq -25V$				1.0	mA
		With Load	5.0 mA $\leq I_O \leq 1.0$A				0.5	
N_O	Noise		$T_A = 25°C$, 10 Hz $\leq f \leq 100$ kHz			25	80	μV/V_O
$\Delta V_I/\Delta V_O$	Ripple Rejection		$f = 2400$ Hz, $V_I = -13V$ $I_O = 350$ mA, $T_J = 25°C$		54	60		dB
V_{DO}	Dropout Voltage		$I_O = 1.0$A, $T_J = 25°C$			1.1	2.3	V
I_{pk}	Peak Output Current		$T_J = 25°C$		1.3	2.1	3.3	A
$\Delta V_O/\Delta T$	Average Temperature Coefficient of Output Voltage		$I_O = 5.0$ mA, $-55°C \leq T_A \leq 125°C$				0.3	mV/°C/V_O
I_{OS}	Output Short Circuit Current		$V_I = -35V$, $T_J = 25°C$				1.2	A

LM7908C
Electrical Characteristics

$0°C \leq T_A \leq 125°C$, $V_I = -14V$, $I_O = 500$ mA, $C_I = 2.0$ μF, $C_O = 1.0$ μF, unless otherwise specified

Symbol	Parameter		Conditions (Note 1)		Min	Typ	Max	Units
V_O	Output Voltage		$T_J = 25°C$		-7.7	-8.0	-8.3	V
$V_{R\ LINE}$	Line Regulation		$T_J = 25°C$	$-10.5V \leq V_I \leq -25V$		6.0	160	mV
				$-11V \leq V_I \leq -17V$		2.0	80	
$V_{R\ LOAD}$	Load Regulation		$T_J = 25°C$	5.0 mA $\leq I_O \leq 1.5A$		12	160	mV
				250 mA $\leq I_O \leq 750$ mA		4.0	80	
V_O	Output Voltage		$-10.5V \leq V_I \leq 23V$, 5.0 mA $\leq I_O \leq 1.0A$, $P \leq 15W$		-7.6		-8.4	V
I_Q	Quiescent Current		$T_J = 25°C$			3.5	7.0	mA
ΔI_Q	Quiescent Current Change	With Line	$-10.5V \leq V_I \leq -25V$				1.0	mA
		With Load	5.0 mA $\leq I_O \leq 1.0A$				0.5	
N_O	Noise		$T_A = 25°C$, 10 Hz $\leq f \leq 100$ kHz			200		μV
$\Delta V_I / \Delta V_O$	Ripple Rejection		$f = 2400$ Hz, $V_I = -13V$ $I_O = 350$ mA, $T_J = 25°C$		54	60		dB
V_{DO}	Dropout Voltage		$I_O = 1.0A$, $T_J = 25°C$			1.1		V
I_{pk}	Peak Output Current		$T_J = 25°C$			2.1		A
$\Delta V_O / \Delta T$	Average Temperature Coefficient of Output Voltage		$I_O = 5.0$ mA, $0°C \leq T_A \leq 125°C$			0.6		mV/°C

Note 1: All characteristics except noise voltage and ripple rejection ratio are measured using pulse techniques ($t_W \leq 10$ ms, duty cycle $\leq 5\%$). Output voltage changes due to changes in internal temperature must be taken into account separately.

Equivalent Circuit

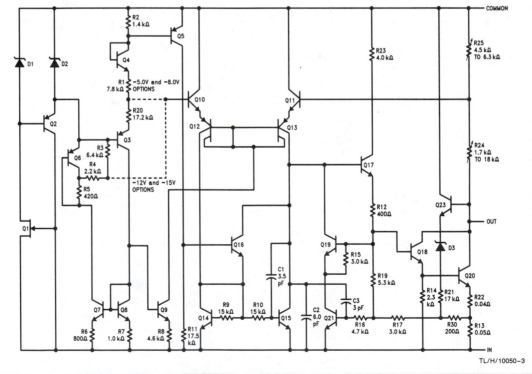

TL/H/10050–3

Typical Performance Characteristics

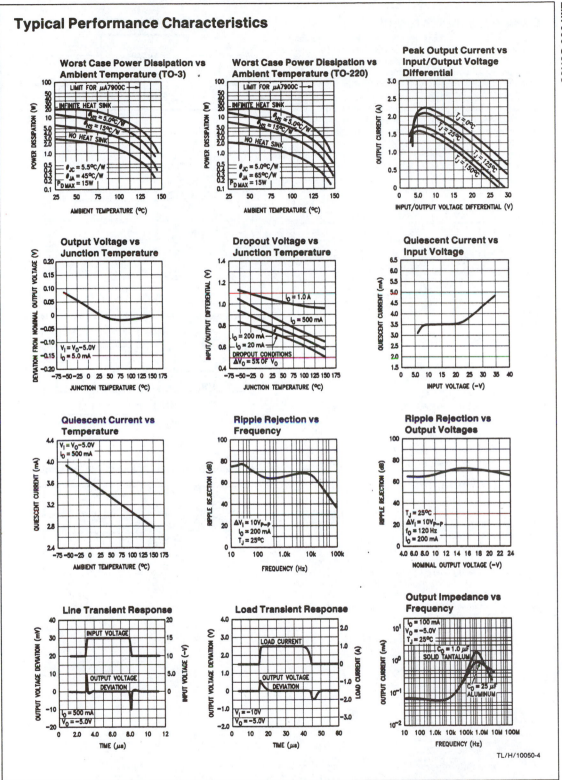

TL/H/10050-4

Design Considerations

The LM7900 fixed voltage regulator series has thermal overload protection from excessive power dissipation, internal short circuit protection which limits the circuit's maximum current, and output transistor safe-area compensation for reducing the output current as the voltage across the pass transistor is increased.

Although the internal power dissipation is limited, the junction temperature must be kept below the maximum specified temperature (150°C for LM7900, 125°C for LM7900C) in order to meet data sheet specifications. To calculate the maximum junction temperature or heat sink required, the following thermal resistance values should be used:

Package	Typ θ_{JC} °C/W	Max θ_{JC} °C/W	Typ θ_{JA} °C/W	Max θ_{JA} °C/W
TO-3	3.5	5.5	40	35
TO-220	3.0	5.0	60	40

$$P_{D\,MAX} = \frac{T_{J\,Max} - T_A}{\theta_{JC} + \theta_{CA}} \text{ or } \frac{T_{J\,Max}\,T_A}{\theta_{JA}}$$

$\theta_{CA} = \theta_{CS} + \theta_{SA}$ (without heat sink)

Solving for T_J:

$T_J = T_A + P_D(\theta_{JC} + \theta_{CA})$ or

$\quad = T_A + P_D\theta_{JA}$ (without heat sink)

Where:

T_J = Junction Temperature

T_A = Ambient Temperature

P_D = Power Dissipation

θ_{JA} = Junction-to-Ambient Thermal Resistance

θ_{JC} = Junction-to-Case Thermal Resistance

θ_{CA} = Case-to-Ambient Thermal Resistance

θ_{CS} = Case-to-Heat Sink Thermal Resistance

θ_{SA} = Heat Sink-to-Ambient Thermal Resistance

Typical Applications

Bypass capacitors are necessary for stable operation of the LM7900 series of regulators over the input voltage and output current ranges. Output bypass capacitors will improve the transient response by the regulator.

The bypass capacitors, (2.0 µF on the input, 1.0 µF on the output) should be ceramic or solid tantalum which have good high frequency characteristics. If aluminum electrolytics are used, their values should be 10 µF or larger. The bypass capacitors should be mounted with the shortest leads, and if possible, directly across the regulator terminals.

Fixed Output Regulator

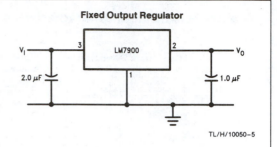

TL/H/10050-5

High Current Voltage Regulator

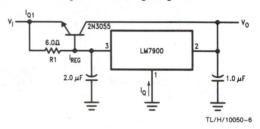

TL/H/10050-6

Output Current HIGH, Foldback Current-Limited

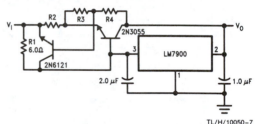

TL/H/10050-7

Output Current HIGH, Short Circuit Protected

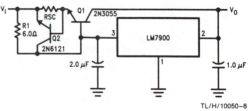

TL/H/10050-8

$$RSC = \frac{V_{BE(Q2)}}{I_{OS}}$$

National Semiconductor

LM317L 3-Terminal Adjustable Regulator

General Description

The LM317L is an adjustable 3-terminal positive voltage regulator capable of supplying 100 mA over a 1.2V to 37V output range. It is exceptionally easy to use and requires only two external resistors to set the output voltage. Further, both line and load regulation are better than standard fixed regulators. Also, the LM317L is available packaged in a standard TO-92 transistor package which is easy to use.

In addition to higher performance than fixed regulators, the LM317L offers full overload protection. Included on the chip are current limit, thermal overload protection and safe area protection. All overload protection circuitry remains fully functional even if the adjustment terminal is disconnected.

Features

- Adjustable output down to 1.2V
- Guaranteed 100 mA output current
- Line regulation typically 0.01%V
- Load regulation typically 0.1%
- Current limit constant with temperature
- Eliminates the need to stock many voltages
- Standard 3-lead transistor package
- 80 dB ripple rejection
- Output is short circuit protected

Normally, no capacitors are needed unless the device is situated more than 6 inches from the input filter capacitors in which case an input bypass is needed. An optional output capacitor can be added to improve transient response. The adjustment terminal can be bypassed to achieve very high ripple rejection ratios which are difficult to achieve with standard 3-terminal regulators.

Besides replacing fixed regulators, the LM317L is useful in a wide variety of other applications. Since the regulator is "floating" and sees only the input-to-output differential voltage, supplies of several hundred volts can be regulated as long as the maximum input-to-output differential is not exceeded.

Also, it makes an especially simple adjustable switching regulator, a programmable output regulator, or by connecting a fixed resistor between the adjustment and output, the LM317L can be used as a precision current regulator. Supplies with electronic shutdown can be achieved by clamping the adjustment terminal to ground which programs the output to 1.2V where most loads draw little current.

The LM317L is available in a standard TO-92 transistor package and the SO-8 package. The LM317L is rated for operation over a −25°C to 125°C range.

Connection Diagram

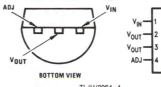

TL/H/9064–4
**Order Number LM317LZ
See NS Package
Number Z03A**

TL/H/9064–5
**Order Number LM317LM
See NS Package
Number M08A**

Typical Applications

1.2V–25V Adjustable Regulator

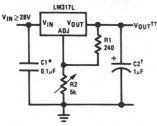

TL/H/9064–1

Full output current not available at high input-output voltages

†Optional—improves transient response

*Needed if device is more than 6 inches from filter capacitors

$$\dagger\dagger V_{OUT} = 1.25V \left(1 + \frac{R2}{R1} \right) + I_{ADJ} (R_2)$$

Fully Protected (Bulletproof) Lamp Driver

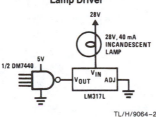

TL/H/9064–2

Lamp Flasher

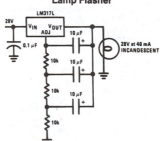

TL/H/9064–3

Output rate—4 flashes per second at 10% duty cycle

LM317L

Absolute Maximum Ratings

If Military/Aerospace specified devices are required, please contact the National Semiconductor Sales Office/Distributors for availability and specifications.

Power Dissipation	Internally Limited
Input-Output Voltage Differential	40V
Operating Junction Temperature Range	−40°C to +125°C

Storage Temperature	−55°C to +150°C
Lead Temperature (Soldering, 4 seconds)	260°C
Output is Short Circuit Protected	
ESD rating to be determined.	

Electrical Characteristics (Note 1)

Parameter	Conditions	Min	Typ	Max	Units
Line Regulation	T_j = 25°C, 3V ≤ (V_{IN} − V_{OUT}) ≤ 40V, I_L ≤ 20 mA (Note 2)		0.01	0.04	%/V
Load Regulation	T_j = 25°C, 5 mA ≤ I_{OUT} ≤ I_{MAX}, (Note 2)		0.1	0.5	%
Thermal Regulation	T_j = 25°C, 10 ms Pulse		0.04	0.2	%/W
Adjustment Pin Current			50	100	µA
Adjustment Pin Current Change	5 mA ≤ I_L ≤ 100 mA 3V ≤ (V_{IN} − V_{OUT}) ≤ 40V, P ≤ 625 mW		0.2	5	µA
Reference Voltage	3V ≤ (V_{IN} − V_{OUT}) ≤ 40V, (Note 3) 5 mA ≤ I_{OUT} ≤ 100 mA, P ≤ 625 mW	1.20	1.25	1.30	V
Line Regulation	3V ≤ (V_{IN} − V_{OUT}) ≤ 40V, I_L ≤ 20 mA (Note 2)		0.02	0.07	%/V
Load Regulation	5 mA ≤ I_{OUT} ≤ 100 mA, (Note 2)		0.3	1.5	%
Temperature Stability	T_{MIN} ≤ T_j ≤ T_{Max}		0.65		%
Minimum Load Current	(V_{IN} − V_{OUT}) ≤ 40V 3V ≤ (V_{IN} − V_{OUT}) ≤ 15V		3.5 1.5	5 2.5	mA
Current Limit	3V ≤ (V_{IN} − V_{OUT}) ≤ 13V (V_{IN} − V_{OUT}) = 40V	100 25	200 50	300 150	mA mA
Rms Output Noise, % of V_{OUT}	T_j = 25°C, 10 Hz ≤ f ≤ 10 kHz		0.003		%
Ripple Rejection Ratio	V_{OUT} = 10V, f = 120 Hz, C_{ADJ} = 0 C_{ADJ} = 10 µF	 66	65 80		dB dB
Long-Term Stability	T_j = 125°C, 1000 Hours		0.3	1	%
Thermal Resistance Junction to Ambient	Z Package 0.4″ Leads Z Package 0.125 Leads SO-8 Package		180 160 165		°C/W °C/W °C/W
Thermal Rating of SO Package			165		°C/W

Note 1: Unless otherwise noted, these specifications apply: −25°C ≤ T_j ≤ 125°C for the LM317L; V_{IN} − V_{OUT} = 5V and I_{OUT} = 40 mA. Although power dissipation is internally limited, these specifications are applicable for power dissipations up to 625 mW. I_{MAX} is 100 mA.

Note 2: Regulation is measured at constant junction temperature, using pulse testing with a low duty cycle. Changes in output voltage due to heating effects are covered under the specification for thermal regulation.

Note 3: Thermal resistance of the TO-92 package is 180°C/W junction to ambient with 0.4″ leads from a PC board and 160°C/W junction to ambient with 0.125″ lead length to PC board.

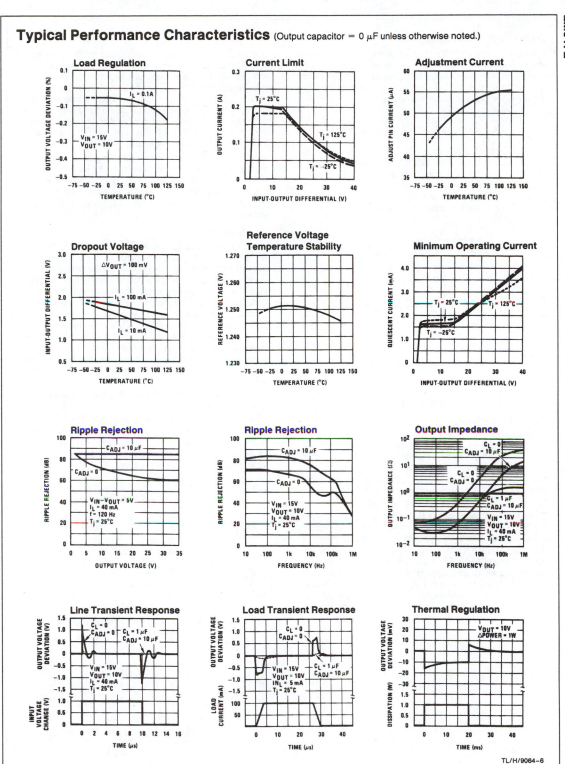

Typical Performance Characteristics (Output capacitor = 0 μF unless otherwise noted.)

TL/H/9064–6

LM317L

Application Hints

In operation, the LM317L develops a nominal 1.25V reference voltage, V_{REF}, between the output and adjustment terminal. The reference voltage is impressed across program resistor R1 and, since the voltage is constant, a constant current I_1 then flows through the output set resistor R2, giving an output voltage of

$$V_{OUT} = V_{REF} \left(1 + \frac{R2}{R1} \right) + I_{ADJ}(R2)$$

Since the 100 μA current from the adjustment terminal represents an error term, the LM317L was designed to minimize I_{ADJ} and make it very constant with line and load changes. To do this, all quiescent operating current is returned to the output establishing a minimum load current requirement. If there is insufficient load on the output, the output will rise.

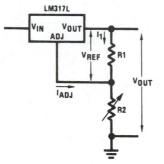

FIGURE 1

TL/H/9064–7

External Capacitors

An input bypass capacitor is recommended in case the regulator is more than 6 inches away from the usual large filter capacitor. A 0.1 μF disc or 1 μF solid tantalum on the input is suitable input bypassing for almost all applications. The device is more sensitive to the absence of input bypassing when adjustment or output capacitors are used, but the above values will eliminate the possiblity of problems.

The adjustment terminal can be bypassed to ground on the LM317L to improve ripple rejection and noise. This bypass capacitor prevents ripple and noise from being amplified as the output voltage is increased. With a 10 μF bypass capacitor 80 dB ripple rejection is obtainable at any output level. Increases over 10 μF do not appreciably improve the ripple rejection at frequencies above 120 Hz. If the bypass capacitor is used, it is sometimes necessary to include protection diodes to prevent the capacitor from discharging through internal low current paths and damaging the device.

In general, the best type of capacitors to use is solid tantalum. *Solid tantalum capacitors have low impedance even at high frequencies.* Depending upon capacitor construction, it takes about 25 μF in aluminum electrolytic to equal 1 μF solid tantalum at high frequencies. Ceramic capacitors are also good at high frequencies; but some types have a large decrease in capacitance at frequencies around 0.5 MHz. For this reason, a 0.01 μF disc may seem to work better than a 0.1 μF disc as a bypass.

Although the LM317L is stable with no output capacitors, like any feedback circuit, certain values of external capacitance can cause excessive ringing. This occurs with values between 500 pF and 5000 pF. A 1 μF solid tantalum (or 25 μF aluminum electrolytic) on the output swamps this effect and insures stability.

Load Regulation

The LM317L is capable of providing extremely good load regulation but a few precautions are needed to obtain maximum performance. The current set resistor connected between the adjustment terminal and the output terminal (usually 240Ω) should be tied directly to the output of the regulator rather than near the load. This eliminates line drops from appearing effectively in series with the reference and degrading regulation. For example, a 15V regulator with 0.05Ω resistance between the regulator and load will have a load regulation due to line resistance of 0.05Ω × I_L. If the set resistor is connected near the load the effective line resistance will be 0.05Ω (1 + R2/R1) or in this case, 11.5 times worse.

Figure 2 shows the effect of resistance between the regulator and 240Ω set resistor.

With the TO-92 package, it is easy to minimize the resistance from the case to the set resistor, by using two separate leads to the output pin. The ground of R2 can be returned near the ground of the load to provide remote ground sensing and improve load regulation.

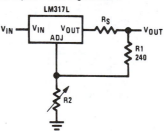

TL/H/9064–8

FIGURE 2. Regulator with Line Resistance in Output Lead

LM317L

Application Hints (Continued)

Thermal Regulation

When power is dissipated in an IC, a temperature gradient occurs across the IC chip affecting the individual IC circuit components. With an IC regulator, this gradient can be especially severe since power dissipation is large. Thermal regulation is the effect of these temperature gradients on output voltage (in percentage output change) per watt of power change in a specified time. Thermal regulation error is independent of electrical regulation or temperature coefficient, and occurs within 5 ms to 50 ms after a change in power dissipation. Thermal regulation depends on IC layout as well as electrical design. The thermal regulation of a voltage regulator is defined as the percentage change of V_{OUT}, per watt, within the first 10 ms after a step of power is applied. The LM317L specification is 0.2%/W, maximum.

In the Thermal Regulation curve at the bottom of the Typical Performance Characteristics page, a typical LM317L's output changes only 7 mV (or 0.07% of $V_{OUT} = -10V$) when a 1W pulse is applied for 10 ms. This performance is thus well inside the specification limit of 0.2%/W $\times$ 1W = 0.2% maximum. When the 1W pulse is ended, the thermal regulation again shows a 7 mV change as the gradients across the LM317L chip die out. Note that the load regulation error of about 14 mV (0.14%) is additional to the thermal regulation error.

Protection Diodes

When external capacitors are used with *any* IC regulator it is sometimes necessary to add protection diodes to pre-vent the capacitors from discharging through low current points into the regulator. Most 10 μF capacitors have low enough internal series resistance to deliver 20A spikes when shorted. Although the surge is short, there is enough energy to damage parts of the IC.

When an output capacitor is connected to a regulator and the input is shorted, the output capacitor will discharge into the output of the regulator. The discharge current depends on the value of the capacitor, the output voltage of the regulator, and the rate of decrease of V_{IN}. In the LM317L, this discharge path is through a large junction that is able to sustain a 2A surge with no problem. This is not true of other types of positive regulators. For output capacitors of 25 μF or less, the LM317L's ballast resistors and output structure limit the peak current to a low enough level so that there is no need to use a protection diode.

The bypass capacitor on the adjustment terminal can discharge through a low current junction. Discharge occurs when *either* the input or output is shorted. Internal to the LM317L is a 50Ω resistor which limits the peak discharge current. No protection is needed for output voltages of 25V or less and 10 μF capacitance. *Figure 3* shows an LM317L with protection diodes included for use with outputs greater than 25V and high values of output capacitance.

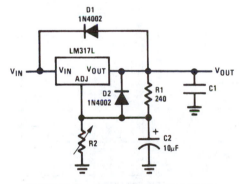

FIGURE 3. Regulator with Protection Diodes

TL/H/9064-9

$$V_{OUT} = 1.25V \left(1 + \frac{R2}{R1}\right) I_{ADJ} R2$$

D1 protects against C1
D2 protects against C2

LM555/LM555C Timer

General Description

The LM555 is a highly stable device for generating accurate time delays or oscillation. Additional terminals are provided for triggering or resetting if desired. In the time delay mode of operation, the time is precisely controlled by one external resistor and capacitor. For astable operation as an oscillator, the free running frequency and duty cycle are accurately controlled with two external resistors and one capacitor. The circuit may be triggered and reset on falling waveforms, and the output circuit can source or sink up to 200 mA or drive TTL circuits.

Features

- Direct replacement for SE555/NE555
- Timing from microseconds through hours
- Operates in both astable and monostable modes

- Adjustable duty cycle
- Output can source or sink 200 mA
- Output and supply TTL compatible
- Temperature stability better than 0.005% per °C
- Normally on and normally off output

Applications

- Precision timing
- Pulse generation
- Sequential timing
- Time delay generation
- Pulse width modulation
- Pulse position modulation
- Linear ramp generator

Schematic Diagram

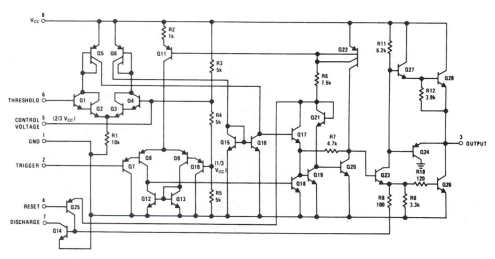

TL/H/7851-1

LM555/LM555C

Absolute Maximum Ratings

If Military/Aerospace specified devices are required, please contact the National Semiconductor Sales Office/Distributors for availability and specifications.

Supply Voltage	+18V
Power Dissipation (Note 1) LM555H, LM555CH	760 mW
Operating Temperature Ranges	
LM555N, LM555CN	1180 mW
LM555C	0°C to +70°C
LM555	−55°C to +125°C
Storage Temperature Range	−65°C to +150°C

Soldering Information
Dual-In-Line Package
Soldering (10 Seconds) 260°C
Small Outline Package
Vapor Phase (60 Seconds) 215°C
Infrared (15 Seconds) 220°C
See AN-450 "Surface Mounting Methods and Their Effect on Product Reliability" for other methods of soldering surface mount devices.

Electrical Characteristics (T_A = 25°C, V_{CC} = +5V to +15V, unless otherwise specified)

Parameter	Conditions	LM555 Min	LM555 Typ	LM555 Max	LM555C Min	LM555C Typ	LM555C Max	Units
Supply Voltage		4.5		18	4.5		16	V
Supply Current	V_{CC} = 5V, R_L = ∞		3	5		3	6	mA
	V_{CC} = 15V, R_L = ∞ (Low State) (Note 2)		10	12		10	15	mA
Timing Error, Monostable								
Initial Accuracy			0.5			1		%
Drift with Temperature	R_A = 1k to 100 kΩ, C = 0.1 μF, (Note 3)		30			50		ppm/°C
Accuracy over Temperature			1.5			1.5		%
Drift with Supply			0.05			0.1		%/V
Timing Error, Astable								
Initial Accuracy			1.5			2.25		%
Drift with Temperature	R_A, R_B = 1k to 100 kΩ, C = 0.1 μF, (Note 3)		90			150		ppm/°C
Accuracy over Temperature			2.5			3.0		%
Drift with Supply			0.15			0.30		%/V
Threshold Voltage			0.667			0.667		x V_{CC}
Trigger Voltage	V_{CC} = 15V	4.8	5	5.2		5		V
	V_{CC} = 5V	1.45	1.67	1.9		1.67		V
Trigger Current			0.01	0.5		0.5	0.9	μA
Reset Voltage		0.4	0.5	1	0.4	0.5	1	V
Reset Current			0.1	0.4		0.1	0.4	mA
Threshold Current	(Note 4)		0.1	0.25		0.1	0.25	μA
Control Voltage Level	V_{CC} = 15V	9.6	10	10.4	9	10	11	V
	V_{CC} = 5V	2.9	3.33	3.8	2.6	3.33	4	V
Pin 7 Leakage Output High			1	100		1	100	nA
Pin 7 Sat (Note 5)								
Output Low	V_{CC} = 15V, I_7 = 15 mA		150			180		mV
Output Low	V_{CC} = 4.5V, I_7 = 4.5 mA		70	100		80	200	mV

LM555/LM555C

Electrical Characteristics $T_A = 25°C$, $V_{CC} = +5V$ to $+15V$, (unless otherwise specified) (Continued)

Parameter	Conditions	LM555			LM555C			Units
		Min	Typ	Max	Min	Typ	Max	
Output Voltage Drop (Low)	$V_{CC} = 15V$							
	$I_{SINK} = 10$ mA		0.1	0.15		0.1	0.25	V
	$I_{SINK} = 50$ mA		0.4	0.5		0.4	0.75	V
	$I_{SINK} = 100$ mA		2	2.2		2	2.5	V
	$I_{SINK} = 200$ mA		2.5			2.5		V
	$V_{CC} = 5V$							
	$I_{SINK} = 8$ mA		0.1	0.25				V
	$I_{SINK} = 5$ mA					0.25	0.35	V
Output Voltage Drop (High)	$I_{SOURCE} = 200$ mA, $V_{CC} = 15V$		12.5			12.5		V
	$I_{SOURCE} = 100$ mA, $V_{CC} = 15V$	13	13.3		12.75	13.3		V
	$V_{CC} = 5V$	3	3.3		2.75	3.3		V
Rise Time of Output			100			100		ns
Fall Time of Output			100			100		ns

Note 1: For operating at elevated temperatures the device must be derated above 25°C based on a +150°C maximum junction temperature and a thermal resistance of 164°c/w (T0-5), 106°c/w (DIP) and 170°c/w (S0-8) junction to ambient.

Note 2: Supply current when output high typically 1 mA less at $V_{CC} = 5V$.

Note 3: Tested at $V_{CC} = 5V$ and $V_{CC} = 15V$.

Note 4: This will determine the maximum value of $R_A + R_B$ for 15V operation. The maximum total ($R_A + R_B$) is 20 MΩ.

Note 5: No protection against excessive pin 7 current is necessary providing the package dissipation rating will not be exceeded.

Note 6: Refer to RETS555X drawing of military LM555H and LM555J versions for specifications.

Connection Diagrams

Metal Can Package

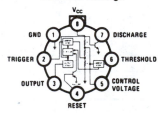

TL/H/7851–2

Top View

Order Number LM555H or LM555CH
See NS Package Number H08C

Dual-In-Line and Small Outline Packages

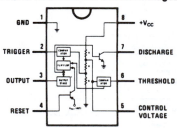

TL/H/7851–3

Top View

Order Number LM555J, LM555CJ,
LM555CM or LM555CN
See NS Package Number J08A, M08A or N08E

LM555/LM555C

Typical Performance Characteristics

Minimum Pulse Width Required for Triggering

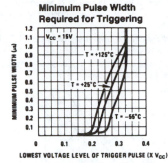

Supply Current vs Supply Voltage

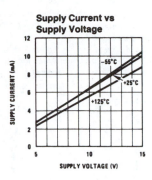

High Output Voltage vs Output Source Current

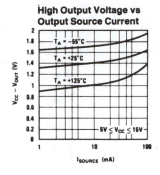

Low Output Voltage vs Output Sink Current

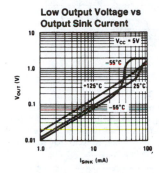

Low Output Voltage vs Output Sink Current

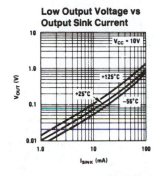

Low Output Voltage vs Output Sink Current

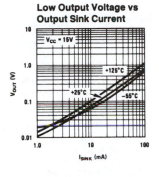

Output Propagation Delay vs Voltage Level of Trigger Pulse

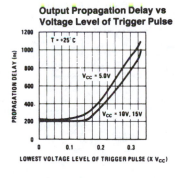

Output Propagation Delay vs Voltage Level of Trigger Pulse

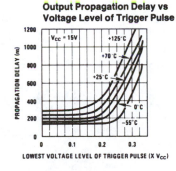

Discharge Transistor (Pin 7) Voltage vs Sink Current

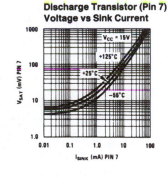

Discharge Transistor (Pin 7) Voltage vs Sink Current

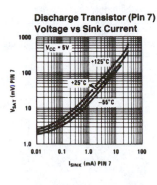

TL/H/7851–4

Applications Information

MONOSTABLE OPERATION

In this mode of operation, the timer functions as a one-shot *(Figure 1)*. The external capacitor is initially held discharged by a transistor inside the timer. Upon application of a negative trigger pulse of less than 1/3 V_{CC} to pin 2, the flip-flop is set which both releases the short circuit across the capacitor and drives the output high.

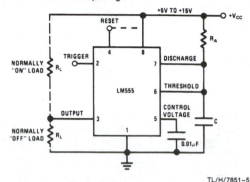

TL/H/7851-5

FIGURE 1. Monostable

The voltage across the capacitor then increases exponentially for a period of t = 1.1 R_A C, at the end of which time the voltage equals 2/3 V_{CC}. The comparator then resets the flip-flop which in turn discharges the capacitor and drives the output to its low state. *Figure 2* shows the waveforms generated in this mode of operation. Since the charge and the threshold level of the comparator are both directly proportional to supply voltage, the timing internal is independent of supply.

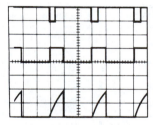

V_{CC} = 5V
TIME = 0.1 ms/DIV.
R_A = 9.1 kΩ
C = 0.01 µF

Top Trace: Input 5V/Div.
Middle Trace: Output 5V/Div.
Bottom Trace: Capacitor Voltage 2V/Div.

TL/H/7851-6

FIGURE 2. Monostable Waveforms

During the timing cycle when the output is high, the further application of a trigger pulse will not effect the circuit. However the circuit can be reset during this time by the application of a negative pulse to the reset terminal (pin 4). The output will then remain in the low state until a trigger pulse is again applied.

When the reset function is not in use, it is recommended that it be connected to V_{CC} to avoid any possibility of false triggering.

Figure 3 is a nomograph for easy determination of R, C values for various time delays.

NOTE: In monostable operation, the trigger should be driven high before the end of timing cycle.

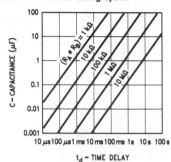

TL/H/7851-7

FIGURE 3. Time Delay

ASTABLE OPERATION

If the circuit is connected as shown in *Figure 4* (pins 2 and 6 connected) it will trigger itself and free run as a multivibrator. The external capacitor charges through R_A + R_B and discharges through R_B. Thus the duty cycle may be precisely set by the ratio of these two resistors.

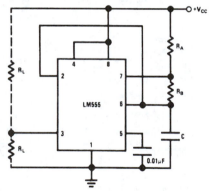

TL/H/7851-8

FIGURE 4. Astable

In this mode of operation, the capacitor charges and discharges between 1/3 V_{CC} and 2/3 V_{CC}. As in the triggered mode, the charge and discharge times, and therefore the frequency are independent of the supply voltage.

Applications Information (Continued)

Figure 5 shows the waveforms generated in this mode of operation.

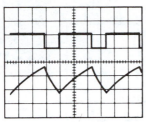

TL/H/7851–9

V_{CC} = 5V   Top Trace: Output 5V/Div.
TIME = 20 μs/DIV.   Bottom Trace: Capacitor Voltage 1V/Div.
R_A = 3.9 kΩ
R_B = 3 kΩ
C = 0.01 μF

FIGURE 5. Astable Waveforms

The charge time (output high) is given by:

$$t_1 = 0.693 (R_A + R_B) C$$

And the discharge time (output low) by:

$$t_2 = 0.693 (R_B) C$$

Thus the total period is:

$$T = t_1 + t_2 = 0.693 (R_A + 2R_B) C$$

The frequency of oscillation is:

$$f = \frac{1}{T} = \frac{1.44}{(R_A + 2 R_B) C}$$

Figure 6 may be used for quick determination of these RC values.

The duty cycle is: $$D = \frac{R_B}{R_A + 2R_B}$$

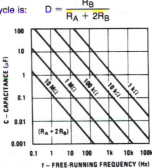

TL/H/7851–10

FIGURE 6. Free Running Frequency

FREQUENCY DIVIDER

The monostable circuit of *Figure 1* can be used as a frequency divider by adjusting the length of the timing cycle. *Figure 7* shows the waveforms generated in a divide by three circuit.

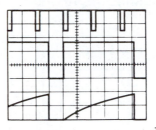

TL/H/7851–11

V_{CC} = 5V   Top Trace: Input 4V/Div.
TIME = 20 μs/DIV.   Middle Trace: Output 2V/Div.
R_A = 9.1 kΩ   Bottom Trace: Capacitor 2V/Div.
C = 0.01 μF

FIGURE 7. Frequency Divider

PULSE WIDTH MODULATOR

When the timer is connected in the monostable mode and triggered with a continuous pulse train, the output pulse width can be modulated by a signal applied to pin 5. *Figure 8* shows the circuit, and in *Figure 9* are some waveform examples.

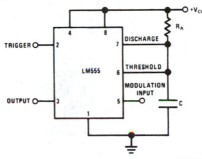

TL/H/7851–12

FIGURE 8. Pulse Width Modulator

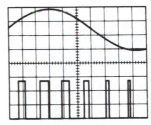

TL/H/7851–13

V_{CC} = 5V   Top Trace: Modulation 1V/Div.
TIME = 0.2 ms/DIV. Bottom Trace: Output Voltage 2V/Div.
R_A = 9.1 kΩ
C = 0.01 μF

FIGURE 9. Pulse Width Modulator

PULSE POSITION MODULATOR

This application uses the timer connected for astable operation, as in *Figure 10,* with a modulating signal again applied to the control voltage terminal. The pulse position varies with the modulating signal, since the threshold voltage and hence the time delay is varied. *Figure 11* shows the waveforms generated for a triangle wave modulation signal.

Applications Information (Continued)

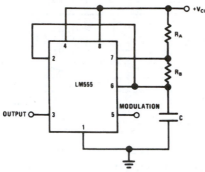

TL/H/7851–14

FIGURE 10. Pulse Position Modulator

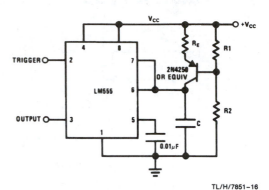

TL/H/7851–16

FIGURE 12

Figure 13 shows waveforms generated by the linear ramp. The time interval is given by:

$$T = \frac{2/3\ V_{CC}\ R_E\ (R_1 + R_2)\ C}{R_1\ V_{CC} - V_{BE}\ (R_1 + R_2)}$$

$$V_{BE} \cong 0.6V$$

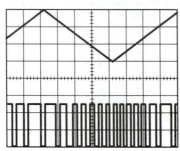

TL/H/7851–15

V_{CC} = 5V   Top Trace: Modulation Input 1V/Div.
TIME = 0.1 ms/DIV.   Bottom Trace: Output 2V/Div.
R_A = 3.9 kΩ
R_B = 3 kΩ
C = 0.01 μF

FIGURE 11. Pulse Position Modulator

LINEAR RAMP

When the pullup resistor, R_A, in the monostable circuit is replaced by a constant current source, a linear ramp is generated. *Figure 12* shows a circuit configuration that will perform this function.

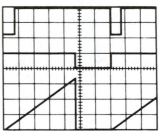

TL/H/7851–17

V_{CC} = 5V   Top Trace: Input 3V/Div.
TIME = 20 μs/DIV.   Middle Trace: Output 5V/Div.
R_1 = 47 kΩ   Bottom Trace: Capacitor Voltage 1V/Div.
R_2 = 100 kΩ
R_E = 2.7 kΩ
C = 0.01 μF

FIGURE 13. Linear Ramp

50% DUTY CYCLE OSCILLATOR

For a 50% duty cycle, the resistors R_A and R_B may be connected as in *Figure 14*. The time period for the out-

LM555/LM555C

Applications Information (Continued)

put high is the same as previous, $t_1 = 0.693\ R_A\ C$. For the output low it is $t_2 =$

$$\left[(R_A\,R_B)/(R_A + R_B)\right] C \; \ell n \left[\frac{R_B - 2R_A}{2R_B - R_A}\right]$$

Thus the frequency of oscillation is $f = \dfrac{1}{t_1 + t_2}$

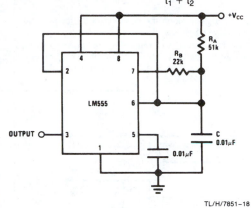

TL/H/7851–18

FIGURE 14. 50% Duty Cycle Oscillator

Note that this circuit will not oscillate if R_B is greater than $1/2\ R_A$ because the junction of R_A and R_B cannot bring pin 2 down to $1/3\ V_{CC}$ and trigger the lower comparator.

ADDITIONAL INFORMATION

Adequate power supply bypassing is necessary to protect associated circuitry. Minimum recommended is 0.1 μF in parallel with 1 μF electrolytic.

Lower comparator storage time can be as long as 10 μs when pin 2 is driven fully to ground for triggering. This limits the monostable pulse width to 10 μs minimum.

Delay time reset to output is 0.47 μs typical. Minimum reset pulse width must be 0.3 μs, typical.

Pin 7 current switches within 30 ns of the output (pin 3) voltage.

National Semiconductor

LM741/LM741A/LM741C/LM741E Operational Amplifier

General Description

The LM741 series are general purpose operational amplifiers which feature improved performance over industry standards like the LM709. They are direct, plug-in replacements for the 709C, LM201, MC1439 and 748 in most applications.

The amplifiers offer many features which make their application nearly foolproof: overload protection on the input and output, no latch-up when the common mode range is exceeded, as well as freedom from oscillations.

The LM741C/LM741E are identical to the LM741/LM741A except that the LM741C/LM741E have their performance guaranteed over a 0°C to +70°C temperature range, instead of −55°C to +125°C.

Schematic and Connection Diagrams (Top Views)

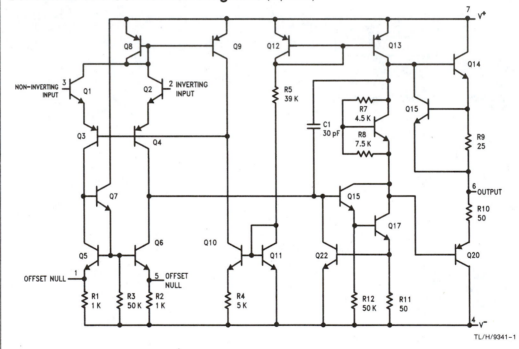

TL/H/9341–1

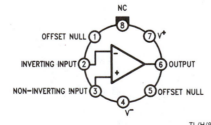

Metal Can Package

TL/H/9341–2

Order Number LM741H, LM741AH, LM741CH or LM741EH
See NS Package Number H08C

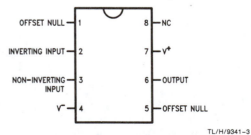

Dual-In-Line or S.O. Package

TL/H/9341–3

Order Number LM741J, LM741AJ, LM741CJ, LM741CM, LM741CN or LM741EN
See NS Package Number J08A, M08A or N08E

Absolute Maximum Ratings

If Military/Aerospace specified devices are required, please contact the National Semiconductor Sales Office/ Distributors for availability and specifications.
(Note 5)

	LM741A	LM741E	LM741	LM741C
Supply Voltage	±22V	±22V	±22V	±18V
Power Dissipation (Note 1)	500 mW	500 mW	500 mW	500 mW
Differential Input Voltage	±30V	±30V	±30V	±30V
Input Voltage (Note 2)	±15V	±15V	±15V	±15V
Output Short Circuit Duration	Continuous	Continuous	Continuous	Continuous
Operating Temperature Range	−55°C to +125°C	0°C to +70°C	−55°C to +125°C	0°C to +70°C
Storage Temperature Range	−65°C to +150°C	−65°C to +150°C	−65°C to +150°C	−65°C to +150°C
Junction Temperature	150°C	100°C	150°C	100°C
Soldering Information				
N-Package (10 seconds)	260°C	260°C	260°C	260°C
J- or H-Package (10 seconds)	300°C	300°C	300°C	300°C
M-Package				
Vapor Phase (60 seconds)	215°C	215°C	215°C	215°C
Infrared (15 seconds)	215°C	215°C	215°C	215°C

See AN-450 "Surface Mounting Methods and Their Effect on Product Reliability" for other methods of soldering surface mount devices.

ESD Tolerance (Note 6)	400V	400V	400V	400V

Electrical Characteristics (Note 3)

Parameter	Conditions	LM741A/LM741E Min	Typ	Max	LM741 Min	Typ	Max	LM741C Min	Typ	Max	Units
Input Offset Voltage	$T_A = 25°C$ $R_S \leq 10\ k\Omega$ $R_S \leq 50\Omega$		0.8	3.0		1.0	5.0		2.0	6.0	mV mV
	$T_{AMIN} \leq T_A \leq T_{AMAX}$ $R_S \leq 50\Omega$ $R_S \leq 10\ k\Omega$			4.0			6.0			7.5	mV mV
Average Input Offset Voltage Drift				15							µV/°C
Input Offset Voltage Adjustment Range	$T_A = 25°C, V_S = \pm 20V$	±10				±15			±15		mV
Input Offset Current	$T_A = 25°C$		3.0	30		20	200		20	200	nA
	$T_{AMIN} \leq T_A \leq T_{AMAX}$			70		85	500			300	nA
Average Input Offset Current Drift				0.5							nA/°C
Input Bias Current	$T_A = 25°C$		30	80		80	500		80	500	nA
	$T_{AMIN} \leq T_A \leq T_{AMAX}$			0.210			1.5			0.8	µA
Input Resistance	$T_A = 25°C, V_S = \pm 20V$	1.0	6.0		0.3	2.0		0.3	2.0		MΩ
	$T_{AMIN} \leq T_A \leq T_{AMAX}$, $V_S = \pm 20V$	0.5									MΩ
Input Voltage Range	$T_A = 25°C$							±12	±13		V
	$T_{AMIN} \leq T_A \leq T_{AMAX}$				±12	±13					V
Large Signal Voltage Gain	$T_A = 25°C, R_L \geq 2\ k\Omega$ $V_S = \pm 20V, V_O = \pm 15V$ $V_S = \pm 15V, V_O = \pm 10V$	50			50	200		20	200		V/mV V/mV
	$T_{AMIN} \leq T_A \leq T_{AMAX}$, $R_L \geq 2\ k\Omega$, $V_S = \pm 20V, V_O = \pm 15V$ $V_S = \pm 15V, V_O = \pm 10V$ $V_S = \pm 5V, V_O = \pm 2V$	32 10			 25			 15			V/mV V/mV V/mV

Electrical Characteristics (Note 3) (Continued)

Parameter	Conditions	LM741A/LM741E Min	Typ	Max	LM741 Min	Typ	Max	LM741C Min	Typ	Max	Units
Output Voltage Swing	$V_S = \pm 20V$ $R_L \geq 10\ k\Omega$ $R_L \geq 2\ k\Omega$	± 16 ± 15									V V
	$V_S = \pm 15V$ $R_L \geq 10\ k\Omega$ $R_L \geq 2\ k\Omega$				± 12 ± 10	± 14 ± 13		± 12 ± 10	± 14 ± 13		V V
Output Short Circuit Current	$T_A = 25°C$ $T_{AMIN} \leq T_A \leq T_{AMAX}$	10 10	25	35 40		25			25		mA mA
Common-Mode Rejection Ratio	$T_{AMIN} \leq T_A \leq T_{AMAX}$ $R_S \leq 10\ k\Omega, V_{CM} = \pm 12V$ $R_S \leq 50\Omega, V_{CM} = \pm 12V$	80	95		70	90		70	90		dB dB
Supply Voltage Rejection Ratio	$T_{AMIN} \leq T_A \leq T_{AMAX}$, $V_S = \pm 20V$ to $V_S = \pm 5V$ $R_S \leq 50\Omega$ $R_S \leq 10\ k\Omega$	86	96		77	96		77	96		dB dB
Transient Response Rise Time Overshoot	$T_A = 25°C$, Unity Gain		0.25 6.0	0.8 20		0.3 5			0.3 5		μs %
Bandwidth (Note 4)	$T_A = 25°C$	0.437	1.5								MHz
Slew Rate	$T_A = 25°C$, Unity Gain	0.3	0.7			0.5			0.5		$V/\mu s$
Supply Current	$T_A = 25°C$					1.7	2.8		1.7	2.8	mA
Power Consumption	$T_A = 25°C$ $V_S = \pm 20V$ $V_S = \pm 15V$		80	150		50	85		50	85	mW mW
LM741A	$V_S = \pm 20V$ $T_A = T_{AMIN}$ $T_A = T_{AMAX}$			165 135							mW mW
LM741E	$V_S = \pm 20V$ $T_A = T_{AMIN}$ $T_A = T_{AMAX}$			150 150							mW mW
LM741	$V_S = \pm 15V$ $T_A = T_{AMIN}$ $T_A = T_{AMAX}$					60 45	100 75				mW mW

Note 1: For operation at elevated temperatures, these devices must be derated based on thermal resistance, and T_j max. (listed under "Absolute Maximum Ratings"). $T_j = T_A + (\theta_{jA}\ P_D)$.

Thermal Resistance	Cerdip (J)	DIP (N)	HO8 (H)	SO-8 (M)
θ_{jA} (Junction to Ambient)	100°C/W	100°C/W	170°C/W	195°C/W
θ_{jC} (Junction to Case)	N/A	N/A	25°C/W	N/A

Note 2: For supply voltages less than $\pm 15V$, the absolute maximum input voltage is equal to the supply voltage.

Note 3: Unless otherwise specified, these specifications apply for $V_S = \pm 15V$, $-55°C \leq T_A \leq +125°C$ (LM741/LM741A). For the LM741C/LM741E, these specifications are limited to $0°C \leq T_A \leq +70°C$.

Note 4: Calculated value from: BW (MHz) = 0.35/Rise Time(μs).

Note 5: For military specifications see RETS741X for LM741 and RETS741AX for LM741A.

Note 6: Human body model, 1.5 kΩ in series with 100 pF.

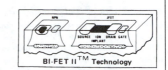

BI-FET II™ Technology

TL081CP Wide Bandwidth JFET Input Operational Amplifier

General Description

The TL081 is a low cost high speed JFET input operational amplifier with an internally trimmed input offset voltage (BI-FET II™ technology). The device requires a low supply current and yet maintains a large gain bandwidth product and a fast slew rate. In addition, well matched high voltage JFET input devices provide very low input bias and offset currents. The TL081 is pin compatible with the standard LM741 and uses the same offset voltage adjustment circuitry. This feature allows designers to immediately upgrade the overall performance of existing LM741 designs.

The TL081 may be used in applications such as high speed integrators, fast D/A converters, sample-and-hold circuits and many other circuits requiring low input offset voltage, low input bias current, high input impedance, high slew rate and wide bandwidth. The devices has low noise and offset voltage drift, but for applications where these requirements are critical, the LF356 is recommended. If maximum supply current is important, however, the TL081C is the better choice.

Features

- Internally trimmed offset voltage   15 mV
- Low input bias current   50 pA
- Low input noise voltage   25 nV/$\sqrt{Hz}$
- Low input noise current   0.01 pA/$\sqrt{Hz}$
- Wide gain bandwidth   4 MHz
- High slew rate   13 V/μs
- Low supply current   1.8 mA
- High input impedance $10^{12}\Omega$
- Low total harmonic distortion $A_V = 10$, $<0.02\%$
  $R_L = 10k$, $V_O = 20$ Vp-p,
  BW $= 20$ Hz-20 kHz
- Low 1/f noise corner   50 Hz
- Fast settling time to 0.01%   2 μs

Typical Connection

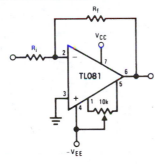

TL/H/8358–1

Simplified Schematic

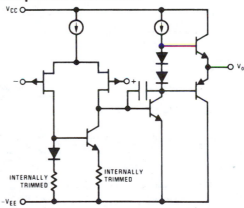

TL/H/8358–2

Connection Diagram

Dual-In-Line Package

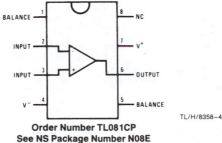

TL/H/8358–4

Order Number TL081CP
See NS Package Number N08E

TL081CP

Absolute Maximum Ratings

If Military/Aerospace specified devices are required, please contact the National Semiconductor Sales Office/Distributors for availability and specifications.

Supply Voltage	± 18V
Power Dissipation (Notes 1 and 6)	670 mW
Operating Temperature Range	0°C to +70°C
$T_{j(MAX)}$	115°C
Differential Input Voltage	± 30V
Input Voltage Range (Note 2)	± 15V
Output Short Circuit Duration	Continuous
Storage Temperature Range	-65°C to $+150$°C
Lead Temp. (Soldering, 10 seconds)	260°C
θ_{jA}	120°C/W
ESD rating to be determined.	

DC Electrical Characteristics (Note 3)

Symbol	Parameter	Conditions	TL081C Min	TL081C Typ	TL081C Max	Units
V_{OS}	Input Offset Voltage	$R_S = 10$ kΩ, $T_A = 25$°C		5	15	mV
		Over Temperature			20	mV
$\Delta V_{OS}/\Delta T$	Average TC of Input Offset Voltage	$R_S = 10$ kΩ		10		μV/°C
I_{OS}	Input Offset Current	$T_j = 25$°C, (Notes 3, 4)		25	100	pA
		$T_j \leq 70$°C			4	nA
I_B	Input Bias Current	$T_j = 25$°C, (Notes 3, 4)		50	200	pA
		$T_j \leq 70$°C			8	nA
R_{IN}	Input Resistance	$T_j = 25$°C		10^{12}		Ω
A_{VOL}	Large Signal Voltage Gain	$V_S = \pm 15$V, $T_A = 25$°C $V_O = \pm 10$V, $R_L = 2$ kΩ	25	100		V/mV
		Over Temperature	15			V/mV
V_O	Output Voltage Swing	$V_S = \pm 15$V, $R_L = 10$ kΩ	± 12	± 13.5		V
V_{CM}	Input Common-Mode Voltage Range	$V_S = \pm 15$V	± 11	$+15$ -12		V V
CMRR	Common-Mode Rejection Ratio	$R_S \leq 10$ kΩ	70	100		dB
PSRR	Supply Voltage Rejection Ratio	(Note 5)	70	100		dB
I_S	Supply Current			1.8	2.8	mA

AC Electrical Characteristics (Note 3)

Symbol	Parameter	Conditions	TL081C Min	TL081C Typ	TL081C Max	Units
SR	Slew Rate	$V_S = \pm 15$V, $T_A = 25$°C		13		V/μs
GBW	Gain Bandwidth Product	$V_S = \pm 15$V, $T_A = 25$°C		4		MHz
e_n	Equivalent Input Noise Voltage	$T_A = 25$°C, $R_S = 100\Omega$, f = 1000 Hz		25		nV/$\sqrt{Hz}$
i_n	Equivalent Input Noise Current	$T_j = 25$°C, f = 1000 Hz		0.01		pA/$\sqrt{Hz}$

Note 1: For operating at elevated temperature, the device must be derated based on a thermal resistance of 120°C/W junction to ambient for N package.

Note 2: Unless otherwise specified the absolute maximum negative input voltage is equal to the negative power supply voltage.

Note 3: These specifications apply for $V_S = \pm 15$V and 0°C $\leq T_A \leq +70$°C. V_{OS}, I_B and I_{OS} are measured at $V_{CM} = 0$.

Note 4: The input bias currents are junction leakage currents which approximately double for every 10°C increase in the junction temperature, T_j. Due to the limited production test time, the input bias currents measured are correlated to junction temperature. In normal operation the junction temperature rises above the ambient temperature as a result of internal power dissipation, P_D. $T_j = T_A + \theta_{jA} P_D$ where θ_{jA} is the thermal resistance from junction to ambient. Use of a heat sink is recommended if input bias current is to be kept to a minimum.

Note 5: Supply voltage rejection ratio is measured for both supply magnitudes increasing or decreasing simultaneously in accordance with common practice from $V_S = \pm 5$V to ± 15V.

Note 6: Max. Power Dissipation is defined by the package characteristics. Operating the part near the Max. Power Dissipation may cause the part to operate outside guaranteed limits.

Typical Performance Characteristics

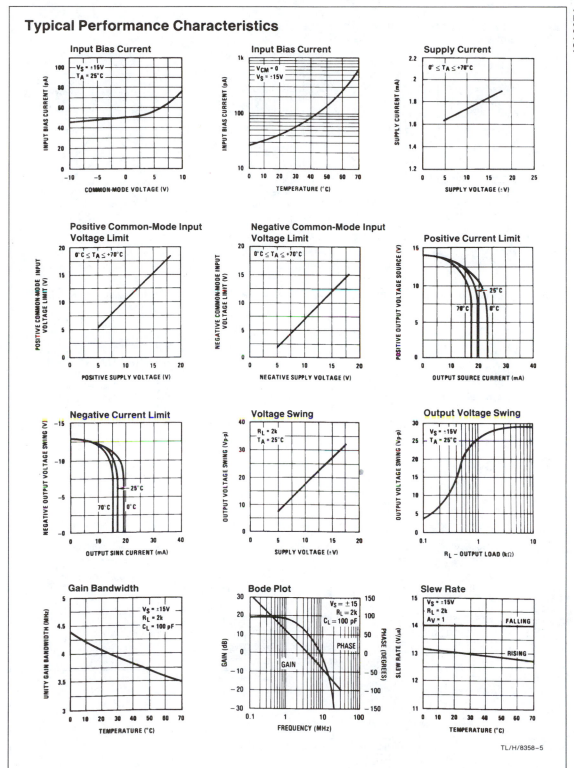

TL/H/8358-5

TL081CP

Typical Performance Characteristics (Continued)

Distortion vs Frequency

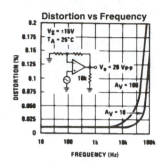

Undistorted Output Voltage Swing

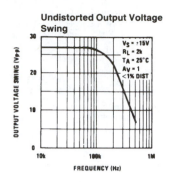

Open Loop Frequency Response

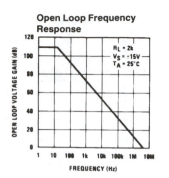

Common-Mode Rejection Ratio

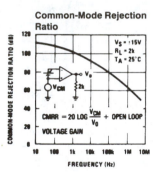

Power Supply Rejection Ratio

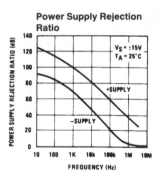

Equivalent Input Noise Voltage

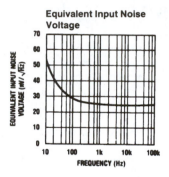

Open Loop Voltage Gain (V/V)

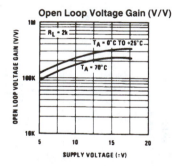

Output Impedance

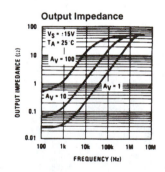

Inverter Settling Time

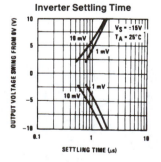

TL/H/8358–6

Pulse Response

Small Signal Inverting

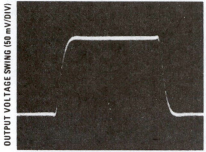

TL/H/8358-7

Small Signal Non-Inverting

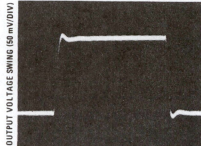

TL/H/8358-13

Large Signal Inverting

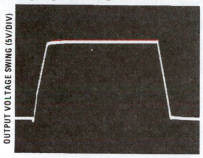

TL/H/8358-14

Large Signal Non-Inverting

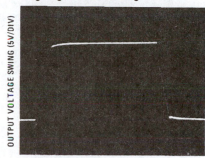

TL/H/8358-15

Current Limit ($R_L = 100\Omega$)

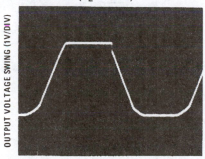

TL/H/8358-16

Application Hints

The TL081 is an op amp with an internally trimmed input offset voltage and JFET input devices (BI-FET II). These JFETs have large reverse breakdown voltages from gate to source and drain eliminating the need for clamps across the inputs. Therefore, large differential input voltages can easily be accommodated without a large increase in input current. The maximum differential input voltage is independent of the supply voltages. However, neither of the input voltages should be allowed to exceed the negative supply as this will cause large currents to flow which can result in a destroyed unit.

Exceeding the negative common-mode limit on either input will force the output to a high state, potentially causing a reversal of phase to the output.

Exceeding the negative common-mode limit on both inputs will force the amplifier output to a high state. In neither case does a latch occur since raising the input back within the

Application Hints (Continued)

common-mode range again puts the input stage and thus the amplifier in a normal operating mode.

Exceeding the positive common-mode limit on a single input will not change the phase of the output; however, if both inputs exceed the limit, the output of the amplifier will be forced to a high state.

The amplifier will operate with a common-mode input voltage equal to the positive supply; however, the gain bandwidth and slew rate may be decreased in this condition. When the negative common-mode voltage swings to within 3V of the negative supply, an increase in input offset voltage may occur.

The TL081 is biased by a zener reference which allows normal circuit operation on ±4V power supplies. Supply voltages less than these may result in lower gain bandwidth and slew rate.

The TL081 will drive a 2 kΩ load resistance to ±10V over the full temperature range of 0°C to +70°C. If the amplifier is forced to drive heavier load currents, however, an increase in input offset voltage may occur on the negative voltage swing and finally reach an active current limit on both positive and negative swings.

Precautions should be taken to ensure that the power supply for the integrated circuit never becomes reversed in polarity or that the unit is not inadvertently installed backwards in a socket as an unlimited current surge through the

resulting forward diode within the IC could cause fusing of the internal conductors and result in a destroyed unit.

Because these amplifiers are JFET rather than MOSFET input op amps they do not require special handling.

As with most amplifiers, care should be taken with lead dress, component placement and supply decoupling in order to ensure stability. For example, resistors from the output to an input should be placed with the body close to the input to minimize "pick-up" and maximize the frequency of the feedback pole by minimizing the capacitance from the input to ground.

A feedback pole is created when the feedback around any amplifier is resistive. The parallel resistance and capacitance from the input of the device (usually the inverting input) to AC ground set the frequency of the pole. In many instances the frequency of this pole is much greater than the expected 3 dB frequency of the closed loop gain and consequently there is negligible effect on stability margin. However, if the feedback pole is less than approximately 6 times the expected 3 dB frequency a lead capacitor should be placed from the output to the input of the op amp. The value of the added capacitor should be such that the RC time constant of this capacitor and the resistance it parallels is greater than or equal to the original feedback pole time constant.

Detailed Schematic

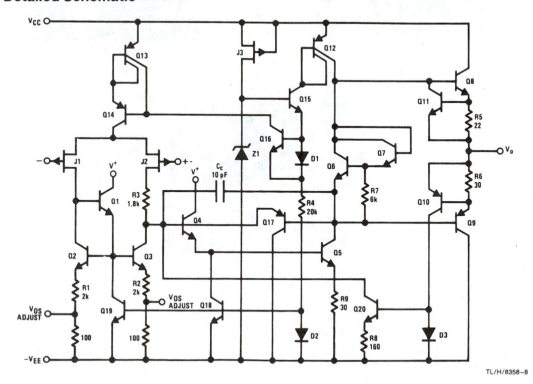

TL/H/8358–8

Typical Applications

Supply Current Indicator/Limiter

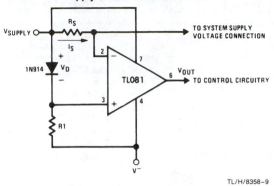

TL/H/8358–9

- V_{OUT} switches high when $R_S I_S > V_D$

Hi-Z$_{IN}$ Inverting Amplifier

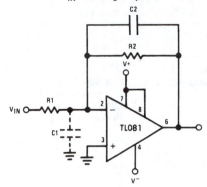

TL/H/8358–10

Parasitic input capacitance C1 $\cong$ (3 pF for TL081 plus any additional layout capacitance) interacts with feedback elements and creates undesirable high frequency pole. To compensate, add C2 such that: $R2C2 \cong R1C1$.

Ultra-Low (or High) Duty Cycle Pulse Generator

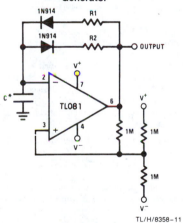

TL/H/8358–11

- $t_{OUTPUT\ HIGH} \approx R1C\ \ell n\ \dfrac{4.8 - 2V_S}{4.8 - V_S}$

- $t_{OUTPUT\ LOW} \approx R2C\ \ell n\ \dfrac{2V_S - 7.8}{V_S - 7.8}$

  where $V_S = V^+ + |V^-|$

*low leakage capacitor

Long Time Integrator

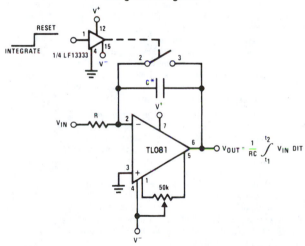

$$V_{OUT} = \frac{1}{RC} \int_{t_1}^{t_2} V_{IN}\ DIT$$

TL/H/8358–12

* Low leakage capacitor
- 50k pot used for less sensitive V_{OS} adjust

Monolithic Precision, Low Power
FET-Input Electrometer Op Amp

AD515A*

FEATURES
Ultralow Bias Current: 75fA max (AD515AL)
                            150fA max (AD515AK)
                            300fA max (AD515AJ)
Low Power: 1.5mA max Quiescent Current
             (0.6mA typ)
Low Offset Voltage: 1.0mV max (AD515AK & L)
Low Drift: 15µV/°C max (AD515AK)
Low Noise: 4µV p-p, 0.1Hz to 10Hz

AD515A PIN CONFIGURATION

GUARD PIN (CONNECTED TO CASE)
TAB

8
OFFSET
NULL   1   7   V+
INVERTING
INPUT   2   6   OUTPUT
NONINVERTING 3   5   OFFSET
INPUT   NULL
              4
OFFSET   V–   OFFSET
NULL   NULL

1   10kΩ   5

4   V–

PRODUCT DESCRIPTION

The AD515A is a monolithic FET-input operational amplifier
with a guaranteed maximum input bias current of 75fA
(AD515AL). The AD515A is a monolithic successor to the
industry standard AD515 electrometer, and will replace the
AD515 in most aplications. The AD515A also delivers laser-
trimmed offset voltage, low drift, low noise and low power, a
combination of features not previously available in ultralow bias
current circuits. All devices are internally compensated, protected
against latch-up and are short circuit protected.

The AD515A's combination of low input bias current, low
offset voltage and low drift optimizes it for a wide variety of
electrometer and very high impedance buffer applications including photo-current detection, vacuum ion-gage measurement,
long-term precision integration and low drift sample/hold applications. This amplifier is also an excellent choice for all forms of
biomedical instrumentation such as pH/pIon sensitive electrodes,
very low current oxygen sensors, and high impedance biological
microprobes. In addition, the low cost and pin compatibility of
the AD515A with standard FET op amps will allow designers to
upgrade the performance of present systems at little or no additional
cost. The $10^{15}\Omega$ common-mode input impedance ensures that
the input bias current is essentially independent of common-mode
voltage.

As with previous electrometer amplifier designs from Analog
Devices, the case is brought out to its own connection (Pin 8) so
that the case can be independently connected to a point at the
same potential as the input, thus minimizing stray leakage to
the case. This feature will also shield the input circuitry from
external noise and supply transients.

The AD515A is available in three versions of bias current and
offset voltage, the "J", "K" and "L"; all are specified for rated

*Covered by Patent No. 4,639,683.

performance from 0 to +70°C and supplied in a hermetically
sealed TO-99 package. The industry standard hybrid version,
AD515, will also be available.

PRODUCT HIGHLIGHTS

1. The AD515A provides subpicoampere bias currents in an
   integrated circuit amplifier.
   - The ultralow input bias currents are specified as the
     maximum measured at either input with the device fully
     warmed up on ±15V supplies at +25°C ambient with no
     heat sink. This parameter is 100% tested.
   - By using ±5V supplies, input bias current can typically
     be brought below 50fA.

2. The input offset voltage on all grades is laser trimmed, typically
   less than 500µV.
   - The offset voltage drift is 15µV/°C maximum on the K
     grade.
   - If additional nulling is desired, the amount required will
     have a minimal effect on offset drift (approximately 3µV/°C
     per mV).

3. The low quiescent current drain of 0.6mA typical and 1.5mA
   maximum, keeps self-heating effects to a minimum and renders
   the AD515A suitable for a wide range of remote probe
   applications.

4. The combination of low input noise voltage and very low
   input noise current is such that for source impedances from
   $1M\Omega$ to $10^{11}\Omega$, the Johnson noise of the source will easily
   dominate the noise characteristic.

5. Every AD515A receives a 24-hour stabilization bake at +150°C,
   to ensure reliability and long-term stability.

SPECIFICATIONS (typical @ +25°C with $V_S = \pm 15V$ dc, unless otherwise specified)

Model	AD515AJ	AD515AK	AD515AL
OPEN-LOOP GAIN[1]			
$V_{OUT} = \pm 10V, R_L \geqslant 2k\Omega$	**20,000V/V min**	**40,000V/V min**	**25,000V/V min**
$R_L \geqslant 10k\Omega$	**40,000V/V min**	**100,000V/V min**	**50,000V/V min**
$T_A =$ min to max $R_L \geqslant 2k\Omega$	**15,000V/V min**	**40,000V/V min**	**25,000V/V min**
OUTPUT CHARACTERISTICS			
Voltage@$R_L = 2k\Omega$, $T_A =$ min to max	**±10V min** (±12V typ)	★	★
@$R_L = 10k\Omega$, $T_A =$ min to max	**±12V min** (±13V typ)	★	★
Load Capacitance[2]	1000pF		
Short-Circuit Current	**10mA min** (20mA typ)		
FREQUENCY RESPONSE			
Unity Gain, Small Signal	1MHz	★	★
Full Power Response	**5kHz min** (50kHz typ)	★	★
Slew Rate Inverting Unity Gain	**0.3V/μs min** (3.0V/μs typ)	★	★
Overload Recovery Inverting Unity Gain	100μs max (2μs typ)	★	★
INPUT OFFSET VOLTAGE[3]	**3.0mV max** (0.4mV typ)	**1.0mV max** (0.4mV typ)	**1.0mV max** (0.4mV typ)
vs. Temperature, $T_A =$ min to max	**50μV/°C max**	**15μV/°C max**	**25μV/°C max**
vs. Supply, $T_A =$ min to max	**400μV/V max** (50μV/V typ)	**100μV/V max**	**200μV/V max**
INPUT BIAS CURRENT			
Either Input[4]	**300fA max**	**150fA max**	**75fA max**
INPUT IMPEDANCE			
Differential $V_{DIFF} = \pm 1V$	$1.6pF\|10^{13}\Omega$	★	★
Common Mode	$0.8pF\|10^{15}\Omega$	★	★
INPUT NOISE			
Voltage, 0.1Hz to 10Hz	4.0μV (p-p)	★	★
$f = 10Hz$	75nV/√Hz	★	★
$f = 100Hz$	55nV/√Hz	★	★
$f = 1kHz$	50nV/√Hz	★	★
Current, 0.1Hz to 10Hz	0.007pA (p-p)	★	★
10Hz to 10kHz	0.01pA rms	★	★
INPUT VOLTAGE RANGE			
Differential	**±20V min**	★	★
Common Mode, $T_A =$ min to max	**±10V min** (+12V, −11 typ)	★	★
Common-Mode Rejection, $V_{IN} = \pm 10V$	**66dB min** (94dB typ)	**80dB min**	**70dB min**
Maximum Safe Input Voltage[5]	$\pm V_S$	★	★
POWER SUPPLY			
Rated Performance	±15V	★	★
Operating	**±5V min** (±18V max)	★	★
Quiescent Current	**1.5mA max** (0.6mA typ)	★	★
TEMPERATURE			
Operating, Rated Performance	0 to +70°C	★	★
Storage	−65°C to +150°C	★	★
PACKAGE OPTION[6]			
TO-99 (H-08A)	AD515AJH	AD515AKH	AD515ALH

NOTES

*Specifications same as AD515AJ.

[1] Open Loop Gain is specified with or without nulling of V_{OS}.

[2] A conservative design would not exceed 750pF of load capacitance.

[3] Input Offset Voltage specifications are guaranteed after 5 minutes of operation at $T_A = +25°C$.

[4] Bias Current specifications are guaranteed after 5 minutes of operation at $T_A = +25°C$. For higher temperatures, the current doubles every +10°C.

[5] If it is possible for the input voltage to exceed the supply voltage, a series protection resistor should be added to limit input current to 0.1mA. The input devices can handle overload currents of 0.1mA indefinitely without damage. See next page.

[6] See Section 20 for package outline information.

Specifications subject to change without notice.

Specifications shown in **boldface** are tested on all production units at final test.

ESD PRECAUTIONS

Charges as high as 4000V readily accumulate on the human body and test equipment and discharge without detection. Therefore, reasonable ESD precautions are recommended to avoid functional damage or performance degradation. Unused devices should be stored in conductive foam or shunts, and the foam should be discharged to the destination socket before devices are removed. For further information on ESD precautions, refer to Analog Devices' ESD Prevention Manual.

Applying the AD515A

LAYOUT AND CONNECTIONS CONSIDERATIONS

The design of very high impedance measurement systems introduces a new level of problems associated with the reduction of leakage paths and noise pickup.

1. A primary consideration in high impedance system designs is to attempt to place the measuring device as near to the signal source as possible. This will minimize current leakage paths, noise pickup and capacitive loading. The AD515A, with its combination of low offset voltage (normally eliminating the need for trimming), low quiescent current (minimal source heating, possible battery operation), internal compensation and small physical size lends itself very nicely to installation at the signal source or inside a probe. Also, as a result of the high load capacitance rating, the AD515A can comfortably drive a long signal cable.

2. The use of guarding techniques is essential to realizing the capability of the ultralow input currents of the AD515A. Guarding is achieved by applying a low impedance bootstrap potential to the outside of the insulation material surrounding the high impedance signal line. This bootstrap potential is held at the same level as that of the high impedance line; therefore, there is no voltage drop across the insulation and, hence, no leakage. The guard will also act as a shield to reduce noise pickup and serves an additional function of reducing the effective capacitance to the input line. The case of the AD515A is brought out separately to Pin 8 so that the case can also be connected to the guard potential. This technique virtually eliminates potential leakage paths across the package insulation, provides a noise shield for the sensitive circuitry and reduces common-mode input capacitance to about 0.8pF. Figure 1 shows a proper printed circuit board layout for input guarding and connecting the case guard. Figures 2 and 3 show guarding connections for typical inverting and noninverting applications. If Pin 8 is not used for guarding, it should be connected to ground or a power supply to reduce noise.

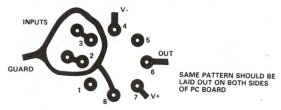

Figure 1. Board Layout for Guarding Inputs with Guarded TO-99 Package

3. Printed circuit board layout and construction is critical for achieving the ultimate in low leakage performance that the AD515A can deliver. The best performance will be realized by using a teflon IC socket for the AD515A; but at least a teflon stand-off should be used for the high impedance lead. If this is not feasible, the input guarding scheme shown in Figure 1 will minimize leakage as much as possible; the guard ring should be applied to both sides of the board. The guard ring is connected to a low impedance potential at the same level as the inputs. High impedance signal lines should not be extended for any unnecessary length on a printed circuit; to minimize noise and leakage, they must be carried in rigid, shielded cables.

4. Another important concern for achieving and maintaining low leakage currents is complete cleanliness of circuit boards and components. Completed assemblies should be washed thoroughly in a low residue solvent such as TMC Freon or high-purity methanol followed by a rinse with deionized water and nitrogen drying. If service is anticipated in a high contaminant or high humidity environment, a high dielectric conformal coating is recommended. All insulation materials except Kel-F or teflon will show rapid degradation of surface leakage at high humidities.

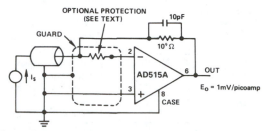

Figure 2. Picoampere Current-to-Voltage Converter Inverting Configuration

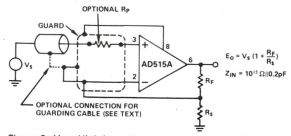

Figure 3. Very High Impedance Noninverting Amplifier

INPUT PROTECTION

The AD515A is guaranteed for a maximum safe input potential equal to the power supply potential.

Many instrumentation situations, such as flame detectors in gas chromatographs, involve measurement of low level currents from high-voltage sources. In such applications, a sensor fault condition may apply a very high potential to the input of the current-to-voltage converting amplifier. This possibility necessitates some form of input protection. Many electrometer type devices, especially CMOS designs, can require elaborate Zener protection schemes which often compromise overall performance. The AD515A requires input protection only if the source is not current limited, and as such is similar to many JFET-input designs. The failure mode would be overheating from excess current rather than voltage breakdown. If the source is not current limited, all that is required is a resistor in series with the affected input terminal so that the maximum overload current is 0.1mA (for example, 1MΩ for a 100V overload). This simple scheme will cause no significant reduction in performance and give complete overload protection. Figures 2 and 3 show proper connections.

COAXIAL CABLE AND CAPACITANCE EFFECTS

If it is not possible to attach the AD515A virtually on top of the signal source, considerable care should be exercised in designing the connecting lines carrying the high impedance signal. Shielded coaxial cable must be used for noise reduction, but use of coaxial cables for high impedance work can add problems from cable leakage, noise and capacitance. Only the best polyethylene or virgin teflon (not reconstituted) should be used to obtain the highest possible insulation resistance.

Cable systems should be made as rigid and vibration free as possible since cable movement can cause noise signals of three types, all significant in high impedance systems. Frictional movement of the shield over the insulation material generates a charge which is sensed by the signal line as a noise voltage. Low noise cable with graphite lubricant such as Amphenol 21-537 will reduce the noise, but short, rigid lines are better. Cable movements will also make small changes in the internal cable capacitance and capacitance to other objects. Since the total charge on these capacitances cannot be changed instantly, a noise voltage results as predicted from: $\Delta V = Q/\Delta C$. Noise voltage is also generated by the motion of a conductor in a magnetic field.

The conductor-to-shield capacitance of coaxial cable is usually about 30pF/foot. Charging this capacitance can cause considerable stretching of high impedance signal rise-time, thus cancelling the low input capacitance feature of the AD515A. There are two ways to circumvent this problem. For inverting signals or low-level current measurements, the signal is carried on the line connected to the inverting input and shielded (guarded) by the ground line as shown in Figure 2. Since the signal is always at virtual ground, no voltage change is required and no capacitances are charged. In many circumstances, this will destabilize the circuit; if so, capacitance from output to inverting input will stabilize the circuit.

Noninverting and buffer situations are more critical since the signal line voltage and therefore charge will change, causing signal delay. This effect can be reduced considerably by connecting the cable shield to a guard potential instead of ground, an option shown in Figure 3. Since such a connection results in positive feedback to the input, the circuit may be destabilized and oscillate. If so, capacitance from positive input to ground must be added to make the net capacitance at Pin 3 positive. This technique can considerably reduce the effective capacitance which must be charged.

Typical Performance Curves

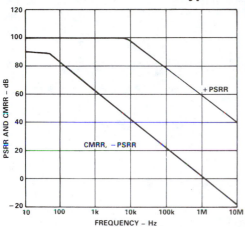

Figure 4. PSRR and CMRR vs. Frequency

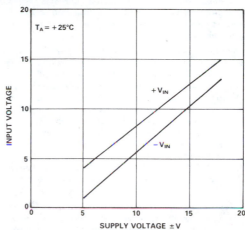

Figure 6. Input Common-Mode Range vs. Supply Voltage

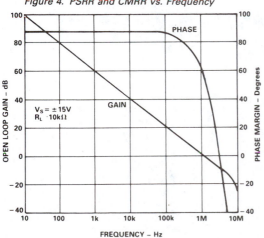

Figure 5. Open Loop Frequency Response

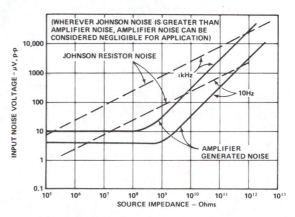

Figure 7. Peak-to-Peak Input Noise Voltage vs. Source Impedance and Bandwidth

AD515A

ELECTROMETER APPLICATION NOTES

The AD515A offers subpicoampere input bias currents available in an integrated circuit package. This design will open up many new application opportunities for measurements from very high impedance and very low current sources. Performing accurate measurements of this sort requires careful attention to detail; the notes given here will aid the user in realizing the full measurement potential of the AD515A and perhaps extending its performance limits.

1. As with all junction FET input devices, the temperature of the FETs themselves is all important in determining the input bias currents. Over the operating temperature range, the input bias currents closely follow a characteristic of doubling every 10°C; therefore, every effort should be made to minimize device operating temperature.

2. The heat dissipation can be reduced initially by careful investigation of the application. First, if it is possible to reduce the required power supplies, this should be done since internal power consumption contributes the largest component of self-heating. To minimize this effect, the quiescent current of the AD515A has been reduced to less than 1mA. Figure 8 shows typical input bias current and quiescent current versus supply voltage.

3. Output loading effects, which are normally ignored, can cause a significant increase in chip temperature and therefore bias current. For example, a 2kΩ load driven at 10V at the output will cause at least an additional 25mW dissipation in the output stage (and some in other stages) over the typical 24mW, thereby at least doubling the effects of self-heating. The results of this form of additional power dissipation are demonstrated in Figure 9, which shows normalized input bias current versus additional power dissipated. Therefore, although many dc performance parameters are specified driving a 2kΩ load, to reduce this additional dissipation, we recommend restricting the load resistance to be at least 10kΩ.

4. Figure 10 shows the AD515A's input current versus differential input voltage. Input current at either terminal stays below a few hundred fA until one input terminal is forced higher than 1 to 1.5V above the other terminal. Input current limits at 30μA under these conditions.

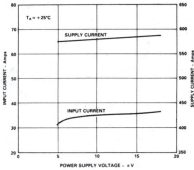

Figure 8. Input Bias Current and Supply Current vs. Supply Voltage

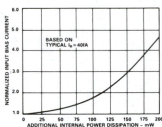

Figure 9. Input Bias Current vs. Additional Power Dissipation

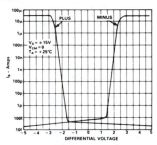

Figure 10. Input Bias Current vs. Differential Input Voltage

AD515A CIRCUIT APPLICATION NOTES

The AD515A is quite simple to apply to a wide variety of applications because of the pretrimmed offset voltage and internal compensation, which minimize required external components and eliminate the need for adjustments to the device itself. The major considerations in applying this device are the external problems of layout and heat control which have already been discussed. In circuit situations employing the use of very high value resistors, such as low level current to voltage converters, electrometer operational amplifiers can be destabilized by a pole created by the small capacitance at the negative input. If this occurs, a capacitor of 2 to 5pF in parallel with the resistor will stabilize the loop. A much larger capacitor may be used if desired to limit bandwidth and thereby reduce wideband noise.

Selection of passive components employed in high impedance situations is critical. High MΩ resistors should be of the carbon film or deposited ceramic oxide to obtain the best in low noise and high stability performance. The best packaging for high MΩ resistors is a glass body sprayed with silicone varnish to minimize humidity effects. These resistors must be handled very carefully to prevent surface contamination. Capacitors for any high impedance or long-term integration situation should be of a polystyrene formulation for optimum performance. Most other types have too low an insulation resistance, or high dielectric absorption.

Unlike situations involving standard operational amplifiers with much higher bias currents, balancing the impedances seen at the input terminals of the AD515A is usually unnecessary and probably undesirable. At the large source impedances where these effects matter, obtaining quality, matched resistors will be difficult. More important, instead of a cancelling effect, as with bias current, the noise voltage of the additional resistor will add by root-sum-of-squares to that of the other resistor thus increasing the total noise by about 40%. Noise currents driving the resistors also add, but in the AD515A are significant only above $10^{11}\Omega$.

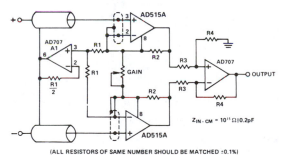

(ALL RESISTORS OF SAME NUMBER SHOULD BE MATCHED ±0.1%)
(BUFFER A1 BOOSTS COMMON MODE Z_{IN} BY DRIVING CABLE SHIELDS AT COMMON MODE VOLTAGE AND NEUTRALIZING CM CAPACITANCE)

Figure 11. Very High Impedance Instrumentation Amplifier

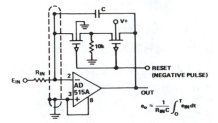

Figure 12. Low Drift Integrator and Low-Leakage Guarded Reset

LOW-LEVEL CURRENT-TO-VOLTAGE CONVERTERS

Figure 2 shows a standard low-level current-to-voltage converter. To obtain higher sensitivity, it is obvious to simply use a higher value feedback resistor. However, high value resistors above $10^9\Omega$ tend to be expensive, large, noisy and unstable. To avoid this, it may be desirable to use a circuit configuration with output gain, as in Figure 13. The drawback is that input errors of offset voltage drift and noise are multiplied by the same gain, but the precision performance of the AD515A makes the tradeoff easier.

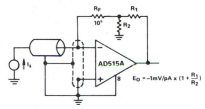

Figure 13. Picoampere to Voltage Converter with Gain

One of the problems with low-level leakage current testing or low-level current transducers (such as Clark oxygen sensors) is finding a way to apply voltage bias to the device while still grounding the device and the bias source. Figure 14 shows a technique in which the desired bias is applied at the noninverting terminal thus forcing that voltage at the inverting terminal. The current is sensed by R_F, and the AD524 instrumentation amplifier converts the floating differential signal to a single-ended output.

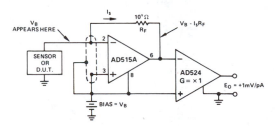

Figure 14. Current-to-Voltage Converters with Grounded Bias and Sensor

LM111/LM211/LM311 Voltage Comparator

LM111/LM211/LM311

General Description

The LM111, LM211 and LM311 are voltage comparators that have input currents nearly a thousand times lower than devices like the LM106 or LM710. They are also designed to operate over a wider range of supply voltages: from standard ±15V op amp supplies down to the single 5V supply used for IC logic. Their output is compatible with RTL, DTL and TTL as well as MOS circuits. Further, they can drive lamps or relays, switching voltages up to 50V at currents as high as 50 mA.

Both the inputs and the outputs of the LM111, LM211 or the LM311 can be isolated from system ground, and the output can drive loads referred to ground, the positive supply or the negative supply. Offset balancing and strobe capability are provided and outputs can be wire OR'ed. Although slower than the LM106 and LM710 (200 ns response time vs

40 ns) the devices are also much less prone to spurious oscillations. The LM111 has the same pin configuration as the LM106 and LM710.

The LM211 is identical to the LM111, except that its performance is specified over a −25°C to +85°C temperature range instead of −55°C to +125°C. The LM311 has a temperature range of 0°C to +70°C.

Features

- Operates from single 5V supply
- Input current: 150 nA max. over temperature
- Offset current: 20 nA max. over temperature
- Differential input voltage range: ±30V
- Power consumption: 135 mW at ±15V

Typical Applications**

Offset Balancing

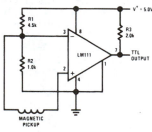

Detector for Magnetic Transducer

Relay Driver with Strobe

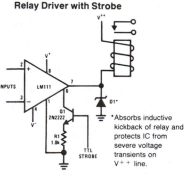

*Absorbs inductive kickback of relay and protects IC from severe voltage transients on V++ line.

Note: Do Not Ground Strobe Pin.

Strobing

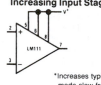

Note: Do Not Ground Strobe Pin. Output is turned off when current is pulled from Strobe Pin.

Increasing Input Stage Current*

*Increases typical common mode slew from 7.0V/μs to 18V/μs.

Digital Transmission Isolator

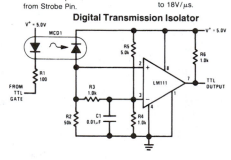

Strobing off Both Input* and Output Stages

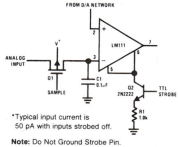

*Typical input current is 50 pA with inputs strobed off.

Note: Do Not Ground Strobe Pin.

Note: Pin connections shown on schematic diagram and typical applications are for H08 metal can package.

TL/H/5704–1

LM111/LM211/LM311

Absolute Maximum Ratings for the LM111/LM211

If Military/Aerospace specified devices are required, please contact the National Semiconductor Sales Office/Distributors for availability and specifications. (Note 7)

Total Supply Voltage (V_{84})	36V
Output to Negative Supply Voltage (V_{74})	50V
Ground to Negative Supply Voltage (V_{14})	30V
Differential Input Voltage	$\pm30V$
Input Voltage (Note 1)	$\pm15V$
Power Dissipation (Note 2)	500 mW
Output Short Circuit Duration	10 sec

Operating Temperature Range LM111	$-55°C$ to $125°C$
LM211	$-25°C$ to $85°C$
Storage Temperature Range	$-65°C$ to $150°C$
Lead Temperature (Soldering, 10 sec)	260°C
Voltage at Strobe Pin	$V^+ -5V$

Soldering Information
Dual-In-Line Package
Soldering (10 seconds) . . . . .260°C
Small Outline Package
Vapor Phase (60 seconds) .215°C
Infrared (15 seconds) . . . . .220°C
See AN-450 "Surface Mounting Methods and Their Effect on Product Reliability" for other methods of soldering surface mount devices.

ESD Rating (Note 8)   300V

Electrical Characteristics for the LM111 and LM211 (Note 3)

Parameter	Conditions	Min	Typ	Max	Units
Input Offset Voltage (Note 4)	$T_A = 25°C$, $R_S \leq 50k$		0.7	3.0	mV
Input Offset Current (Note 4)	$T_A = 25°C$		4.0	10	nA
Input Bias Current	$T_A = 25°C$		60	100	nA
Voltage Gain	$T_A = 25°C$	40	200		V/mV
Response Time (Note 5)	$T_A = 25°C$		200		ns
Saturation Voltage	$V_{IN} \leq -5$ mV, $I_{OUT} = 50$ mA $T_A = 25°C$		0.75	1.5	V
Strobe ON Current (Note 6)	$T_A = 25°C$	2.0	3.0	5.0	mA
Output Leakage Current	$V_{IN} \geq 5$ mV, $V_{OUT} = 35V$ $T_A = 25°C$, $I_{STROBE} = 3$ mA		0.2	10	nA
Input Offset Voltage (Note 4)	$R_S \leq 50$ k			4.0	mV
Input Offset Current (Note 4)				20	nA
Input Bias Current				150	nA
Input Voltage Range	$V^+ = 15V$, $V^- = -15V$, Pin 7 Pull-Up May Go To 5V	-14.5	13.8,-14.7	13.0	V
Saturation Voltage	$V^+ \geq 4.5V$, $V^- = 0$ $V_{IN} \leq -6$ mV, $I_{OUT} \leq 8$ mA		0.23	0.4	V
Output Leakage Current	$V_{IN} \geq 5$ mV, $V_{OUT} = 35V$		0.1	0.5	μA
Positive Supply Current	$T_A = 25°C$		5.1	6.0	mA
Negative Supply Current	$T_A = 25°C$		4.1	5.0	mA

Note 1: This rating applies for ±15 supplies. The positive input voltage limit is 30V above the negative supply. The negative input voltage limit is equal to the negative supply voltage or 30V below the positive supply, whichever is less.

Note 2: The maximum junction temperature of the LM111 is 150°C, while that of the LM211 is 110°C. For operating at elevated temperatures, devices in the H08 package must be derated based on a thermal resistance of 165°C/W, junction to ambient, or 20°C/W, junction to case. The thermal resistance of the dual-in-line package is 110°C/W, junction to ambient.

Note 3: These specifications apply for $V_S = \pm15V$ and Ground pin at ground, and $-55°C \leq T_A \leq +125°C$, unless otherwise stated. With the LM211, however, all temperature specifications are limited to $-25°C \leq T_A \leq +85°C$. The offset voltage, offset current and bias current specifications apply for any supply voltage from a single 5V supply up to $\pm15V$ supplies.

Note 4: The offset voltages and offset currents given are the maximum values required to drive the output within a volt of either supply with a 1 mA load. Thus, these parameters define an error band and take into account the worst-case effects of voltage gain and input impedance.

Note 5: The response time specified (see definitions) is for a 100 mV input step with 5 mV overdrive.

Note 6: This specification gives the range of current which must be drawn from the strobe pin to ensure the output is properly disabled. Do not short the strobe pin to ground; it should be current driven at 3 to 5 mA.

Note 7: Refer to RETS111X for the LM111H, LM111J and LM111J-8 military specifications.

Note 8: Human body model, 1.5 kΩ in series with 100 pF.

Absolute Maximum Ratings for the LM311

If Military/Aerospace specified devices are required, please contact the National Semiconductor Sales Office/Distributors for availability and specifications.

Total Supply Voltage (V_{84})	36V
Output to Negative Supply Voltage V_{74})	40V
Ground to Negative Supply Voltage V_{14})	30V
Differential Input Voltage	±30V
Input Voltage (Note 1)	±15V
Power Dissipation (Note 2)	500 mW
ESD Rating (Note 7)	300V

Output Short Circuit Duration	10 sec
Operating Temperature Range	0° to 70°C
Storage Temperature Range	−65°C to 150°C
Lead Temperature (soldering, 10 sec)	260°C
Voltage at Strobe Pin	$V^+ - 5V$

Soldering Information
Dual-In-Line Package
Soldering (10 seconds) . . . . .260°C
Small Outline Package
Vapor Phase (60 seconds) . . . . .215°C
Infrared (15 seconds) . . . . .220°C

See AN-450 "Surface Mounting Methods and Their Effect on Product Reliability" for other methods of soldering surface mount devices.

Electrical Characteristics for the LM311 (Note 3)

Parameter	Conditions	Min	Typ	Max	Units
Input Offset Voltage (Note 4)	$T_A = 25°C$, $R_S \leq 50k$		2.0	7.5	mV
Input Offset Current (Note 4)	$T_A = 25°C$		6.0	50	nA
Input Bias Current	$T_A = 25°C$		100	250	nA
Voltage Gain	$T_A = 25°C$	40	200		V/mV
Response Time (Note 5)	$T_A = 25°C$		200		ns
Saturation Voltage	$V_{IN} \leq -10$ mV, $I_{OUT} = 50$ mA $T_A = 25°C$		0.75	1.5	V
Strobe ON Current	$T_A = 25°C$	1.5	3.0		mA
Output Leakage Current	$V_{IN} \geq 10$ mV, $V_{OUT} = 35V$ $T_A = 25°C$, $I_{STROBE} = 3$ mA $V^- = V_{GRND} = -5V$		0.2	50	nA
Input Offset Voltage (Note 4)	$R_S \leq 50K$			10	mV
Input Offset Current (Note 4)				70	nA
Input Bias Current				300	nA
Input Voltage Range		−14.5	13.8, −14.7	13.0	V
Saturation Voltage	$V^+ \geq 4.5V$, $V^- = 0$ $V_{IN} \leq -10$ mV, $I_{OUT} \leq 8$ mA		0.23	0.4	V
Positive Supply Current	$T_A = 25°C$		5.1	7.5	mA
Negative Supply Current	$T_A = 25°C$		4.1	5.0	mA

Note 1: This rating applies for ±15V supplies. The positive input voltage limit is 30V above the negative supply. The negative input voltage limit is equal to the negative supply voltage or 30V below the positive supply, whichever is less.

Note 2: The maximum junction temperature of the LM311 is 110°C. For operating at elevated temperature, devices in the H08 package must be derated based on a thermal resistance of 165°C/W, junction to ambient, or 20°C/W, junction to case. The thermal resistance of the dual-in-line package is 100°C/W, junction to ambient.

Note 3: These specifications apply for $V_S = \pm 15V$ and the Ground pin at ground, and 0°C < T_A < +70°C, unless otherwise specified. The offset voltage, offset current and bias current specifications apply for any supply voltage from a single 5V supply up to ±15V supplies.

Note 4: The offset voltages and offset currents given are the maximum values required to drive the output within a volt of either supply with 1 mA load. Thus, these parameters define an error band and take into account the worst-case effects of voltage gain and input impedance.

Note 5: The response time specified (see definitions) is for a 100 mV input step with 5 mV overdrive.

Note 6: This specification gives the range of current which must be drawn from the strobe pin to ensure the output is properly disabled. Do not short the strobe pin to ground; it should be current driven at 3 to 5 mA.

Note 7: Human body model, 1.5 kΩ in series with 100 pF.

LM111/LM211

LM111/LM211 Typical Performance Characteristics

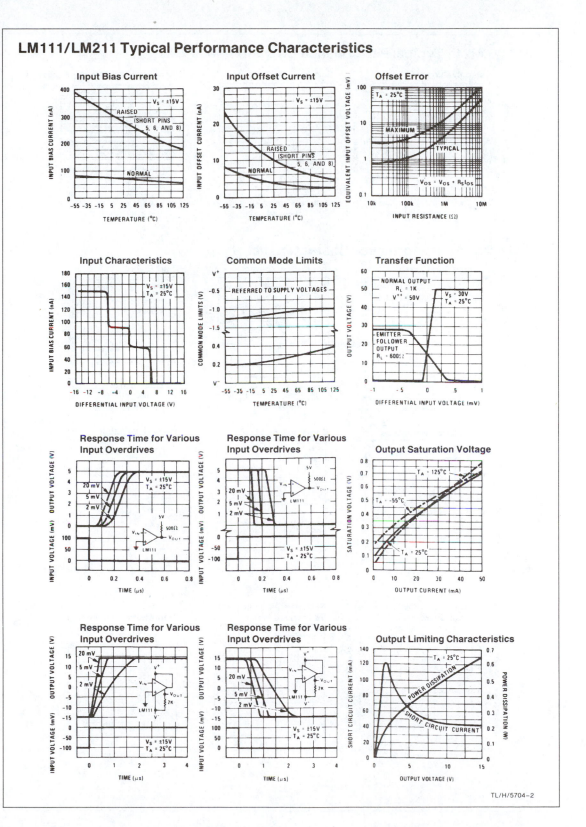

TL/H/5704–2

LM111/LM211 Typical Performance Characteristics (Continued)

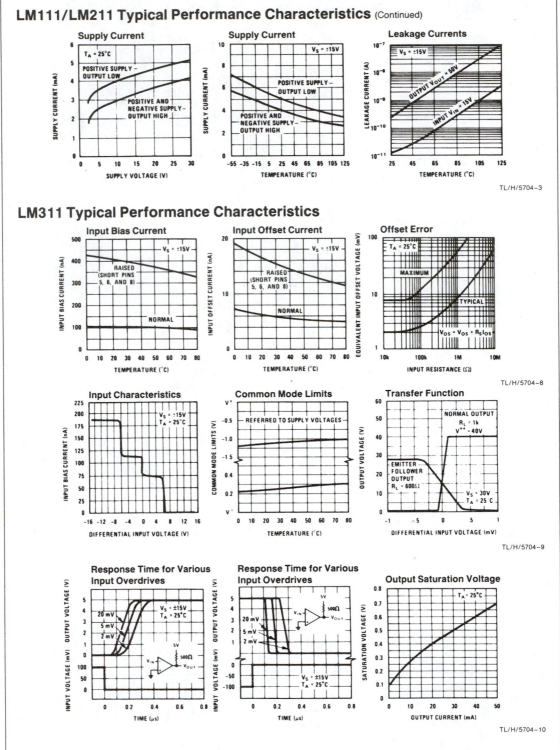

TL/H/5704-3

LM311 Typical Performance Characteristics

TL/H/5704-8

TL/H/5704-9

TL/H/5704-10

LM311 Typical Performance Characteristics (Continued)

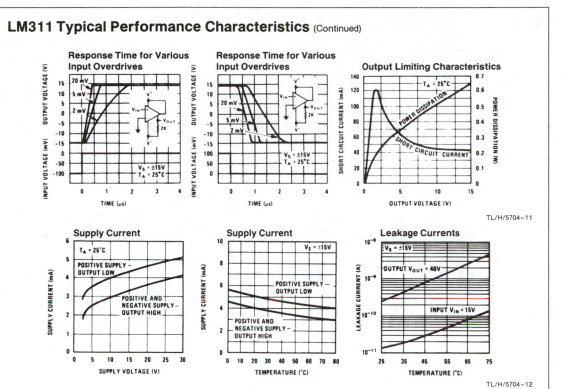

TL/H/5704–11

TL/H/5704–12

Schematic Diagram*

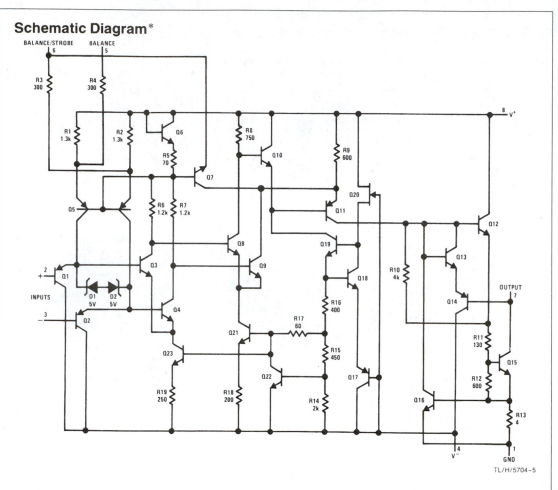

TL/H/5704–5

Connection Diagrams*

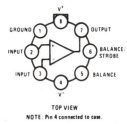

Metal Can Package

TOP VIEW
NOTE: Pin 4 connected to case.

Order Number LM111H,
LM211H or LM311H
See NS Package Number H08C

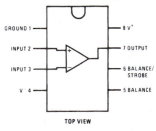

Dual-In-Line Package

TOP VIEW

Order Number LM111J-8, LM211J-8,
LM311J-8, LM311M or LM311N
See NS Package Number J08A,
M08A or N08E

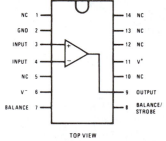

Dual-In-Line Package

TOP VIEW

TL/H/5704–6

Order Number LM111J, LM211J,
LM311J or LM311N-14
See NS Number Package
J14A or N14A

*Pin connections shown on schematic diagram are for H08 package.

ANALOG
DEVICES

Precision Instrumentation Amplifier

AD524

FEATURES
Low Noise: 0.3µV p-p 0.1Hz to 10Hz
Low Nonlinearity: 0.003% (G = 1)
High CMRR: 120dB (G = 1000)
Low Offset Voltage: 50µV
Low Offset Voltage Drift: 0.5µV/°C
Gain Bandwidth Product: 25MHz
Pin Programmable Gains of 1, 10, 100, 1000
Input Protection, Power On – Power Off
No External Components Required
Internally Compensated
MIL-STD-883B, Chips, and Plus Parts Available
16-Pin Ceramic DIP Package and 20-Terminal
   Leadless Chip Carriers Available

AD524 FUNCTIONAL BLOCK DIAGRAM

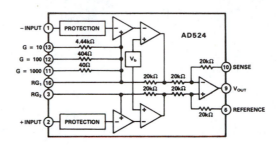

PRODUCT DESCRIPTION
The AD524 is a precision monolithic instrumentation amplifier designed for data acquisition applications requiring high accuracy under worst-case operating conditions. An outstanding combination of high linearity, high common mode rejection, low offset voltage drift, and low noise makes the AD524 suitable for use in many data acquisition systems.

The AD524 has an output offset voltage drift of less than 25µV/°C, input offset voltage drift of less than 0.5µV/°C, CMR above 90dB at unity gain (120dB at G = 1000) and maximum nonlinearity of 0.003% at G = 1. In addition to the outstanding dc specifications the AD524 also has a 25MHz gain bandwidth product (G = 100). To make it suitable for high speed data acquisition systems the AD524 has an output slew rate of 5V/µs and settles in 15µs to 0.01% for gains of 1 to 100.

As a complete amplifier the AD524 does not require any external components for fixed gains of 1, 10, 100 and 1,000. For other gain settings between 1 and 1000 only a single resistor is required. The AD524 input is fully protected for both power on and power off fault conditions.

The AD524 IC instrumentation amplifier is available in four different versions of accuracy and operating temperature range. The economical "A" grade, the low drift "B" grade and lower drift, higher linearity "C" grade are specified from −25°C to +85°C. The "S" grade guarantees performance to specification over the extended temperature range −55°C to +125°C. Devices are available in a 16-pin ceramic DIP package and a 20-terminal leadless chip carrier.

PRODUCT HIGHLIGHTS
1. The AD524 has guaranteed low offset voltage, offset voltage drift and low noise for precision high gain applications.

2. The AD524 is functionally complete with pin programmable gains of 1, 10, 100 and 1000, and single resistor programmable for any gain.

3. Input and output offset nulling terminals are provided for very high precision applications and to minimize offset voltage changes in gain ranging applications.

4. The AD524 is input protected for both power on and power off fault conditions.

5. The AD524 offers superior dynamic performance with a gain bandwidth product of 25MHz, full power response of 75kHz and a settling time of 15µs to 0.01% of a 20V step (G = 100).

SPECIFICATIONS (@ $V_S = \pm15V$, $R_L = 2k\Omega$ and $T_A = +25°C$ unless otherwise specified)

Model	AD524A Min	Typ	Max	AD524B Min	Typ	Max	AD524C Min	Typ	Max	AD524S Min	Typ	Max	Units
GAIN													
Gain Equation (External Resistor Gain Programming)		$\left[\frac{40,000}{R_G}+1\right]\pm20\%$			$\left[\frac{40,000}{R_G}+1\right]\pm20\%$			$\left[\frac{40,000}{R_G}+1\right]\pm20\%$			$\left[\frac{40,000}{R_G}+1\right]\pm20\%$		
Gain Range (Pin Programmable)	1 to 1000			1 to 1000			1 to 1000			1 to 1000			
Gain Error													
G = 1			±0.05			±0.03			±0.02			±0.05	%
G = 10			±0.25			±0.15			±0.1%			±0.25	%
G = 100			±0.5			±0.35			±0.25			±0.5	%
G = 1000			±2.0			±1.0			±0.5			±2.0	%
Nonlinearity													
G = 1			±0.01			±0.005			±0.003			±0.01	%
G = 10, 100			±0.01			±0.005			±0.003			±0.01	%
G = 1000			±0.01			±0.01			±0.01			±0.01	%
Gain vs. Temperature													
G = 1		5			5			5			5		ppm/°C
G = 10		15			10			10			10		ppm/°C
G = 100		35			25			25			25		ppm/°C
G = 1000		100			50			50			50		ppm/°C
VOLTAGE OFFSET (May be Nulled)													
Input Offset Voltage		250			100			50			100		μV
vs. Temperature		2			0.75			0.5			2.0		μV/°C
Output Offset Voltage		5			3			2.0			3.0		mV
vs. Temperature		100			50			25			50		μV/°C
Offset Referred to the Input vs. Supply													
G = 1	70			75			80			75			dB
G = 10	85			95			100			95			dB
G = 100	95			105			110			105			dB
G = 1000	100			110			115			110			dB
INPUT CURRENT													
Input Bias Current			±50			±25			±15			±50	nA
vs. Temperature		±100			±100			±100			±100		pA/°C
Input Offset Current			±35			±15			±10			±35	nA
vs. Temperature		±100			±100			±100			±100		pA/°C
INPUT													
Input Impedance													
Differential Resistance		10^9			10^9			10^9			10^9		Ω
Differential Capacitance		10			10			10			10		pF
Common Mode Resistance		10^9			10^9			10^9			10^9		Ω
Common Mode Capacitance		10			10			10			10		pF
Input Voltage Range													
Max Differ. Input Linear (V_{DL})[1]		±10			±10			±10			±10		V
Max Common Mode Linear (V_{CM})		$12V-\left(\frac{G}{2}\times V_D\right)$			$12V-\left(\frac{G}{2}\times V_D\right)$			$12V-\left(\frac{G}{2}\times V_D\right)$			$12V-\left(\frac{G}{2}\times V_D\right)$		V
Common Mode Rejection dc to 60Hz with 1kΩ Source Imbalance													
G = 1	70			75			80			70			dB
G = 10	90			95			100			90			dB
G = 100	100			105			110			100			dB
G = 1000	110			115			120			110			dB
OUTPUT RATING													
V_{OUT}, $R_L = 2k\Omega$		±10			±10			±10			±10		V
DYNAMIC RESPONSE													
Small Signal −3dB													
G = 1		1			1			1			1		MHz
G = 10		400			400			400			400		kHz
G = 100		150			150			150			150		kHz
G = 1000		25			25			25			25		kHz
Slew Rate		5.0			5.0			5.0			5.0		V/μs
Settling Time to 0.01%, 20V Step													
G = 1 to 100		15			15			15			15		μs
G = 1000		75			75			75			75		μs
NOISE													
Voltage Noise, 1kHz													
R.T.I.		7			7			7			7		nV/√Hz
R.T.O.		90			90			90			90		nV/√Hz
R.T.I., 0.1 to 10Hz													
G = 1		15			15			15			15		μV p-p
G = 10		2			2			2			2		μV p-p
G = 100, 1000		0.3			0.3			0.3			0.3		μV p-p
Current Noise 0.1Hz to 10Hz		60			60			60			60		pA p-p

AD524

Model	AD524A Min	AD524A Typ	AD524A Max	AD524B Min	AD524B Typ	AD524B Max	AD524C Min	AD524C Typ	AD524C Max	AD524S Min	AD524S Typ	AD524S Max	Units
SENSE INPUT													
R_{IN}		20			20			20			20		kΩ ± 20%
I_{IN}		15			15			15			15		μA
Voltage Range	± 10			± 10			± 10			± 10			V
Gain to Output		1			1			1			1		%
REFERENCE INPUT													
R_{IN}		40			40			40			40		kΩ ± 20%
I_{IN}		15			15			15			15		μA
Voltage Range	± 10			± 10			10			10			V
Gain to Output		1			1			1			1		%
TEMPERATURE RANGE													
Specified Performance	- 25		+ 85	25		+ 85	25		+ 85	55		+ 125	°C
Storage	- 65		+ 150	65		+ 150	65		+ 150	- 65		+ 150	°C
POWER SUPPLY													
Power Supply Range	± 6	± 15	± 18	± 6	± 15	± 18	± 6	± 15	± 18	± 6	± 15	± 18	V
Quiescent Current		3.5	5.0		3.5	5.0		3.5	5.0		3.5	5.0	mA
PACKAGE OPTIONS[2]													
16-Pin Ceramic (D-16)		AD524AD			AD524BD			AD524CD			AD524SD		
LCC DIP (E-20A)		AD524AE			AD524BE			AD524CE			AD524SE		

NOTES

[1]V_{DL} is the maximum differential input voltage at G = 1 for specified nonlinearity.
V_{DL} at other gains = 10V/G.
V_D = Actual differential input voltage.
   Example: G = 10, V_D = 0.50.
   V_{CM} = 12V − (10/2 × 0.50V) = 9.5V
[2]See Section 20 for package outline information.
Specifications subject to change without notice.

All min and max specifications are guaranteed. Specifications shown in **boldface** are tested on all production units at final electrical test. Results from those tests are used to calculate outgoing quality levels.

ABSOLUTE MAXIMUM RATINGS*

Supply Voltage . . . . ± 18V
Internal Power Dissipation 450mW
Input Voltage,
(Either Input Simultaneously) . . . . . . $|V_{IN}| + |V_S| < 36V$
Output Short Circuit Duration . . . . Indefinite
Storage Temperature Range
   D . . . . −65°C to +150°C
   E . . . . −65°C to +150°C
Operating Temperature Range
   AD524A/B/C . . . . −25°C to +85°C
   AD524S . . . . −55°C to +125°C
Lead Temperature Range (Soldering 60 seconds) . . +300°C

NOTES
*Stresses above those listed under "Absolute Maximum Ratings" may cause permanent damage to the device. This is a stress rating only and functional operation of the device at these or any other conditions above those indicated in the operational section of this specification is not implied. Exposure to absolute maximum rating conditions for extended periods may affect device reliability.

CONNECTION DIAGRAMS

Ceramic (D) Package

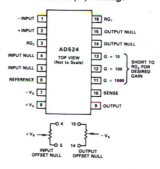

METALIZATION PHOTOGRAPH
Contact factory for latest dimensions.
Dimensions shown in inches and (mm).

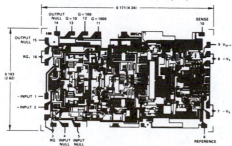

Leadless Chip Carrier (E) Package

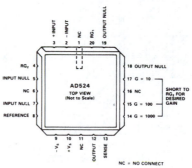

Typical Characteristics

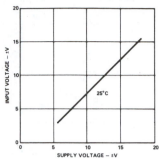

Figure 1. Input Voltage Range vs. Supply Voltage, G = 1

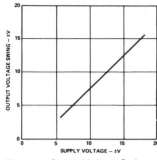

Figure 2. Output Voltage Swing vs. Supply Voltage

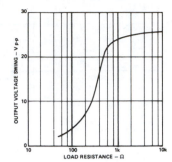

Figure 3. Output Voltage Swing vs. Load Resistance

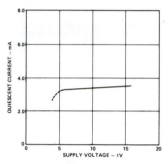

Figure 4. Quiescent Current vs. Supply Voltage

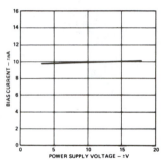

Figure 5. Input Bias Current vs. Supply Voltage

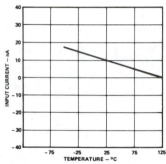

Figure 6. Input Bias Current vs. Temperature

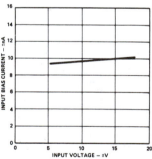

Figure 7. Input Bias Current vs. CMV

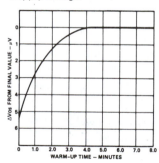

Figure 8. Offset Voltage, RTI, Turn On Drift

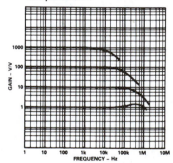

Figure 9. Gain vs. Frequency

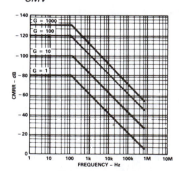

Figure 10. CMRR vs. Frequency RTI, Zero to 1k Source Imbalance

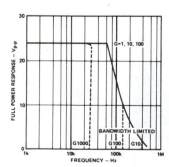

Figure 11. Large Signal Frequency Response

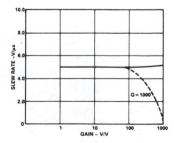

Figure 12. Slew Rate vs. Gain

Typical Characteristics — AD524

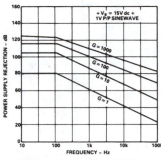

Figure 13. Positive PSRR vs. Frequency

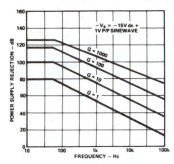

Figure 14. Negative PSRR vs. Frequency

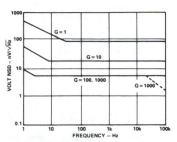

Figure 15. RTI Noise Spectral Density vs. Gain

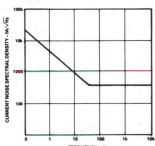

Figure 16. Input Current Noise

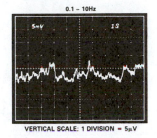

Figure 17. Low Frequency Noise – G = 1 (System Gain = 1000)

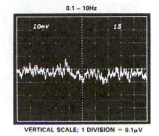

Figure 18. Low Frequency Noise – G = 1000 (System Gain = 100,000)

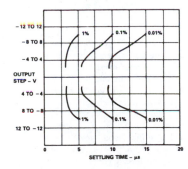

Figure 19. Settling Time Gain = 1

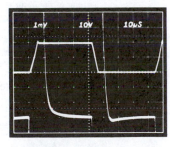

Figure 20. Large Signal Pulse Response and Settling Time – G = 1

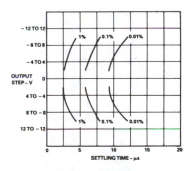

Figure 21. Settling Time Gain = 10

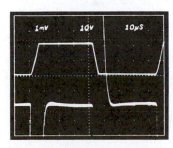

Figure 22. Large Signal Pulse Response and Settling Time G = 10

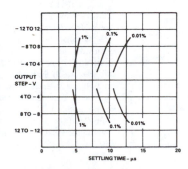

Figure 23. Settling Time Gain = 100

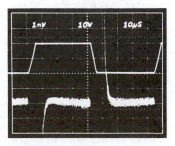

Figure 24. Large Signal Pulse Response and Settling Time G = 100

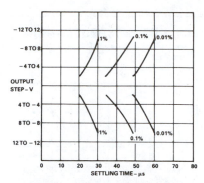

Figure 25. Settling Time Gain = 1000

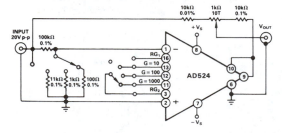

Figure 26. Large Signal Pulse Response and Settling Time G = 1000

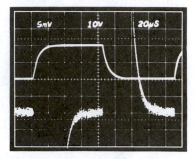

Figure 27. Settling Time Test Circuit

Theory of Operation

The AD524 is a monolithic instrumentation amplifier based on the classic 3 op amp circuit. The advantage of monolithic construction is the closely matched components that enhance the performance of the input preamp. The preamp section develops the programmed gain by the use of feedback concepts. The programmed gain is developed by varying the value of R_G (smaller values increase the gain) while the feedback forces the collector currents Q1, Q2, Q3 and Q4 to be constant which impresses the input voltage across R_G.

As R_G is reduced to increase the programmed gain, the transconductance of the input preamp increases to the transconductance of the input transistors. This has three important advantages. First, this approach allows the circuit to achieve a very high open loop gain of 3×10^8 at a programmed gain of 1000 thus reducing gain related errors to a negligible 30ppm. Second, the

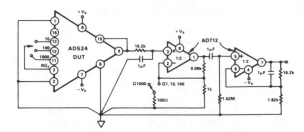

Figure 28. Noise Test Circuit

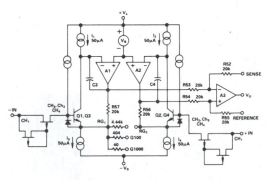

Figure 29. Simplified Circuit of Amplifier; Gain is Defined as $((R56 + R57)/(R_G)) + 1$. For a Gain of 1, R_G is an Open Circuit

gain bandwith product which is determined by C3 or C4 and the input transconductance, reaches 25MHz. Third, the input voltage noise reduces to a value determined by the collector current of the input transistors for an RTI noise of $7nV/\sqrt{Hz}$ at G = 1000.

INPUT PROTECTION

As interface amplifiers for data acquisition systems, instrumentation amplifiers are often subjected to input overloads, i.e., voltage levels in excess of the full scale for the selected gain range. At low gains, 10 or less, the gain resistor acts as a current limiting element in series with the inputs. At high gains the lower value of R_G will not adequately protect the inputs from excessive currents. Standard practice would be to place series limiting resistors in each input, but to limit input current to below 5mA with a full differential overload (36V) would require over 7k of resistance which would add $10nV\sqrt{Hz}$ of noise. To provide both input protection and low noise a special series protect FET was used.

A unique FET design was used to provide a bidirectional current limit, thereby, protecting against both positive and negative overloads. Under nonoverload conditions, three channels CH_2, CH_3, CH_4, act as a resistance ($\approx 1k\Omega$) in series with the input as before. During an overload in the positive direction, a fourth channel, CH_1, acts as a small resistance ($\approx 3k\Omega$) in series with the gate, which draws only the leakage current, and the FET limits I_{DSS}. When the FET enhances under a negative overload, the gate current must go through the small FET formed by CH_1 and when this FET goes into saturation, the gate current is limited and the main FET will go into controlled enhancement. The bidirectional limiting holds the maximum input current to 3mA over the 36V range.

Applying the AD524

INPUT OFFSET AND OUTPUT OFFSET

Voltage offset specifications are often considered a figure of merit for instrumentation amplifiers. While initial offset may be adjusted to zero, shifts in offset voltage due to temperature variations will cause errors. Intelligent systems can often correct for this factor with an auto-zero cycle, but there are many small-signal high-gain applications that don't have this capability.

Voltage offset and drift comprise two components each; input and output offset and offset drift. Input offset is that component of offset that is directly proportional to gain i.e., input offset as measured at the output at G = 100 is 100 times greater than at G = 1. Output offset is independent of gain. At low gains, output offset drift is dominant, while at high gains input offset drift dominates. Therefore, the output offset voltage drift is normally specified as drift at G = 1 (where input effects are insignificant), while input offset voltage drift is given by drift specification at a high gain (where output offset effects are negligible). All input-related numbers are referred to the input (RTI) which is to say that the effect on the output is "G" times larger. Voltage offset vs. power supply is also specified at one or more gain settings and is also RTI.

By separating these errors, one can evaluate the total error independent of the gain setting used. In a given gain configuration both errors can be combined to give a total error referred to the input (R.T.I.) or output (R.T.O.) by the following formula:

Total Error R.T.I. = input error + (output error/gain)

Total Error R.T.O. = (Gain × input error) + output error

As an illustration, a typical AD524 might have a +250μV output offset and a −50μV input offset. In a unity gain configuration, the *total* output offset would be 200μV or the sum of the two. At a gain of 100, the output offset would be −4.75mV or: +250μV + 100(−50μV) = −4.75mV.

The AD524 provides for both input and output offset adjustment. This simplifies very high precision applications and minimize offset voltage changes in switched gain applications. In such applications the input offset is adjusted first at the highest programmed gain, then the output offset is adjusted at G = 1.

GAIN

The AD524 has internal high accuracy pretrimmed resistors for pin programmable gain of 1, 10, 100 and 1000. One of the preset gains can be selected by pin strapping the appropriate gain terminal and RG$_2$ together (for G = 1 RG$_2$ is not connected).

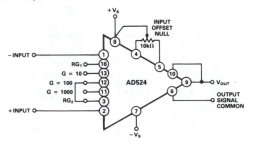

Figure 30. Operating Connections for G = 100

The AD524 can be configured for gains other than those that are internally preset; there are two methods to do this. The first method uses just an external resistor connected between pins 3 and 16 which programs the gain according to the formula $R_G = \frac{40k}{G-1}$ (see Figure 31). For best results R_G should be a precision resistor with a low temperature coefficient. An external R_G affects both gain accuracy and gain drift due to the mismatch between it and the internal thin-film resistors. Gain accuracy is determined by the tolerance of the external R_G and the absolute accuracy of the internal resistors (± 20%). Gain drift is determined by the mismatch of the temperature coefficient of R_G and the temperature coefficient of the internal resistors (− 50ppm/°C typ).

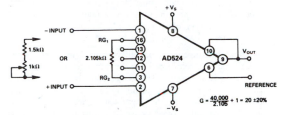

Figure 31. Operating Connections for G = 20

The second technique uses the internal resistors in parallel with an external resistor (Figure 32). This technique minimizes the gain adjustment range and reduces the effects of temperature coefficient sensitivity.

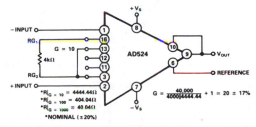

Figure 32. Operating Connections for G = 20, Low Gain T.C. Technique

The AD524 may also be configured to provide gain in the output stage. Figure 33 shows an H pad attenuator connected to the reference and sense lines of the AD524. R1, R2 and R3 should be made as low as possible to minimize the gain variation and reduction of CMRR. Varying R2 will precisely set the gain without affecting CMRR. CMRR is determined by the match of R1 and R3.

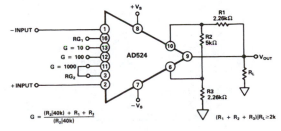

Figure 33. Gain of 2000

Output Gain	R2	R1,R3	Nominal Gain
2	5kΩ	2.26kΩ	2.02
5	1.05kΩ	2.05kΩ	5.01
10	1kΩ	4.42kΩ	10.1

Table I. Output Gain Resistor Values

INPUT BIAS CURRENTS

Input bias currents are those currents necessary to bias the input transistors of a dc amplifier. Bias currents are an additional source of input error and must be considered in an total error budget. The bias currents when multiplied by the source resistance appear as an offset voltage. What is of concern in calculating bias current errors is the change in bias current with respect to signal voltage and temperature. Input offset current is the difference between the two input bias currents. The effect of offset current is an input offset voltage whose magnitude is the offset current times the source impedance imbalance.

Although instrumentation amplifiers have differential inputs, there must be a return path for the bias currents. If this is not provided, those currents will charge stray capacitances, causing the output to drift uncontrollably or to saturate. Therefore, when amplifying "floating" input sources such as transformers and thermocouples, as well as ac-coupled sources, there must still be a dc path from each input to ground.

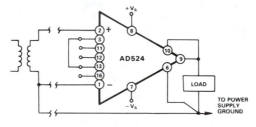

a. Transformer Coupled

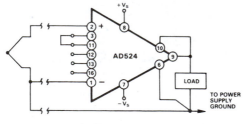

b. Thermocouple

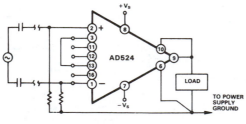

c. AC Coupled

Figure 34. Indirect Ground Returns for Bias Currents

COMMON-MODE REJECTION

Common-mode rejection is a measure of the change in output voltage when both inputs are changed equal amounts. These specifications are usually given for a full-range input voltage change and a specified source imbalance. "Common-Mode Rejection Ratio" (CMRR) is a ratio expression while "Common-Mode Rejection" (CMR) is the logarithm of that ratio. For example, a CMRR of 10,000 corresponds to a CMR of 80dB.

In an instrumentation amplifier, ac common-mode rejection is only as good as the differential phase shift. Degradation of ac common-mode rejection is caused by unequal drops across differing track resistances and a differential phase shift due to varied stray capacitances or cable capacitances. In many applications shielded cables are used to minimize noise. This technique can create common mode rejection errors unless the shield is properly driven. Figures 35 and 36 shows active data guards which are configured to improve ac common mode rejection by "bootstrapping" the capacitances of the input cabling, thus minimizing differential phase shift.

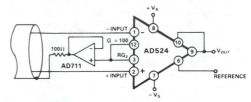

Figure 35. Shield Driver, G ≥ 100

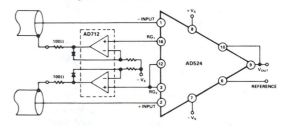

Figure 36. Differential Shield Driver

GROUNDING

Many data-acquisition components have two or more ground pins which are not connected together within the device. These grounds must be tied together at one point, usually at the system power-supply ground. Ideally, a single solid ground would be desirable. However, since current flows through the ground wires and etch stripes of the circuit cards, and since these paths

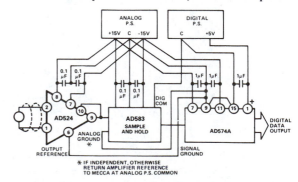

Figure 37. Basic Grounding Practice

AD524

have resistance and inductance, hundreds of millivolts can be generated between the system ground point and the data acquisition components. Separate ground returns should be provided to minimize the current flow in the path from the sensitive points to the system ground point. In this way supply currents and logic-gate return currents are not summed into the same return path as analog signals where they would cause measurement errors.

Since the output voltage is developed with respect to the potential on the reference terminal an instrumentation amplifier can solve many grounding problems.

SENSE TERMINAL

The sense terminal is the feedback point for the instrument amplifier's output amplifier. Normally it is connected to the instrument amplifier output. If heavy load currents are to be drawn through long leads, voltage drops due to current flowing through lead resistance can cause errors. The sense terminal can be wired to the instrument amplifier at the load thus putting the IxR drops "inside the loop" and virtually eliminating this error source.

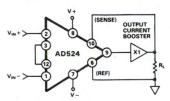

Figure 38. AD524 Instrumentation Amplifier with Output Current Booster

Typically, IC instrumentation amplifiers are rated for a full ± 10 volt output swing into 2kΩ. In some applications, however, the need exists to drive more current into heavier loads. Figure 38 shows how a high-current booster may be connected "inside the loop" of an instrumentation amplifier to provide the required current boost without significantly degrading overall performance. Nonlinearities, offset and gain inaccuracies of the buffer are minimized by the loop gain of the IA output amplifier. Offset drift of the buffer is similarly reduced.

REFERENCE TERMINAL

The reference terminal may be used to offset the output by up to ± 10V. This is useful when the load is "floating" or does not share a ground with the rest of the system. It also provides a direct means of injecting a precise offset. It must be remembered that the total output swing is ± 10 volts to be shared between signal and reference offset.

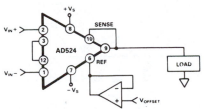

Figure 39. Use of Reference Terminal to Provide Output Offset

When the IA is of the three-amplifier configuration it is necessary that nearly zero impedance be presented to the reference terminal.

Any significant resistance from the reference terminal to ground increases the gain of the noninverting signal path thereby upsetting the common-mode rejection of the IA.

In the AD524 a reference source resistance will unbalance the CMR trim by the ratio of $20k\Omega/R_{REF}$. For example, if the reference source impedance is 1Ω, CMR will be reduced to 86dB ($20k\Omega/1\Omega = 86dB$). An operational amplifier may be used to provide that low impedance reference point as shown in Figure 39. The input offset voltage characteristics of that amplifier will add directly to the output offset voltage performance of the instrumentation amplifier.

An instrumentation amplifier can be turned into a voltage-to-current converter by taking advantage of the sense and reference terminals as shown in Figure 40.

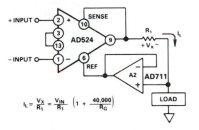

$$I_L = \frac{V_X}{R_1} = \frac{V_{IN}}{R_1}\left(1 + \frac{40,000}{R_G}\right)$$

Figure 40. Voltage-to-Current Converter

By establishing a reference at the "low" side of a current setting resistor, an output current may be defined as a function of input voltage, gain and the value of that resistor. Since only a small current is demanded at the input of the buffer amplifier A_2, the forced current I_L will largely flow through the load. Offset and drift specifications of A_2 must be added to the output offset and drift specifications of the IA.

PROGRAMMABLE GAIN

Figure 41 shows the AD524 being used as a software programmable gain amplifier. Gain switching can be accomplished with mechanical switches such as DIP switches or reed relays. It should be noted that the "on" resistance of the switch in series with the internal gain resistor becomes part of the gain equation and will have an effect on gain accuracy.

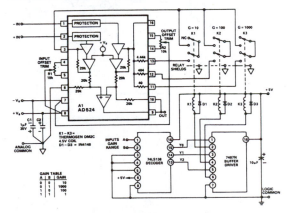

Figure 41. 3 Decade Gain Programmable Amplifier

The AD524 can also be connected for gain in the output stage. Figure 42 shows an AD547 used as an active attenuator in the output amplifier's feedback loop. The active attenuation presents a very low impedance to the feedback resistors therefore minimizing the common rejection ratio degradation.

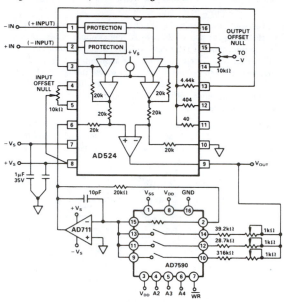

Figure 42. Programmable Output Gain

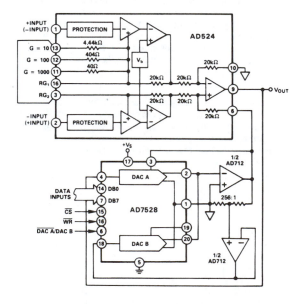

Figure 43. Programmable Output Gain Using a DAC

Another method for developing the switching scheme is to use a DAC. The AD7528 dual DAC which acts essentially as a pair of switched resistive attenuators having high analog linearity and symmetrical bipolar transmission is ideal in this application. The multiplying DAC's advantage is that it can handle inputs of either polarity or zero without affecting the programmed gain. The circuit shown uses an AD7528 to set the gain (DAC A) and to perform a fine adjustment (DAC B).

AUTO-ZERO CIRCUITS

In many applications it is necessary to provide very accurate data in high gain configurations. At room temperature the offset effects can be nulled by the use of offset trimpots. Over the operating temperature range, however, offset nulling becomes a problem. The circuit of Figure 44 show a CMOS DAC operating in the bipolar mode and connected to the reference terminal to provide software controllable offset adjustments.

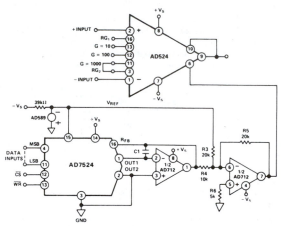

In many applications complex software algorithms for auto-zero applications are not available. For those applications Figure 45 provides a hardware solution.

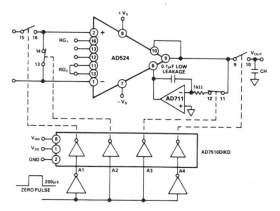

Figure 45. Auto-Zero Circuit

Error Budget Analysis — AD524

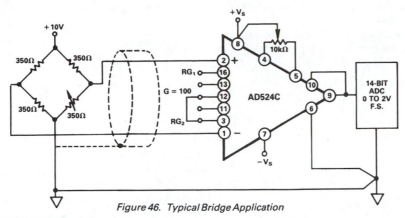

Figure 46. Typical Bridge Application

ERROR BUDGET ANALYSIS

To illustrate how instrumentation amplifier specifications are applied, we will now examine a typical case where an AD524 is required to amplify the output of an unbalanced transducer. Figure 46 shows a differential transducer, unbalanced by 100Ω, supplying a 0 to 20mV signal to an AD524C. The output of the IA feeds a 14-bit A to D converter with a 0 to 2 volt input voltage range. The operating temperature range is −25°C to +85°C. Therefore, the largest change in temperature ΔT within the operating range is from ambient to +85°C (85°C − 25°C =60°C).

In many applications, differential linearity and resolution are of prime importance. This would be so in cases where the absolute value of a variable is less important than changes in value. In these applications, only the irreducible errors (45ppm = 0.004%) are significant. Furthermore, if a system has an intelligent processor monitoring the A to D output, the addition of a auto-gain/auto-zero cycle will remove all reducible errors and may eliminate the requirement for initial calibration. This will also reduce errors to 0.004%.

Error Source	AD524C Specifications	Calculation	Effect on Absolute Accuracy at T_A = 25°C	Effect on Absolute Accuracy at T_A = 85°C	Effect on Resolution
Gain Error	±0.25%	±0.25% = 2500ppm	2500ppm	2500ppm	–
Gain Instability	25ppm	(25ppm/°C)(60°C) = 1500ppm	–	1500ppm	–
Gain Nonlinearity	±0.003%	±0.003% = 30ppm	–	–	30ppm
Input Offset Voltage	±50μV, RTI	±50μV/20mV = ±2500ppm	2500ppm	2500ppm	–
Input Offset Voltage Drift	±0.5μV/°C	(±0.5μV/°C)(60°C) = 30μV 30μV/20mV = 1500ppm	–	1500ppm	–
Output Offset Voltage[1]	±2.0mV	±2.0mV/20mV = 1000ppm	1000ppm	1000ppm	–
Output Offset Voltage Drift[1]	±25μV/°C	(±25μV/°C)(60°C) = 1500μV 1500μV/20mV = 750ppm	–	750ppm	–
Bias Current – Source Imbalance Error	±15nA	(±15nA)(100Ω) = 1.5μV 1.5μV/20mV = 75ppm	75ppm	75ppm	–
Bias Current – Source Imbalance Drift	±100pA/°C	(±100pA/°C)(100Ω)(60°C) = 0.6μV 0.6μV/20mV = 30ppm	–	30ppm	–
Offset Current – Source Imbalance Error	±10nA	(±10nA)(100Ω) = 1μV 1μV/20mV = 50ppm	50ppm	50ppm	–
Offset Current – Source Imbalance Drift	±100pA/°C	(100pA/°C)(100Ω)(60°C) = 0.6μV 0.6μV/20mV = 30ppm	–	30ppm	–
Offset Current – Source Resistance – Error	±10nA	(10nA)(175Ω) = 3.5μV 3.5μV/20mV = 87.5ppm	87.5ppm	87.5ppm	–
Offset Current – Source Resistance – Drift	±100pA/°C	(100pA/°C)(175Ω)(60°C) = 1μV 1μV/20mV = 50ppm	–	50ppm	–
Common Mode Rejection 5V dc	115dB	115dB = 1.8ppm × 5V = 8.8μV 8.8μV/20mV = 444ppm	444ppm	444ppm	–
Noise, RTI (0.1–10Hz)	0.3μV p-p	0.3μV p-p/20mV = 15ppm	–	–	15ppm
		Total Error	6656.5ppm	10516.5ppm	45ppm

[1]Output offset voltage and output offset voltage drift are given as RTI figures.

Table II. Error Budget Analysis of AD524CD in Bridge Application

Figure 47 shows a simple application, in which the variation of the cold-junction voltage of a Type J thermocouple–iron(+)–constantan– is compensated for by a voltage developed in series by the temperature-sensitive output current of an AD590 semiconductor temperature sensor.

The circuit is calibrated by adjusting R_T for proper output voltage with the measuring junction at a known reference tem-

perature and the circuit near 25°C. If resistors with low tempcos are used, compensation accuracy will be to within ± 0.5°C, for temperatures between + 15°C and + 35°C. Other thermocouple types may be accommodated with the standard resistance values shown in the table. For other ranges of ambient temperature, the equation in the figure may be solved for the optimum values of R_T and R_A.

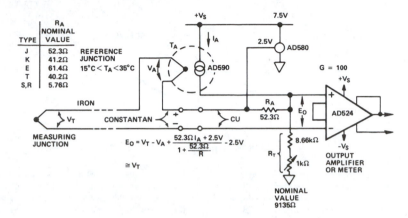

Figure 47. Cold-Junction Compensation

The microprocessor controlled data acquisition system shown in Figure 48 includes both auto-zero and auto-gain capability. By dedicating two of the differential inputs, one to ground and one to the A/D reference, the proper program calibration cycles can eliminate both initial accuracy errors and accuracy errors over temperature. The auto-zero cycle, in this application, converts a

number that appears to be ground and then writes that same number (8 bit) to the AD7524 which eliminates the zero error since its output has an inverted scale. The auto-gain cycle converts the A/D reference and compares it with full scale. A multiplicative correction factor is then computed and applied to subsequent readings.

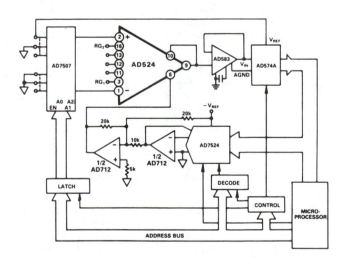

Figure 48. Microprocessor Controlled Data Acquisition System

 **National
Semiconductor**

54LS00/DM54LS00/DM74LS00
Quad 2-Input NAND Gates

General Description
This device contains four independent gates each of which performs the logic NAND function.

Features
■ Alternate Military/Aerospace device (54LS00) is available. Contact a National Semiconductor Sales Office/Distributor for specifications.

Connection Diagram

Dual-In-Line Package

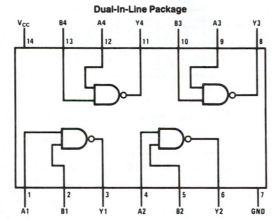

TL/F/6439-1

**Order Number 54LS00DMQB, 54LS00FMQB, 54LS00LMQB, DM54LS00J, DM54LS00W, DM74LS00M or DM74LS00N
See NS Package Number E20A, J14A, M14A, N14A or W14B**

Function Table

$$Y = \overline{AB}$$

Inputs		Output
A	**B**	**Y**
L	L	H
L	H	H
H	L	H
H	H	L

H = High Logic Level
L = Low Logic Level

Absolute Maximum Ratings (Note)

If Military/Aerospace specified devices are required, please contact the National Semiconductor Sales Office/Distributors for availability and specifications.

Supply Voltage	7V
Input Voltage	7V
Operating Free Air Temperature Range	
DM54LS and 54LS	$-55°C$ to $+125°C$
DM74LS	$0°C$ to $+70°C$
Storage Temperature Range	$-65°C$ to $+150°C$

Note: The "Absolute Maximum Ratings" are those values beyond which the safety of the device cannot be guaranteed. The device should not be operated at these limits. The parametric values defined in the "Electrical Characteristics" table are not guaranteed at the absolute maximum ratings. The "Recommended Operating Conditions" table will define the conditions for actual device operation.

Recommended Operating Conditions

Symbol	Parameter	DM54LS00			DM74LS00			Units
		Min	Nom	Max	Min	Nom	Max	
V_{CC}	Supply Voltage	4.5	5	5.5	4.75	5	5.25	V
V_{IH}	High Level Input Voltage	2			2			V
V_{IL}	Low Level Input Voltage			0.7			0.8	V
I_{OH}	High Level Output Current			-0.4			-0.4	mA
I_{OL}	Low Level Output Current			4			8	mA
T_A	Free Air Operating Temperature	-55		125	0		70	°C

Electrical Characteristics over recommended operating free air temperature range (unless otherwise noted)

Symbol	Parameter	Conditions		Min	Typ (Note 1)	Max	Units
V_I	Input Clamp Voltage	V_{CC} = Min, I_I = -18 mA				-1.5	V
V_{OH}	High Level Output Voltage	V_{CC} = Min, I_{OH} = Max, V_{IL} = Max	DM54	2.5	3.4		V
			DM74	2.7	3.4		
V_{OL}	Low Level Output Voltage	V_{CC} = Min, I_{OL} = Max, V_{IH} = Min	DM54		0.25	0.4	V
			DM74		0.35	0.5	
		I_{OL} = 4 mA, V_{CC} = Min	DM74		0.25	0.4	
I_I	Input Current @ Max Input Voltage	V_{CC} = Max, V_I = 7V				0.1	mA
I_{IH}	High Level Input Current	V_{CC} = Max, V_I = 2.7V				20	μA
I_{IL}	Low Level Input Current	V_{CC} = Max, V_I = 0.4V				-0.36	mA
I_{OS}	Short Circuit Output Current	V_{CC} = Max (Note 2)	DM54	-20		-100	mA
			DM74	-20		-100	
I_{CCH}	Supply Current with Outputs High	V_{CC} = Max			0.8	1.6	mA
I_{CCL}	Supply Current with Outputs Low	V_{CC} = Max			2.4	4.4	mA

Switching Characteristics at V_{CC} = 5V and T_A = 25°C (See Section 1 for Test Waveforms and Output Load)

Symbol	Parameter	R_L = 2 kΩ				Units
		C_L = 15 pF		C_L = 50 pF		
		Min	Max	Min	Max	
t_{PLH}	Propagation Delay Time Low to High Level Output	3	10	4	15	ns
t_{PHL}	Propagation Delay Time High to Low Level Output	3	10	4	15	ns

Note 1: All typicals are at V_{CC} = 5V, T_A = 25°C.

Note 2: Not more than one output should be shorted at a time, and the duration should not exceed one second.

 **National Semiconductor**

54LS04/DM54LS04/DM74LS04 Hex Inverting Gates

General Description

This device contains six independent gates each of which performs the logic INVERT function.

Features

■ Alternate Military/Aerospace device (54LS04) is available. Contact a National Semiconductor Sales Office/Distributor for specifications.

Connection Diagram

Dual-In-Line Package

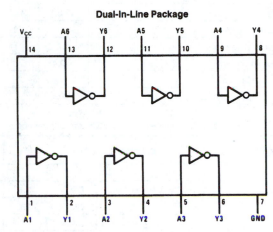

TL/F/6345–1

Order Number 54LS04DMQB, 54LS04FMQB, 54LS04LMQB, DM54LS04J, DM54LS04W, DM74LS04M or DM74LS04N
See NS Package Number E20A, J14A, M14A, N14A or W14B

Function Table

$$Y = \overline{A}$$

Input	Output
A	Y
L	H
H	L

H = High Logic Level
L = Low Logic Level

LS04

Absolute Maximum Ratings (Note)

If Military/Aerospace specified devices are required, please contact the National Semiconductor Sales Office/Distributors for availability and specifications.

Supply Voltage	7V
Input Voltage	7V
Operating Free Air Temperature Range	
DM54LS and 54LS	−55°C to +125°C
DM74LS	0°C to +70°C
Storage Temperature Range	−65°C to +150°C

Note: The "Absolute Maximum Ratings" are those values beyond which the safety of the device cannot be guaranteed. The device should not be operated at these limits. The parametric values defined in the "Electrical Characteristics" table are not guaranteed at the absolute maximum ratings. The "Recommended Operating Conditions" table will define the conditions for actual device operation.

Recommended Operating Conditions

Symbol	Parameter	DM54LS04			DM74LS04			Units
		Min	Nom	Max	Min	Nom	Max	
V_{CC}	Supply Voltage	4.5	5	5.5	4.75	5	5.25	V
V_{IH}	High Level Input Voltage	2			2			V
V_{IL}	Low Level Input Voltage			0.7			0.8	V
I_{OH}	High Level Output Current			−0.4			−0.4	mA
I_{OL}	Low Level Output Current			4			8	mA
T_A	Free Air Operating Temperature	−55		125	0		70	°C

Electrical Characteristics over recommended operating free air temperature range (unless otherwise noted)

Symbol	Parameter	Conditions		Min	Typ (Note 1)	Max	Units
V_I	Input Clamp Voltage	V_{CC} = Min, I_I = −18 mA				−1.5	V
V_{OH}	High Level Output Voltage	V_{CC} = Min, I_{OH} = Max, V_{IL} = Max	DM54	2.5	3.4		V
			DM74	2.7	3.4		
V_{OL}	Low Level Output Voltage	V_{CC} = Min, I_{OL} = Max, V_{IH} = Min	DM54		0.25	0.4	V
			DM74		0.35	0.5	
		I_{OL} = 4 mA, V_{CC} = Min	DM74		0.25	0.4	
I_I	Input Current @ Max Input Voltage	V_{CC} = Max, V_I = 7V				0.1	mA
I_{IH}	High Level Input Current	V_{CC} = Max, V_I = 2.7V				20	μA
I_{IL}	Low Level Input Current	V_{CC} = Max, V_I = 0.4V				−0.36	mA
I_{OS}	Short Circuit Output Current	V_{CC} = Max (Note 2)	DM54	−20		−100	mA
			DM74	−20		−100	
I_{CCH}	Supply Current with Outputs High	V_{CC} = Max			1.2	2.4	mA
I_{CCL}	Supply Current with Outputs Low	V_{CC} = Max			3.6	6.6	mA

Switching Characteristics at V_{CC} = 5V and T_A = 25°C (See Section 1 for Test Waveforms and Output Load)

Symbol	Parameter	R_L = 2 kΩ				Units
		C_L = 15 pF		C_L = 50 pF		
		Min	Max	Min	Max	
t_{PLH}	Propagation Delay Time Low to High Level Output	3	10	4	15	ns
t_{PHL}	Propagation Delay Time High to Low Level Output	3	10	4	15	ns

Note 1: All typicals are at V_{CC} = 5V, T_A = 25°C.

Note 2: Not more than one output should be shorted at a time, and the duration should not exceed one second.

54LS08/DM54LS08/DM74LS08 Quad 2-Input AND Gates

General Description
This device contains four independent gates each of which performs the logic AND function.

Features
■ Alternate Military/Aerospace device (54LS08) is available. Contact a National Semiconductor Sales Office/Distributor for specifications.

Connection Diagram

Dual-In-Line Package

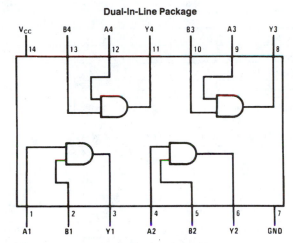

TL/F/6347–1

Order Number 54LS08DMQB, 54LS08FMQB, 54LS08LMQB, DM54LS08J, DM54LS08W, DM74LS08M or DM74LS08N
See NS Package Number E20A, J14A, M14A, N14A or W14B

Function Table

Y = AB

Inputs		Output
A	B	Y
L	L	L
L	H	L
H	L	L
H	H	H

H = High Logic Level
L = Low Logic Level

LS08

Absolute Maximum Ratings (Note)

If Military/Aerospace specified devices are required, please contact the National Semiconductor Sales Office/Distributors for availability and specifications.

Supply Voltage	7V
Input Voltage	7V
Operating Free Air Temperature Range	
DM54LS and 54LS	−55°C to +125°C
DM74LS	0°C to +70°C
Storage Temperature Range	−65°C to +150°C

Note: The "Absolute Maximum Ratings" are those values beyond which the safety of the device cannot be guaranteed. The device should not be operated at these limits. The parametric values defined in the "Electrical Characteristics" table are not guaranteed at the absolute maximum ratings. The "Recommended Operating Conditions" table will define the conditions for actual device operation.

Recommended Operating Conditions

Symbol	Parameter	DM54LS08			DM74LS08			Units
		Min	Nom	Max	Min	Nom	Max	
V_{CC}	Supply Voltage	4.5	5	5.5	4.75	5	5.25	V
V_{IH}	High Level Input Voltage	2			2			V
V_{IL}	Low Level Input Voltage			0.7			0.8	V
I_{OH}	High Level Output Current			−0.4			−0.4	mA
I_{OL}	Low Level Output Current			4			8	mA
T_A	Free Air Operating Temperature	−55		125	0		70	°C

Electrical Characteristics over recommended operating free air temperature range (unless otherwise noted)

Symbol	Parameter	Conditions		Min	Typ (Note 1)	Max	Units
V_I	Input Clamp Voltage	V_{CC} = Min, I_I = −18 mA				−1.5	V
V_{OH}	High Level Output Voltage	V_{CC} = Min, I_{OH} = Max, V_{IH} = Min	DM54	2.5	3.4		V
			DM74	2.7	3.4		
V_{OL}	Low Level Output Voltage	V_{CC} = Min, I_{OL} = Max, V_{IL} = Max	DM54		0.25	0.4	V
			DM74		0.35	0.5	
		I_{OL} = 4 mA, V_{CC} = Min	DM74		0.25	0.4	
I_I	Input Current @ Max Input Voltage	V_{CC} = Max, V_I = 7V				0.1	mA
I_{IH}	High Level Input Current	V_{CC} = Max, V_I = 2.7V				20	µA
I_{IL}	Low Level Input Current	V_{CC} = Max, V_I = 0.4V				−0.36	mA
I_{OS}	Short Circuit Output Current	V_{CC} = Max (Note 2)	DM54	−20		−100	mA
			DM74	−20		−100	
I_{CCH}	Supply Current with Outputs High	V_{CC} = Max			2.4	4.8	mA
I_{CCL}	Supply Current with Outputs Low	V_{CC} = Max			4.4	8.8	mA

Switching Characteristics at V_{CC} = 5V and T_A = 25°C (See Section 1 for Test Waveforms and Output Load)

Symbol	Parameter	R_L = 2 kΩ				Units
		C_L = 15 pF		C_L = 50 pF		
		Min	Max	Min	Max	
t_{PLH}	Propagation Delay Time Low to High Level Output	4	13	6	18	ns
t_{PHL}	Propagation Delay Time High to Low Level Output	3	11	5	18	ns

Note 1: All typicals are at V_{CC} = 5V, T_A = 25°C.

Note 2: Not more than one output should be shorted at a time, and the duration should not exceed one second.

LS47

National Semiconductor

54LS47/DM74LS47
BCD to 7-Segment Decoder/Driver

General Description

The 'LS47 accepts four lines of BCD (8421) input data, generates their complements internally and decodes the data with seven AND/OR gates having open-collector outputs to drive indicator segments directly. Each segment output is guaranteed to sink 24 mA in the ON (LOW) state and withstand 15V in the OFF (HIGH) state with a maximum leakage current of 250 μA. Auxiliary inputs provided blanking, lamp test and cascadable zero-suppression functions. Also see the 'LS47 data sheet.

Features

- Open-collector outputs
- Drive indicator segments directly
- Cascadable zero-suppression capability
- Lamp test input

Connection Diagram

Dual-In-Line Package

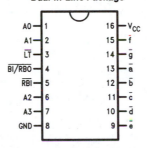

TL/F/9817–1

**Order Number 54LS47DMQB, 54LS47FMQB,
DM74LS47M or DM74LS47N
See NS Package Number J16A, M16A, N16E or W16A**

Pin Names	Description
A0–A3	BCD Inputs
$\overline{\text{RBI}}$	Ripple Blanking Input (Active LOW)
$\overline{\text{LT}}$	Lamp Test Input (Active LOW)
$\overline{\text{BI}}/\overline{\text{RBO}}$	Blanking Input (Active LOW) or
	Ripple Blanking Output (Active LOW)
$\overline{\text{a}}$–$\overline{\text{g}}$	*Segment Outputs (Active LOW)

*OC—Open Collector

Absolute Maximum Ratings (Note)

If Military/Aerospace specified devices are required, please contact the National Semiconductor Sales Office/Distributors for availability and specifications.

Supply Voltage	7V
Input Voltage	7V
Operating Free Air Temperature Range	
54LS	$-55°C$ to $+125°C$
DM74LS	$0°C$ to $+70°C$
Storage Temperature Range	$-65°C$ to $+150°C$

Note: *The "Absolute Maximum Ratings" are those values beyond which the safety of the device cannot be guaranteed. The device should not be operated at these limits. The parametric values defined in the "Electrical Characteristics" table are not guaranteed at the absolute maximum ratings. The "Recommended Operating Conditions" table will define the conditions for actual device operation.*

Recommended Operating Conditions

Symbol	Parameter	54LS47			DM74SL47			Units
		Min	Nom	Max	Min	Nom	Max	
V_{CC}	Supply Voltage	4.5	5	5.5	4.75	5	5.25	V
V_{IH}	High Level Input Voltage	2			2			V
V_{IL}	Low Level Input Voltage			0.7			0.8	V
I_{OH}	High Level Output Current @ 15V = V_{OH}*			-50			-250	μA
I_{OL}	Low Level Output Current			12			24	mA
T_A	Free Air Operating Temperature	-55		125	0		70	°C

*OFF state at $\bar{a}$–$\bar{g}$.

Electrical Characteristics

Over recommended operating free air temperature range (unless otherwise noted)

Symbol	Parameter	Conditions		Min	Typ (Note 1)	Max	Units
V_I	Input Clamp Voltage	V_{CC} = Min, I_I = -18 mA				-1.5	V
V_{OH}	High Level Output Voltage	V_{CC} = Min, I_{OH} = Max, V_{IL} = Max	54LS	2.4			V
			DM74	2.7	3.4		
V_{OL}	Low Level Output Voltage	V_{CC} = Min, I_{OL} = Max, V_{IH} = Min	54LS			0.4	V
			DM74		0.35	0.5	
		I_{OL} = 4 mA, V_{CC} = Min	DM74		0.25	0.4	
I_I	Input Current @ Max Input Voltage	V_{CC} = Max, V_I = 10V				100	μA
I_{IH}	High Level Input Current	V_{CC} = Max, V_I = 2.7V				20	μA
I_{IL}	Low Level Input Current	V_{CC} = Max, V_I = 0.4V				-0.4	mA
I_{OS}	Short Circuit Output Current	V_{CC} = Max (Note 2), I_{OS} at $\overline{BI/RBO}$	54LS	-0.3		-2.0	mA
			DM74	-0.3		-2.0	
I_{CC}	Supply Current	V_{CC} = Max				13	mA

Note 1: All typicals are at V_{CC} = 5V, T_A = 25°C.

Note 2: Not more than one output should be shorted at a time, and the duration should not exceed one second.

LS47

Switching Characteristics at $V_{CC} = +5.0V$, $T_A = +25°C$

Symbol	Parameter	Conditions	$R_L = 665\Omega$ $C_L = 15$ pF		Units
			Min	Max	
t_{PLH} t_{PHL}	Propagation Delay An to $\overline{a}-\overline{g}$			100 100	ns
t_{PLH} t_{PHL}	Propagation Delay $\overline{RBI}$ to $\overline{a}-\overline{g}$*			100 100	ns

*$\overline{LT}$ = HIGH, A0–A3 = LOW

Functional Description

The 'LS47 decodes the input data in the pattern indicated in the Truth Table and the segment identification illustration. If the input data is decimal zero, a LOW signal applied to the $\overline{RBI}$ blanks the display and causes a multidigit display. For example, by grounding the $\overline{RBI}$ of the highest order decoder and connecting its $\overline{BI}/\overline{RBO}$ to $\overline{RBI}$ of the next lowest order decoder, etc., leading zeros will be suppressed. Similarly, by grounding $\overline{RBI}$ of the lowest order decoder and connecting its $\overline{BI}/\overline{RBO}$ to $\overline{RBI}$ of the next highest order decoder, etc., trailing zeros will be suppressed. Leading and trailing zeros can be suppressed simultaneously by using external gates, i.e.: by driving $\overline{RBI}$ of a intermediate decoder from an OR gate whose inputs are $\overline{BI}/\overline{RBO}$ of the next highest and lowest order decoders. $\overline{BI}/\overline{RBO}$ also serves as an unconditional blanking input. The internal NAND gate that generates the $\overline{RBO}$ signal has a resistive pull-up, as opposed to a totem pole, and thus $\overline{BI}/\overline{RBO}$ can be forced LOW by external means, using wired-collector logic. A LOW signal thus applied to $\overline{BI}/\overline{RBO}$ turns off all segment outputs. This blanking feature can be used to control display intensity by varying the duty cycle of the blanking signal. A LOW signal applied to $\overline{LT}$ turns on all segment outputs, provided that $\overline{BI}/\overline{RBO}$ is not forced LOW.

Logic Diagram

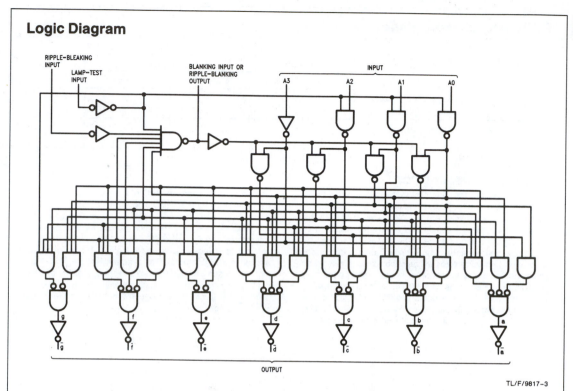

TL/F/9817-3

Numerical Designations—Resultant Displays

TL/F/9817-4

LS47

Logic Symbol

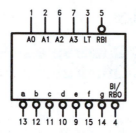

TL/F/9817–2

V_{CC} = Pin 16
GND = Pin 8

Truth Table

Decimal or Function	Inputs							Outputs							Note
	$\overline{LT}$	$\overline{RBI}$	A3	A2	A1	A0	$\overline{BI/RBO}$	$\bar{a}$	$\bar{b}$	$\bar{c}$	$\bar{d}$	$\bar{e}$	$\bar{f}$	$\bar{g}$	
0	H	H	L	L	L	L	H	L	L	L	L	L	L	H	1
1	H	X	L	L	L	H	H	H	L	L	H	H	H	H	1
2	H	X	L	L	H	L	H	L	L	H	L	L	H	L	
3	H	X	L	L	H	H	H	L	L	L	L	H	H	L	
4	H	X	L	H	L	L	H	H	L	L	H	H	L	L	
5	H	X	L	H	L	H	H	L	H	L	L	H	L	L	
6	H	X	L	H	H	L	H	H	H	L	L	L	L	L	
7	H	X	L	H	H	H	H	L	L	L	H	H	H	H	
8	H	X	H	L	L	L	H	L	L	L	L	L	L	L	
9	H	X	H	L	L	H	H	L	L	L	H	H	L	L	
10	H	X	H	L	H	L	H	H	H	H	L	L	H	L	
11	H	X	H	L	H	H	H	H	H	L	L	H	H	L	
12	H	X	H	H	L	L	H	H	L	H	H	H	L	L	
13	H	X	H	H	L	H	H	L	H	H	L	H	L	L	
14	H	X	H	H	H	L	H	H	H	H	L	L	L	L	
15	H	X	H	H	H	H	H	H	H	H	H	H	H	H	
$\overline{BI}$	X	X	X	X	X	X	L	H	H	H	H	H	H	H	2
$\overline{RBI}$	H	L	L	L	L	L	L	H	H	H	H	H	H	H	3
$\overline{LT}$	L	X	X	X	X	X	H	L	L	L	L	L	L	L	4

Note 1: $\overline{BI/RBO}$ is wire-AND logic serving as blanking input ($\overline{BI}$) and/or ripple-blanking output ($\overline{RBO}$). The blanking out ($\overline{BI}$) must be open or held at a HIGH level when output functions 0 through 15 are desired, and ripple-blanking input ($\overline{RBI}$) must be open or at a HIGH level if blanking or a decimal 0 is not desired. X = input may be HIGH or LOW.

Note 2: When a LOW level is applied to the blanking input (forced condition) all segment outputs go to a HIGH level regardless of the state of any other input condition.

Note 3: When ripple-blanking input ($\overline{RBI}$) and inputs A0, A1, A2 and A3 are LOW level, with the lamp test input at HIGH level, all segment outputs go to a HIGH level and the ripple-blanking output ($\overline{RBO}$) goes to a LOW level (response condition).

Note 4: When the blanking input/ripple-blanking output ($\overline{BI/RBO}$) is open or held at a HIGH level, and a LOW level is applied to lamp test input, all segment outputs go to a LOW level.

CD4001M/CD4001C Quadruple 2-Input NOR Gate
CD4011M/CD4011C Quadruple 2-Input NAND Gate

General Description

The CD4001M/CD4001C, CD4011M/CD4011C are monolithic complementary MOS (CMOS) quadruple two-input NOR and NAND gate integrated circuits. N- and P-channel enhancement mode transistors provide a symmetrical circuit with output swings essentially equal to the supply voltage. This results in high noise immunity over a wide supply voltage range. No DC power other than that caused by leakage current is consumed during static conditions. All inputs are protected against static discharge and latching conditions.

Features

- Wide supply voltage range   3.0V to 15V
- Low power   10 nW (typ.)
- High noise immunity   0.45 V_{DD} (typ.)

Connection Diagrams

Dual-In-Line Package

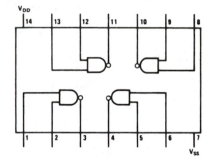

TL/F/5938–1

Top View
CD4011M/CD4011C

Dual-In-Line Package

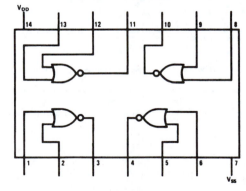

TL/F/5938–2

Top View
CD4001M/CD4001C

Order Number CD4001* or CD4011*
*Please look into Section 8, Appendix D for availability of various package types.

Absolute Maximum Ratings (Note 1)

If Military/Aerospace specified devices are required, contact the National Semiconductor Sales Office/ Distributors for availability and specifications.

Voltage on any Pin $V_{SS} - 0.3V$ to $V_{DD} + 0.3V$

Operating Temperature Range
     CD4001M, CD4011M $-55°C$ to $+125°C$
     CD4001C, CD4011C $-40°C$ to $+85°C$

Storage Temperature Range $-65°C$ to $+150°C$
Power Dissipation (P_D)
     Dual-In-Line   700 mW
     Small Outline   500 mW
Operating V_{DD} Range $V_{SS} + 3.0V$ to $V_{SS} + 15V$
Lead Temp.(Soldering, 10 sec.)   260°C

DC Electrical Characteristics CD4001M, CD4011M

Symbol	Parameter	Conditions	Limits							Units
			$-55°C$		$+25°C$			$+125°C$		
			Min	Max	Min	Typ	Max	Min	Max	
I_L	Quiescent Device Current	$V_{DD} = 5.0V$		0.05		0.001	0.05		3.0	μA
		$V_{DD} = 10V$		0.1		0.001	0.1		6.0	μA
P_D	Quiescent Device Dissipation/Package	$V_{DD} = 5.0V$		0.25		0.005	0.25		15	μW
		$V_{DD} = 10V$		1.0		0.01	1.0		60	μW
V_{OL}	Output Voltage low Level	$V_{DD} = 5.0V, V_I = V_{DD}, I_O = 0A$		0.05		0	0.05		0.05	V
		$V_{DD} = 10V, V_I = V_{DD}, I_O = 0A$		0.05		0	0.05		0.05	V
V_{OH}	Output Voltage High Level	$V_{DD} = 5.0V, V_I = V_{SS}, I_O = 0A$	4.95		4.95	5.0		4.95		V
		$V_{DD} = 10V, V_I = V_{SS}, I_O = 0A$	9.95		9.95	10		9.95		V
V_{NL}	Noise Immunity (All Inputs)	$V_{DD} = 5.0V, V_O = 3.6V, I_O = 0A$	1.5		1.5	2.25		1.4		V
		$V_{DD} = 10V, V_O = 7.2V, I_O = 0A$	3.0		3.0	4.5		2.9		V
V_{NH}	Noise Immunity (All Inputs)	$V_{DD} = 5.0V, V_O = 0.95V, I_O = 0A$	1.4		1.5	2.25		1.5		V
		$V_{DD} = 10V, V_O = 2.9V, I_O = 0A$	2.9		3.0	4.5		3.0		V
I_DN	Output Drive Current N-Channel (4001) (Note 2)	$V_{DD} = 5.0V, V_O = 0.4V, V_I = V_{DD}$	0.5		0.40	1.0		0.28		mA
		$V_{DD} = 10V, V_O = 0.5V, V_I = V_{DD}$	1.1		0.9	2.5		0.65		mA
I_DP	Output Drive Current P-Channel (4001) (Note 2)	$V_{DD} = 5.0V, V_O = 2.5V, V_I = V_{SS}$	-0.62		-0.5	-2.0		-0.35		mA
		$V_{DD} = 10V, V_O = 9.5V, V_I = V_{SS}$	-0.62		-0.5	-1.0		-0.35		mA
I_DN	Output Drive Current N-Channel (4011) (Note 2)	$V_{DD} = 5.0V, V_O = 0.4V, V_I = V_{DD}$	0.31		0.25	0.5		0.175		mA
		$V_{DD} = 10V, V_O = 0.5V, V_I = V_{DD}$	0.63		0.5	0.6		0.35		mA
I_DP	Output Drive Current P-Channel (4011) (Note 2)	$V_{DD} = 5.0V, V_O = 2.5V, V_I = V_{SS}$	-0.31		-0.25	-0.5		-0.175		mA
		$V_{DD} = 10V, V_O = 9.5V, V_I = V_{SS}$	-0.75		-0.6	-1.2		-0.4		mA
I_I	Input Current					10				pA

Note 1: "Absolute Maximum Ratings" are those values beyond which the safety of the device cannot be guaranteed. Except for "Operating Temperature Range" they are not meant to imply that the devices should be operated at these limits. The table of "Electrical Characteristics" provides conditions for actual device operation.

Note 2: I_DN and I_DP are tested one output at a time.

CD4013BM/CD4013BC

National Semiconductor

CD4013BM/CD4013BC Dual D Flip-Flop

General Description

The CD4013B dual D flip-flop is a monolithic complementary MOS (CMOS) integrated circuit constructed with N- and P-channel enhancement mode transistors. Each flip-flop has independent data, set, reset, and clock inputs and "Q" and "Q̄" outputs. These devices can be used for shift register applications, and by connecting "Q̄" output to the data input, for counter and toggle applications. The logic level present at the "D" input is transferred to the Q output during the positive-going transition of the clock pulse. Setting or resetting is independent of the clock and is accomplished by a high level on the set or reset line respectively.

Features

- Wide supply voltage range 3.0V to 15V
- High noise immunity 0.45 V_{DD} (typ.)
- Low power TTL compatibility fan out of 2 driving 74L or 1 driving 74LS

Applications

- Automotive
- Data terminals
- Instrumentation
- Medical electronics
- Alarm system
- Industrial electronics
- Remote metering
- Computers

Connection Diagram

Dual-In-Line Package

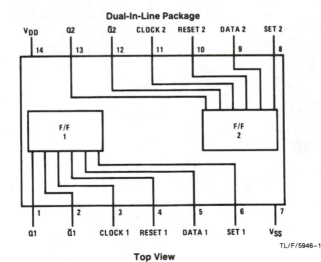

Top View

TL/F/5946-1

Order Number CD4013B*

Please see Section 8, Appendix D for availability of various package types.

Truth Table

CL[†]	D	R	S	Q	Q̄
⌐	0	0	0	0	1
⌐	1	0	0	1	0
⌐	x	0	0	Q	Q̄
x	x	1	0	0	1
x	x	0	1	1	0
x	x	1	1	1	1

No change
† = Level change
x = Don't care case

Absolute Maximum Ratings (Notes 1 & 2)

If Military/Aerospace specified devices are required, contact the National Semiconductor Sales Office/Distributors for availability and specifications.

DC Supply Voltage (V_{DD}) -0.5 V_{DC} to $+18$ V_{DC}
Input Voltage (V_{IN}) -0.5 V_{DC} to $V_{DD} +0.5$ V_{DC}
Storage Temp. Range (T_S) $-65°C$ to $+150°C$
Power Dissipation (P_D)
Dual-In-Line   700 mW
Small Outline   500 mW
Lead Temperature (T_L)
(Soldering, 10 seconds)   260°C

Recommended Operating Conditions (Note 2)

DC Supply Voltage (V_{DD}) $+3$ V_{DC} to $+15$ V_{DC}
Input Voltage (V_{IN}) 0 V_{DC} to V_{DD} V_{DC}
Operating Temperature Range (T_A)
CD4013BM $-55°C$ to $+125°C$
CD4013BC $-40°C$ to $+85°C$

DC Electrical Characteristics CD4013BM (Note 2)

Symbol	Parameter	Conditions	$-55°C$ Min	$-55°C$ Max	$+25°C$ Min	$+25°C$ Typ	$+25°C$ Max	$+125°C$ Min	$+125°C$ Max	Units
I_{DD}	Quiescent Device Current	$V_{DD}=5V$, $V_{IN}=V_{DD}$ or V_{SS}		1.0			1.0		30	µA
		$V_{DD}=10V$, $V_{IN}=V_{DD}$ or V_{SS}		2.0			2.0		60	µA
		$V_{DD}=15V$, $V_{IN}=V_{DD}$ or V_{SS}		4.0			4.0		120	µA
V_{OL}	Low Level Output Voltage	$\|I_O\|<1.0$ µA, $V_{DD}=5V$		0.05			0.05		0.05	V
		$V_{DD}=10V$		0.05			0.05		0.05	V
		$V_{DD}=15V$		0.05			0.05		0.05	V
V_{OH}	High Level Output Voltage	$\|I_O\|<1.0$ µA, $V_{DD}=5V$	4.95		4.95			4.95		V
		$V_{DD}=10V$	9.95		9.95			9.95		V
		$V_{DD}=15V$	14.95		14.95			14.95		V
V_{IL}	Low Level Input Voltage	$\|I_O\|<1.0$ µA, $V_{DD}=5V$, $V_O=0.5V$ or $4.5V$		1.5			1.5		1.5	V
		$V_{DD}=10V$, $V_O=1.0V$ or $9.0V$		3.0			3.0		3.0	V
		$V_{DD}=15V$, $V_O=1.5V$ or $13.5V$		4.0			4.0		4.0	V
V_{IH}	High Level Input Voltage	$\|I_O\|<1.0$ µA, $V_{DD}=5V$, $V_O=0.5V$ or $4.5V$	3.5		3.5			3.5		V
		$V_{DD}=10V$, $V_O=1.0V$ or $9.0V$	7.0		7.0			7.0		V
		$V_{DD}=15V$, $V_O=1.5V$ or $13.5V$	11.0		11.0			11.0		V
I_{OL}	Low Level Output Current (Note 3)	$V_{DD}=5V$, $V_O=0.4V$	0.64		0.51	0.88		0.36		mA
		$V_{DD}=10V$, $V_O=0.5V$	1.6		1.3	2.25		0.9		mA
		$V_{DD}=15V$, $V_O=1.5V$	4.2		3.4	8.8		2.4		mA
I_{OH}	High Level Output Current (Note 3)	$V_{DD}=5V$, $V_O=4.6V$	-0.64		-0.51	-0.88		-0.36		mA
		$V_{DD}=10V$, $V_O=9.5V$	-1.6		-1.3	-2.25		-0.9		mA
		$V_{DD}=15V$, $V_O=13.5V$	-4.2		-3.4	-8.8		-2.4		mA
I_{IN}	Input Current	$V_{DD}=15V$, $V_{IN}=0V$		-0.1		-10^{-5}	-0.1		-1.0	µA
		$V_{DD}=15V$, $V_{IN}=15V$		0.1		10^{-5}	0.1		1.0	µA

DC Electrical Characteristics CD4013BC (Note 2)

Symbol	Parameter	Conditions	$-40°C$ Min	$-40°C$ Max	$+25°C$ Min	$+25°C$ Typ	$+25°C$ Max	$+85°C$ Min	$+85°C$ Max	Units
I_{DD}	Quiescent Device Current	$V_{DD}=5V$, $V_{IN}=V_{DD}$ or V_{SS}		4.0			4.0		30	µA
		$V_{DD}=10V$, $V_{IN}=V_{DD}$ or V_{SS}		8.0			8.0		60	µA
		$V_{DD}=15V$, $V_{IN}=V_{DD}$ or V_{SS}		16.0			16.0		120	µA
V_{OL}	Low Level Output Voltage	$\|I_O\|<1.0$ µA, $V_{DD}=5V$		0.05			0.05		0.05	V
		$V_{DD}=10V$		0.05			0.05		0.05	V
		$V_{DD}=15V$		0.05			0.05		0.05	V
V_{OH}	High Level Output Voltage	$\|I_O\|<1.0$ µA, $V_{DD}=5V$	4.95		4.95			4.95		V
		$V_{DD}=10V$	9.95		9.95			9.95		V
		$V_{DD}=15V$	14.95		14.95			14.95		V
V_{IL}	Low Level Input Voltage	$\|I_O\|<1.0$ µA, $V_{DD}=5V$, $V_O=0.5V$ or $4.5V$		1.5			1.5		1.5	V
		$V_{DD}=10V$, $V_O=1.0V$ or $9.0V$		3.0			3.0		3.0	V
		$V_{DD}=15V$, $V_O=1.5V$ or $13.5V$		4.0			4.0		4.0	V

DC Electrical Characteristics CD4013BC (Note 2) (Continued)

Symbol	Parameter	Conditions	−40°C		+25°C			+85°C		Units
			Min	Max	Min	Typ	Max	Min	Max	
V_{IH}	High Level Input Voltage	$\lvert I_O \rvert < 1.0\ \mu A$								
		$V_{DD} = 5V, V_O = 0.5V$ or $4.5V$	3.5		3.5			3.5		V
		$V_{DD} = 10V, V_O = 1.0V$ or $9.0V$	7.0		7.0			7.0		V
		$V_{DD} = 15V, V_O = 1.5V$ or $13.5V$	11.0		11.0			11.0		V
I_{OL}	Low Level Output Current (Note 3)	$V_{DD} = 5V, V_O = 0.4V$	0.52		0.44	0.88		0.36		mA
		$V_{DD} = 10V, V_O = 0.5V$	1.3		1.1	2.25		0.9		mA
		$V_{DD} = 15V, V_O = 1.5V$	3.6		3.0	8.8		2.4		mA
I_{OH}	High Level Output Current (Note 3)	$V_{DD} = 5V, V_O = 4.6V$	−0.52		−0.44	−0.88		−0.36		mA
		$V_{DD} = 10V, V_O = 9.5V$	−1.3		−1.1	−2.25		−0.9		mA
		$V_{DD} = 15V, V_O = 13.5V$	−3.6		−3.0	−8.8		−2.4		mA
I_{IN}	Input Current	$V_{DD} = 15V, V_{IN} = 0V$		−0.3		-10^{-5}	−0.3		−1.0	μA
		$V_{DD} = 15V, V_{IN} = 15V$		0.3		10^{-5}	0.3		1.0	μA

Note 1: "Absolute Maximum Ratings" are those values beyond which the safety of the device cannot be guaranteed, they are not meant to imply that the devices should be operated at these limits. The tables of "Recommended Operating Conditions" and "Electrical Characteristics" provide conditions for actual device operation.

Note 2: $V_{SS} = 0V$ unless otherwise specified.

Note 3: I_{OH} and I_{OL} are measured one output at a time.

AC Electrical Characteristics* $T_A = 25°C$, $C_L = 50$ pF, $R_L = 200k$, unless otherwise noted

Symbol	Parameter	Conditions	Min	Typ	Max	Units
CLOCK OPERATION						
t_{PHL}, t_{PLH}	Propagation Delay Time	$V_{DD} = 5V$		200	350	ns
		$V_{DD} = 10V$		80	160	ns
		$V_{DD} = 15V$		65	120	ns
t_{THL}, t_{TLH}	Transition Time	$V_{DD} = 5V$		100	200	ns
		$V_{DD} = 10V$		50	100	ns
		$V_{DD} = 15V$		40	80	ns
t_{WL}, t_{WH}	Minimum Clock Pulse Width	$V_{DD} = 5V$		100	200	ns
		$V_{DD} = 10V$		40	80	ns
		$V_{DD} = 15V$		32	65	ns
t_{RCL}, t_{FCL}	Maximum Clock Rise and Fall Time	$V_{DD} = 5V$			15	μs
		$V_{DD} = 10V$			10	μs
		$V_{DD} = 15V$			5	μs
t_{SU}	Minimum Set-Up Time	$V_{DD} = 5V$		20	40	ns
		$V_{DD} = 10V$		15	30	ns
		$V_{DD} = 15V$		12	25	ns
f_{CL}	Maximum Clock Frequency	$V_{DD} = 5V$	2.5	5		MHz
		$V_{DD} = 10V$	6.2	12.5		MHz
		$V_{DD} = 15V$	7.6	15.5		MHz
SET AND RESET OPERATION						
$t_{PHL(R)}$, $t_{PLH(S)}$	Propagation Delay Time	$V_{DD} = 5V$		150	300	ns
		$V_{DD} = 10V$		65	130	ns
		$V_{DD} = 15V$		45	90	ns
$t_{WH(R)}$, $t_{WH(S)}$	Minimum Set and Reset Pulse Width	$V_{DD} = 5V$		90	180	ns
		$V_{DD} = 10V$		40	80	ns
		$V_{DD} = 15V$		25	50	ns
C_{IN}	Average Input Capacitance	Any Input		5	7.5	pF

*AC Parameters are guaranteed by DC correlated testing.

Schematic Diagram

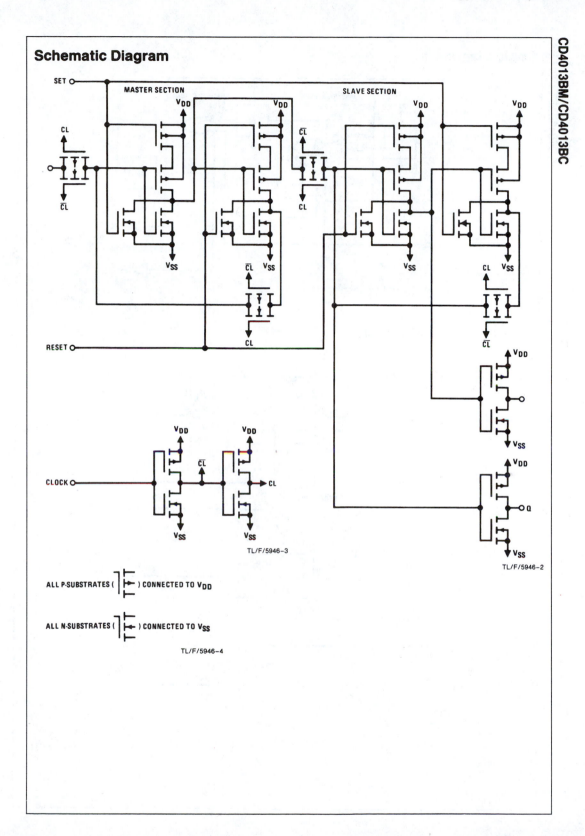

TL/F/5946–3

ALL P-SUBSTRATES ( ) CONNECTED TO V$_{DD}$

ALL N-SUBSTRATES ( ) CONNECTED TO V$_{SS}$

TL/F/5946–4

TL/F/5946–2

CD4013BM/CD4013BC

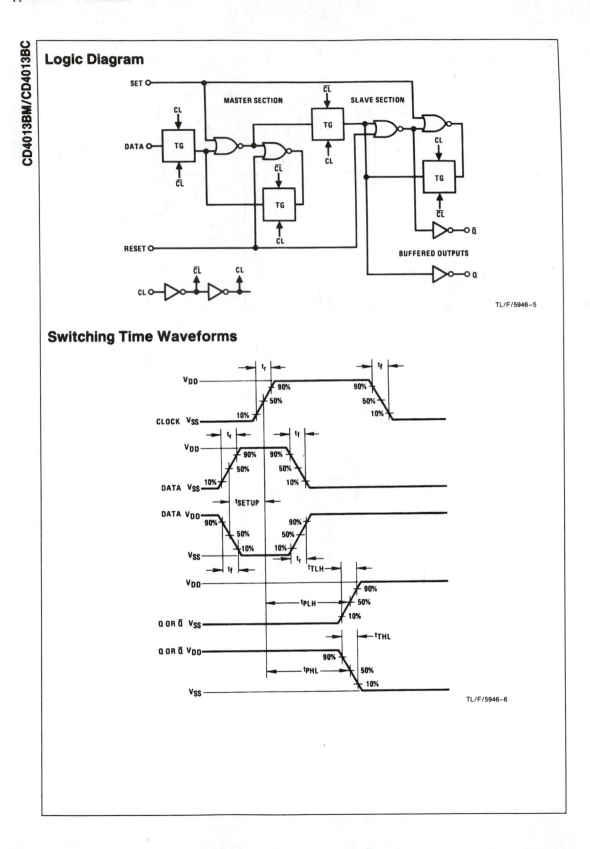

CD4013BM/CD4013BC

Logic Diagram

TL/F/5946-5

Switching Time Waveforms

TL/F/5946-6

CD4020BM/CD4020BC/CD4040BM/CD4040BC/CD4060BM/CD4060BC

National Semiconductor

CD4020BM/CD4020BC
14-Stage Ripple Carry Binary Counters
CD4040BM/CD4040BC
12-Stage Ripple Carry Binary Counters
CD4060BM/CD4060BC
14-Stage Ripple Carry Binary Counters

General Description

The CD4020BM/CD4020BC, CD4060BM/CD4060BC are 14-stage ripple carry binary counters, and the CD4040BM/CD4040BC is a 12-stage ripple carry binary counter. The counters are advanced one count on the negative transition of each clock pulse. The counters are reset to the zero state by a logical "1" at the reset input independent of clock.

Features

- Wide supply voltage range — 1.0V to 15V
- High noise immunity — 0.45 V_{DD} (typ.)
- Low power TTL compatibility — Fan out of 2 driving 74L or 1 driving 74LS
- Medium speed operation — 8 MHz typ. at V_{DD} = 10V
- Schmitt trigger clock input

Connection Diagrams

Dual-In-Line Package
CD4020BM/CD4020BC

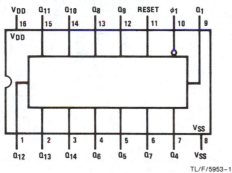

Top View

TL/F/5953–1

Order Number CD4020B*, CD4040B* or CD4060B*

*Please look into Section 8, Appendix D for availability of various package types.

Dual-In-Line Package
CD4040BM/CD4040BC

Dual-In-Line Package
CD4060BM/CD4060BC

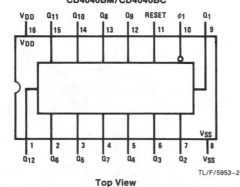

Top View

TL/F/5953–2

Top View

TL/F/5953–3

Absolute Maximum Ratings (Notes 1 and 2)

If Military/Aerospace specified devices are required, contact the National Semiconductor Sales Office/ Distributors for availability and specifications.

Supply Voltage (V_{DD})	$-0.5V$ to $+18V$
Input Voltage (V_{IN})	$-0.5V$ to $V_{DD} + 0.5V$
Storage Temperature Range (T_S)	$-65°C$ to $+150°C$
Package Dissipation (P_D)	
Dual-In-Line	700 mW
Small Outline	500 mW
Lead Temperature (T_L)	
(Soldering, 10 seconds)	260°C

Recommended Operating Conditions

Supply Voltage (V_{DD})	$+3V$ to $+15V$
Input Voltage (V_{IN})	0V to V_{DD}
Operating Temperature Range (T_A)	
CD40XXBM	$-55°C$ to $+125°C$
CD40XXBC	$-40°C$ to $+85°C$

DC Electrical Characteristics CD40XXBM (Note 2)

Symbol	Parameter	Conditions	$-55°C$ Min	$-55°C$ Max	$+25°C$ Min	$+25°C$ Typ	$+25°C$ Max	$+125°C$ Min	$+125°C$ Max	Units
I_{DD}	Quiescent Device Current	$V_{DD}=5V$, $V_{IN}=V_{DD}$ or V_{SS}		5			5		150	μA
		$V_{DD}=10V$, $V_{IN}=V_{DD}$ or V_{SS}		10			10		300	μA
		$V_{DD}=15V$, $V_{IN}=V_{DD}$ or V_{SS}		20			20		600	μA
V_{OL}	Low Level Output Voltage	$V_{DD}=5V$		0.05		0	0.05		0.05	V
		$V_{DD}=10V$		0.05		0	0.05		0.05	V
		$V_{DD}=15V$		0.05		0	0.05		0.05	V
V_{OH}	High Level Output Voltage	$V_{DD}=5V$	4.95		4.95	5		4.95		V
		$V_{DD}=10V$	9.95		9.95	10		9.95		V
		$V_{DD}=15V$	14.95		14.95	15		14.95		V
V_{IL}	Low Level Input Voltage	$V_{DD}=5V$, $V_O=0.5V$ or $4.5V$		1.5		2	1.5		1.5	V
		$V_{DD}=10V$, $V_O=1.0V$ or $9.0V$		3.0		4	3.0		3.0	V
		$V_{DD}=15V$, $V_O=1.5V$ or $13.5V$		4.0		6	4.0		4.0	V
V_{IH}	High Level Input Voltage	$V_{DD}=5V$, $V_O=0.5V$ or $4.5V$	3.5		3.5	3		3.5		V
		$V_{DD}=10V$, $V_O=1.0V$ or $9.0V$	7.0		7.0	6		7.0		V
		$V_{DD}=15V$, $V_O=1.5V$ or $13.5V$	11.0		11.0	9		11.0		V
I_{OL}	Low Level Output Current (See Note 3)	$V_{DD}=5V$, $V_O=0.4V$	0.64		0.51	0.88		0.36		mA
		$V_{DD}=10V$, $V_O=0.5V$	1.6		1.3	2.25		0.9		mA
		$V_{DD}=15V$, $V_O=1.5V$	4.2		3.4	8.8		2.4		mA
I_{OH}	High Level Output Current (See Note 3)	$V_{DD}=5V$, $V_O=4.6V$	-0.64		-0.51	-0.88		-0.36		mA
		$V_{DD}=10V$, $V_O=9.5V$	-1.6		-1.3	-2.25		-0.9		mA
		$V_{DD}=15V$, $V_O=13.5V$	-4.2		-3.4	-8.8		-2.4		mA
I_{IN}	Input Current	$V_{DD}=15V$, $V_{IN}=0V$		-0.10	-10^{-5}		-0.10		-1.0	μA
		$V_{DD}=15V$, $V_{IN}=15V$		0.10	10^{-5}		0.10		1.0	μA

Note 1: "Absolute Maximum Ratings" are those values beyond which the safety of the device cannot be guaranteed. They are not meant to imply that the devices should be operated at these limits. The tables of "Recommended Operating Conditions" and "Electrical Characteristics" provide conditions for actual device operation.

Note 2: $V_{SS}=0V$ unless otherwise specified.

Note 3: Data does not apply to oscillator points ϕ_0 and $\overline{\phi_0}$ of CD4060BM/CD4060BC. I_{OH} and I_{OL} are tested one output at a time.

DC Electrical Characteristics 40XXBC (Note 2)

Symbol	Parameter	Conditions	$-40°C$ Min	$-40°C$ Max	$+25°C$ Min	$+25°C$ Typ	$+25°C$ Max	$+85°C$ Min	$+85°C$ Max	Units
I_{DD}	Quiescent Device Current	$V_{DD}=5V$, $V_{IN}=V_{DD}$ or V_{SS}		20			20		150	μA
		$V_{DD}=10V$, $V_{IN}=V_{DD}$ or V_{SS}		40			40		300	μA
		$V_{DD}=15V$, $V_{IN}=V_{DD}$ or V_{SS}		80			80		600	μA
V_{OL}	Low Level Output Voltage	$V_{DD}=5V$		0.05		0	0.05		0.05	V
		$V_{DD}=10V$		0.05		0	0.05		0.05	V
		$V_{DD}=15V$		0.05		0	0.05		0.05	V

CD4020BM/CD4020BC/CD4040BM/CD4040BC/CD4060BM/CD4060BC

DC Electrical Characteristics 40XXBC (Note 2) (Continued)

Symbol	Parameter	Conditions	−40°C		+25°C			+85°C		Units
			Min	Max	Min	Typ	Max	Min	Max	
V_{OH}	High Level Output Voltage	V_{DD} = 5V	4.95		4.95	5		4.95		V
		V_{DD} = 10V	9.95		9.95	10		9.95		V
		V_{DD} = 15V	14.95		14.95	15		14.95		V
V_{IL}	Low Level Input Voltage	V_{DD} = 5V, V_O = 0.5V or 4.5V		1.5		2	1.5		1.5	V
		V_{DD} = 10V, V_O = 1.0V or 9.0V		3.0		4	3.0		3.0	V
		V_{DD} = 15V, V_O = 1.5V or 13.5V		4.0		6	4.0		4.0	V
V_{IH}	High Level Input Voltage	V_{DD} = 5V, V_O = 0.5V or 4.5V	3.5		3.5	3		3.5		V
		V_{DD} = 10V, V_O = 1.0V or 9.0V	7.0		7.0	6		7.0		V
		V_{DD} = 15V, V_O = 1.5V or 13.5V	11.0		11.0	9		11.0		V
I_{OL}	Low Level Output Current (See Note 3)	V_{DD} = 5V, V_O = 0.4V	0.52		0.44	0.88		0.36		mA
		V_{DD} = 10V, V_O = 0.5V	1.3		1.1	2.25		0.9		mA
		V_{DD} = 15V, V_O = 1.5V	3.6		3.0	8.8		2.4		mA
I_{OH}	High Level Output Current (See Note 3)	V_{DD} = 5V, V_O = 4.6V	−0.52		−0.44	−0.88		−0.36		mA
		V_{DD} = 10V, V_O = 9.5V	−1.3		−1.1	−2.25		−0.9		mA
		V_{DD} = 15V, V_O = 13.5V	−3.6		−3.0	−8.8		−2.4		mA
I_{IN}	Input Current	V_{DD} = 15V, V_{IN} = 0V		−0.30		-10^{-5}	−0.30		−1.0	μA
		V_{DD} = 15V, V_{IN} = 15V		0.30		10^{-5}	0.30		1.0	μA

AC Electrical Characteristics* CD4020BM/CD4020BC, CD4040BM/CD4040BC

T_A = 25°C, C_L = 50 pF, R_L = 200k, t_r = t_f = 20 ns, unless otherwise noted

Symbol	Parameter	Conditions	Min	Typ	Max	Units
t_{PHL1}, t_{PLH1}	Propagation Delay Time to Q_1	V_{DD} = 5V		250	550	ns
		V_{DD} = 10V		100	210	ns
		V_{DD} = 15V		75	150	ns
t_{PHL}, t_{PLH}	Interstage Propagation Delay Time from Q_n to Q_{n+1}	V_{DD} = 5V		150	330	ns
		V_{DD} = 10V		60	125	ns
		V_{DD} = 15V		45	90	ns
t_{THL}, t_{TLH}	Transition Time	V_{DD} = 5V		100	200	ns
		V_{DD} = 10V		50	100	ns
		V_{DD} = 15V		40	80	ns
t_{WL}, t_{WH}	Minimum Clock Pulse Width	V_{DD} = 5V		125	335	ns
		V_{DD} = 10V		50	125	ns
		V_{DD} = 15V		40	100	ns
t_{rCL}, t_{fCL}	Maximum Clock Rise and Fall Time	V_{DD} = 5V			No Limit	ns
		V_{DD} = 10V			No Limit	ns
		V_{DD} = 15V			No Limit	ns
f_{CL}	Maximum Clock Frequency	V_{DD} = 5V	1.5	4		MHz
		V_{DD} = 10V	4	10		MHz
		V_{DD} = 15V	5	12		MHz
$t_{PHL(R)}$	Reset Propagation Delay	V_{DD} = 5V		200	450	ns
		V_{DD} = 10V		100	210	ns
		V_{DD} = 15V		80	170	ns
$t_{WH(R)}$	Minimum Reset Pulse Width	V_{DD} = 5V		200	450	ns
		V_{DD} = 10V		100	210	ns
		V_{DD} = 15V		80	170	ns
C_{in}	Average Input Capacitance	Any Input		5	7.5	pF
C_{pd}	Power Dissipation Capacitance			50		pF

*AC Parameters are guaranteed by DC correlated testing.

AC Electrical Characteristics* CD4060BM/CD4060BC

$T_A = 25°C$, $C_L = 50$ pF, $R_L = 200$k, $t_r = t_f = 20$ ns, unless otherwise noted

Symbol	Parameter	Conditions	Min	Typ	Max	Units
t_{PHL4}, t_{PLH4}	Propagation Delay Time to Q_4	$V_{DD} = 5V$		550	1300	ns
		$V_{DD} = 10V$		250	525	ns
		$V_{DD} = 15V$		200	400	ns
t_{PHL}, t_{PLH}	Interstage Propagation Delay Time from Q_n to Q_{n+1}	$V_{DD} = 5V$		150	330	ns
		$V_{DD} = 10V$		60	125	ns
		$V_{DD} = 15V$		45	90	ns
t_{THL}, t_{TLH}	Transition Time	$V_{DD} = 5V$		100	200	ns
		$V_{DD} = 10V$		50	100	ns
		$V_{DD} = 15V$		40	80	ns
t_{WL}, t_{WH}	Minimum Clock Pulse Width	$V_{DD} = 5V$		170	500	ns
		$V_{DD} = 10V$		65	170	ns
		$V_{DD} = 15V$		50	125	ns
t_{rCL}, t_{fCL}	Maximum Clock Rise and Fall Time	$V_{DD} = 5V$			No Limit	ns
		$V_{DD} = 10V$			No Limit	ns
		$V_{DD} = 15V$			No Limit	ns
f_{CL}	Maximum Clock Frequency	$V_{DD} = 5V$	1	3		MHz
		$V_{DD} = 10V$	3	8		MHz
		$V_{DD} = 15V$	4	10		MHz
$t_{PHL(R)}$	Reset Propagation Delay	$V_{DD} = 5V$		200	450	ns
		$V_{DD} = 10V$		100	210	ns
		$V_{DD} = 15V$		80	170	ns
$t_{WH(R)}$	Minimum Reset Pulse Width	$V_{DD} = 5V$		200	450	ns
		$V_{DD} = 10V$		100	210	ns
		$V_{DD} = 15V$		80	170	ns
C_{in}	Average Input Capacitance	Any Input		5	7.5	pF
C_{pd}	Power Dissipation Capacitance			50		pF

*AC Parameters are guaranteed by DC correlated testing.

CD4060B Typical Oscillator Connections

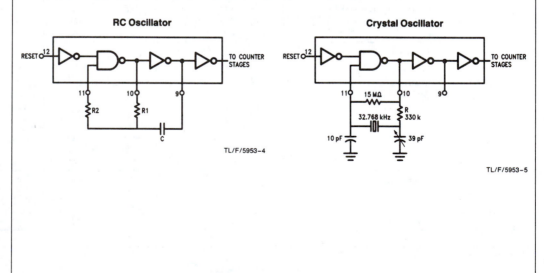

RC Oscillator

TL/F/5953-4

Crystal Oscillator

TL/F/5953-5

Schematic Diagrams

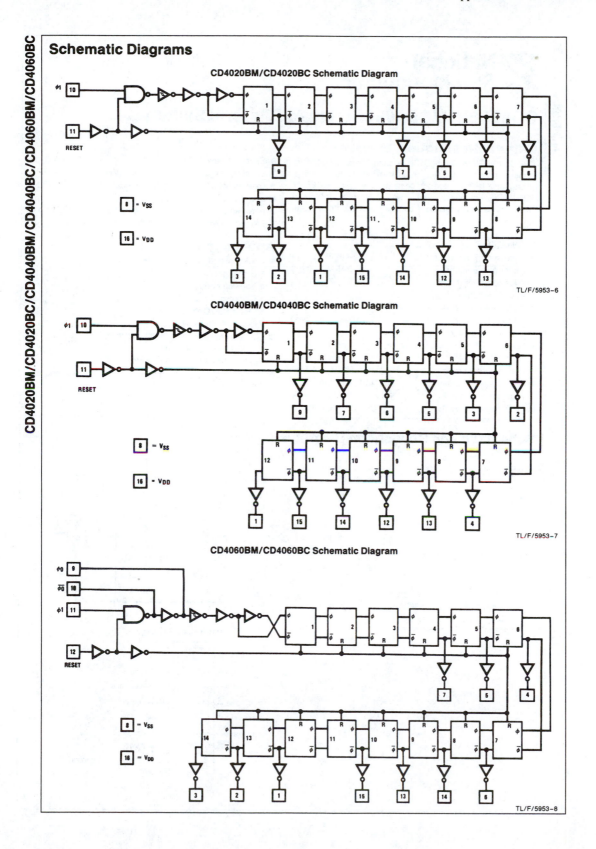

CD4020BM/CD4020BC Schematic Diagram

CD4040BM/CD4040BC Schematic Diagram

CD4060BM/CD4060BC Schematic Diagram

TL/F/5953-6

TL/F/5953-7

TL/F/5953-8

CD4020BM/CD4020BC/CD4040BM/CD4040BC/CD4060BM/CD4060BC

CD4066BM/CD4066BC

CD4066BM/CD4066BC Quad Bilateral Switch

General Description

The CD4066BM/CD4066BC is a quad bilateral switch intended for the transmission or multiplexing of analog or digital signals. It is pin-for-pin compatible with CD4016BM/CD4016BC, but has a much lower "ON" resistance, and "ON" resistance is relatively constant over the input-signal range.

Features

■ Wide supply voltage range   3V to 15V
■ High noise immunity   0.45 V_{DD} (typ.)
■ Wide range of digital and ±7.5 V_{PEAK}
  analog switching
■ "ON" resistance for 15V operation   80Ω
■ Matched "ON" resistance $\Delta R_{ON} = 5\Omega$ (typ.)
  over 15V signal input
■ "ON" resistance flat over peak-to-peak signal range
■ High "ON"/"OFF"   65 dB (typ.)
  output voltage ratio   @ $f_{is} = 10$ kHz, $R_L = 10$ kΩ
■ High degree linearity   0.1% distortion (typ.)
  High degree linearity   @ $f_{is} = 1$ kHz, $V_{is} = 5V_{p-p}$,
  High degree linearity $V_{DD} - V_{SS} = 10V$, $R_L = 10$ kΩ

■ Extremely low "OFF"   0.1 nA (typ.)
  switch leakage   @ $V_{DD} - V_{SS} = 10V$, $T_A = 25°C$
■ Extremely high control input impedance $10^{12}\Omega$(typ.)
■ Low crosstalk -50 dB (typ.)
  between switches   @ $f_{is} = 0.9$ MHz, $R_L = 1$ kΩ
■ Frequency response, switch "ON"   40 MHz (typ.)

Applications

■ Analog signal switching/multiplexing
  • Signal gating
  • Squelch control
  • Chopper
  • Modulator/Demodulator
  • Commutating switch
■ Digital signal switching/multiplexing
■ CMOS logic implementation
■ Analog-to-digital/digital-to-analog conversion
■ Digital control of frequency, impedance, phase, and analog-signal-gain

Schematic and Connection Diagrams

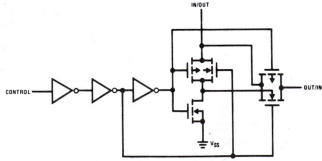

Order Number CD4066B*

*Please look into Section 8, Appendix D
for availability of various package types.

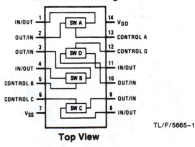

Dual-In-Line Package

Top View

TL/F/5665–1

Absolute Maximum Ratings (Notes 1 & 2)

If Military/Aerospace specified devices are required, contact the National Semiconductor Sales Office/Distributors for availability and specifications.

Supply Voltage (V_{DD})	$-0.5V$ to $+18V$
Input Voltage (V_{IN})	$-0.5V$ to $V_{DD}+0.5V$
Storage Temperature Range (T_S)	$-65°C$ to $+150°C$
Power Dissipation (P_D)	
Dual-In-Line	700 mW
Small Outline	500 mW
Lead Temperature (T_L)	
(Soldering, 10 seconds)	300°C

Recommended Operating Conditions (Note 2)

Supply Voltage (V_{DD})	3V to 15V
Input Voltage (V_{IN})	0V to V_{DD}
Operating Temperature Range (T_A)	
CD4066BM	$-55°C$ to $+125°C$
CD4066BC	$-40°C$ to $+85°C$

DC Electrical Characteristics CD4066BM (Note 2)

Symbol	Parameter	Conditions	$-55°C$ Min	$-55°C$ Max	$+25°C$ Min	$+25°C$ Typ	$+25°C$ Max	$+125°C$ Min	$+125°C$ Max	Units
I_{DD}	Quiescent Device Current	$V_{DD}=5V$		0.25		0.01	0.25		7.5	μA
		$V_{DD}=10V$		0.5		0.01	0.5		15	μA
		$V_{DD}=15V$		1.0		0.01	1.0		30	μA

SIGNAL INPUTS AND OUTPUTS

Symbol	Parameter	Conditions	$-55°C$ Min	$-55°C$ Max	$+25°C$ Min	$+25°C$ Typ	$+25°C$ Max	$+125°C$ Min	$+125°C$ Max	Units
R_{ON}	"ON" Resistance	$R_L=10\ k\Omega$ to $\frac{V_{DD}-V_{SS}}{2}$ $V_C=V_{DD}, V_{IS}=V_{SS}$ to V_{DD}								
		$V_{DD}=5V$		800		270	1050		1300	Ω
		$V_{DD}=10V$		310		120	400		550	Ω
		$V_{DD}=15V$		200		80	240		320	Ω
ΔR_{ON}	Δ"ON" Resistance Between any 2 of 4 Switches	$R_L=10\ k\Omega$ to $\frac{V_{DD}-V_{SS}}{2}$ $V_C=V_{DD}, V_{IS}=V_{SS}$ to V_{DD}								
		$V_{DD}=10V$				10				Ω
		$V_{DD}=15V$				5				Ω
I_{IS}	Input or Output Leakage Switch "OFF"	$V_C=0$ $V_{IS}=15V$ and $0V$, $V_{OS}=0V$ and $15V$		± 50		± 0.1	± 50		± 500	nA

CONTROL INPUTS

Symbol	Parameter	Conditions	$-55°C$ Min	$-55°C$ Max	$+25°C$ Min	$+25°C$ Typ	$+25°C$ Max	$+125°C$ Min	$+125°C$ Max	Units
V_{ILC}	Low Level Input Voltage	$V_{IS}=V_{SS}$ and V_{DD} $V_{OS}=V_{DD}$ and V_{SS} $I_{IS}=\pm 10\ \mu A$								
		$V_{DD}=5V$		1.5		2.25	1.5		1.5	V
		$V_{DD}=10V$		3.0		4.5	3.0		3.0	V
		$V_{DD}=15V$		4.0		6.75	4.0		4.0	V
V_{IHC}	High Level Input Voltage	$V_{DD}=5V$	3.5		3.5	2.75		3.5		V
		$V_{DD}=10V$ (see note 6)	7.0		7.0	5.5		7.0		V
		$V_{DD}=15V$	11.0		11.0	8.25		11.0		V
I_{IN}	Input Current	$V_{DD}-V_{SS}=15V$ $V_{DD}\geq V_{IS}\geq V_{SS}$ $V_{DD}\geq V_C\geq V_{SS}$		± 0.1		$\pm 10^{-5}$	± 0.1		± 1.0	μA

DC Electrical Characteristics CD4066BC (Note 2)

Symbol	Parameter	Conditions	$-40°C$ Min	$-40°C$ Max	$+25°C$ Min	$+25°C$ Typ	$+25°C$ Max	$+85°C$ Min	$+85°C$ Max	Units
I_{DD}	Quiescent Device Current	$V_{DD}=5V$		1.0		0.01	1.0		7.5	μA
		$V_{DD}=10V$		2.0		0.01	2.0		15	μA
		$V_{DD}=15V$		4.0		0.01	4.0		30	μA

DC Electrical Characteristics (Continued) CD4066BC (Note 2)

Symbol	Parameter	Conditions	−40°C		+25°C			+85°C		Units
			Min	Max	Min	Typ	Max	Min	Max	
SIGNAL INPUTS AND OUTPUTS										
R_{ON}	"ON" Resistance	$R_L = 10\ k\Omega$ to $\frac{V_{DD}-V_{SS}}{2}$ $V_C = V_{DD}$, V_{SS} to V_{DD}								
		$V_{DD} = 5V$		850		270	1050		1200	Ω
		$V_{DD} = 10V$		330		120	400		520	Ω
		$V_{DD} = 15V$		210		80	240		300	Ω
ΔR_{ON}	Δ"ON" Resistance Between Any 2 of 4 Switches	$R_L = 10\ k\Omega$ to $\frac{V_{DD}-V_{SS}}{2}$ $V_{CC} = V_{DD}$, $V_{IS} = V_{SS}$ to V_{DD}								
		$V_{DD} = 10V$				10				Ω
		$V_{DD} = 15V$				5				Ω
I_{IS}	Input or Output Leakage Switch "OFF"	$V_C = 0$		±50		±0.1	±50		±200	nA
CONTROL INPUTS										
V_{ILC}	Low Level Input Voltage	$V_{IS} = V_{SS}$ and V_{DD} $V_{OS} = V_{DD}$ and V_{SS} $I_{IS} = \pm 10\mu A$								
		$V_{DD} = 5V$		1.5		2.25	1.5		1.5	V
		$V_{DD} = 10V$		3.0		4.5	3.0		3.0	V
		$V_{DD} = 15V$		4.0		6.75	4.0		4.0	V
V_{IHC}	High Level Input Voltage	$V_{DD} = 5V$	3.5		3.5	2.75		3.5		V
		$V_{DD} = 10V$ (See note 6)	7.0		7.0	5.5		7.0		V
		$V_{DD} = 15V$	11.0		11.0	8.25		11.0		V
I_{IN}	Input Current	$V_{DD} - V_{SS} = 15V$ $V_{DD} \geq V_{IS} \geq V_{SS}$ $V_{DD} \geq V_C \geq V_{SS}$		±0.3		$\pm 10^{-5}$	±0.3		±1.0	μA

AC Electrical Characteristics* $T_A = 25°C$, $t_r = t_f = 20$ ns and $V_{SS} = 0V$ unless otherwise noted

Symbol	Parameter	Conditions	Min	Typ	Max	Units
t_{PHL}, t_{PLH}	Propagation Delay Time Signal Input to Signal Output	$V_C = V_{DD}$, $C_L = 50$ pF, (*Figure 1*) $R_L = 200k$				
		$V_{DD} = 5V$		25	55	ns
		$V_{DD} = 10V$		15	35	ns
		$V_{DD} = 15V$		10	25	ns
t_{PZH}, t_{PZL}	Propagation Delay Time Control Input to Signal Output High Impedance to Logical Level	$R_L = 1.0\ k\Omega$, $C_L = 50$ pF, (*Figures 2* and *3*)				
		$V_{DD} = 5V$			125	ns
		$V_{DD} = 10V$			60	ns
		$V_{DD} = 15V$			50	ns
t_{PHZ}, t_{PLZ}	Propagation Delay Time Control Input to Signal Output Logical Level to High Impedance	$R_L = 1.0\ k\Omega$, $C_L = 50$ pF, (*Figures 2* and *3*)				
		$V_{DD} = 5V$			125	ns
		$V_{DD} = 10V$			60	ns
		$V_{DD} = 15V$			50	ns
	Sine Wave Distortion	$V_C = V_{DD} = 5V$, $V_{SS} = -5V$ $R_L = 10\ k\Omega$, $V_{IS} = 5V_{p\text{-}p}$, $f = 1$ kHz, (*Figure 4*)		0.1		%
	Frequency Response-Switch "ON" (Frequency at −3 dB)	$V_C = V_{DD} = 5V$, $V_{SS} = -5V$, $R_L = 1\ k\Omega$, $V_{IS} = 5V_{p\text{-}p}$, $20\ Log_{10}\ V_{OS}/V_{OS}$ (1 kHz) − dB, (*Figure 4*)		40		MHz

AC Electrical Characteristics* (Continued) $T_A = 25°C$, $t_r = t_f = 20$ ns and $V_{SS} = 0V$ unless otherwise noted

Symbol	Parameter	Conditions	Min	Typ	Max	Units
	Feedthrough — Switch "OFF" (Frequency at -50 dB)	$V_{DD} = 5.0V$, $V_{CC} = V_{SS} = -5.0V$, $R_L = 1$ kΩ, $V_{IS} = 5.0V_{p-p}$, 20 Log$_{10}$, $V_{OS}/V_{IS} = -50$ dB, *(Figure 4)*		1.25		
	Crosstalk Between Any Two Switches (Frequency at -50 dB)	$V_{DD} = V_{C(A)} = 5.0V$; $V_{SS} = V_{C(B)} = 5.0V$, R_L1 kΩ, $V_{IS(A)} = 5.0$ V$_{p-p}$, 20 Log$_{10}$, $V_{OS(B)}/V_{IS(A)} = -50$ dB *(Figure 5)*		0.9		MHz
	Crosstalk; Control Input to Signal Output	$V_{DD} = 10V$, $R_L = 10$ kΩ, $R_{IN} = 1.0$ kΩ, $V_{CC} = 10V$ Square Wave, $C_L = 50$ pF *(Figure 6)*		150		mV$_{p-p}$
	Maximum Control Input	$R_L = 1.0$ kΩ, $C_L = 50$ pF, *(Figure 7)* $V_{OS(f)} = \frac{1}{2} V_{OS}(1.0$ kHz$)$				
		$V_{DD} = 5.0V$		6.0		MHz
		$V_{DD} = 10V$		8.0		MHz
		$V_{DD} = 15V$		8.5		MHz
C_{IS}	Signal Input Capacitance			8.0		pF
C_{OS}	Signal Output Capacitance	$V_{DD} = 10V$		8.0		pF
C_{IOS}	Feedthrough Capacitance	$V_C = 0V$		0.5		pF
C_{IN}	Control Input Capacitance			5.0	7.5	pF

*AC Parameters are guaranteed by DC correlated testing.

Note 1: "Absolute Maximum Ratings" are those values beyond which the safety of the device cannot be guaranteed. They are not meant to imply that the devices should be operated at these limits. The tables of "Recommended Operating Conditions" and "Electrical Characteristics" provide conditions for actual device operation.

Note 2: $V_{SS} = 0V$ unless otherwise specified.

Note 3: These devices should not be connected to circuits with the power "ON".

Note 4: In all cases, there is approximately 5 pF of probe and jig capacitance in the output; however, this capacitance is included in C_L wherever it is specified.

Note 5: V_{IS} is the voltage at the in/out pin and V_{OS} is the voltage at the out/in pin. V_C is the voltage at the control input.

Note 6: Conditions for V_{IHC}: a) $V_{IS} = V_{DD}$, $I_{OS} =$ standard B series I_{OH} b) $V_{IS} = 0V$, $I_{OL} =$ standard B series I_{OL}.

AC Test Circuits and Switching Time Waveforms

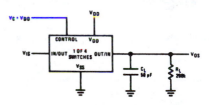
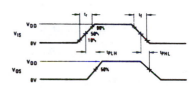

FIGURE 1. t_{PHL}, t_{PLH} Propagation Delay Time Signal Input to Signal Output

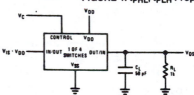
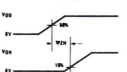
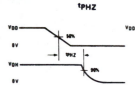

FIGURE 2. t_{PZH}, t_{PHZ} Propagation Delay Time Control to Signal Output

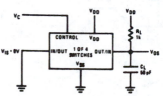
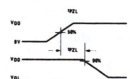
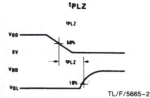

TL/F/5665-2

FIGURE 3. t_{PZL}, t_{PLZ} Propagtion Delay Time Control to Signal Output

AC Test Circuits and Switching Time Waveforms (Continued)

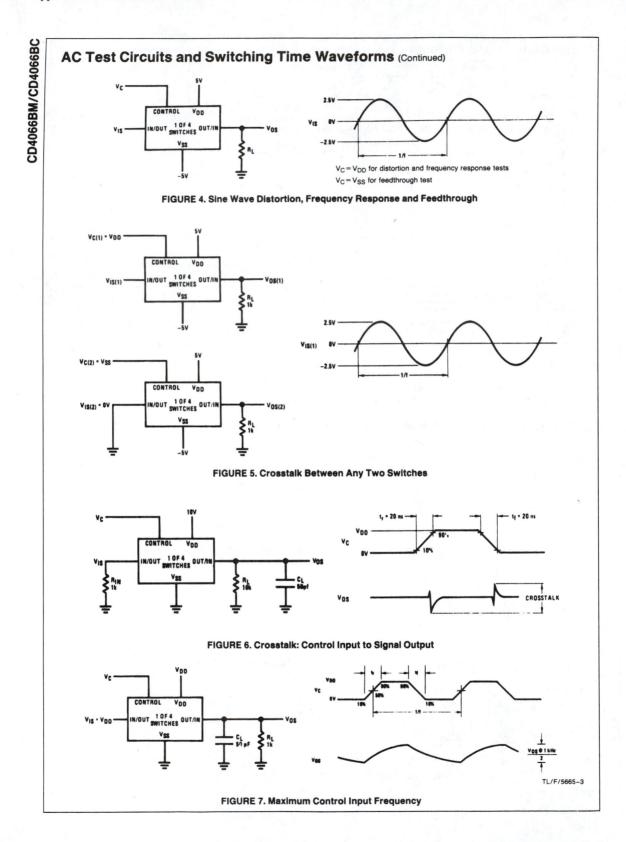

FIGURE 4. Sine Wave Distortion, Frequency Response and Feedthrough

$V_C = V_{DD}$ for distortion and frequency response tests
$V_C = V_{SS}$ for feedthrough test

FIGURE 5. Crosstalk Between Any Two Switches

FIGURE 6. Crosstalk: Control Input to Signal Output

FIGURE 7. Maximum Control Input Frequency

TL/F/5665–3

Typical Performance Characteristics

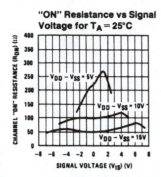

"ON" Resistance vs Signal Voltage for $T_A = 25°C$

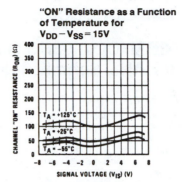

"ON" Resistance as a Function of Temperature for $V_{DD} - V_{SS} = 15V$

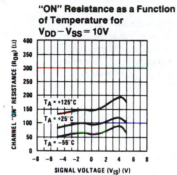

"ON" Resistance as a Function of Temperature for $V_{DD} - V_{SS} = 10V$

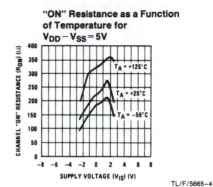

"ON" Resistance as a Function of Temperature for $V_{DD} - V_{SS} = 5V$

TL/F/5665–4

Special Considerations

In applications where separate power sources are used to drive V_{DD} and the signal input, the V_{DD} current capability should exceed V_{DD}/R_L (R_L = effective external load of the 4 CD4066BM/CD4066BC bilateral switches). This provision avoids any permanent current flow or clamp action of the V_{DD} supply when power is applied or removed from CD4066BM/CD4066BC.

In certain applications, the external load-resistor current may include both V_{DD} and signal-line components. To avoid

drawing V_{DD} current when switch current flows into terminals 1, 4, 8 or 11, the voltage drop across the bidirectional switch must not exceed 0.6V at $T_A \leq 25°C$, or 0.4V at $T_A > 25°C$ (calculated from R_{ON} values shown).

No V_{DD} current will flow through R_L if the switch current flows into terminals 2, 3, 9 or 10.

CD4093BM/CD4093BC Quad 2-Input NAND Schmitt Trigger

General Description

The CD4093B consists of four Schmitt-trigger circuits. Each circuit functions as a 2-input NAND gate with Schmitt-trigger action on both inputs. The gate switches at different points for positive and negative-going signals. The difference between the positive ($V_T{}^+$) and the negative voltage ($V_T{}^-$) is defined as hysteresis voltage (V_H).

All outputs have equal source and sink currents and conform to standard B-series output drive (see Static Electrical Characteristics).

Features

- Wide supply voltage range   3.0V to 15V
- Schmitt-trigger on each input
  with no external components
- Noise immunity greater than 50%

- Equal source and sink currents
- No limit on input rise and fall time
- Standard B-series output drive
- Hysteresis voltage (any input) T_A = 25°C

Typical	V_{DD} = 5.0V	V_H = 1.5V
	V_{DD} = 10V	V_H = 2.2V
	V_{DD} = 15V	V_H = 2.7V
Guaranteed		V_H = 0.1 V_{DD}

Applications

- Wave and pulse shapers
- High-noise-environment systems
- Monostable multivibrators
- Astable multivibrators
- NAND logic

Connection Diagram

Dual-In-Line Package

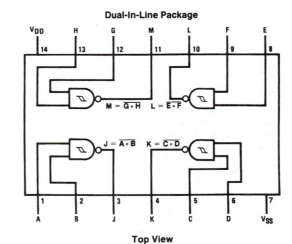

$M = \overline{G \cdot H}$ $L = \overline{E \cdot F}$

$J = \overline{A \cdot B}$ $K = \overline{C \cdot D}$

TL/F/5982–1

Top View

Order Number CD4093B*

*Please look into Section 8, Appendix D for availability of various package types.

CD4093BM/CD4093BC

Absolute Maximum Ratings (Notes 1 & 2)

If Military/Aerospace specified devices are required, contact the National Semiconductor Sales Office/Distributors for availability and specifications.

DC Supply Voltage (V_{DD}) -0.5 to $+18$ V_{DC}
Input Voltage (V_{IN}) -0.5 to V_{DD} $+0.5$ V_{DC}
Storage Temperature Range (T_S) $-65°C$ to $+150°C$
Power Dissipation (P_D)
  Dual-In-Line 700 mW
  Small Outline 500 mW
Lead Temperature (T_L)
  (Soldering, 10 seconds) 260°C

Recommended Operating Conditions (Note 2)

DC Supply Voltage (V_{DD}) 3 to 15 V_{DC}
Input Voltage (V_{IN}) 0 to V_{DD} V_{DC}
Operating Temperature Range (T_A)
  CD4093BM $-55°C$ to $+125°C$
  CD4093BC $-40°C$ to $+85°C$

DC Electrical Characteristics CD4093BM (Note 2)

Symbol	Parameter	Conditions	−55°C		+25°C			+125°C		Units
			Min	Max	Min	Typ	Max	Min	Max	
I_{DD}	Quiescent Device Current	$V_{DD}=5V$		0.25			0.25		7.5	µA
		$V_{DD}=10V$		0.5			0.5		15.0	µA
		$V_{DD}=15V$		1.0			1.0		30.0	µA
V_{OL}	Low Level Output Voltage	$V_{IN}=V_{DD}, \|I_O\|<1$ µA								
		$V_{DD}=5V$		0.05	0	0.05		0.05	V
		$V_{DD}=10V$		0.05	0	0.05		0.05	V
		$V_{DD}=15V$		0.05	0	0.05		0.05	V
V_{OH}	High Level Output Voltage	$V_{IN}=V_{SS}, \|I_O\|<1$ µA								
		$V_{DD}=5V$	4.95		4.95	5		4.95		V
		$V_{DD}=10V$	9.95		9.95	10		9.95		V
		$V_{DD}=15V$	14.95		14.95	15		14.95		V
V_T^-	Negative-Going Threshold Voltage (Any Input)	$\|I_O\|<1$ µA								
		$V_{DD}=5V, V_O=4.5V$	1.3	2.25	1.5	1.8	2.25	1.5	2.3	V
		$V_{DD}=10V, V_O=9V$	2.85	4.5	3.0	4.1	4.5	3.0	4.65	V
		$V_{DD}=15V, V_O=13.5V$	4.35	6.75	4.5	6.3	6.75	4.5	6.9	V
V_T^+	Positive-Going Threshold Voltage (Any Input)	$\|I_O\|<1$ µA								
		$V_{DD}=5V, V_O=0.5V$	2.75	3.65	2.75	3.3	3.5	2.65	3.5	V
		$V_{DD}=10V, V_O=1V$	5.5	7.15	5.5	6.2	7.0	5.35	7.0	V
		$V_{DD}=15V, V_O=1.5V$	8.25	10.65	8.25	9.0	10.5	8.1	10.5	V
V_H	Hysteresis ($V_T^+ - V_T^-$) (Any Input)	$V_{DD}=5V$	0.5	2.35	0.5	1.5	2.0	0.35	2.0	V
		$V_{DD}=10V$	1.0	4.30	1.0	2.2	4.0	0.70	4.0	V
		$V_{DD}=15V$	1.5	6.30	1.5	2.7	6.0	1.20	6.0	V
I_{OL}	Low Level Output Current (Note 3)	$V_{IN}=V_{DD}$								
		$V_{DD}=5V, V_O=0.4V$	0.64		0.51	0.88		0.36		mA
		$V_{DD}=10V, V_O=0.5V$	1.6		1.3	2.25		0.9		mA
		$V_{DD}=15V, V_O=1.5V$	4.2		3.4	8.8		2.4		mA
I_{OH}	High Level Output Current (Note 3)	$V_{IN}=V_{SS}$								
		$V_{DD}=5V, V_O=4.6V$	−0.64		0.51	−0.88		−0.36		mA
		$V_{DD}=10V, V_O=9.5V$	−1.6		−1.3	−2.25		−0.9		mA
		$V_{DD}=15V, V_O=13.5V$	−4.2		−3.4	−8.8		−2.4		mA
I_{IN}	Input Current	$V_{DD}=15V, V_{IN}=0V$		−0.1		-10^{-5}	−0.1		−1.0	µA
		$V_{DD}=15V, V_{IN}=15V$		0.1		10^{-5}	0.1		1.0	µA

Note 1: "Absolute Maximum Ratings" are those values beyond which the safety of the device cannot be guaranteed; they are not meant to imply that the devices should be operated at these limits. The table of "Recommended Operating Conditions" and "Electrical Characteristics" provides conditions for actual device operation.

Note 2: $V_{SS}=0V$ unless otherwise specified.

Note 3: I_{OH} and I_{OL} are tested one output at a time.

DC Electrical Characteristics CD4093BC (Note 2)

Symbol	Parameter	Conditions	−40°C Min	−40°C Max	+25°C Min	+25°C Typ	+25°C Max	+85°C Min	+85°C Max	Units
I_{DD}	Quiescent Device Current	$V_{DD} = 5V$		1.0			1.0		7.5	μA
		$V_{DD} = 10V$		2.0			2.0		15.0	μA
		$V_{DD} = 15V$		4.0			4.0		30.0	μA
V_{OL}	Low Level Output Voltage	$V_{IN} = V_{DD}, |I_O| < 1\,\mu A$								
		$V_{DD} = 5V$		0.05		0	0.05		0.05	V
		$V_{DD} = 10V$		0.05		0	0.05		0.05	V
		$V_{DD} = 15V$		0.05		0	0.05		0.05	V
V_{OH}	High Level Output Voltage	$V_{IN} = V_{SS}, |I_O| < 1\,\mu A$								
		$V_{DD} = 5V$	4.95		4.95	5		4.95		V
		$V_{DD} = 10V$	9.95		9.95	10		9.95		V
		$V_{DD} = 15V$	14.95		14.95	15		14.95		V
V_T^-	Negative-Going Threshold Voltage (Any Input)	$|I_O| < 1\,\mu A$								
		$V_{DD} = 5V, V_O = 4.5V$	1.3	2.25	1.5	1.8	2.25	1.5	2.3	V
		$V_{DD} = 10V, V_O = 9V$	2.85	4.5	3.0	4.1	4.5	3.0	4.65	V
		$V_{DD} = 15V, V_O = 13.5V$	4.35	6.75	4.5	6.3	6.75	4.5	6.9	V
V_T^+	Positive-Going Threshold Voltage (Any Input)	$|I_O| < 1\,\mu A$								
		$V_{DD} = 5V, V_O = 0.5V$	2.75	3.6	2.75	3.3	3.5	2.65	3.5	V
		$V_{DD} = 10V, V_O = 1V$	5.5	7.15	5.5	6.2	7.0	5.35	7.0	V
		$V_{DD} = 15V, V_O = 1.5V$	8.25	10.65	8.25	9.0	10.5	8.1	10.5	V
V_H	Hysteresis $(V_T^+ - V_T^-)$ (Any Input)	$V_{DD} = 5V$	0.5	2.35	0.5	1.5	2.0	0.35	2.0	V
		$V_{DD} = 10V$	1.0	4.3	1.0	2.2	4.0	0.70	4.0	V
		$V_{DD} = 15V$	1.5	6.3	1.5	2.7	6.0	1.20	6.0	V
I_{OL}	Low Level Output Current (Note 3)	$V_{IN} = V_{DD}$								
		$V_{DD} = 5V, V_O = 0.4V$	0.52		0.44	0.88		0.36		mA
		$V_{DD} = 10V, V_O = 0.5V$	1.3		1.1	2.25		0.9		mA
		$V_{DD} = 15V, V_O = 1.5V$	3.6		3.0	8.8		2.4		mA
I_{OH}	High Level Output Current (Note 3)	$V_{IN} = V_{SS}$								
		$V_{DD} = 5V, V_O = 4.6V$	−0.52		0.44	−0.88		−0.36		mA
		$V_{DD} = 10V, V_O = 9.5V$	−1.3		−1.1	−2.25		−0.9		mA
		$V_{DD} = 15V, V_O = 13.5V$	−3.6		−3.0	−8.8		−2.4		mA
I_{IN}	Input Current	$V_{DD} = 15V, V_{IN} = 0V$		−0.3		-10^{-5}	−0.3		−1.0	μA
		$V_{DD} = 15V, V_{IN} = 15V$		0.3		10^{-5}	0.3		1.0	μA

AC Electrical Characteristics*

$T_A = 25°C$, $C_L = 50$ pF, $R_L = 200k$, Input t_r, $t_f = 20$ ns, unless otherwise specified

Symbol	Parameter	Conditions	Min	Typ	Max	Units
t_{PHL}, t_{PLH}	Propagation Delay Time	$V_{DD} = 5V$		300	450	ns
		$V_{DD} = 10V$		120	210	ns
		$V_{DD} = 15V$		80	160	ns
t_{THL}, t_{TLH}	Transition Time	$V_{DD} = 5V$		90	145	ns
		$V_{DD} = 10V$		50	75	ns
		$V_{DD} = 15V$		40	60	ns
C_{IN}	Input Capacitance	(Any Input)		5.0	7.5	pF
C_{PD}	Power Dissipation Capacitance	(Per Gate)		24		pF

*AC Parameters are guaranteed by DC correlated testing.

Note 2: $V_{SS} = 0V$ unless otherwise specified.

Note 3: I_{OH} and I_{OL} are tested one output at a time.

Typical Applications

Gated Oscillator

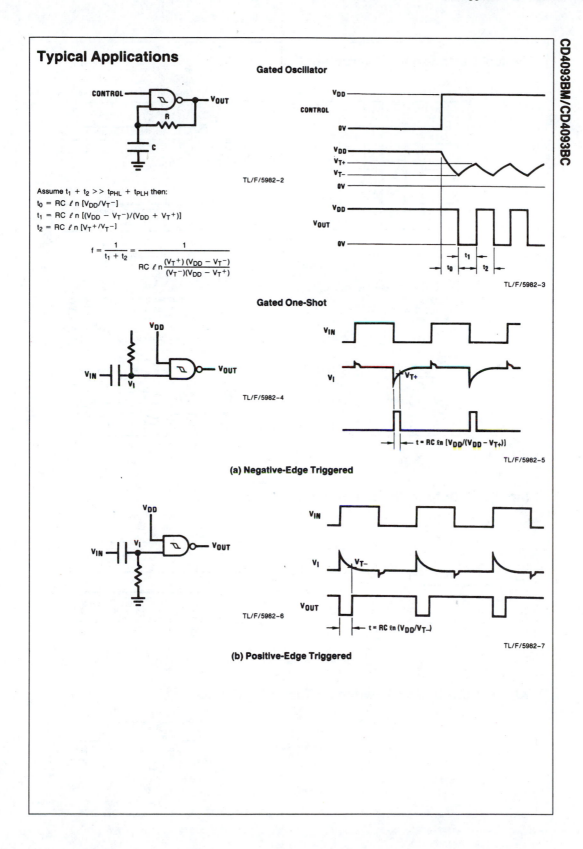

TL/F/5982-2

Assume $t_1 + t_2 >> t_{PHL} + t_{PLH}$ then:

$t_0 = RC \, \ell n \, [V_{DD}/V_T{}^-]$

$t_1 = RC \, \ell n \, [(V_{DD} - V_T{}^-)/(V_{DD} + V_T{}^+)]$

$t_2 = RC \, \ell n \, [V_T{}^+/V_T{}^-]$

$$f = \frac{1}{t_1 + t_2} = \frac{1}{RC \, \ell n \, \dfrac{(V_T{}^+) \, (V_{DD} - V_T{}^-)}{(V_T{}^-)(V_{DD} - V_T{}^+)}}$$

TL/F/5982-3

Gated One-Shot

TL/F/5982-4

$t = RC \, \ell n \, [V_{DD}/(V_{DD} - V_{T+})]$

TL/F/5982-5

(a) Negative-Edge Triggered

TL/F/5982-6

$t = RC \, \ell n \, (V_{DD}/V_{T-})$

TL/F/5982-7

(b) Positive-Edge Triggered

CD4093BM/CD4093BC

Typical Performance Characteristics

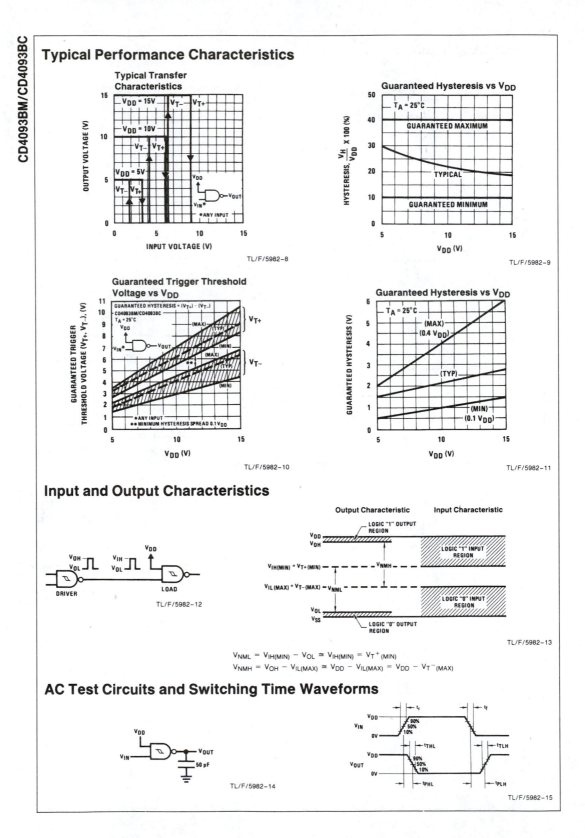

Typical Transfer Characteristics

TL/F/5982-8

Guaranteed Hysteresis vs V$_{DD}$

TL/F/5982-9

Guaranteed Trigger Threshold Voltage vs V$_{DD}$

TL/F/5982-10

Guaranteed Hysteresis vs V$_{DD}$

TL/F/5982-11

Input and Output Characteristics

TL/F/5982-12

TL/F/5982-13

$$V_{NML} = V_{IH(MIN)} - V_{OL} \cong V_{IH(MIN)} = V_T{}^+ (MIN)$$
$$V_{NMH} = V_{OH} - V_{IL(MAX)} \cong V_{DD} - V_{IL(MAX)} = V_{DD} - V_T{}^- (MAX)$$

AC Test Circuits and Switching Time Waveforms

TL/F/5982-14

TL/F/5982-15

National Semiconductor

CD4528BM/CD4528BC Dual Monostable Multivibrator

General Description

The CD4528B is a dual monostable multivibrator. Each device is retriggerable and resettable. Triggering can occur from either the rising or falling edge of an input pulse, resulting in an output pulse over a wide range of widths. Pulse duration and accuracy are determined by external timing components Rx and Cx.

Features

- Wide supply voltage range   3.0V to 18V
- Separate reset available
- Quiescent current = 5.0 nA/package (typ.) at 5.0 V_{DC}
- Diode protection on all inputs
- Triggerable from leading or trailing edge pulse
- Capable of driving two low-power TTL loads or one low-power Schottky TTL load over the rated temperature range

Connection Diagrams

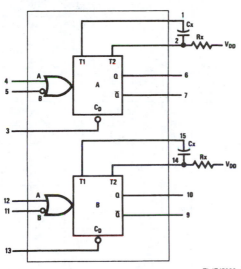

TL/F/5998-1

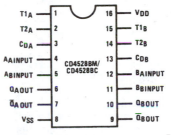

TL/F/5998-2

Dual-In-Line Package

Top View

Order Number CD4528B*

*Please look into Section 8, Appendix D for availability of various package types.

Truth Table

Inputs			Outputs	
Clear	A	B	Q	Q̄
L	X	X	L	H
X	H	X	L	H
X	X	L	L	H
H	L	↓	⊓	⊔
H	↑	H	⊓	⊔

H = High Level
L = Low Level
↑ = Transition from Low to High
↓ = Transition from High to Low
⊓ = One High Level Pulse
⊔ = One Low Level Pulse
X = Irrelevant

Absolute Maximum Ratings (Notes 1 & 2)

If Military/Aerospace specified devices are required, contact the National Semiconductor Sales Office/Distributors for availability and specifications.

DC Supply Voltage (V_{DD}) $-0.5 V_{DC}$ to $+18 V_{DC}$
Input Voltage, All Inputs (V_{IN}) $-0.5 V_{DC}$ to $V_{DD} +0.5 V_{DC}$
Storage Temperature
   Range (T_S) $-65°C$ to $+150°C$
Power Dissipation (P_D)
   Dual-In-Line 700 mW
   Small Outline 500 mW
Lead Temperature (T_L)
   (Soldering, 10 seconds) 260°C

Recommended Operating Conditions (Note 2)

DC Supply Voltage (V_{DD}) 3V to 15V
Input Voltage (V_{IN}) 0V to V_{DD} V_{DC}
Operating Temperature Range (T_A)
   CD4528BM $-55°C$ to $+125°C$
   CD4528BC $-40°C$ to $+85°C$

DC Electrical Characteristics CD4528BM (Note 2)

Symbol	Parameter	Conditions	−55°C Min	−55°C Max	+25°C Min	+25°C Typ	+25°C Max	+125°C Min	+125°C Max	Units
I_{DD}	Quiescent Device Current	V_{DD} = 5V		5	0.005		5		150	µA
		V_{DD} = 10V		10	0.010		10		300	µA
		V_{DD} = 15V		20	0.015		20		600	µA
V_{OL}	Low Level Output Voltage	V_{DD} = 5V		0.05			0.05		0.05	V
		V_{DD} = 10V		0.05			0.05		0.05	V
		V_{DD} = 15V		0.05			0.05		0.05	V
V_{OH}	High Level Output Voltage	V_{DD} = 5V	4.95		4.95	5.0		4.95		V
		V_{DD} = 10V	9.95		9.95	10.0		9.95		V
		V_{DD} = 15V	14.95		14.95	15.0		14.95		V
V_{IL}	Low Level Input Voltage	V_{DD} = 5V, V_O = 0.5V or 4.5V		1.5		2.25	1.5		1.5	V
		V_{DD} = 10V, V_O = 1V or 9V		3.0		4.50	3.0		3.0	V
		V_{DD} = 15V, V_O = 1.5V or 13.5V		4.0		6.75	4.0		4.0	V
V_{IH}	High Level Input Voltage	V_{DD} = 5V, V_O = 0.5V or 4.5V	3.5		3.5	2.75		3.5		V
		V_{DD} = 10V, V_O = 1V or 9V	7.0		7.0	5.50		7.0		V
		V_{DD} = 15V, V_O = 1.5V or 13.5V	11.0		11.0	8.25		11.0		V
I_{OL}	Low Level Output Current (Note 3)	V_{DD} = 5V, V_O = 0.4V	0.64		0.51	0.88		0.36		mA
		V_{DD} = 10V, V_O = 0.5V	1.6		1.3	2.25		0.9		mA
		V_{DD} = 15V, V_O = 1.5V	4.2		3.4	8.8		2.4		mA
I_{OH}	High Level Output Current (Note 3)	V_{DD} = 5V, V_O = 4.6V	−0.25		−0.2	−0.36		−0.14		mA
		V_{DD} = 10V, V_O = 9.5V	−0.62		−0.5	−0.9		−0.35		mA
		V_{DD} = 15V, V_O = 13.5V	−1.8		−1.5	−3.5		−1.1		mA
I_{IN}	Input Current	V_{DD} = 15V, V_{IN} = 0V		−0.1	-10^{-5}		−0.1		−1.0	µA
		V_{DD} = 15V, V_{IN} = 15V		0.1		10^{-5}	0.1		1.0	µA

Note 1: "Absolute Maximum Ratings" are those values beyond which the safety of the device cannot be guaranteed. Except for "Operating Temperature Range", they are not meant to imply that the devices should be operated at these limits. The table of "Electrical Characteristics" provides conditions for actual device operation.

Note 2: V_{SS} = 0V unless otherwise specified.

Note 3: I_{OH} and I_{OL} are tested one output at a time.

DC Electrical Characteristics CD4528BC (Note 2)

Symbol	Parameter	Conditions	−40°C		+25°C			+85°C		Units
			Min	Max	Min	Typ	Max	Min	Max	
I_{DD}	Quiescent Device Current	$V_{DD} = 5V$		20		0.005	20		150	μA
		$V_{DD} = 10V$		40		0.010	40		300	μA
		$V_{DD} = 15V$		80		0.015	80		600	μA
V_{OL}	Low Level Output Voltage	$V_{DD} = 5V$		0.05			0.05		0.05	V
		$V_{DD} = 10V$		0.05			0.05		0.05	V
		$V_{DD} = 15V$		0.05			0.05		0.05	V
V_{OH}	High Level Output Voltage	$V_{DD} = 5V$	4.95		4.95	5.0		4.95		V
		$V_{DD} = 10V$	9.95		9.95	10.0		9.95		V
		$V_{DD} = 15V$	14.95		14.95	15.0		14.95		V
V_{IL}	Low Level Input Voltage	$V_{DD} = 5V, V_O = 0.5V$ or $4.5V$		1.5		2.25	1.5		1.5	V
		$V_{DD} = 10V, V_O = 1V$ or $9V$		3.0		4.50	3.0		3.0	V
		$V_{DD} = 15V, V_O = 1.5V$ or $13.5V$		4.0		6.75	4.0		4.0	V
V_{IH}	High Level Input Voltage	$V_{DD} = 5V, V_O = 0.5V$ or $4.5V$	3.5		3.5	2.75		3.5		V
		$V_{DD} = 10V, V_O = 1V$ or $9V$	7.0		7.0	5.50		7.0		V
		$V_{DD} = 15V, V_O = 1.5V$ or $13.5V$	11.0		11.0	8.25		11.0		V
I_{OL}	Low Level Output Current (Note 3)	$V_{DD} = 5V, V_O = 0.4V$	0.52		0.44	0.88		0.36		mA
		$V_{DD} = 10V, V_O = 0.5V$	1.3		1.1	2.25		0.9		mA
		$V_{DD} = 15V, V_O = 1.5V$	3.6		3.0	8.8		2.4		mA
I_{OH}	High Level Output Current (Note 3)	$V_{DD} = 5V, V_O = 4.6V$	−0.2		−0.16	−0.36		−0.12		mA
		$V_{DD} = 10V, V_O = 9.5V$	−0.5		−0.4	−0.9		−0.3		mA
		$V_{DD} = 15V, V_O = 13.5V$	−1.4		−1.2	−3.5		−1.0		mA
I_{IN}	Input Current	$V_{DD} = 15V, V_{IN} = 0V$		−0.3		-10^{-5}	−0.3		−1.0	μA
		$V_{DD} = 15V, V_{IN} = 15V$		0.3		10^{-5}	0.3		1.0	μA

Note 1: "Absolute Maximum Ratings" are those values beyond which the safety of the device cannot be guaranteed. Except for "Operating Temperature Range", they are not meant to imply that the devices should be operated at these limits. The table of "Electrical Characteristics" provides conditions for actual device operation.

Note 2: $V_{SS} = 0V$ unless otherwise specified.

Note 3: I_{OH} and I_{OL} are tested one output at a time.

AC Electrical Characteristics* CD4528BM

$T_A = 25°C$, $C_L = 50$ pF, $R_L = 200$ kΩ, Input $t_r = t_f = 20$ ns, unless otherwise specified

Parameter	Conditions	Min	Typ	Max	Units
Output Rise Time	$t_r = (3.0$ ns/pF$) C_L + 30$ ns, $V_{DD} = 5.0$V		180	400	ns
	$t_r = (1.5$ ns/pF$) C_L + 15$ ns, $V_{DD} = 10.0$V		90	200	ns
	$t_r = (1.1$ ns/pF$) C_L + 10$ ns, $V_{DD} = 15.0$V		65	160	ns
Output Fall Time	$t_f = (1.5$ ns/pF$) C_L + 25$ ns, $V_{DD} = 5.0$V		100	200	ns
	$t_f = (0.75$ ns/pF$) C_L + 12.5$ ns, $V_{DD} = 10$V		50	100	ns
	$t_f = (0.55$ ns/pF$) C_L + 9.5$ ns, $V_{DD} = 15.0$V		35	80	ns
Turn-Off, Turn-On Delay A or B to Q or $\overline{Q}$ Cx = 15 pF, Rx = 5.0 kΩ	t_{PLH}, $t_{PHL} = (1.7$ ns/pF$) C_L + 240$ ns, $V_{DD} = 5.0$V		230	500	ns
	t_{PLH}, $t_{PHL} = (0.66$ ns/pF$) C_L + 8$ ns, $V_{DD} = 10.0$V		100	250	ns
	t_{PLH}, $t_{PHL} = (0.5$ ns/pF$) C_L + 65$ ns, $V_{DD} = 15.0$V		65	150	ns
Turn-Off, Turn-On Delay A or B to Q or $\overline{Q}$ Cx = 100 pF, Rx = 10 kΩ	t_{PLH}, $t_{PHL} = (1.7$ ns/pF$) C_L + 620$ ns, $V_{DD} = 5.0$V		230	500	ns
	t_{PLH}, $t_{PHL} = (0.66$ ns/pF$) C_L + 257$ ns, $V_{DD} = 10.0$V		100	250	ns
	t_{PLH}, $t_{PHL} = (0.5$ ns/pF$) C_L + 185$ ns, $V_{DD} = 15.0$V		65	150	ns
Minimum Input Pulse Width A or B Cx = 15 pF, Rx = 5.0 kΩ	$V_{DD} = 5.0$V		60	150	ns
	$V_{DD} = 10.0$V		20	50	ns
	$V_{DD} = 15$V		20	50	ns
Cx = 1000 pF, Rx = 10 kΩ	$V_{DD} = 5.0$V		60	150	ns
	$V_{DD} = 10.0$V		20	50	ns
	$V_{DD} = 15.0$V		20	50	ns
Output Pulse Width Q or $\overline{Q}$ For Cx < 0.01 μF (See Graph for Appropriate V_{DD} Level) Cx = 15 pF, Rx = 5.0 kΩ	$V_{DD} = 5.0$V		550		ns
	$V_{DD} = 10.0$V		350		ns
	$V_{DD} = 15.0$V		300		ns
For Cx > 0.01 μF Use $PW_{out} = 0.2$ Rx Cx ln $[V_{DD} - V_{SS}]$ Cx = 10,000 pF, Rx = 10 kΩ	$V_{DD} = 5.0$V	15	29	45	μs
	$V_{DD} = 10.0$V	10	37	90	μs
	$V_{DD} = 15.0$V	15	42	95	μs
Pulse Width Match between Circuits in the Same Package Cx = 10,000 pF, Rx = 10 kΩ	$V_{DD} = 5.0$V		6	25	%
	$V_{DD} = 10.0$V		8	35	%
	$V_{DD} = 15.0$V		8	35	%
Reset Propagation Delay, t_{PLH}, t_{PHL} Cx = 15 pF, Rx = 5.0 kΩ	$V_{DD} = 5.0$V		325	600	ns
	$V_{DD} = 10.0$V		90	225	ns
	$V_{DD} = 15.0$V		60	170	ns
Cx = 1000 pF, Rx = 10 kΩ	$V_{DD} = 5.0$V		7.0		μs
	$V_{DD} = 10.0$V		6.7		μs
	$V_{DD} = 15.0$V		6.7		μs
Minimum Retrigger Time Cx = 15 pF, Rx = 5.0 kΩ	$V_{DD} = 5.0$V		0		ns
	$V_{DD} = 10.0$V		0		ns
	$V_{DD} = 15.0$V		0		ns
Cx = 1000 pF, Rx = 10 kΩ	$V_{DD} = 5.0$V		0		ns
	$V_{DD} = 10.0$V		0		ns
	$V_{DD} = 15.0$V		0		ns

*AC parameters are guaranteed by DC correlated testing.

CD4528BM/CD4528BC

Logic Diagrams (½ of Device Shown)

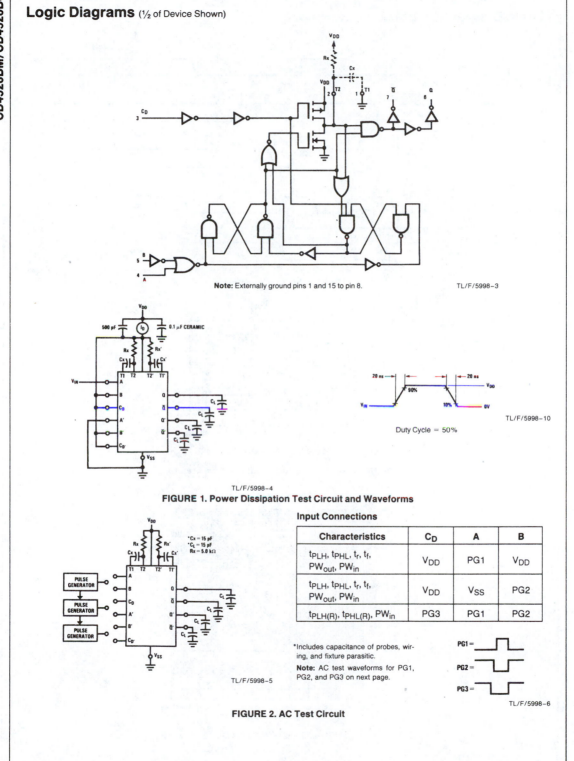

Note: Externally ground pins 1 and 15 to pin 8.

TL/F/5998–3

FIGURE 1. Power Dissipation Test Circuit and Waveforms

TL/F/5998–4

TL/F/5998–10

Duty Cycle = 50%

Input Connections

Characteristics	C_D	A	B
t_{PLH}, t_{PHL}, t_r, t_f, PW_{out}, PW_{in}	V_{DD}	PG1	V_{DD}
t_{PLH}, t_{PHL}, t_r, t_f, PW_{out}, PW_{in}	V_{DD}	V_{SS}	PG2
$t_{PLH(R)}$, $t_{PHL(R)}$, PW_{in}	PG3	PG1	PG2

*Includes capacitance of probes, wiring, and fixture parasitic.

Note: AC test waveforms for PG1, PG2, and PG3 on next page.

PG1 =

PG2 =

PG3 =

TL/F/5998–6

*Cx = 15 pF
*CL = 15 pF
Rx = 5.0 kΩ

TL/F/5998–5

FIGURE 2. AC Test Circuit

Logic Diagrams (½ of Device Shown) (Continued)

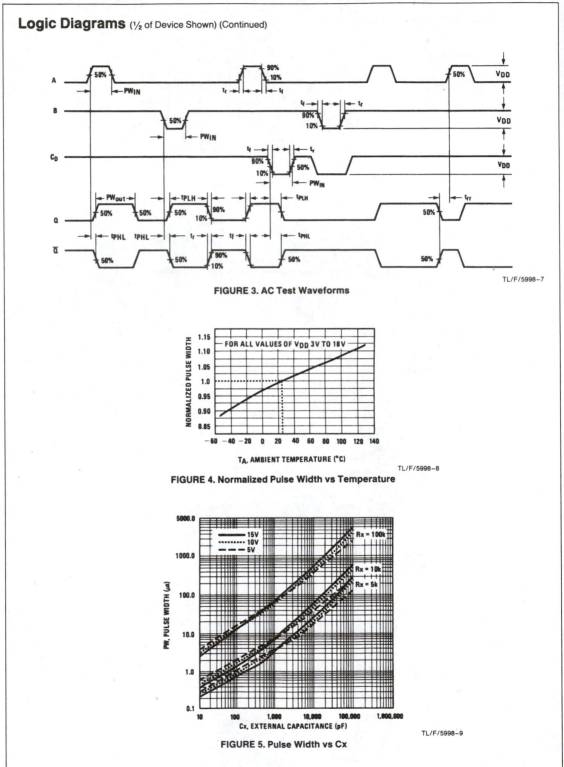

TL/F/5998–7

FIGURE 3. AC Test Waveforms

TL/F/5998–8

FIGURE 4. Normalized Pulse Width vs Temperature

TL/F/5998–9

FIGURE 5. Pulse Width vs Cx

National Semiconductor

DAC0800/DAC0801/DAC0802 8-Bit Digital-to-Analog Converters

General Description

The DAC0800 series are monolithic 8-bit high-speed current-output digital-to-analog converters (DAC) featuring typical settling times of 100 ns. When used as a multiplying DAC, monotonic performance over a 40 to 1 reference current range is possible. The DAC0800 series also features high compliance complementary current outputs to allow differential output voltages of 20 Vp-p with simple resistor loads as shown in *Figure 1*. The reference-to-full-scale current matching of better than ±1 LSB eliminates the need for full-scale trims in most applications while the nonlinearities of better than ±0.1% over temperature minimizes system error accumulations.

The noise immune inputs of the DAC0800 series will accept TTL levels with the logic threshold pin, V_{LC}, grounded. Changing the V_{LC} potential will allow direct interface to other logic families. The performance and characteristics of the device are essentially unchanged over the full ±4.5V to ±18V power supply range; power dissipation is only 33 mW with ±5V supplies and is independent of the logic input states.

The DAC0800, DAC0802, DAC0800C, DAC0801C and DAC0802C are a direct replacement for the DAC-08, DAC-08A, DAC-08C, DAC-08E and DAC-08H, respectively.

Features

- Fast settling output current   100 ns
- Full scale error   ±1 LSB
- Nonlinearity over temperature   ±0.1%
- Full scale current drift   ±10 ppm/°C
- High output compliance   −10V to +18V
- Complementary current outputs
- Interface directly with TTL, CMOS, PMOS and others
- 2 quadrant wide range multiplying capability
- Wide power supply range   ±4.5V to ±18V
- Low power consumption   33 mW at ±5V
- Low cost

Typical Applications

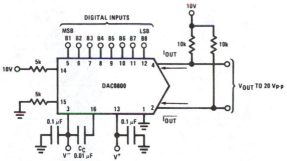

TL/H/5686−1

FIGURE 1. ±20 V_{P-P} Output Digital-to-Analog Converter (Note 4)

Ordering Information

Non-Linearity	Temperature Range	Order Numbers				
		J Package (J16A)*		N Package (N16A)*		SO Package (M16A)
±0.1% FS	−55°C ≤ T_A ≤ +125°C	DAC0802LJ	DAC-08AQ			
±0.1% FS	0°C ≤ T_A ≤ +70°C	DAC0802LCJ	DAC-08HQ	DAC0802LCN	DAC-08HP	DAC0802LCM
±0.19% FS	−55°C ≤ T_A ≤ +125°C	DAC0800LJ	DAC-08Q			
±0.19% FS	0°C ≤ T_A ≤ +70°C	DAC0800LCJ	DAC-08EQ	DAC0800LCN	DAC-08EP	DAC0800LCM
±0.39% FS	0°C ≤ T_A ≤ +70°C	DAC0801LCJ	DAC-08CQ	DAC0801LCN	DAC-08CP	DAC0801LCM

*Devices may be ordered by using either order number.

DAC0800/DAC0801/DAC0802

Absolute Maximum Ratings (Note 1)

If Military/Aerospace specified devices are required, please contact the National Semiconductor Sales Office/Distributors for availability and specifications.

Supply Voltage (V$^+$ − V$^-$)	±18V or 36V
Power Dissipation (Note 2)	500 mW
Reference Input Differential Voltage (V14 to V15)	V$^-$ to V$^+$
Reference Input Common-Mode Range (V14, V15)	V$^-$ to V$^+$
Reference Input Current	5 mA
Logic Inputs	V$^-$ to V$^-$ plus 36V
Analog Current Outputs (V$_S^-$ = −15V)	4.25 mA
ESD Susceptibility (Note 3)	TBD V
Storage Temperature	−65°C to +150°C

Lead Temp. (Soldering, 10 seconds)	
Dual-In-Line Package (plastic)	260°C
Dual-In-Line Package (ceramic)	300°C
Surface Mount Package	
Vapor Phase (60 seconds)	215°C
Infrared (15 seconds)	220°C

Operating Conditions (Note 1)

	Min	Max	Units
Temperature (T$_A$)			
DAC0802L	−55	+125	°C
DAC0800L	−55	+125	°C
DAC0800LC	0	+70	°C
DAC0801LC	0	+70	°C
DAC0802LC	0	+70	°C

Electrical Characteristics
The following specifications apply for V$_S$ = ±15V, I$_{REF}$ = 2 mA and T$_{MIN}$ ≤ T$_A$ ≤ T$_{MAX}$ unless otherwise specified. Output characteristics refer to both I$_{OUT}$ and $\overline{I_{OUT}}$.

Symbol	Parameter	Conditions	DAC0802L/ DAC0802LC			DAC0800L/ DAC0800LC			DAC0801LC			Units
			Min	Typ	Max	Min	Typ	Max	Min	Typ	Max	
	Resolution		8	8	8	8	8	8	8	8	8	Bits
	Monotonicity		8	8	8	8	8	8	8	8	8	Bits
	Nonlinearity				±0.1			±0.19			±0.39	%FS
t$_s$	Settling Time	To ±½ LSB, All Bits Switched "ON" or "OFF", T$_A$ = 25°C		100	135					100	150	ns
		DAC0800L					100	135				ns
		DAC0800LC					100	150				ns
tPLH, tPHL	Propagation Delay	T$_A$ = 25°C										
	Each Bit			35	60		35	60		35	60	ns
	All Bits Switched			35	60		35	60		35	60	ns
TCI$_{FS}$	Full Scale Tempco			±10	±50		±10	±50		±10	±80	ppm/°C
V$_{OC}$	Output Voltage Compliance	Full Scale Current Change <½ LSB, R$_{OUT}$ > 20 MΩ Typ	−10		18	−10		18	−10		18	V
I$_{FS4}$	Full Scale Current	V$_{REF}$ = 10.000V, R14 = 5.000 kΩ R15 = 5.000 kΩ, T$_A$ = 25°C	1.984	1.992	2.000	1.94	1.99	2.04	1.94	1.99	2.04	mA
I$_{FSS}$	Full Scale Symmetry	I$_{FS4}$ − I$_{FS2}$		±0.5	±4.0		±1	±8.0		±2	±16	µA
I$_{ZS}$	Zero Scale Current			0.1	1.0		0.2	2.0		0.2	4.0	µA
I$_{FSR}$	Output Current Range	V$^-$ = −5V	0	2.0	2.1	0	2.0	2.1	0	2.0	2.1	mA
		V$^-$ = −8V to −18V	0	2.0	4.2	0	2.0	4.2	0	2.0	4.2	mA
V$_{IL}$	Logic Input Levels	V$_{LC}$ = 0V										
	Logic "0"				0.8			0.8			0.8	V
V$_{IH}$	Logic "1"		2.0			2.0			2.0			V
I$_{IL}$	Logic Input Current	V$_{LC}$ = 0V										
	Logic "0"	−10V ≤ V$_{IN}$ ≤ +0.8V		−2.0	−10		−2.0	−10		−2.0	−10	µA
I$_{IH}$	Logic "1"	2V ≤ V$_{IN}$ ≤ +18V		0.002	10		0.002	10		0.002	10	µA
V$_{IS}$	Logic Input Swing	V$^-$ = −15V	−10		18	−10		18	−10		18	V
V$_{THR}$	Logic Threshold Range	V$_S$ = ±15V	−10		13.5	−10		13.5	−10		13.5	V
I$_{15}$	Reference Bias Current			−1.0	−3.0		−1.0	−3.0		−1.0	−3.0	µA
dI/dt	Reference Input Slew Rate	*(Figure 12)*	4.0	8.0		4.0	8.0		4.0	8.0		mA/µs
PSSI$_{FS+}$	Power Supply Sensitivity	4.5V ≤ V$^+$ ≤ +18V		0.0001	0.01		0.0001	0.01		0.0001	0.01	%/%
PSSI$_{FS-}$		−4.5V ≤ V$^-$ ≤ 18V I$_{REF}$ = 1mA		0.0001	0.01		0.0001	0.01		0.0001	0.01	%/%
I+	Power Supply Current	V$_S$ = ±5V, I$_{REF}$ = 1 mA		2.3	3.8		2.3	3.8		2.3	3.8	mA
I−				−4.3	−5.8		−4.3	−5.8		−4.3	−5.8	mA
I+		V$_S$ = 5V, −15V, I$_{REF}$ = 2 mA		2.4	3.8		2.4	3.8		2.4	3.8	mA
I−				−6.4	−7.8		−6.4	−7.8		−6.4	−7.8	mA
I+		V$_S$ = ±15V, I$_{REF}$ = 2 mA		2.5	3.8		2.5	3.8		2.5	3.8	mA
I−				−6.5	−7.8		−6.5	−7.8		−6.5	−7.8	mA

Electrical Characteristics (Continued)

The following specifications apply for $V_S = \pm 15V$, $I_{REF} = 2$ mA and $T_{MIN} \leq T_A \leq T_{MAX}$ unless otherwise specified. Output characteristics refer to both I_{OUT} and $\overline{I_{OUT}}$.

Symbol	Parameter	Conditions	DAC0802L/ DAC0802LC			DAC0800L/ DAC0800LC			DAC0801LC			Units
			Min	Typ	Max	Min	Typ	Max	Min	Typ	Max	
P_D	Power Dissipation	$\pm 5V$, $I_{REF} = 1$ mA		33	48		33	48		33	48	mW
		$5V$, $-15V$, $I_{REF} = 2$ mA		108	136		108	136		108	136	mW
		$\pm 15V$, $I_{REF} = 2$ mA		135	174		135	174		135	174	mW

Note 1: Absolute Maximum Ratings indicate limits beyond which damage to the device may occur. DC and AC electrical specifications do not apply when operating the device beyond its specified operating conditions.

Note 2: The maximum junction temperature of the DAC0800, DAC0801 and DAC0802 is 125°C. For operating at elevated temperatures, devices in the Dual-In-Line J package must be derated based on a thermal resistance of 100°C/W, junction-to-ambient, 175°C/W for the molded Dual-In-Line N package and 100°C/W for the Small Outline M package.

Note 3: Human body model, 100 pF discharged through a 1.5 kΩ resistor.

Note 4: Pin-out numbers for the DAC080X represent the Dual-In-Line package. The Small Outline package pin-out differs from the Dual-In-Line package.

Connection Diagrams

Dual-In-Line Package

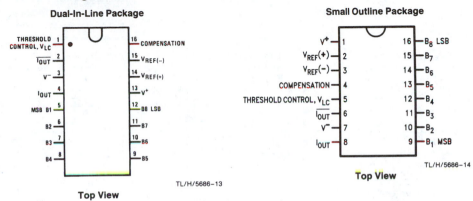

Top View

TL/H/5686–13

Small Outline Package

Top View

TL/H/5686–14

See Ordering Information

Block Diagram (Note 4)

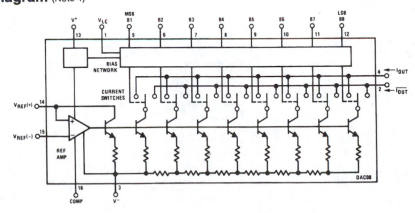

TL/H/5686–2

DAC0800/DAC0801/DAC0802

Typical Performance Characteristics

Full Scale Current vs Reference Current

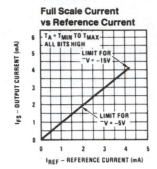

LSB Propagation Delay Vs I$_{FS}$

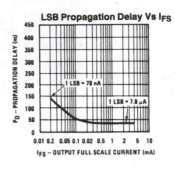

Reference Input Frequency Response

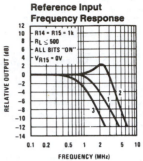

Curve 1: $C_C = 15$ pF, $V_{IN} = 2$ Vp-p centered at 1V.
Curve 2: $C_C = 15$ pF, $V_{IN} = 50$ mVp-p centered at 200 mV.
Curve 3: $C_C = 0$ pF, $V_{IN} = 100$ mVp-p at 0V and applied through 50 Ω connected to pin 14.2V applied to R14.

Reference Amp Common-Mode Range

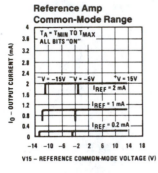

Note. Positive common-mode range is always (V+) − 1.5V

Logic Input Current vs Input Voltage

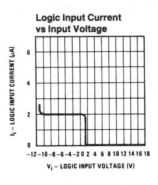

V$_{TH}$ − V$_{LC}$ vs Temperature

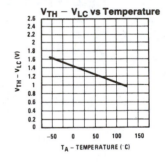

Output Current vs Output Voltage (Output Voltage Compliance)

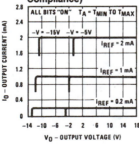

Output Voltage Compliance vs Temperature

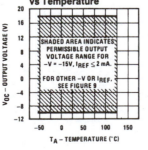

Bit Transfer Characteristics

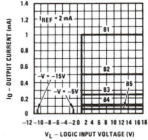

TL/H/5686–3

Note. B1–B8 have identical transfer characteristics. Bits are fully switched with less than ½ LSB error, at less than ±100 mV from actual threshold. These switching points are guaranteed to lie between 0.8 and 2V over the operating temperature range (V$_{LC}$ = 0V).

Typical Performance Characteristics (Continued)

Power Supply Current vs +V

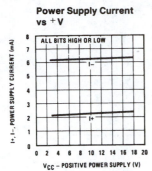

Power Supply Current vs −V

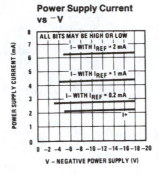

Power Supply Current vs Temperature

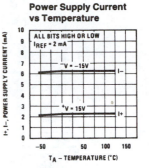

TL/H/5686–4

Equivalent Circuit

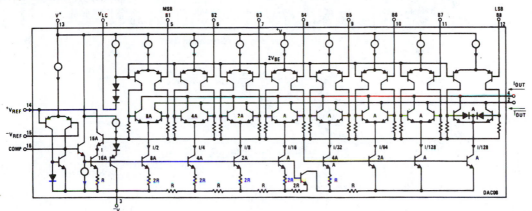

TL/H/5686–15

FIGURE 2

Typical Applications (Continued)

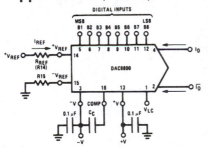

TL/H/5686–5

$$I_{FS} \approx \frac{+V_{REF}}{R_{REF}} \times \frac{255}{256}$$

$I_O + \overline{I_O} = I_{FS}$ for all logic states

For fixed reference, TTL operation, typical values are:

V_{REF} = 10.000V

R_{REF} = 5.000k

$R15 \approx R_{REF}$

C_C = 0.01 μF

V_{LC} = 0V (Ground)

FIGURE 3. Basic Positive Reference Operation (Note 4)

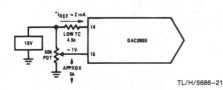

TL/H/5686–21

FIGURE 4. Recommended Full Scale Adjustment Circuit (Note 4)

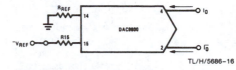

TL/H/5686–16

$$I_{FS} \approx \frac{-V_{REF}}{R_{REF}} \times \frac{255}{256}$$

Note. R_{REF} sets I_{FS}; R15 is for bias current cancellation

FIGURE 5. Basic Negative Reference Operation (Note 4)

Typical Applications (Continued)

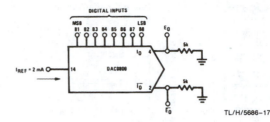

TL/H/5686–17

	B1	B2	B3	B4	B5	B6	B7	B8	I_O mA	$\overline{I_O}$ mA	E_O	$\overline{E_O}$
Full Scale	1	1	1	1	1	1	1	1	1.992	0.000	−9.960	0.000
Full Scale − LSB	1	1	1	1	1	1	1	0	1.984	0.008	−9.920	−0.040
Half Scale + LSB	1	0	0	0	0	0	0	1	1.008	0.984	−5.040	−4.920
Half Scale	1	0	0	0	0	0	0	0	1.000	0.992	−5.000	−4.960
Half Scale − LSB	0	1	1	1	1	1	1	1	0.992	1.000	−4.960	−5.000
Zero Scale + LSB	0	0	0	0	0	0	0	1	0.008	1.984	−0.040	−9.920
Zero Scale	0	0	0	0	0	0	0	0	0.000	1.992	0.000	−9.960

FIGURE 6. Basic Unipolar Negative Operation (Note 4)

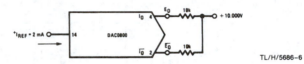

TL/H/5686–6

	B1	B2	B3	B4	B5	B6	B7	B8	E_O	$\overline{E_O}$
Pos. Full Scale	1	1	1	1	1	1	1	1	−9.920	+10.000
Pos. Full Scale − LSB	1	1	1	1	1	1	1	0	−9.840	+9.920
Zero Scale + LSB	1	0	0	0	0	0	0	1	−0.080	+0.160
Zero Scale	1	0	0	0	0	0	0	0	0.000	+0.080
Zero Scale − LSB	0	1	1	1	1	1	1	1	+0.080	0.000
Neg. Full Scale + LSB	0	0	0	0	0	0	0	1	+9.920	−9.840
Neg. Full Scale	0	0	0	0	0	0	0	0	+10.000	−9.920

FIGURE 7. Basic Bipolar Output Operation (Note 4)

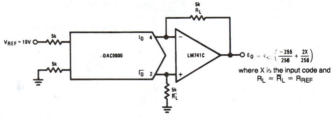

$$E_O = V_{ref}\left(\frac{-255}{256} + \frac{2X}{256}\right)$$

where X is the input code and $R_L = \overline{R_L} = R_{REF}$

TL/H/5686–18

If $R_L = \overline{R_L}$ within ±0.05%, output is symmetrical about ground

	B1	B2	B3	B4	B5	B6	B7	B8	E_O
Pos. Full Scale	1	1	1	1	1	1	1	1	+9.960
Pos. Full Scale − LSB	1	1	1	1	1	1	1	0	+9.880
(+)Zero Scale	1	0	0	0	0	0	0	0	+0.040
(−)Zero Scale	0	1	1	1	1	1	1	1	−0.040
Neg. Full Scale + LSB	0	0	0	0	0	0	0	1	−9.880
Neg. Full Scale	0	0	0	0	0	0	0	0	−9.960

FIGURE 8. Symmetrical Offset Binary Operation (Note 4)

DAC0800/DAC0801/DAC0802

Typical Applications (Continued)

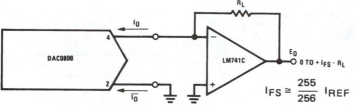

For complementary output (operation as negative logic DAC), connect inverting input of op amp to $\overline{I_O}$ (pin 2), connect I_O (pin 4) to ground.

$$I_{FS} \cong \frac{255}{256} I_{REF}$$

TL/H/5686–19

FIGURE 9. Positive Low Impedance Output Operation (Note 4)

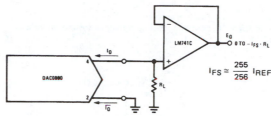

For complementary output (operation as a negative logic DAC) connect non-inverting input of op am to $\overline{I_O}$ (pin 2); connect I_O (pin 4) to ground.

$$I_{FS} \cong \frac{255}{256} I_{REF}$$

TL/H/5686–20

FIGURE 10. Negative Low Impedance Output Operation (Note 4)

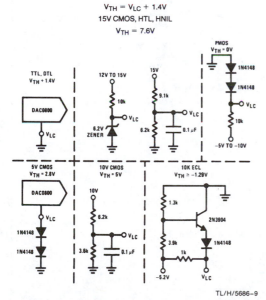

$V_{TH} = V_{LC} + 1.4V$

15V CMOS, HTL, HNIL

$V_{TH} = 7.6V$

TL/H/5686–9

Note. Do not exceed negative logic input range of DAC.

FIGURE 11. Interfacing with Various Logic Families

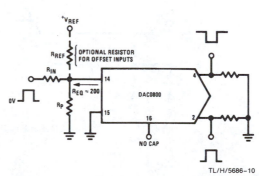

Typical values: $R_{IN} = 5k$, $+V_{IN} = 10V$

TL/H/5686–10

FIGURE 12. Pulsed Reference Operation (Note 4)

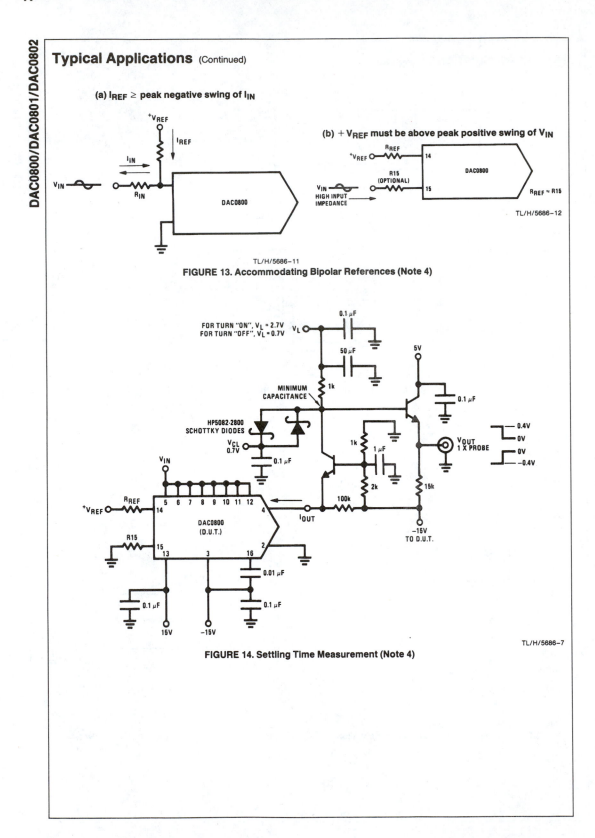

Typical Applications (Continued)

DAC0800/DAC0801/DAC0802

(a) $I_{REF} \geq$ peak negative swing of I_{IN}

(b) $+V_{REF}$ must be above peak positive swing of V_{IN}

TL/H/5686–12

TL/H/5686–11

FIGURE 13. Accommodating Bipolar References (Note 4)

FOR TURN "ON", V_L = 2.7V
FOR TURN "OFF", V_L = 0.7V

TL/H/5686–7

FIGURE 14. Settling Time Measurement (Note 4)

Typical Applications (Continued)

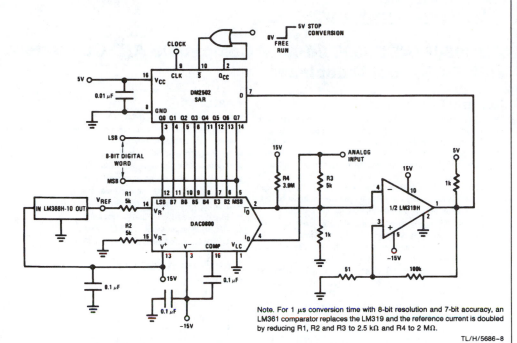

Note. For 1 μs conversion time with 8-bit resolution and 7-bit accuracy, an LM361 comparator replaces the LM319 and the reference current is doubled by reducing R1, R2 and R3 to 2.5 kΩ and R4 to 2 MΩ.

TL/H/5686-8

FIGURE 15. A Complete 2 μs Conversion Time, 8-Bit A/D Converter (Note 4)

ADC0808/ADC0809 8-Bit μP Compatible A/D Converters with 8-Channel Multiplexer

General Description

The ADC0808, ADC0809 data acquisition component is a monolithic CMOS device with an 8-bit analog-to-digital converter, 8-channel multiplexer and microprocessor compatible control logic. The 8-bit A/D converter uses successive approximation as the conversion technique. The converter features a high impedance chopper stabilized comparator, a 256R voltage divider with analog switch tree and a successive approximation register. The 8-channel multiplexer can directly access any of 8-single-ended analog signals.

The device eliminates the need for external zero and full-scale adjustments. Easy interfacing to microprocessors is provided by the latched and decoded multiplexer address inputs and latched TTL TRI-STATE® outputs.

The design of the ADC0808, ADC0809 has been optimized by incorporating the most desirable aspects of several A/D conversion techniques. The ADC0808, ADC0809 offers high speed, high accuracy, minimal temperature dependence, excellent long-term accuracy and repeatability, and consumes minimal power. These features make this device ideally suited to applications from process and machine control to consumer and automotive applications. For 16-channel multiplexer with common output (sample/hold port) see ADC0816 data sheet. (See AN-247 for more information.)

Features

- Easy interface to all microprocessors
- Operates ratiometrically or with 5 V_{DC} or analog span adjusted voltage reference
- No zero or full-scale adjust required
- 8-channel multiplexer with address logic
- 0V to 5V input range with single 5V power supply
- Outputs meet TTL voltage level specifications
- Standard hermetic or molded 28-pin DIP package
- 28-pin molded chip carrier package
- ADC0808 equivalent to MM74C949
- ADC0809 equivalent to MM74C949-1

Key Specifications

- Resolution .... 8 Bits
- Total Unadjusted Error .... $\pm\frac{1}{2}$ LSB and ± 1 LSB
- Single Supply .... 5 V_{DC}
- Low Power .... 15 mW
- Conversion Time .... 100 μs

Block Diagram

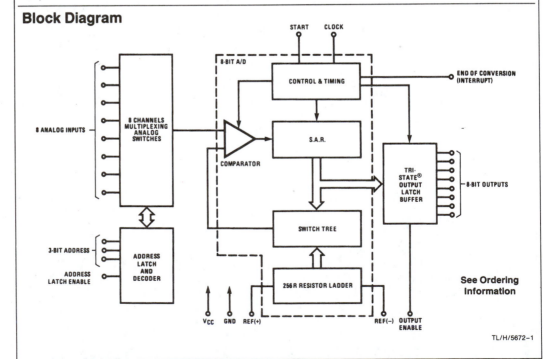

TL/H/5672-1

Absolute Maximum Ratings (Notes 1 & 2)

If Military/Aerospace specified devices are required, please contact the National Semiconductor Sales Office/Distributors for availability and specifications.

Supply Voltage (V_{CC}) (Note 3)	6.5V
Voltage at Any Pin	$-0.3V$ to ($V_{CC}+0.3V$)
Except Control Inputs	
Voltage at Control Inputs	$-0.3V$ to $+15V$
(START, OE, CLOCK, ALE, ADD A, ADD B, ADD C)	
Storage Temperature Range	$-65°C$ to $+150°C$
Package Dissipation at $T_A=25°C$	875 mW
Lead Temp. (Soldering, 10 seconds)	
Dual-In-Line Package (plastic)	260°C
Dual-In-Line Package (ceramic)	300°C
Molded Chip Carrier Package	
Vapor Phase (60 seconds)	215°C
Infrared (15 seconds)	220°C
ESD Susceptibility (Note 11)	400V

Operating Conditions (Notes 1 & 2)

Temperature Range (Note 1)	$T_{MIN}\leq T_A\leq T_{MAX}$
ADC0808CJ	$-55°C\leq T_A\leq +125°C$
ADC0808CCJ, ADC0808CCN,	
ADC0809CCN	$-40°C\leq T_A\leq +85°C$
ADC0808CCV, ADC0809CCV	$-40°C\leq T_A\leq +85°C$
Range of V_{CC} (Note 1)	4.5 V_{DC} to 6.0 V_{DC}

Electrical Characteristics

Converter Specifications: $V_{CC}=5$ $V_{DC}=V_{REF+}$, $V_{REF(-)}=$ GND, $T_{MIN}\leq T_A\leq T_{MAX}$ and $f_{CLK}=640$ kHz unless otherwise stated.

Symbol	Parameter	Conditions	Min	Typ	Max	Units
	ADC0808 Total Unadjusted Error (Note 5)	25°C			$\pm\frac{1}{2}$	LSB
		T_{MIN} to T_{MAX}			$\pm\frac{3}{4}$	LSB
	ADC0809 Total Unadjusted Error (Note 5)	0°C to 70°C			± 1	LSB
		T_{MIN} to T_{MAX}			$\pm 1\frac{1}{4}$	LSB
	Input Resistance	From Ref(+) to Ref(−)	1.0	2.5		kΩ
	Analog Input Voltage Range	(Note 4) V(+) or V(−)	GND-0.10		$V_{CC}+0.10$	V_{DC}
$V_{REF(+)}$	Voltage, Top of Ladder	Measured at Ref(+)		V_{CC}	$V_{CC}+0.1$	V
$\frac{V_{REF(+)}+V_{REF(-)}}{2}$	Voltage, Center of Ladder		$V_{CC}/2-0.1$	$V_{CC}/2$	$V_{CC}/2+0.1$	V
$V_{REF(-)}$	Voltage, Bottom of Ladder	Measured at Ref(−)	-0.1	0		V
I_{IN}	Comparator Input Current	$f_c=640$ kHz, (Note 6)	-2	± 0.5	2	μA

Electrical Characteristics

Digital Levels and DC Specifications: ADC0808CJ 4.5V$\leq V_{CC}\leq 5.5V$, $-55°C\leq T_A\leq +125°C$ unless otherwise noted ADC0808CCJ, ADC0808CCN, ADC0808CCV, ADC0809CCN and ADC0809CCV, 4.75$\leq V_{CC}\leq 5.25V$, $-40°C\leq T_A\leq +85°C$ unless otherwise noted

Symbol	Parameter	Conditions	Min	Typ	Max	Units
ANALOG MULTIPLEXER						
$I_{OFF(+)}$	OFF Channel Leakage Current	$V_{CC}=5V$, $V_{IN}=5V$, $T_A=25°C$		10	200	nA
		T_{MIN} to T_{MAX}			1.0	μA
$I_{OFF(-)}$	OFF Channel Leakage Current	$V_{CC}=5V$, $V_{IN}=0$, $T_A=25°C$	-200	-10		nA
		T_{MIN} to T_{MAX}	-1.0			μA

ADC0808/ADC0809

Electrical Characteristics (Continued)

Digital Levels and DC Specifications: ADC0808CJ $4.5V \leq V_{CC} \leq 5.5V$, $-55°C \leq T_A \leq +125°C$ unless otherwise noted ADC0808CCJ, ADC0808CCN, ADC0808CCV, ADC0809CCN and ADC0809CCV, $4.75 \leq V_{CC} \leq 5.25V$, $-40°C \leq T_A \leq +85°C$ unless otherwise noted

Symbol	Parameter	Conditions	Min	Typ	Max	Units
CONTROL INPUTS						
$V_{IN(1)}$	Logical "1" Input Voltage		$V_{CC}-1.5$			V
$V_{IN(0)}$	Logical "0" Input Voltage				1.5	V
$I_{IN(1)}$	Logical "1" Input Current (The Control Inputs)	$V_{IN}=15V$			1.0	μA
$I_{IN(0)}$	Logical "0" Input Current (The Control Inputs)	$V_{IN}=0$		-1.0		μA
I_{CC}	Supply Current	$f_{CLK}=640$ kHz		0.3	3.0	mA
DATA OUTPUTS AND EOC (INTERRUPT)						
$V_{OUT(1)}$	Logical "1" Output Voltage	$I_O=-360 \mu A$	$V_{CC}-0.4$			V
$V_{OUT(0)}$	Logical "0" Output Voltage	$I_O=1.6$ mA			0.45	V
$V_{OUT(0)}$	Logical "0" Output Voltage EOC	$I_O=1.2$ mA			0.45	V
I_{OUT}	TRI-STATE Output Current	$V_O=5V$ / $V_O=0$		-3	3	μA / μA

Electrical Characteristics

Timing Specifications $V_{CC}=V_{REF(+)}=5V$, $V_{REF(-)}=GND$, $t_r=t_f=20$ ns and $T_A=25°C$ unless otherwise noted.

Symbol	Parameter	Conditions	Min	Typ	Max	Units
t_{WS}	Minimum Start Pulse Width	(Figure 5)		100	200	ns
t_{WALE}	Minimum ALE Pulse Width	(Figure 5)		100	200	ns
t_s	Minimum Address Set-Up Time	(Figure 5)		25	50	ns
t_H	Minimum Address Hold Time	(Figure 5)		25	50	ns
t_D	Analog MUX Delay Time From ALE	$R_S=0\Omega$ (Figure 5)		1	2.5	μS
t_{H1}, t_{H0}	OE Control to Q Logic State	$C_L=50$ pF, $R_L=10k$ (Figure 8)		125	250	ns
t_{1H}, t_{0H}	OE Control to Hi-Z	$C_L=10$ pF, $R_L=10k$ (Figure 8)		125	250	ns
t_c	Conversion Time	$f_c=640$ kHz, (Figure 5) (Note 7)	90	100	116	μS
f_c	Clock Frequency		10	640	1280	kHz
t_{EOC}	EOC Delay Time	(Figure 5)	0		$8+2 \mu S$	Clock Periods
C_{IN}	Input Capacitance	At Control Inputs		10	15	pF
C_{OUT}	TRI-STATE Output Capacitance	At TRI-STATE Outputs, (Note 12)		10	15	pF

Note 1: Absolute Maximum Ratings indicate limits beyond which damage to the device may occur. DC and AC electrical specifications do not apply when operating the device beyond its specified operating conditions.

Note 2: All voltages are measured with respect to GND, unless otehwise specified.

Note 3: A zener diode exists, internally, from V_{CC} to GND and has a typical breakdown voltage of 7 V_{DC}.

Note 4: Two on-chip diodes are tied to each analog input which will forward conduct for analog input voltages one diode drop below ground or one diode drop greater than the V_{CC}n supply. The spec allows 100 mV forward bias of either diode. This means that as long as the analog V_{IN} does not exceed the supply voltage by more than 100 mV, the output code will be correct. To achieve an absolute $0V_{DC}$ to $5V_{DC}$ input voltage range will therefore require a minimum supply voltage of 4.900 V_{DC} over temperature variations, initial tolerance and loading.

Note 5: Total unadjusted error includes offset, full-scale, linearity, and multiplexer errors. See Figure 3. None of these A/Ds requires a zero or full-scale adjust. However, if an all zero code is desired for an analog input other than 0.0V, or if a narrow full-scale span exists (for example: 0.5V to 4.5V full-scale) the reference voltages can be adjusted to achieve this. See Figure 13.

Note 6: Comparator input current is a bias current into or out of the chopper stabilized comparator. The bias current varies directly with clock frequency and has little temperature dependence (Figure 6). See paragraph 4.0.

Note 7: The outputs of the data register are updated one clock cycle before the rising edge of EOC.

Note 8: Human body model, 100 pF discharged through a 1.5 kΩ resistor.

Functional Description

Multiplexer. The device contains an 8-channel single-ended analog signal multiplexer. A particular input channel is selected by using the address decoder. Table I shows the input states for the address lines to select any channel. The address is latched into the decoder on the low-to-high transition of the address latch enable signal.

TABLE I

SELECTED	ADDRESS LINE		
ANALOG CHANNEL	C	B	A
IN0	L	L	L
IN1	L	L	H
IN2	L	H	L
IN3	L	H	H
IN4	H	L	L
IN5	H	L	H
IN6	H	H	L
IN7	H	H	H

CONVERTER CHARACTERISTICS

The Converter

The heart of this single chip data acquisition system is its 8-bit analog-to-digital converter. The converter is designed to give fast, accurate, and repeatable conversions over a wide range of temperatures. The converter is partitioned into 3 major sections: the 256R ladder network, the successive approximation register, and the comparator. The converter's digital outputs are positive true.

The 256R ladder network approach *(Figure 1)* was chosen over the conventional R/2R ladder because of its inherent monotonicity, which guarantees no missing digital codes. Monotonicity is particularly important in closed loop feedback control systems. A non-monotonic relationship can cause oscillations that will be catastrophic for the system. Additionally, the 256R network does not cause load variations on the reference voltage.

The bottom resistor and the top resistor of the ladder network in *Figure 1* are not the same value as the remainder of the network. The difference in these resistors causes the output characteristic to be symmetrical with the zero and full-scale points of the transfer curve. The first output transition occurs when the analog signal has reached $+\frac{1}{2}$ LSB and succeeding output transitions occur every 1 LSB later up to full-scale.

The successive approximation register (SAR) performs 8 iterations to approximate the input voltage. For any SAR type converter, n-iterations are required for an n-bit converter. *Figure 2* shows a typical example of a 3-bit converter. In the ADC0808, ADC0809, the approximation technique is extended to 8 bits using the 256R network.

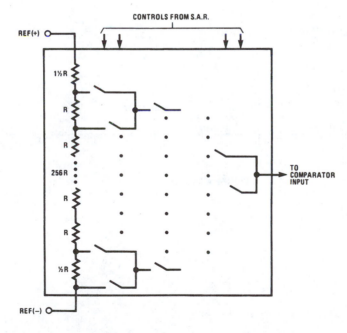

TL/H/5672-2

FIGURE 1. Resistor Ladder and Switch Tree

Functional Description (Continued)

The A/D converter's successive approximation register (SAR) is reset on the positive edge of the start conversion (SC) pulse. The conversion is begun on the falling edge of the start conversion pulse. A conversion in process will be interrupted by receipt of a new start conversion pulse. Continuous conversion may be accomplished by tying the end-of-conversion (EOC) output to the SC input. If used in this mode, an external start conversion pulse should be applied after power up. End-of-conversion will go low between 0 and 8 clock pulses after the rising edge of start conversion.

The most important section of the A/D converter is the comparator. It is this section which is responsible for the ultimate accuracy of the entire converter. It is also the comparator drift which has the greatest influence on the repeatability of the device. A chopper-stabilized comparator provides the most effective method of satisfying all the converter requirements.

The chopper-stabilized comparator converts the DC input signal into an AC signal. This signal is then fed through a high gain AC amplifier and has the DC level restored. This technique limits the drift component of the amplifier since the drift is a DC component which is not passed by the AC amplifier. This makes the entire A/D converter extremely insensitive to temperature, long term drift and input offset errors.

Figure 4 shows a typical error curve for the ADC0808 as measured using the procedures outlined in AN-179.

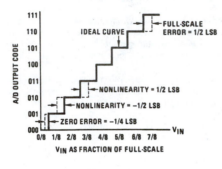

FIGURE 2. 3-Bit A/D Transfer Curve

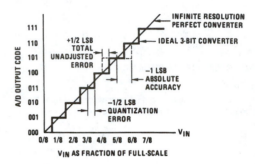

FIGURE 3. 3-Bit A/D Absolute Accuracy Curve

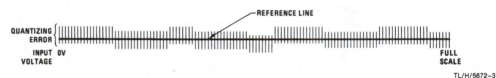

FIGURE 4. Typical Error Curve

TL/H/5672–3

Connection Diagrams

Dual-In-Line Package

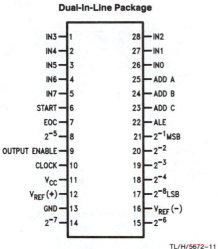

TL/H/5672–11

**Order Number ADC0808CCN, ADC0809CCN,
ADC0808CCJ or ADC0808CJ
See NS Package J28A or N28A**

Molded Chip Carrier Package

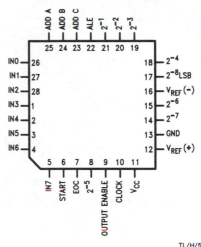

TL/H/5672–12

**Order Number ADC0808CCV or ADC0809CCV
See NS Package V28A**

Timing Diagram

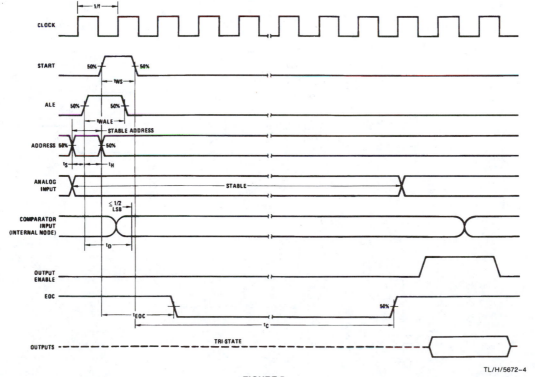

TL/H/5672–4

FIGURE 5

ADC0808/ADC0809

Typical Performance Characteristics

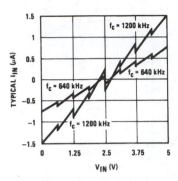

FIGURE 6. Comparator I_{IN} vs V_{IN}
($V_{CC} = V_{REF} = 5V$)

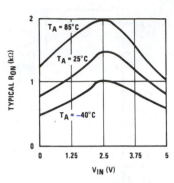

TL/H/5672–5

FIGURE 7. Multiplexer R_{ON} vs V_{IN}
($V_{CC} = V_{REF} = 5V$)

TRI-STATE Test Circuits and Timing Diagrams

t_{1H}, t_{H1}

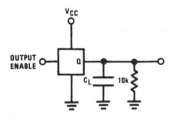

t_{1H}, $C_L = 10$ pF

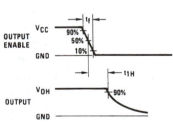

t_{H1}, $C_L = 50$ pF

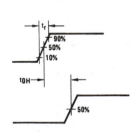

t_{0H}, t_{H0}

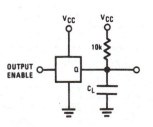

t_{0H}, $C_L = 10$ pF

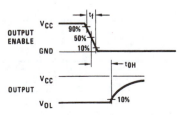

t_{H0}, $C_L = 50$ pF

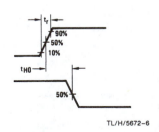

TL/H/5672–6

FIGURE 8

Applications Information

OPERATION

1.0 RATIOMETRIC CONVERSION

The ADC0808, ADC0809 is designed as a complete Data Acquisition System (DAS) for ratiometric conversion systems. In ratiometric systems, the physical variable being measured is expressed as a percentage of full-scale which is not necessarily related to an absolute standard. The voltage input to the ADC0808 is expressed by the equation

$$\frac{V_{IN}}{V_{fs} - V_Z} = \frac{D_X}{D_{MAX} - D_{MIN}} \qquad (1)$$

V_{IN} = Input voltage into the ADC0808
V_{fs} = Full-scale voltage
V_Z = Zero voltage
D_X = Data point being measured
D_{MAX} = Maximum data limit
D_{MIN} = Minimum data limit

A good example of a ratiometric transducer is a potentiometer used as a position sensor. The position of the wiper is directly proportional to the output voltage which is a ratio of the full-scale voltage across it. Since the data is represented as a proportion of full-scale, reference requirements are greatly reduced, eliminating a large source of error and cost for many applications. A major advantage of the ADC0808, ADC0809 is that the input voltage range is equal to the supply range so the transducers can be connected directly across the supply and their outputs connected directly into the multiplexer inputs, *(Figure 9)*.

Ratiometric transducers such as potentiometers, strain gauges, thermistor bridges, pressure transducers, etc., are suitable for measuring proportional relationships; however, many types of measurements must be referred to an absolute standard such as voltage or current. This means a system reference must be used which relates the full-scale voltage to the standard volt. For example, if $V_{CC} = V_{REF} = 5.12V$, then the full-scale range is divided into 256 standard steps. The smallest standard step is 1 LSB which is then 20 mV.

2.0 RESISTOR LADDER LIMITATIONS

The voltages from the resistor ladder are compared to the selected into 8 times in a conversion. These voltages are coupled to the comparator via an analog switch tree which is referenced to the supply. The voltages at the top, center and bottom of the ladder must be controlled to maintain proper operation.

The top of the ladder, Ref(+), should not be more positive than the supply, and the bottom of the ladder, Ref(−), should not be more negative than ground. The center of the ladder voltage must also be near the center of the supply because the analog switch tree changes from N-channel switches to P-channel switches. These limitations are automatically satisfied in ratiometric systems and can be easily met in ground referenced systems.

Figure 10 shows a ground referenced system with a separate supply and reference. In this system, the supply must be trimmed to match the reference voltage. For instance, if a 5.12V is used, the supply should be adjusted to the same voltage within 0.1V.

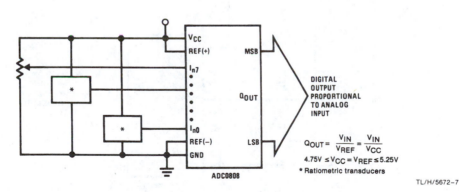

FIGURE 9. Ratiometric Conversion System

Applications Information (Continued)

The ADC0808 needs less than a milliamp of supply current so developing the supply from the reference is readily accomplished. In *Figure 11* a ground referenced system is shown which generates the supply from the reference. The buffer shown can be an op amp of sufficient drive to supply the milliamp of supply current and the desired bus drive, or if a capacitive bus is driven by the outputs a large capacitor will supply the transient supply current as seen in *Figure 12*. The LM301 is overcompensated to insure stability when loaded by the 10 μF output capacitor.

The top and bottom ladder voltages cannot exceed V_{CC} and ground, respectively, but they can be symmetrically less than V_{CC} and greater than ground. The center of the ladder voltage should always be near the center of the supply. The sensitivity of the converter can be increased, (i.e., size of the LSB steps decreased) by using a symmetrical reference system. In *Figure 13*, a 2.5V reference is symmetrically centered about $V_{CC}/2$ since the same current flows in identical resistors. This system with a 2.5V reference allows the LSB bit to be half the size of a 5V reference system.

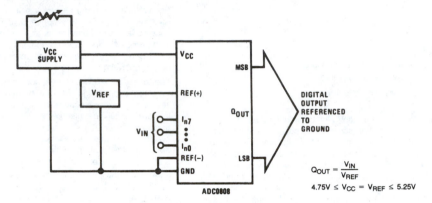

$$Q_{OUT} = \frac{V_{IN}}{V_{REF}}$$

$$4.75V \leq V_{CC} = V_{REF} \leq 5.25V$$

FIGURE 10. Ground Referenced Conversion System Using Trimmed Supply

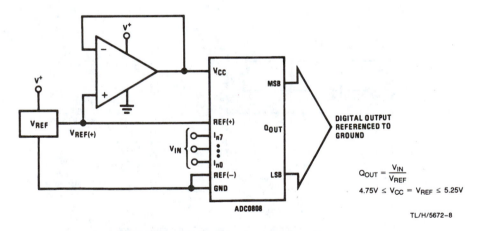

$$Q_{OUT} = \frac{V_{IN}}{V_{REF}}$$

$$4.75V \leq V_{CC} = V_{REF} \leq 5.25V$$

TL/H/5672-8

FIGURE 11: Ground Referenced Conversion System with Reference Generating V_{CC} Supply

Applications Information (Continued)

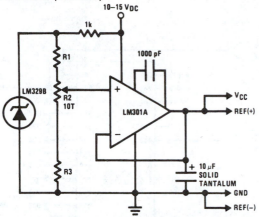

FIGURE 12. Typical Reference and Supply Circuit

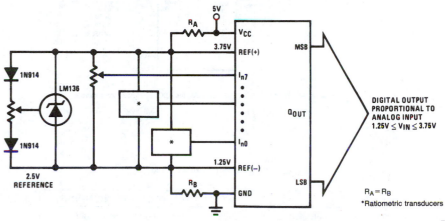

TL/H/5672–9

FIGURE 13. Symmetrically Centered Reference

3.0 CONVERTER EQUATIONS

The transition between adjacent codes N and N + 1 is given by:

$$V_{IN} = \left\{ (V_{REF(+)} - V_{REF(-)}) \left[\frac{N}{256} + \frac{1}{512} \right] \pm V_{TUE} \right\} + V_{REF(-)} \quad (2)$$

The center of an output code N is given by:

$$V_{IN} \left\{ (V_{REF(+)} - V_{REF(-)}) \left[\frac{N}{256} \right] \pm V_{TUE} \right\} + V_{REF(-)} \quad (3)$$

The output code N for an arbitrary input are the integers within the range:

$$N = \frac{V_{IN} - V_{REF(-)}}{V_{REF(+)} - V_{REF(-)}} \times 256 \pm \text{Absolute Accuracy} \quad (4)$$

where: V_{IN} = Voltage at comparator input

$V_{REF(+)}$ = Voltage at Ref(+)

$V_{REF(-)}$ = Voltage at Ref(−)

V_{TUE} = Total unadjusted error voltage (typically $V_{REF(+)} \div 512$)

4.0 ANALOG COMPARATOR INPUTS

The dynamic comparator input current is caused by the periodic switching of on-chip stray capacitances. These are connected alternately to the output of the resistor ladder/ switch tree network and to the comparator input as part of the operation of the chopper stabilized comparator.

The average value of the comparator input current varies directly with clock frequency and with V_{IN} as shown in *Figure 6.*

If no filter capacitors are used at the analog inputs and the signal source impedances are low, the comparator input current should not introduce converter errors, as the transient created by the capacitance discharge will die out before the comparator output is strobed.

If input filter capacitors are desired for noise reduction and signal conditioning they will tend to average out the dynamic comparator input current. It will then take on the characteristics of a DC bias current whose effect can be predicted conventionally.

ADC0808/ADC0809

Typical Application

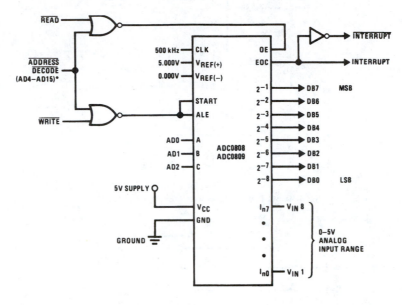

TL/H/5672–10

*Address latches needed for 8085 and SC/MP interfacing the ADC0808 to a microprocessor

MICROPROCESSOR INTERFACE TABLE

PROCESSOR	READ	WRITE	INTERRUPT (COMMENT)
8080	MEMR	MEMW	INTR (Thru RST Circuit)
8085	RD	WR	INTR (Thru RST Circuit)
Z-80	RD	WR	INT (Thru RST Circuit, Mode 0)
SC/MP	NRDS	NWDS	SA (Thru Sense A)
6800	VMA•φ2•R/W	VMA•φ•R/W	IRQA or IRQB (Thru PIA)

Ordering Information

TEMPERATURE RANGE		−40°C to +85°C			−55°C to +125°C
Error	± ½ LSB Unadjusted	ADC0808CCN	ADC0808CCV	ADC0808CCJ	ADC0808CJ
	± 1 LSB Unadjusted	ADC0809CCN	ADC0809CCV		
Package Outline		N28A Molded DIP	V28A Molded Chip Carrier	J28A Ceramic DIP	J28A Ceramic DIP

Very Fast, Complete
12-Bit A/D Converter
AD578

FEATURES

Performance

Complete 12-Bit A/D Converter with Reference and Clock
Fast Conversion: 3μs (max)
Buried Zener Reference for Long Term Stability and Low Gain T.C.: ±30ppm/°C max
Max Nonlinearity: <±0.012%
No Missing Codes Over Temperature
Low Power: 875mW
Hermetic Package Available
Available to MIL-STD-883

Versatility

Positive-True Parallel or Serial Logic Outputs
Short Cycle Capability
Precision +10V Reference for External Applications
Adjustable Internal Clock
"Z" Models for ±12V Supplies

AD578 FUNCTIONAL BLOCK DIAGRAM

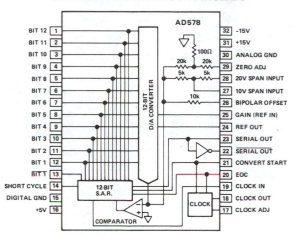

GENERAL DESCRIPTION

The AD578 is a high speed 12-bit successive approximation analog-to-digital converter that includes an internal clock, reference and comparator. Its hybrid IC design utilizes MSI digital and linear monolithic chips in conjunction with a 12-bit monolithic DAC to provide superior performance and versatility with IC size, price and reliability.

Important performance characteristics of the AD578 include a maximum linearity error at +25°C of ±0.012%, maximum gain temperature coefficient of ±30ppm/°C, typical power dissipation of 875mW and maximum conversion time of 3μs.

The fast conversion speeds of 3μs (L grade) 4.5μs (K, T grades) and 6μs (J, S grades) make the AD578 an excellent choice in a variety of applications where system throughput rates from 166kHz to 333kHz are required. In addition, it may be short cycled to obtain faster conversion speeds at lower resolutions.

The design of the AD578 includes scaling resistors that provide analog input signal ranges of ±5V, ±10V, 0 to +10V or 0 to +20V. Adding flexibility and value is the +10V precision reference which can be used for external applications.

The AD578 is available with either the polymer seal (N) for use in benign environmental applications or hermetic solder-seal (D) for more harsh or rigorous surroundings. Both are contained in a 32-pin side-brazed, ceramic DIP.

The AD578S, T are available processed to MIL-STD-883 Level B, Method 5008.

PRODUCT HIGHLIGHTS

1. The AD578 is a complete 12-bit A/D converter. No external components are required to perform a conversion.

2. The fast conversion rate of the AD578 makes it an excellent choice for high speed data acquisition and digital signal processing applications.

3. The internal buried zener reference is laser trimmed to 10.00V ±1.0% and ±15ppm/°C typical T.C. The reference is available for external use and can provide up to 1mA.

4. The scaling resistors are included on the monolithic DAC for exceptional thermal tracking.

5. The component count is minimized, resulting in low bond wire and chip count and high MTBF.

6. Short cycle and external clock capabilities are provided for applications requiring faster conversion speeds and/or lower resolutions.

7. The integrated package construction provides high quality and reliability with small size and weight.

SPECIFICATIONS (typical @ +25°C, ±15V and +5V unless otherwise noted)

Model	AD578J	AD578K	AD578L	AD578SD[1]	AD578TD[1]
RESOLUTION	12 Bits	*	*	*	*
ANALOG INPUTS					
Voltage Ranges					
Bipolar	±5.0V, ±10V	*	*	*	*
Unipolar	0 to +10V, 0 to +20V	*	*	*	*
Input Impedance					
0 to +10V, ±5V	5kΩ	*	*	*	*
±10V, 0 to +20V	10kΩ	*	*	*	*
DIGITAL INPUTS					
Convert Command[2]	1LSTTL Load	*	*	*	*
Clock Input	1LSTTL Load	*	*	*	*
TRANSFER CHARACTERISTICS					
Gain Error[3,4]	±0.1% FSR, ±0.25% FSR max	*	*	*	*
Unipolar Offset[4]	±0.1% FSR, ±0.25% FSR max	*	*	*	*
Bipolar Error[4,5]	±0.1% FSR, ±0.25% FSR max	*	*	*	*
Linearity Error, 25°C	±1/2LSB max	*	*	*	*
T_{min} to T_{max}	±3/4LSB	*	*	±3/4LSB max	±3/4LSB max
DIFFERENTIAL LINEARITY ERROR (Minimum resolution for which no missing codes are guaranteed)					
+25°C	12 Bits	*	*	*	*
T_{min} to T_{max}	12 Bits	*	*	*	*
POWER SUPPLY SENSITIVITY					
+15V ±10%	0.005%/%ΔV_S max	*	*	*	*
−15V ±10%	0.005%/%ΔV_S max	*	*	*	*
+5V ±10%	0.005%/%ΔV_S max	*	*	*	*
TEMPERATURE COEFFICIENTS					
Gain	±15ppm/°C typ	*	*	*	*
	±30ppm/°C max	*	*	±50ppm/°C max	±30ppm/°C max
Unipolar Offset	±3ppm/°C typ	*	*	*	*
	±10ppm/°C max	*	*	±15ppm/°C max	±10ppm/°C max
Bipolar Offset	±8ppm/°C typ	*	*	*	*
	±20ppm/°C max	*	*	±25ppm/°C max	±20ppm/°C max
Differential Linearity	±2ppm/°C typ	*	*	*	*
CONVERSION TIME[6,7,8](max)	6.0μs	4.5μs	3μs	6.0μs	4.5μs
PARALLEL OUTPUTS					
Unipolar Code	Binary	*	*	*	*
Bipolar Code	Offset Binary/Two's Complement	*	*	*	*
Output Drive	2LSTTL Loads	*	*	*	*
SERIAL OUTPUTS (NRZ FORMAT)					
Unipolar Code	Binary/Complementary Binary	*	*	*	*
Bipolar Code	Offset Binary/Comp. Offset Binary	*	*	*	*
Output Drive	2LSTTL Loads	*	*	*	*
END OF CONVERSION (EOC)	Logic "1" During Conversion	*	*	*	*
Output Drive	8LSTTL Loads	*	*	*	*
INTERNAL CLOCK[8]					
Output Drive	2LSTTL Loads	*	*	*	*
INTERNAL REFERENCE					
Voltage	10.000 ±100mV	*	*	*	*
Drift	±12ppm/°C, ±20ppm/°C max	*	*	*	*
External Current	±1mA max	*	*	*	*
POWER SUPPLY REQUIREMENTS[9]					
Range for Rated Accuracy	4.75 to 5.25 and ±13.5 to ±16.5	*	*	*	*
Supply Current +15V	3mA typ, 8mA max	*	*	*	*
−15V	22mA typ, 35mA max	*	*	*	*
+5V	100mA typ, 140mA max	*	*	*	*
Power Dissipation	875mW typ	*	*	*	*
TEMPERATURE RANGE					
Operating	0 to +70°C	*	*	−55°C to +125°C	−55°C to +125°C
Storage	−55°C to +150°C	*	*	−65°C to +150°C	−65°C to +150°C

NOTES
[1] Available to MIL-STD-883, Level B. See ADI Military Products Databook for detail specifications.
[2] Positive pulse 200ns wide (min) leading edge (0 to 1) resets outputs. Trailing edge initiates conversion.
[3] With 50Ω, 1% fixed resistor in place of gain adjust potentiometer.
[4] Adjustable to zero.
[5] With 50Ω, 1% resistor between Ref Out and Bipolar Offset (Pins 24 & 26).
[6] Conversion time is defined as the time between the falling edge of convert start and the falling edge of the EOC.
[7] Each grade is specified at the conversion speed shown.
[8] Externally adjustable by a resistor or capacitor (see Figure 7).
[9] For "Z" models order AD578ZJ, ZK, ZL (±11.6V to ±16.5V).
*Specifications same as AD578J.
Specifications subject to change without notice.

AD578

THEORY OF OPERATION

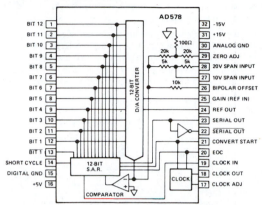

Figure 1. AD578 Functional Diagram and Pinout

The AD578 is a complete pretrimmed 12-bit A/D converter which requires no external components to provide the successive-approximation analog-to-digital conversion function. A block diagram of the AD578 is shown in Figure 1.

When the control section is commanded to initiate a conversion it enables the clock and resets the successive-approximation register (SAR). The SAR, timed by the clock, sequences through the conversion cycle and returns an end-of-convert flag to the control section. The control section disables the clock and brings the output status flag low. The parallel data bits become valid on the rising edge of the clock pulse starting with t_1 and ending with t_{12} (Figure 2), and accurately represent the input signal to within ±1/2LSB.

The temperature-compensated buried Zener reference provides the primary voltage reference to the DAC and guarantees excellent stability with both time and temperature. The reference is trimmed to 10.00 volts ±1.0%, it is buffered and can supply up to 1.0mA to an external load in addition to the current required to drive the reference input resistor (0.5mA) and bipolar offset resistor (1mA). The thin-film application resistors are trimmed to match the full scale output current of the DAC. There are two 5kΩ input scaling resistors to allow either a 10 volt or 20 volt span. The 10kΩ bipolar offset resistor is grounded for unipolar operation or connected to the 10 volt reference for bipolar operation.

UNIPOLAR CALIBRATION

The AD578 is intended to have a nominal 1/2LSB offset so that the exact analog input for a given code will be in the middle of that code (halfway between the transitions to the codes above and below it). Thus, when properly calibrated, the first transition (from 0000 0000 0000 to 0000 0000 0001) will occur for an input level of +1/2LSB (1.22mV for 10V range).

If pin 26 is connected to pin 30, the unit will behave in this manner, within specifications. Refer to Table I and Figure 3 for further clarification. If the offset trim (R1) is used, it should be trimmed as above, although a different offset can be set for a particular system requirement. This circuit will give approximately ±25mV of offset trim range.

The full scale trim is done by applying a signal 1 1/2LSB below the nominal full scale (9.9963V for a 10V range). Trim R2 to give the last transition (1111 1111 1110 to 1111 1111 1111).

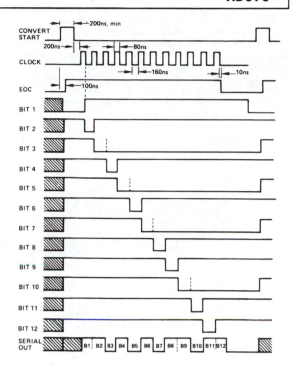

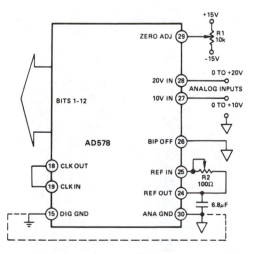

CLOCK
INTERNAL: CONNECT CLOCK OUT (18) TO CLOCK IN (19)
EXTERNAL: CONNECT EXTERNAL CLOCK TO CLOCK IN (19)
CLOCK SHOULD BE AT LEAST 30% DUTY CYCLE WITH
MINIMUM PERIOD, T_{MIN} OF 100ns.

NOTE
[1] THE RISING EDGE OF CONVERT START PULSE RESETS THE MSB TO ZERO,
AND THE LSBs TO ONE. THE TRAILING EDGE INITIATES CONVERSION.

Figure 2. AD578 3μs Timing Diagram

Figure 3. Unipolar Input Connections

BIPOLAR OPERATION

The connections for bipolar ranges are shown in Figure 4. Again, as for the unipolar ranges, if the offset and gain specifications are sufficient the 100Ω trimmer shown can be replaced by a 50Ω ±1% fixed resistor. The analog input is applied as for the unipolar ranges. Bipolar calibration is similar to unipolar calibration. First, a signal 1/2LSB above negative full scale (−4.9988V for the ±5V range) is applied, and R1 is trimmed to give the first transition (0000 0000 0000 to 0000 0000 0001). Then a signal 1 1/2LSB below positive full scale (+4.9963V for the ±5V range) is applied and R2 trimmed to give the last transition (1111 1111 1110 to 1111 1111 1111).

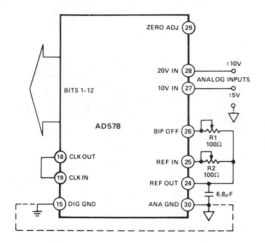

Figure 4. Bipolar Input Connections

LAYOUT CONSIDERATION

Many data-acquisition components have two or more ground pins which are not connected together within the device. These "grounds" are usually referred to as the Logic Power Return, Analog Common (Analog Power Return), and Analog Signal Ground. These grounds must be tied together at one point, usually at the system power-supply ground. Ideally, a single solid ground would be desirable. However, since current flows through the ground wires and etch stripes of the circuit cards, and since these paths have resistance and inductance, hundreds of millivolts can be generated between the system ground point

and the ground pin of the AD578. Separate ground returns should be provided to minimize the current flow in the path from sensitive points to the system ground point. In this way supply currents and logic-gate return currents are not summed into the same return path as analog signals where they would cause measurement errors.

Each of the AD578's supply terminals should be capacitively decoupled as close to the AD578 as possible. A large value capacitor such as 10μF in parallel with a 0.1μF capacitor is usually sufficient. Analog supplies are bypassed to the Analog Power Return pin and the logic supply is bypassed to the Digital GND pin.

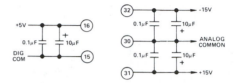

Figure 5. Basic Grounding Practice

To minimize noise the reference output (pin 24) should be decoupled by a 6.8μF capacitor to pin 30.

CLOCK RATE CONTROL

The internal clock is preset to a nominal conversion time of 5.6μs. It can be adjusted for either faster or slower conversions. For faster conversion connect the appropriate 1% resistor between pin 17 and pin 18 and short pin 18 to pin 19.

For slower conversions connect a capacitor between pin 15 and pin 17.

The curves in Figure 6 characterize the conversion time for a given resistor or capacitor connection.

Note: 12-bit operation with no missing codes is not guaranteed when operating in this mode if a particular grade's conversion speed specification has been exceeded.

Short Cycle Input – A Short Cycle Input, pin 14, permits the timing cycle shown in Figure 2 to be terminated after any number of desired bits has been converted, allowing somewhat shorter conversion times in applications not requiring full 12-bit resolution. Short cycle pin connections and associated maximum 12-, 10-, and 8-bit conversion times are summarized in Table II.

Analog Input – Volts (Center of Quantization Interval)				Digital Output Code (Binary For Unipolar Ranges; Offset Binary for Bipolar Ranges)	
0 to +10V Range	0 to +20V Range	−5V to +5V Range	−10V to +10V Range	B1 (MSB)	B12 (LSB)
+9.9976	+19.9951	+4.9976	+9.9951	1 1 1 1 1 1 1 1 1 1 1 1	
+9.9952	+19.9902	+4.9952	+9.9902	1 1 1 1 1 1 1 1 1 1 1 0	
•	•	•	•	•	
•	•	•	•		
+5.0024	+10.0049	+0.0024	+0.0049	1 0 0 0 0 0 0 0 0 0 0 1	
+5.0000	+10.0000	+0.0000	+0.0000	1 0 0 0 0 0 0 0 0 0 0 0	
•	•	•	•	•	
•	•	•	•		
+0.0024	+0.0051	−4.9976	−9.9951	0 0 0 0 0 0 0 0 0 0 0 1	
+0.0000	+0.0000	−5.0000	−10.0000	0 0 0 0 0 0 0 0 0 0 0 0	

Table I. Digital Output Codes vs. Analog Input for Unipolar and Bipolar Ranges

AD578

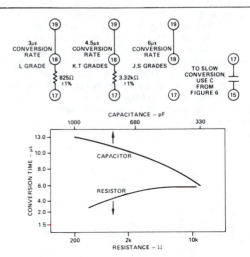

Figure 6. Conversion Times vs. R or C Values

Resolution (Bits)	12	10	8
Connect Pin 14 to Pin	16	2	4
Conversion Speed (μs)	3	2.5	2

Table II. Short Cycle Connections

External Clock – An external clock may be connected directly to the clock input, pin 19. When operating in this mode, the convert start should be held high for a minimum of one clock period in order to reset the SAR and synchronize the conversion cycle. A positive going pulse width of 100 to 200 nanoseconds will provide a continuous string of conversions that start on the first rising edge of the external clock after the EOC output has gone low.

External Buffer Amplifier – In applications where the AD578 is to be driven from high impedance sources or directly from an analog multiplexer a fast slewing, wideband op amp like the AD711 should be used.

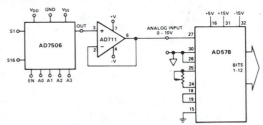

Figure 7. Input Buffer

MICROPROCESSOR INTERFACING

The 3μs conversion times of the AD578 suggests several different methods of interface to microprocessors. In systems where the AD578 is used for high sampling rates on a single signal which is to be digitally processed, CPU-controlled conversion may be inefficient due to the slow cycle times of most microprocessors. It is generally preferable to perform conversions independently, inserting the resultant digital data directly into memory. This can be done using direct memory access (DMA) which is totally transparent to the CPU. Interface to user-designed DMA hardware

is facilitated by the guaranteed data validity on the falling edge of the EOC signal.

In many multichannel data acquisition systems, the processor spends a good deal of time waiting for the ADC to complete its cycle. Converters with total conversion times of 25μs to 100μs are not slow enough to justify use of interrupts, nor fast enough to finish converting during one instruction and are usually timed out with loops, or continuously polled for status. The AD578 allows the microprocessor to time out the converter with just a few dummy instructions. For example, an 8085 system running at a 5MHz clock rate will time out an AD578 by pushing a register pair onto the stack and popping the same pair back off the stack. Such a time-out routine only occupies two bytes of program memory but requires 22 clock cycles (4.4μs). The time saved by not having to wait for the converter allows the processor to run much more efficiently particularly in multichannel systems.

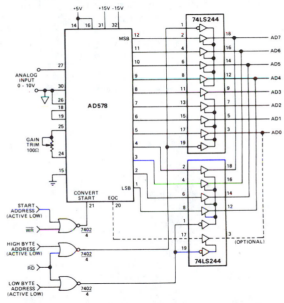

Figure 8. AD578–8085A Interface Connections

Clearly, 12 bits of data must be broken up for interface to an 8-bit wide data bus. There are two possible formats: right-justified and left-justified. In a right-justified system, the least-significant 8 bits occupy one byte and the four MSBs reside in the low nybble of another byte. This format is useful when the data from the ADC is being treated as a binary number between 0 and 4095. The left-justified format supplies the eight most-significant bits in one byte and the 4LSBs in the high nybble of another byte. The data now represents the fractional binary number relating the analog signal to the full-scale voltage. An advantage to this organization is that the most-significant eight bits can be read by the processor as a coarse indication of the true signal value. The full 12-bit word can then be read only when all 12 bits are needed. This allows faster and more efficient control of a process.

Figure 8 shows a typical connection to an 8085-type bus, using left-justified data format for unipolar inputs. Status polling is optional, and can be read simultaneously with the 4LSBs. If it is desired to right-justify the data, pins 1 through 12 of the AD578 should be reversed, as well as the connections to the data bus and high and low byte address signals.

When dealing with bipolar inputs ($\pm 5V$, $\pm 10V$ ranges), using the MSB directly yields an offset binary-coded output. If two's complement coding is desired, it can be produced by substituting $\overline{MSB}$ (pin 13) for the MSB. This facilitates arithmetic operations which are subsequently performed on the ADC output data.

SAMPLED DATA SYSTEMS
The conversion speed of the AD578 allows accurate digitization of high frequency signals and high throughput rates in multi-channel data acquisition systems. The AD578LD, for example,

is capable of a full accuracy conversion in $3\mu s$. In order to benefit from this high speed, a fast sample-hold amplifier (SHA) such as the HTC-0300 is required. This SHA has an acquisition time to 0.01% of approximately 300ns, so that a complete sample-convert-acquire cycle can be accomplished in approximately $4\mu s$. This means a sample rate of 250kHz can be realized, allowing a signal with no frequency components above 125kHz to be sampled with no loss of information. Note that the EOC signal from the AD578 places the SHA in the hold mode in advance of the actual start of the conversion cycle, and releases the SHA from the HOLD mode only after completion of the conversion. After allowing at least 300ns for the SHA to acquire the next analog value, the converter can again be started.

AD578 ORDERING GUIDE*

	Conversion Speed	Temperature Range	Package Option[1]
AD578JN(JD)	$6.0\mu s$	0 to +70°C	Solder Seal (DH-32B)
AD578KN(KD)	$4.5\mu s$	0 to +70°C	Solder Seal (DH-32B)
AD578LN(LD)	$3.0\mu s$	0 to +70°C	Solder Seal (DH-32B)
AD578SD	$6.0\mu s$	−55°C to +125°C	Solder Seal (DH-32B)
AD578SD/883B	$6.0\mu s$	−55°C to +125°C	Solder Seal (DH-32B)
AD578TD/883B	$4.5\mu s$	−55°C to +125°C	Solder Seal (DH-32B)

*For $\pm 12V$ operation "Z" version order: AD578ZJN, . . .

[1] See Section 14 for package outline information.

Index*

*Page numbers in *italics* refer to illustrations; page numbers followed by (t) refer to tables